Right Prism

h = height; p = perimeter of base;
B = area of base
Lateral area: $LA = ph$
Surface area: $SA = LA + 2B$
Volume: $V = Bh$

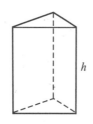

Sphere

r = radius
Surface area: $SA = 4\pi r^2$
Volume: $V = \dfrac{4}{3}\pi r^3$

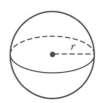

Cube

e = edge
Surface area: $SA = 6e^2$
Volume: $V = e^3$

Regular Pyramid

h = height; p = perimeter of base;
ℓ = slant height; B = area of base

Lateral area: $LA = \dfrac{1}{2}p\ell$

Surface area: $SA = LA + B$ or $\dfrac{1}{2}p\ell + B$

Volume: $V = \dfrac{1}{3}Bh$

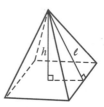

Right Circular Cylinder

r = radius; h = height
Lateral area: $LA = ph$ or $2\pi rh$
Surface area: $SA = LA + 2B$ or $2\pi rh + 2\pi r^2$
Volume: $V = Bh$ or $\pi r^2 h$

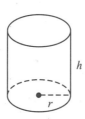

Right Circular Cone

r = radius; h = height; ℓ = slant height
Lateral area: $LA = \dfrac{1}{2}p\ell$ or $\pi r\ell$
Surface area: $SA = LA + B$ or $\pi r\ell + \pi r^2$
Volume: $V = \dfrac{1}{3}Bh$ or $\dfrac{1}{3}\pi r^2 h$

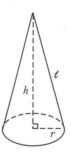

Essentials of Geometry for College Students

Second Edition

Margaret L. Lial
American River College

Barbara A. Brown
Anoka-Ramsey Community College

Arnold R. Steffensen
Northern Arizona University

L. Murphy Johnson
Northern Arizona University

PEARSON

Addison
Wesley

Boston   San Francisco   New York
London   Toronto   Sydney   Tokyo   Singapore   Madrid
Mexico City   Munich   Paris   Cape Town   Hong Kong   Montreal

Publisher: Greg Tobin
Editor in Chief: Maureen O'Connor
Assistant Editor: Katie Nopper
Production Supervisor: Ron Hampton
Production Services: Pre-Press Company
Marketing Manager: Jay Jenkins
Marketing Coordinator: Lindsay Skay
Prepress Supervisor: Caroline Fell
Manufacturing Buyer: Hugh Crawford
Text and Cover Design: Dennis Schaefer
Cover Photograph: David Chmielewski/Corbis

Photo Credits

p. 1, 5, PhotoDisc; p. 2, St. Andrew's University; p. 4, St. Andrew's University; p. 5, St. Andrew's University; p. 11, Getty Thinkstock; p. 21, NASA; p. 44, 50, Corbis; p. 45, Alinari/Art Resource, NY; p. 63, Digital Vision; p. 67, Coloradospringskoa.com; p. 76, St. Andrew's University; p. 84, Corbis; p. 102, St. Andrew's University; p. 114, St. Andrew's University; pp. 127, 153, PhotoDisc; p. 138, PhotoDisc Blue; p. 140, Corbis RF; p. 144, PhotoDisc; p. 145, Digital Vision; p. 146, PhotoDisc; p. 175, 181, Rubberball Productions; p. 194, PhotoDisc; p. 201, PhotoDisc; pp. 215, 232, Radar sat International; pp. 219, 262, Brand X Pictures; p. 222, Corbis; p. 225, Corbis; p. 241, PhotoDisc; p. 247, St. Andrew's University; p. 248, Sara Anderson; p. 255, Corbis; p. 275, NASA; p. 277, Corbis RF; p. 300, NASA; p. 305, Corbis RF; p. 313, St. Andrew's University; p. 331, PhotoDisc; p. 339, Australian Sports Commission; p. 342, Australian Sports Commission; p. 375, Beth Anderson; p. 375, Beth Anderson; p. 376, Beth Anderson; p. 381, PhotoDisc; p. 392, PhotoDisc; p. 398, Digital Vision; p. 402, safeshopper.com; p. 405, St. Andrew's University; p. 406, Digital Vision; p. 412, product; p. 413, Corbis rights managed; p. 413, Beth Anderson; p. 414, NASA; p. 425, PhotoDisc; p. 426, Corbis; p. 427, St. Andrew's University; p. 437, Corbis; p. 438, Brand X Pictures; p. 441, PhotoDisc; p. 457, PhotoDisc; p. 461, Corbis rights managed; p. 477, St. Andrew's University; p. 476, PhotoDisc

Library of Congress Cataloging-in-Publication Data
Essentials of geometry for college students.—2nd ed./
Margaret L. Lial . . . [et al.].
    p. cm.
  Includes index.
  ISBN 0-201-74882-7
  1. Geometry.   I. Lial, Margaret L.

QA453.L5 2003
516—dc21   2002043782

2 3 4 5 6 7 8 9 10—CRW—07 06 05 04

Contents

Preface

Essentials of Geometry for College Students, Second Edition, is designed to provide students with the sound foundation in geometry that is necessary to pursue further courses in college mathematics. It is written for college students who have no previous experience with plane Euclidean geometry and for those who need a refresher in the subject. A background in introductory algebra and a scientific calculator are the only prerequisites. The content of this book is suited to prospective teachers as well as to students of geometry.

Many of the popular features of the previous edition have been maintained, while new material inspired by the AMATYC (American Association of Two-Year Colleges) *Crossroads in Mathematics, Standards for Introductory College Mathematics before Calculus*, and the NCTM (National Council of Teachers of Mathematics) *Curriculum and Evaluation Standards of School Mathematics* has been added. The AMATYC standards on geometry, modeling, reasoning, communication, technology, and deductive proof have been addressed. Some of the features that support these standards include the *Student Activity* and *Technology Connection,* which use geometry software. Found in both the text and exercises, these features are used to investigate and enhance the understanding of theorems. Also included are writing exercises and an introduction of flowchart proofs.

The goal of the second edition was to create a student-friendly textbook that implements some of the AMATYC and NCTM standards. Thus, the features that support this goal are readable, yet accurately worded, explanations that do not sacrifice rigor, motivational small group activities, explorations with today's technology, sound pedagogy and use of color, ample examples and exercises, and comprehensive chapter reviews and tests. Caution notes for the student, a glossary, and *all* odd answers in the back of the textbook are also hallmarks of this text. Optional elements of the textbook include group and technology activities, flowchart proofs, and writing exercises. The instructor may choose to use these features as appropriate without losing continuity in the text.

New Features in the Second Edition

- *Technology Connection* boxes have been added to take advantage of the technology that is currently available to teach mathematics. There are over thirty activities, most of which include using geometry software like Geometer's Sketchpad. These activities are written in a standardized format to allow the use of any geometry software. Also included are several activities that utilize a graphing calculator. They can serve as a regular thread throughout the course, as small group work, or as extra credit. They allow students to explore ideas, make conjectures, and test those conjectures.
- The *Student Activity* feature incorporates the use of everyday items like tissue paper, cardboard tubes, scissors, pipe cleaners, and envelopes. Also used are protractors, compasses, and scientific calculators. There are over thirty activities designed to explain and explore different theorems.
- Learning objectives have been added at the beginning of each section and are restated at the objective's introduction in the text.
- Three cumulative reviews, covering chapters 1–3, chapters 4–7, and chapters 8–10, are provided.
- Writing exercises that encourage students to communicate mathematical ideas using appropriate vocabulary to improve understanding are included with all exercise sets. These exercises are indicated with the icon.
- Since many students have difficulty with two-column proofs, an introduction to flowchart proofs has been added. Flowchart proofs provide an alternative technique to expand students' reasoning skills as they learn to develop mathematical arguments.
- In-text Caution and Note boxes warn students about common errors and emphasize important concepts.
- Section 1.5 on formalizing a proof has been greatly enhanced, and the "thinking backwards" problem-solving technique is used to develop a plan for a proof. This technique is explained in detail using everyday language.
- Less emphasis is placed on students developing proofs of the numbered theorems. Most of these proofs appear in the appropriate section of the text rather than within the exercises.
- *All* the odd answers are now included in the back of the text.
- The book has been reorganized into ten chapters rather than twelve. The chapter previously titled Parallel Lines and Polygons has been separated to form two chapters: Chapter 3, Parallel Lines and Polygons; and Chapter 4, Quadrilaterals. Introduction to proofs, inequalities of triangles, and inequalities of circles, loci, and concurrency have all been integrated into other chapters.

Retained Features

- Informal, yet carefully worded, explanations have remained. The emphasis is on a readable text that students can understand without sacrificing rigor.
- Color is used to highlight important information. Definitions, postulates, theorems, and corollaries are set off in colored boxes for increased emphasis. Text figures and constructions utilize color to clarify concepts presented. Examples present important steps and helpful side comments in color.
- To demonstrate the usefulness and practicality of geometry, applications are integrated throughout the examples and exercises in the text. Each chapter introduction also features a relevant applied problem that is then solved later in the chapter.
- The examples include detailed, step-by-step solutions and side annotations in color.
- Each section of the text ends with a comprehensive set of exercises ranging from basic to more challenging problems, including proofs and applications. Where appropriate, algebra is used as a tool to solve geometric problems.
- Throughout each section, practice exercises with their answers can be found. This integral element keeps students involved with the presentation by allowing them to immediately check their understanding of the material.
- To provide ample opportunities for review, a comprehensive set of chapter review exercises and a practice test are included at the end of each chapter. Answers to *all* of these problems appear at the back of the book.
- The text features many figures, sketches and photographs of noted mathematicians, and other art to generate interest and further motivate students.
- In addition to the chapter review exercises and the practice tests for each chapter, there are comprehensive chapter summaries that include key terms and symbols and a helpful list of proof techniques and summary tables where applicable.

Supplements

For the Student

- *Student's Solutions Manual* (ISBN: 0-321-17353-8) by Barbara A. Brown, Anoka-Ramsey Community College, contains complete, step-by-step solutions to the odd-numbered text exercises and practice problems, as well as all chapter review, practice test, and cumulative review problems.

For the Instructor

- *Instructor's Solutions Manual* (ISBN: 0-321-17355-4) by Barbara A. Brown, contains step-by-step solutions to all section exercises, practice exercises, chapter review exercises, and practice tests.
- *Instructor's Resource Guide/Printed Test Bank* (ISBN: 0-321-17356-2) by Sarah Kueffer, Anoka-Ramsey Community College; and Barbara A. Brown, contains one multiple-choice and two open-answer tests for each chapter, two final examinations, and the answer keys for all tests, as well as teaching notes and solutions for the *Student Activity* feature, transparency masters to prepare overhead acetate transparencies for classroom presentations, and a correlation guide.

Acknowledgments

We are indebted to the following reviewers who provided countless beneficial suggestions and criticisms during the writing of the text:

Marcella Cremer, Richland Community College
Edith Hayes, Texas Woman's University
Stephanie Haynes, Davis and Elkins College
Sue Leland, Montana Technical College
Carol A. Marinas, Barry University
Nicholas Martin, Shepherd College
Kathy Nickell, College of Du Page
Craig Roberts, Southeast Missouri State University
Lois Schuppig, College of Mount Saint Joseph
Mike Scroggins, Lewis and Clark Community College
Rick Silvey, Saint Mary College
Bettie Truitt, Black Hawk College

Special thanks goes to Sarah Kueffer, Anoka-Ramsey Community College, whose expertise was invaluable in the writing of the Technology Connection feature and who reviewed the material for this edition; thanks also to Vince Koehler and Rick Silvey for reviewing the main text as well. We also thank Judy Martinez, who typed the solutions manuals, and Sheri Minkner, who created the art and served as art editor for all of the supplements, for their many hours of hard work. We appreciate the efforts of Bettie Truitt and Lois Schuppig in reviewing the supplements. Last, but not least, thanks to the team at Addison-Wesley who helped make this revision a success: Maureen O'Connor, Ron Hampton, Dennis Schaefer, Jolene Lehr, and Katie Nopper in editorial and production, and Dona Kenly, Jay Jenkins, and Lindsay Skay in marketing.

This book is dedicated to my husband, Bud Brown. This textbook would not be a reality without his unwavering support and encouragement.

Barbara A. Brown
Anoka-Ramsey Community College

To the Student

The word *geometry* stems from the Greek words *geo*, meaning Earth, and *metry*, meaning measure. Geometry originated in Egypt where periodic flooding of the Nile River made it necessary to resurvey the flood plains. As time passed, "Earth measure" took on a broader meaning and involved measuring many things related to the earth. Eventually, it came to be realized that geometry is more than just a collection of facts used for measurement; it is a system in which these facts are related in a precise and logical manner.

There are many answers to the question "Why study geometry?" Unfortunately, some students will reply "Because it's a required course." If we take this one step further and ask "Why is geometry required?" we discover the real answers to the original question.

First, *we use geometry as a tool in many areas of mathematics* as well as in practical situations. For example, you may have already used some of the simple geometry formulas, such as those for finding areas, perimeters, and volumes.

A second reason for studying geometry is *to gain knowledge of symmetry and proportion* that will enable you to appreciate the inherent beauty in man-made art and architecture as well as in nature. For example, the lasting beauty of a building such as the Parthenon on the Acropolis of Athens results from the notion of the *golden ratio*. (See Chapter 5.) The golden ratio, which was known by the early Greeks, gives the most visually appealing dimensions for a rectangle. Or what could be more impressive geometrically than the honeycomb of a bee? (See Chapter 3.)

A third reason for studying geometry is that *geometric figures are used in most areas of mathematics to help us "picture" complex concepts*. For example, in beginning algebra, you were encouraged to draw sketches showing the pertinent information given in many "word problems."

On another less important but interesting level, a fourth reason for knowing geometry involves *solving puzzles*. Most of us are intrigued by simple puzzles and enjoy the challenge of trying to solve them. Actually the thought processes we follow when attempting to find a solution are more valuable than the solution itself. This fact will be illustrated with some classic puzzles in the exercises following Section 1.1.

The fifth, and perhaps most important, reason for studying geometry involves *recognizing geometry as an axiomatic system*. Every branch of mathematics is an axiomatic system, but studying geometry gives us one of the best ways to learn more about these systems and the thought processes used in them. In fact, we will start our study of geometry with inductive and deductive reasoning in Section 1.1.

We hope that you will see the value of geometry more clearly as you proceed through the course. To help you, we begin each chapter with an example of a practical application that is solved later in the chapter. We also include a variety of other examples and exercises of a practical nature throughout the text.

Margaret L. Lial
Barbara A. Brown
Arnold R. Steffensen
L. Murphy Johnson

Foundations of Geometry

B ecause the study of geometry requires an under-standing of the way we think, we begin this chapter by discussing inductive and deductive reasoning. This discussion leads us to consider some basic geometric terms such as *point*, *line*, *plane*, and *angle*.

Throughout our presentation, we emphasize ways that you can use geometry to solve applied problems. One application of the concepts presented in this chapter follows and is discussed in Example 3 of Section 1.1.

AN APPLICATION

The following warranty is given with a new CD player:

"This CD player is warranted for one year from the date of purchase against defects in materials or workmanship. During this period, any such defects will be repaired, or the CD player will be replaced at the company's option without charge. This warranty is void in the case of misuse or negligence."

Assume the premises stated in this warranty and answer the following questions.

(a) Bill purchased a CD player, and when he got home and opened the box, he discovered that the case was cracked. What could Bill conclude?

(b) Beth purchased a CD player, and fourteen months later, the digital display burned out. What could Beth conclude?

(c) Jamie purchased a CD player, and two months later, left it outside after listening to it while sunbathing. That night it rained, and the next day the CD player would not play. What could Jamie conclude?

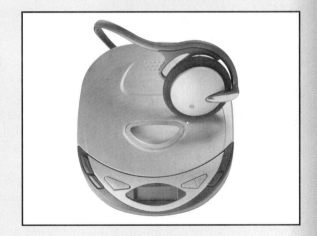

1

1.1   Inductive and Deductive Reasoning

OBJECTIVES

1. Define inductive reasoning.
2. Define an axiomatic system.
3. Define deductive reasoning.

To understand the proofs used in geometry, we begin by considering the two basic ways that we reason or think. The first can be illustrated by studying the following list of numbers:

$$4, 11, 18, 25.$$

What is the next number in this list? Most of us will try to discover what 4, 11, 18, and 25 have in common, and shortly realize that $4 + 7 = 11$, $11 + 7 = 18$, $18 + 7 = 25$, so it would seem that each number after 4 is obtained from the one in front of it by adding 7. As a result, we would probably conclude that the next number should be 32 because $25 + 7 = 32$. Our reasoning process involved considering several *specific* observations, and based on these, we formulated the *general* conclusion that the list will continue in the same pattern if we always add 7 to one number to obtain the next.

OBJECTIVE 1   Define inductive reasoning.

George Cantor (1845–1918)

German mathematician George Cantor is credited with creating a new area of mathematics, set theory, in about 1875. Some of his ideas were viewed as radical by other mathematicians of the day, and a tremendous controversy developed. Today, Cantor's work with sets serves as a foundation for many areas in mathematics, and his efforts are so appreciated that he is referred to as the "Father of Set Theory."

> **DEFINITION: Inductive Reasoning**
>
> We use **inductive reasoning** when we reach a general conclusion based on a limited collection of specific observations.

Natural and social scientists frequently use inductive reasoning. When a laboratory experiment is performed several times with the same result, the physicist might form a general conclusion based on this experimentation. A sociologist might collect information from a limited number of people and attempt to draw a general conclusion about the total population. Also, when a new medicine is tested on a sample of several hundred people, the test results might lead a scientist to draw conclusions about the drug's effectiveness. In all these cases, a general conclusion is drawn from specific observations.

Does inductive reasoning always lead to the same conclusion? The answer is no, not always. If you return to our earlier example of four numbers, in which we concluded that the next number must be 32, you will see that another conclusion is also possible. The next number in the series can be 1, followed by 8, 15, 22, and so forth. This sequence gives the dates of the Thursdays in the year 2001, starting with Thursday, January 4, 2001!

As you can see, although inductive reasoning is widely used, there are no guarantees the conclusion drawn is always correct or that it is the only possible conclusion. The primary flaw with inductive reasoning is that we cannot be sure of what will generally be true based on a limited number of cases.

EXAMPLE 1 Use inductive reasoning to determine the next element in each list. Remember, there might be more than one answer.

(a) 2, 4, 8, 16, 32

We might conclude that the next number is 64 because each number after the first is twice the one before it.

(b) o, t, t, f, f, s, s, e

This one is a bit more difficult. This is a list of the first letters in the words *one, two, three, four,* and so forth. Thus, the next letter would be *n*, the first letter in the word *nine.*

PRACTICE EXERCISE 1

Use inductive reasoning to determine the next element in each list.

(a) 1, 4, 9, 16, 25

(b) 

ANSWERS ON PAGE 6

Although inductive reasoning might not always lead to the same conclusion, it is still an important process, one that is used even by mathematicians. For example, consider the following observations:

$$1 + 1 = 2, \quad 1 + 3 = 4, \quad 3 + 3 = 6, \quad 3 + 5 = 8,$$
$$5 + 1 = 6, \quad 7 + 3 = 10.$$

After examining these equations, a mathematician might conclude that the sum of two odd integers is always an even integer. Certainly these six observations do not prove that this statement is true. What if we listed several hundred such observations with the same results? Would that have proved the statement? The mathematician would say you cannot prove a general statement by giving any number of specific cases, unless, of course, the number of specific cases is limited. The mathematician usually requires a different kind of proof, which is the basis of an *axiomatic system.*

OBJECTIVE 2 Define an axiomatic system.

DEFINITION: Axiomatic System

An **axiomatic system** consists of four parts:
1. undefined terms, 2. definitions, 3. axioms or postulates, and 4. theorems.

Undefined terms are the starting points in a system. Every statement we make is composed of words that have meaning to us. It is impossible to truly define every term because definitions are also formed with words that have meaning. Some terms must be assumed in order to go forward. For example, in geometry the word *set* is an undefined term. Intuitively, we recognize a set to be a collection, or group, or bunch of objects, but this is a definition of *set* only

Euclid (about 300 B.C.)

Euclid, a teacher at the University of Alexandria, collected all known geometry facts into a text called *Elements*. *Elements* contained a systematic and logical arrangement of geometry and was divided into thirteen chapters called *books*. The text was unique in that it began with a few basic assumptions and logically derived everything else from them.

if the terms *collection*, *group*, or *bunch* are also defined. We could continue to find synonyms for the word *set*, but in doing so, we would not be expanding the system. Because we have an intuitive understanding of the word *set*, we simply accept it as an undefined term and use it as a building block for the system.

Definitions are statements that give meaning to new terms that will be used in a system. The words used to form a definition are either undefined terms or previously defined terms. **Postulates**, or **axioms**, are statements about undefined terms and definitions that are accepted as true without verification or proof; they also serve as starting points in a system. In geometry, the word *postulate* is used most often.

Finally, when we have undefined terms, definitions, and postulates, we are ready to begin building the system by deducing results called *theorems*. A **theorem** is a statement that we can prove by using definitions, postulates, and the rules of deduction and logic. Many theorems can be expressed as "*If . . . then*" statements. The phrase that follows the word *if* includes the given information, or the **hypothesis** (plural—**hypotheses**) of the theorem. The phrase following the word *then* includes the statement we are to verify, or the **conclusion** of the theorem.

OBJECTIVE 3 **Define deductive reasoning.** We often begin with a hypothesis, and in a step-by-step manner, obtain other statements by using undefined terms, definitions, postulates, or previously proved theorems, until we reach a conclusion. This process illustrates *deductive reasoning*.

> **DEFINITION: *Deductive Reasoning***
> We use **deductive reasoning** when we reach a specific conclusion based on a collection of generally accepted assumptions.

Let's look at an example of a simple axiomatic system.

EXAMPLE 2 Consider the following axiomatic system.

Undefined terms: happy, pleasant, person
Definition: Terri is a happy person.
Postulate: Every happy person is pleasant.

Using this information, we can state a theorem.

Theorem: Terri is pleasant.

Although we have not yet discussed the procedures for using deductive reasoning to write a proof of a theorem, informally, we might present the following "proof."

PROOF: Because Terri is a happy person, and every happy person is pleasant, it follows that Terri, as one of the happy persons, must also be pleasant.

Using a general formula to find the area of a specific geometric figure is a good example of deductive reasoning. Remember, when we reason deductively, we start with one or more **premises** (undefined terms, definitions, axioms or postulates, or previously proved theorems) and attempt to arrive at a conclusion that logically follows if the premises are accepted. The next example uses deductive reasoning to solve the application given in the chapter introduction.

EXAMPLE 3 Assume the premises stated in the warranty in the margin and answer the following.

(a) Bill purchased a CD player, and when he got home and opened the box, he discovered that the case was cracked. What could Bill conclude?

    Because Bill had just purchased the CD player, one year had not gone by. He had not been negligent, nor had he misused the CD player. He concluded that the company would either repair or replace the CD player.

(b) Beth purchased a CD player, and fourteen months later, the digital display burned out. What could Beth conclude?

    Because fourteen months exceeds the time period of the warranty, Beth concluded that the company would not be required by the warranty to repair or replace the CD player.

(c) Jamie purchased a CD player, and two months later, she left it outside after listening to it while sunbathing. That night it rained, and the next day the CD player would not play. What could Jamie conclude?

    Although the CD player became defective during the warranty period, the defect was due to her negligence. The company would not be required by the warranty to repair or to replace it.

The next example gives practice in recognizing the two types of reasoning—inductive and deductive.

EXAMPLE 4 Determine if each conclusion follows logically from the premises, and state whether the reasoning is inductive or deductive.

(a) *Premise:* My coat is tan.
Premise: Bob's coat is tan.
Premise: Di's coat is tan.
Conclusion: All coats are tan.

    Because we are reasoning from three specific examples and drawing a general conclusion, the process involves inductive reasoning. It is obvious that the conclusion does not logically follow from the premises.

(b) *Premise:* All athletes are in good physical condition.
Premise: Shelly is an athlete.
Conclusion: Shelly is in good physical condition.

    In this case, if we accept the premises, then the conclusion logically follows. This is an example of deductive reasoning. Notice that the conclusion may be true or false; we don't actually know if Shelly is or is not in good condition. However, this is beside the point because the conclusion logically follows from the premises. Certainly, if the premises are true and a conclusion follows, then the conclusion is also true. In mathematics, we assume premises are true so that deduced conclusions are also true statements.

The following warranty is given with a new CD player:

"This CD player is warranted for one year from the date of purchase against defects in materials or workmanship. During this period, any such defects will be repaired, or the CD player will be replaced at the company's option without charge. This warranty is void in the case of misuse or negligence."

Aristotle (384–322 B.C.)

Aristotle, a Greek logician and philosopher, is credited with being the first to systematically study the logic and reasoning used in everyday life. He was a student of Plato and eventually became a tutor of Alexander the Great. The logic of Aristotle forms the basis for how we learn to reason deductively. It also serves as a foundation for the more formalized, symbolic logic studied and used by mathematicians today.

PRACTICE EXERCISE 2

Determine if each conclusion follows logically from the premises, and state whether the reasoning is *inductive* or *deductive*.

(a) *Premise:* This year is leap year.
   Conclusion: Next year will not be leap year.

(b) *Premise:* The Boston Celtics won the NBA Championship in the years 1960 to 1966.
   Premise: The Boston Celtics won the NBA Championship in the years 1968 to 1969.
   Conclusion: The Celtics won the NBA Championship every year
      during the 1960s.   **ANSWERS ON PAGE 6**

A **fallacy** is a conclusion that does not necessarily follow from the premises.

EXAMPLE 5   Consider the following premises.

Premise: A coach tells his team, "If we are to win tonight, then we will have to play very hard."

Premise: The team played extremely hard.

   What can we conclude? Some of you might conclude that "the team won," but this is *not* correct based on the premises. This is a fallacy. The team was told that *If* they win *then* they will have played hard, but nothing was said about what would happen *If* they played hard. In fact, you can see that even if the team members played their hearts out, the other team might have been far superior and defeated them. In this case, a conclusion does not logically follow from the premises. However, if the second premise were replaced with the following:

Premise: The team won the game.

We could then logically conclude that the team played hard.

Answers to Practice Exercises

1. **(a)** One answer is 36 since each number is the square of the numbers 1, 2, 3, 4, and 5. That is, $1 = 1^2$, $4 = 2^2$, $9 = 3^2$, $16 = 4^2$, $25 = 5^2$, and $36 = 6^2$.   **(b)** The next element is ⚆ since each element is one of the numbers 1, 2, 3, 4, and 5 placed next to its mirror image.   **2.** **(a)** Using deductive reasoning, the conclusion follows logically from the given premise and the accepted but unstated premise that leap year occurs every four years.   **(b)** The conclusion does not necessarily follow from the premises. It was reached by inductive reasoning. It is a false statement because in 1967 the Philadelphia 76ers won the NBA Championship.

Use inductive reasoning in Exercises 1–12 to determine the next element in each list.

1. 3, 8, 13, 18, 23

2. 12, 7, 2, −3, −8

3. 1, 3, 9, 27, 81

4. 1, 5, 25, 125

5. 1, −2, 4, −8, 16

6. 40, −20, 10, −5

7. 1, 1, 2, 3, 5, 8, 13, 21

8. 1, 3, 4, 7, 11, 18, 29

9. S, M, T, W, T, F

10. J, F, M, A, M

11. Z, Y, X, W, V

12. $\dfrac{1}{6}, \dfrac{1}{3}, \dfrac{1}{2}, \dfrac{2}{3}$

13. Consider the following statements. Use inductive reasoning to complete the given statement. Test your statement using other numbers.

$3 + 7 = 10 \qquad -5 + 3 = -2$

$-1 + 9 = 8 \qquad -11 + 5 = 16$

$-19 + (-15) = -34$

The sum of two odd numbers is _____.

14. What is the purpose of having undefined terms in an axiomatic system?

15. What is the difference between a postulate and a definition?

16. What is the difference between a postulate and an axiom?

17. What is the difference between a postulate and a theorem?

18. Give a definition of the word *happy*. What synonyms did you use in your definition? If you used a dictionary of synonyms to find those for *happy*, look up the meaning of each synonym. Continuing in this manner, what will eventually happen?

19. Many theorems are "*If . . . then . . .*" statements. What is the information that follows *If* called? What is the information that follows *then* called?

20. Explain the difference between inductive and deductive reasoning.

21. In the *Declaration of Independence*, the statement is made that "All men are created equal." Is this statement a postulate or a theorem?

22. What do you suppose would happen if two axiomatic systems had the same undefined terms and definitions but different postulates?

23. In a democracy such as that of the United States, we assume that the government exists to serve the people. In a dictatorship, it is assumed that the people exist to serve the government. What happens when two countries that support different structures of government try to negotiate?

In Exercises 24–35, determine if each conclusion follows logically from the premises and state whether the reasoning is inductive or deductive.

24. *Premise:* If you are a mathematics major, then you can compute discounts on sale items.
Premise: Becky is a mathematics major.
Conclusion: Becky can compute discounts on sale items.

25. *Premise:* If you are a mathematics major, then you can compute discounts on sale items.
Premise: Becky can compute discounts on sale items.
Conclusion: Becky is a mathematics major.

26. *Premise:* If you are a home buyer, then you make payments.
Premise: Doug makes payments.
Conclusion: Doug is a home buyer.

27. *Premise:* If you are a home buyer, then you make payments.
Premise: Doug is a home buyer.
Conclusion: Doug makes payments.

28. *Premise:* It rained on Tuesday.
    Premise: It rained on Wednesday.
    Conclusion: It will rain on Thursday.

29. *Premise:* Last year, I won money in Las Vegas.
    Premise: The year before last, I won money in Las Vegas.
    Conclusion: I will win money in Las Vegas this year.

30. *Premise:* If I buy a car, then it will be a Buick.
    Premise: If I receive a check from my father, then I will buy a car.
    Premise: I received a check from my father.
    Conclusion: I will buy a Buick.

31. *Premise:* If you are going to be an engineer, then you will study mathematics.
    Premise: If you study mathematics, then you will get a good job.
    Premise: Roy is going to be an engineer.
    Conclusion: Roy will get a good job.

32. *Premise:* If you are an ogg, then you are an arg.
    Premise: If you are a pon, then you are an ogg.
    Conclusion: If you are a pon, then you are an arg.

33. *Premise:* If it is a frog, then it is green.
    Premise: If it hops, then it is a frog.
    Conclusion: If it hops, then it is green.

34. Explain why
    Conclusion: You are an arg.
    Conclusion does not logically follow from the premises in Exercise 32.

35. Explain why
    Conclusion: It is green.
    Conclusion does not logically follow from the premises in Exercise 33.

The puzzles in Exercises 36–47 are classic examples and a certain amount of deductive reasoning is required to solve them. Some of these puzzles are quite challenging, so don't be discouraged if you have trouble finding the solution immediately. Ideally they will make you think a bit and, along the way, provide a bit of entertainment.

36. Mary has two U.S. coins in her purse. Together they total 55¢. One is not a nickel. What are the two coins? [Hint: We did *not* say that neither is a nickel.]

37. A young dog and an older dog are in the backyard. The young dog is the older dog's daughter but the older dog is not the young dog's mother. Explain.

38. If you take 5 apples from 8 apples, what do you have?

39. A rancher had 20 cattle. All but 12 died. How many did he have left?

40. We know there are 12 one-cent stamps in a dozen, but how many two-cent stamps are in a dozen?

41. A museum fired an archaeologist who claimed she found a coin dated 300 B.C. Why?

42. What arithmetic symbol can be placed between 2 and 3 to form a number greater than 2 and less than 3?

43. The number of marbles in a jar doubles every minute and is full in 10 minutes. When was the jar half full?

44. An inflatable raft will carry at maximum 200 lb. How can a man weighing 200 lb and his two daughters, each of whom weighs 100 lb, use the raft to reach an island?

45. How many times can you subtract 5 from 25?

46. A judge wishing to convict a defendant puts two pieces of paper in a hat. He tells the jury that if the defendant draws the piece marked "guilty" he will be convicted, but if he draws the piece marked "innocent" he will be set free. The hitch is that the judge wrote "guilty" on both pieces of paper. But when the crafty defendant showed the jury one piece of paper, the judge was forced to let him go free. How did the defendant outwit the judge?

47. You have 3 sacks, each containing 3 coins. Two of the sacks contain real coins and each coin weighs 1 lb. The third contains counterfeit coins, and each weighs 1 lb 1 oz. A scale is available, but it can be used one time and one time only to obtain a particular measure of weight. How might you use the scale to determine which sack contains the counterfeit coins? [Note: You cannot add or subtract coins to a total because any change of reading up or down on the scale will cause it to zero out.]

The puzzles in Exercises 48–55 have all been attributed to Englishman Henry Ernest Dudeney (1857–1930), called by some the greatest puzzle writer of all time.

48. A block of wood in the shape of a cube, shown in the accompanying figure, is to be sawed into two pieces with one cut so that each resulting piece has a surface in the shape of a regular hexagon (a six-sided figure with sides of equal length). How should the cut be made? [Hint: A cube has six surfaces, and all must be cut to create the two pieces.]

Exercise 48

Exercise 49

49. A bug is sitting on the surface of a solid cube of wood at point A as shown in the figure. If it wants to get to the opposite corner at point B by the shortest possible route, show the path it should take along the surface of the cube.

50. A square piece of cardboard, 8 inches on a side and marked like a checkerboard, is cut into four pieces as shown in Figure 50. The pieces are reassembled to form a rectangle like the one shown in Figure 51. The original square contains 64 little squares whereas the rectangle contains 65 little squares. Where does the extra square come from?

Exercise 50

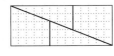

Exercise 51

51. Rearrange the four pieces of the square in Exercise 50 to form a new figure that contains only 63 little squares. In this case, we seem to have lost a little square, whereas, in the preceding exercise, we seem to have gained a square. Explain what happened to this square.

52. Consider the six matches arranged to form a regular hexagon as shown below. Take three more matches and arrange the nine to show another regular six-sided figure. The matches cannot be placed on top of one another, cannot be broken, and there should be no loose ends when you are finished.

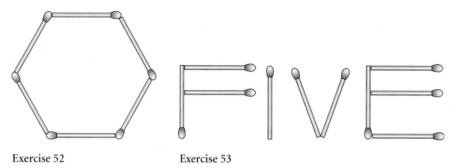

Exercise 52   Exercise 53

53. Consider ten matches arranged to form the word FIVE as shown above. Remove seven of the matches so that what is left is four.

54. Consider the following six "postulates."

  a. Art, Bev, Cheryl, Dot, and Ed all attend the same college where one is a freshman, one is a sophomore, one is a junior, one is a senior, and one is a graduate student.

  b. Art, Bev, and Cheryl have not yet completed their undergraduate work.

  c. Bev is one year ahead of Ed.

  d. Art is not a freshman.

  e. Ed is not a freshman.

  f. Art is in a higher class than Ed.

  Can you discover theorems that give each student's class in college? Write the theorems in complete sentences.

55. Consider the following eight "postulates."

  a. Smith, Jones, and Rodriquez are the engineer, brakeman, and fireman on a train, not necessarily in that order.

  b. Riding on the train are three passengers with the same last names as the crewmembers, identified as passenger Smith, passenger Jones, and passenger Rodriquez in the following statements.

  c. The brakeman lives in Denver.

  d. Passenger Rodriquez lives in San Francisco.

  e. Passenger Jones long ago forgot all the algebra that he learned in high school.

  f. The passenger with the same name as the brakeman lives in New York.

  g. The brakeman and one of the passengers, a professor of mathematical physics, attend the same health club.

  h. Smith beat the fireman in a game of tennis at a court near their homes.

  Can you discover a theorem that tells the name of the engineer? The brakeman? The fireman? Write the theorems in complete sentences.

56. Write a paragraph about a situation in which you have observed or used inductive reasoning. Explain why you believe it is inductive reasoning.

1.2 Points, Lines, and Planes

OBJECTIVES

1. State undefined terms.
2. State postulates about points, lines, planes, and real numbers.

Have you ever looked closely at the screen on a television set? The surface is composed of thousands of small dots (called pixels) that glow when hit by an electron beam. When viewed from a distance, the individual dots blend together forming the picture. Each dot serves as a model for a point, one of the simplest geometric figures that we study.

OBJECTIVE 1   State undefined terms.   In Section 1.1 we discussed axiomatic systems and discovered that such systems consist of undefined terms, definitions, postulates (or axioms), and theorems. The undefined terms provide a starting point for developing an axiomatic system. In geometry, we begin with the following undefined terms:

set,   point,   line,   plane.

Any attempt to define these terms would require more words that are undefined. Intuitively, we have some idea of their meaning. A **set** is a collection or group of objects. A **point** can be thought of as an object that determines a position but that has no dimension (length, width, or height). We can symbolize a point with a dot and label it with a capital letter, such as the point A shown in Figure 1.1(a).

A **line** can be thought of as a set of points in a one-dimensional straight figure that extends in opposite directions without ending. Figure 1.1(b) shows a representation of a line passing through the two points B and C. The arrowheads indicate that the line continues in that direction without ending. We symbolize a line such as this using $\overleftrightarrow{BC}$ or $\overleftrightarrow{CB}$, or when appropriate, by using a lowercase letter such as ℓ.

Finally, a **plane** is a set of points on a flat surface, such as the face of a blackboard, having two dimensions and extending without boundary. We often represent a plane as shown in Figure 1.1(c) and symbolize planes using a script letter such as $\mathcal{P}$.

> **DEFINITION: Space and Geometric Figures**
> The set of all points is called **space**. Any set of points, lines, or planes in space is called a **geometric figure**.

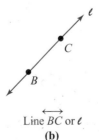

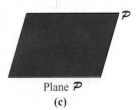

A
•
Point A
(a)

Line $\overleftrightarrow{BC}$ or ℓ
(b)

Plane $\mathcal{P}$
(c)

Figure 1.1

OBJECTIVE 2 **State postulates about points, lines, planes, and real numbers.** We can think of geometry as the study of properties of geometric figures. Some of these properties must be assumed in the form of postulates.

POSTULATE 1.1

Given any two distinct points in space, there is exactly one line that passes through them.

Intuitively, Postulate 1.1 tells us that we can draw one and only one straight line through two different points.

POSTULATE 1.2

Given any three distinct points in space not on the same line, there is exactly one plane that passes through them.

Figure 1.2(a) shows the plane containing the three points A, B, and C. These points are **coplanar**, lie in the same plane. Notice that A, B, and C are not on the same line. If three points such as D, E, and F are on the same line, as in Figure 1.2(b), we can see that more than one plane can contain them.

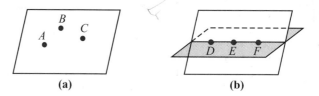

(a)   (b)

Figure 1.2 Points in Planes

Student Activity

On a piece of paper, place 5 points. Label them A, B, C, D, and E as in the drawing shown at the right.
Next, fold the paper so that point E is on the fold.
Open the paper slightly and think of the paper as two planes meeting at the fold. Answer the following questions. There may be more than one correct answer.
1. Name three coplanar points.
2. Name three noncoplanar points.
3. Name a point in both planes.

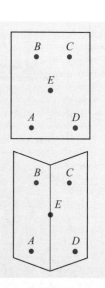

> **POSTULATE 1.3**
>
> The line determined by any two distinct points in a plane is also contained in the plane.

From Postulate 1.3, we see that when we draw lines between points in a given plane, we always remain in the plane. The next postulate guarantees that there are more points in space than those found in any given plane.

> **POSTULATE 1.4**
>
> No plane contains all points in space.

From the previous postulate, space contains at least four points that are not all in the same plane.

The next postulate has numerous applications in algebra.

> **POSTULATE 1.5   *Ruler Postulate***
>
> There is a one-to-one correspondence between the set of all points on a line and the set of all real numbers.

If we draw a line, select a point on it (called the **origin**), mark off equal units in both directions, and label these points with integers, the result is called a **number line**. (See Figure 1.3.) The number corresponding to a given point on the line is called the **coordinate** of the point, and when we identify a point with a given real number, we are **plotting** the point associated with the number.

Origin

$$-3\ -2\ -1\ \ 0\ \ 1\ \ 2\ \ 3$$

Figure 1.3 Number Line

EXAMPLE 1   Plot the points associated with the real numbers $\frac{1}{2}$, $-\frac{3}{4}$, $\sqrt{2}$, and 2.5 on a number line.

Start with a number line like the one in Figure 1.3. Because $\frac{1}{2}$ is halfway between 0 and 1, we locate the point midway between them. The points corresponding to $-\frac{3}{4}$ and 2.5 are determined similarly. We can find the approximate location of the point corresponding to $\sqrt{2}$ by recognizing that $\sqrt{2}$ is approximately 1.4. The four points are plotted in Figure 1.4.

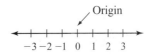

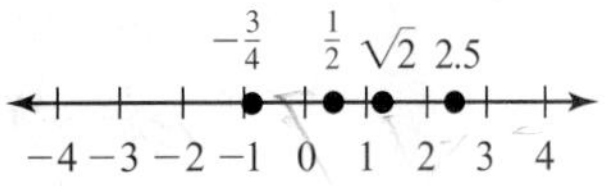

$$-4\ -3\ -2\ -1\ \ 0\ \ 1\ \ 2\ \ 3\ \ 4$$

Figure 1.4 Points on a Number Line

PRACTICE EXERCISE ❶

Approximate the coordinates of points A, B, C, and D as shown on the following number line. (The answers are approximations, so your answers may differ slightly.)

ANSWERS ON PAGE 15

Because there are infinitely many real numbers, there are infinitely many points on any given line. This conclusion is actually a theorem, a statement that can be proved. When proving certain theorems in geometry, we will use some of the following postulates from algebra.

Note Algebraic laws and geometric postulates are both statements accepted as true without needing to be proved. For this reason, the words *law* and *postulate* (when referring to algebraic laws) are used interchangeably throughout this book.

POSTULATE 1.6 The Reflexive Law

Any quantity is equal to itself. $(x = x)$

POSTULATE 1.7 The Symmetric Law

If x and y are any two quantities and $x = y$, then $y = x$.

POSTULATE 1.8 The Transitive Law

If x, y, and z are any three quantities with $x = y$ and $y = z$, then $x = z$.

POSTULATE 1.9 The Addition-Subtraction Law

If w, x, y, and z are any four quantities with $w = x$ and $y = z$, then $w + y = x + z$ and $w - y = x - z$.

The addition-subtraction law states that if equal quantities are added to or subtracted from equal quantities, the results are also equal.

POSTULATE 1.10 The Multiplication-Division Law

If w, x, y, and z are any four quantities with $w = x$ and $y = z$, then $wy = xz$ and $\dfrac{w}{y} = \dfrac{x}{z}$ (provided $y \neq 0$ and $z \neq 0$).

The multiplication-division law states that if equal quantities are multiplied or divided by equal quantities (division by zero excluded), the results are also equal.

> **POSTULATE 1.11   The Substitution Law**
>
> If x and y are any two quantities with $x = y$, then x can be substituted for y in any expression containing y.

> **POSTULATE 1.12   The Distributive Law**
>
> If x, y, and z are any three quantities, then $x(y + z) = xy + xz$.

Postulates 1.6–1.12 are used extensively when solving algebraic equations.

Answer to Practice Exercise

1. A: 3; B: -2; C: $-\dfrac{1}{2}$; D: $\dfrac{7}{4}$

1.2 Exercises

FOR EXTRA HELP: 📖 Student's Solutions Manual   Tutor Center   Addison-Wesley Math Tutor Center

1. How many lines can be drawn between two distinct points?
2. How many planes are determined by three distinct points that are not on the same line?
3. If distinct points A and B are in plane $\mathcal{P}$, and point C is on the line determined by A and B, what can be said about C relative to $\mathcal{P}$?
4. If $\mathcal{P}$ is a plane and A is a point in space, must A be on $\mathcal{P}$?

In Exercises 5–8, A, B, and C are distinct points, ℓ and m are lines, and $\mathcal{P}$ is a plane. Name the postulate illustrated by each statement.

5. If A and B are on ℓ, and A and B are on m, then $m = \ell$.
6. If A and B are on ℓ, ℓ is in $\mathcal{P}$, and C is on ℓ, then C is in $\mathcal{P}$.
7. There is a point A such that A is not in $\mathcal{P}$.
8. If A and B are on ℓ, A is on m, and $\ell \neq m$, then B is not on m.
9. Construct a number line and plot the points associated with the real numbers $3, \dfrac{7}{8}, -\dfrac{1}{3}, -1.5,$ and $\sqrt{3}$.
10. Construct a number line and plot the points associated with the real numbers $-3, \dfrac{1}{4}, -\dfrac{4}{3}, 3.2,$ and $-\sqrt{5}$.

In Exercises 11 and 12, approximate the coordinates of points A, B, C, and D on the given number lines.

11.

12.

In Exercises 13–17, complete each statement using the specified postulate.

13. Reflexive law: $5 = $ __?__.

14. Transitive law: If $a = b$ and $b = 3$, then $a = $ __?__.

15. Addition-subtraction law: If $a = b$, then $a + 7 = b + $ __?__.

16. Multiplication-division law: If $2x = 6$, then $x = $ __?__.

17. Symmetric law: If $a = -3$, then $-3 = $ __?__.

Answer *true* or *false* in Exercises 18–22. If the statement is false, explain why.

18. By the reflexive law, $-2 = -2$.

19. If $w = 7$ and $7 = x$, then $w = x$ by the symmetric law.

20. If $x + 1 = 8$, then $x = 7$ by the addition-subtraction law.

21. If $\frac{1}{3}y = 2$, then $y = 6$ by the addition-subtraction law.

22. If $8 = x$ then $x = 8$ by the reflexive law.

In Exercises 23–24, give the postulate that supports each indicated step in the solution of the equation.

23. Solve $3x + 2 = 4 + 5x$.

Statements	*Reasons*
1. $3x + 2 = 4 + 5x$	1. Given
2. $3x + 2 - 4 = 4 - 4 + 5x$	2. _____
3. $3x - 2 = 5x$	3. Simplify
4. $3x - 3x - 2 = 5x - 3x$	4. _____
5. $-2 = 2x$	5. Simplify
6. $\frac{1}{2}(-2) = \frac{1}{2}(2x)$	6. _____
7. $-1 = x$	7. Simplify
8. $x = -1$	8. _____
Check: -1 in $3x + 2 = 4 + 5x$	
9. $3(-1) + 2 = 4 + 5(-1)$	9. _____
10. $-1 = -1$	10. Simplify

24. Solve $\frac{2}{3}x + 1 = x$.

Statements	Reasons
1. $\frac{2}{3}x + 1 = x$	1. Given
2. $3\left[\frac{2}{3}x + 1\right] = 3x$	2. _____
3. $2x + 3 = 3x$	3. _____
4. $2x - 2x + 3 = 3x - 2x$	4. _____
5. $3 = x$	5. Simplify
6. $x = 3$	6. _____

Check: 3 in $\frac{2}{3}x + 1 = x$

7. $\frac{2}{3}(3) + 1 = 3$	7. _____
8. $3 = 3$	8. Simplify

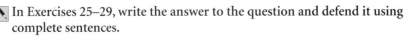 In Exercises 25–29, write the answer to the question and defend it using complete sentences.

25. Can a given point exist on two distinct lines? On five distinct lines?
26. Can two distinct points both exist on two distinct lines?
27. Can a given line exist in two distinct planes? In five distinct planes?
28. Can two given lines with no point in common exist in the same plane?
29. Explain why stools are sometimes made with three legs rather than four, to provide greater stability.

(1.3)   Segments, Rays, and Angles

OBJECTIVES

1. Define line segment and ray.
2. Define angle.
3. Identify special angles.

A line can be thought of as a geometric figure consisting of points that extend infinitely far in opposite directions. We now consider two new figures that are parts of a line. Another undefined term, *between*, describes the position of points on a line in relation to two given points on that line. Intuitively, points on the colored portion of the line in Figure 1.5 are said to be *between* points A and B.

Figure 1.5   Points between A and B

OBJECTIVE 1   Define line segment and ray.

> **DEFINITION: Line Segment**
>
> Let A and B be two distinct points on a line. The geometric figure consisting of all points between A and B, including A and B, is called a **line segment** or **segment**, denoted by $\overline{AB}$ or $\overline{BA}$. The points A and B are called **endpoints** of $\overline{AB}$.

Recall that the line determined by distinct points A and B is denoted by $\overleftrightarrow{AB}$ or $\overleftrightarrow{BA}$, which distinguishes it from the segment $\overline{AB}$ or $\overline{BA}$. The **length** of segment $\overline{AB}$ is the distance between the endpoints A and B and is denoted by AB.

Technology Connection

Geometry software will be needed.

1. Draw a line segment and measure it. Name it $\overline{AC}$.
2. Locate a point B on this line segment.
3. Measure $\overline{AB}$ and $\overline{BC}$.
4. Does $AC = AB + BC$, $BC = AC - AB$, and $AB = AC - BC$?

Compare these results with Postulate 1.13.

The next postulate provides a way to find the length of a segment that is made up of other segments. Refer to Figure 1.6a.

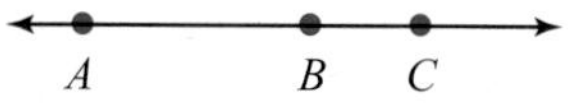

> **POSTULATE 1.13   Segment Addition Postulate**
>
> Let A, B, and C be three points on the same line with B between A and C. Then $AC = AB + BC$, $BC = AC - AB$, and $AB = AC - BC$.

Figure 1.6a   Segment Addition

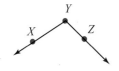

Figure 1.6b

In Figure 1.6a, points A, B, and C are said to be **collinear**, that is, they are on the same line. In this example, B is between A and C. This can be symbolized by writing A-B-C. The segment addition postulate can be applied to many problems whose solutions also require basic algebra.

In Figure 1.6b, points X, Y, and Z are not collinear.

EXAMPLE 1

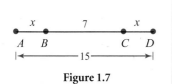

Figure 1.7

In Figure 1.7, $AD = 15$, $BC = 7$, and $AB = CD$. Find AB.

Let $AB = x$. Then because $AB = CD$, $CD = x$. By extending the segment addition postulate, we have

$$AD = AB + BC + CD.$$

Substitute 15 for AD, 7 for BC, and x for both AB and CD to obtain

$$15 = x + 7 + x.$$

$15 = 2x + 7$	Collect like terms
$8 = 2x$	Subtract 7 from both sides
$4 = x$	Divide both sides by 2

Thus, $AB = 4$.

PRACTICE EXERCISE 1

Use the figure below to find the value of x.

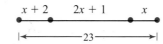

ANSWERS ON PAGE 25

DEFINITION: Ray

Let A and B be two distinct points on a line. The geometric figure consisting of the point A together with all points on $\overleftrightarrow{AB}$ on the same side of A as B is called a **ray**, denoted by $\overrightarrow{AB}$. The point A is called the **endpoint** of $\overrightarrow{AB}$.

Figure 1.8 Ray $\overrightarrow{AB}$

In Figure 1.8, the ray $\overrightarrow{AB}$ is shown in color. It includes point A and all the points to the right of A. Unlike the notations for segments and lines, the rays $\overrightarrow{AB}$ and $\overrightarrow{BA}$ are different. A ray has only one endpoint and it is always written first. Taken together, rays $\overrightarrow{AB}$ and $\overrightarrow{BA}$ make up the line $\overleftrightarrow{AB}$.

Note

SUMMARY

SYMBOL	WORDS	FIGURE
$\overleftrightarrow{AB}$	Line AB	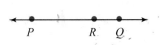 A ———•———————•———→ B
$\overrightarrow{AB}$	Ray AB	A •———————•———→ B
$\overline{AB}$	Line segment AB	A •———————————• B
AB	Length of segment AB	A number

EXAMPLE 2

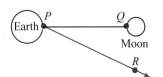

Figure 1.9

Use Figure 1.9 to answer each question.

(a) Is Q on $\overline{PR}$?   (a) No

(b) Is Q on $\overleftrightarrow{PR}$?   (b) Yes

(c) Is Q on $\overrightarrow{PR}$?   (c) Yes

(d) Is Q on $\overrightarrow{RP}$?   (d) No

(e) What are the endpoints of $\overline{PR}$?   (e) P and R

(f) What are the endpoints of $\overrightarrow{PR}$?   (f) Only P

(g) What are the endpoints of $\overleftrightarrow{PR}$?   (g) There are no endpoints.

(h) Are $\overline{PQ}$ and $\overline{QP}$ the same?   (h) Yes

(i) Are $\overrightarrow{PQ}$ and $\overrightarrow{QP}$ the same?   (i) No

(j) If $PQ = 5$ cm and $RQ = 2$ cm, what is the length of $\overline{PR}$?   (j) 3 cm

Scientists have measured the distance between Earth and the Moon to within a few centimeters. A laser beam is directed from point P and reflected back to Earth from a mirror that was left by the Apollo astronauts at point Q. The time it takes for the beam to return to Earth can be measured and used to calculate the distance from Earth to the Moon. The two points P and Q, along with the laser beam, provide a model of a segment. A laser beam directed into space from P through R serves as a model for a ray.

PRACTICE EXERCISE 2

Draw two points X and Y and place point Z on $\overleftrightarrow{XY}$ but not on $\overline{XY}$. Use your figure to answer the following questions.

(a) Is Z on $\overrightarrow{YX}$?

(b) Is Z on $\overline{YX}$?

(c) Is Y on $\overrightarrow{ZX}$?

(d) What are the endpoints of $\overrightarrow{XY}$?

(e) What are the endpoints of $\overline{XY}$?

(f) Are $\overrightarrow{XY}$ and $\overrightarrow{YX}$ the same?

(g) If $ZX = 3$ cm and $XY = 5$ cm, what is the length of $\overline{ZY}$?

ANSWERS ON PAGE 25

OBJECTIVE 2   **Define angle.**   The *angle* is one of the most important figures studied in geometry.

> **DEFINITION: *Angle***
>
> An **angle** is a geometric figure consisting of two rays that share a common endpoint, called the **vertex** of the angle. The rays are called **sides** of the angle.

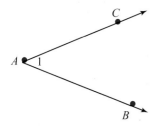

Figure 1.10 ∠*CAB*

Angles are named in three ways, using these letters, using one letter, or using a number. Consider the angle formed by rays $\overrightarrow{AC}$ and $\overrightarrow{AB}$ in Figure 1.10. We use the three points on the angle, A, B, and C, and call the angle ∠*CAB* or ∠*BAC* (the symbol ∠ is read "angle"). Notice in both cases the vertex, A, is *always* written between the other two points. When no confusion can arise, we simply name the angle ∠*A*, using only the vertex point. The third possibility is to write a number such as 1 in the position shown in Figure 1.10 and call the angle ∠1.

We are familiar with measuring the length of a segment using a suitable unit of measure such as inch, centimeter, foot, or meter. In order to measure an angle, we need a measuring unit. The most common unit is the **degree** (°). An angle with measure 0° is formed by two coinciding rays such as $\overrightarrow{AB}$ and $\overrightarrow{AC}$ in Figure 1.11.

Figure 1.11 ∠*BAC* has Measure 0°

If we rotate ray $\overrightarrow{AB}$ in a counterclockwise direction from ray $\overrightarrow{AC}$ in Figure 1.11, the two rays form larger and larger angles as shown in Figure 1.12.

Using degrees as a unit of angular measure originated with the ancient Sumerians. The Sumerians thought that it took Earth 360 days to revolve around the Sun. They assumed Earth's orbit was a circle and that Earth traveled at a constant speed. Thus, to traverse completely around a circle required 360 units of time (days) so that in 1 day, Earth would travel $\frac{1}{360}$ of a circle. The Sumerians defined the measure of the angle formed by $\frac{1}{360}$ of a circle as a degree.

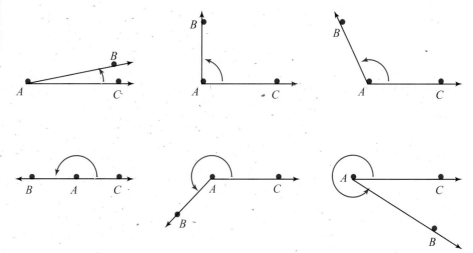

Figure 1.12 Angles

If $\overrightarrow{AB}$ in Figure 1.11 is allowed to rotate completely around until it coincides with ray $\overrightarrow{AC}$ again, the resulting angle is said to measure 360°. Thus, an angle of measure 1° is formed by making $\frac{1}{360}$ of a complete rotation. An angle of measure 1° is shown in Figure 1.13.

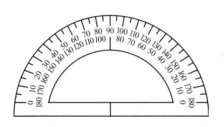

Figure 1.13   Angle Measuring 1°

The approximate measure of an angle in degrees can be found by using a protractor shown in Figure 1.14.

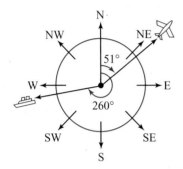

The navigator on an airplane or a ship uses angles measured in degrees to give the direction of travel of the craft. The angle between due north and the direction of travel, measured in degrees in a clockwise direction, is called the *navigational direction* or *course* of the craft. The airplane in the figure is flying on a course of 51° while the ship is sailing a course of 260°.

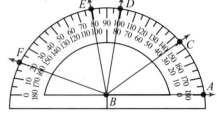

Figure 1.14   Protractor   **Figure 1.15   Measuring Angles**

Note The protractors pictured have dual scales. When the protractor is placed along one side of the angle, use the scale on that same side that starts at "0" to read the measure.

Figure 1.15 shows how a protractor is used to measure various angles. $\angle ABC$ has measure 35°, $\angle ABD$ has measure 80°, $\angle ABE$ has measure 100°, and $\angle ABF$ has measure 160°. Rather than say "$\angle ABC$ has measure 35°," we simply write $m\angle ABC = 35°$.

In some practical applications, angles must be measured with greater precision. One degree is divided into 60 equal parts called minutes ($'$), and one minute is divided into 60 equal parts called seconds ($''$). Thus, the measure of an angle might be given as $55°28'48''$.

OBJECTIVE 3   **Identify special angles.**   Certain angles are given special names.

> **DEFINITION: Special Angles**
> An angle whose sides form a line is called a **straight angle** and has measure 180°. An angle with measure 90° is called a **right angle**. An angle with measure between 0° and 90° is called an **acute angle**. An angle with measure between 90° and 180° is called an **obtuse angle**.

Note All definitions are reversible. For example, the definition says if an angle measures 90° then it is a right angle. The reverse is also true by the definition: If an angle is a right angle, it measures 90°.

EXAMPLE 3   Use Figure 1.16 to answer each question.

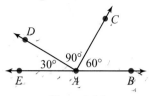

Figure 1.16

(a) Is $\angle BAE$ a straight angle?   Yes   (b) Is $\angle EAD$ an acute angle?   Yes
(c) Is $\angle CAE$ a right angle?   No   (d) Is $\angle DAB$ an obtuse angle?   Yes

Two acute angles or two obtuse angles need not be equal. However, right angles are *always* equal in measure. **Equal angles** are angles with the same measure.

> **DEFINITION: Complementary and Supplementary Angles**
> Two angles whose measures total 90° are called **complementary angles** and each is called the **complement** of the other. Two angles whose measures total 180° are called **supplementary angles**, and each angle is called the **supplement** of the other.

EXAMPLE 4   (a) If $m\angle A = 35°$ and $m\angle B = 55°$, then $\angle A$ and $\angle B$ are complementary because 35° + 55° = 90°.

(b) If $m\angle C = 120°15'45''$ and $m\angle D = 59°44'15''$, then because

$$120°15'45''$$
$$+\ \ 59°44'15''$$
$$\overline{\ 179°59'60''} \text{ but } 60'' = 1'$$
$$= 179°60' \quad \text{ but } 60' = 1°$$
$$= 180°$$

$\angle C$ and $\angle D$ are supplementary angles.

(c) If $m\angle P = 20°$, $m\angle Q = 30°$, and $m\angle R = 40°$, although 20° + 30° + 40° = 90°, we do not call the angles complementary. The definition of complementary angles involves only two angles (not three or more) that add up to 90°.

PRACTICE EXERCISE 3

(a) If $m\angle A = 27°$ and $\angle A$ and $\angle B$ are complementary, find the measure of $\angle B$.
(b) If $m\angle P = 74°26'52''$ and $\angle P$ and $\angle Q$ are supplementary, find the measure of $\angle Q$.

ANSWERS ON PAGE 25

EXAMPLE 5   If $m\angle A = (2x)°$, $m\angle B = (x - 6)°$, and $\angle A$ and $\angle B$ are complementary, find x.

Because $\angle A$ and $\angle B$ are complementary, $m\angle A + m\angle B = 90°$. Substituting, we have

$$2x + (x - 6) = 90$$
$$2x + x - 6 = 90$$
$$3x - 6 = 90$$
$$3x = 96$$
$$x = 32$$

PRACTICE EXERCISE 4

If $m\angle P = (2y - 9)°$, $m\angle Q = (7y)°$, and $\angle P$ and $\angle Q$ are supplementary, find y.

ANSWERS ON PAGE 25

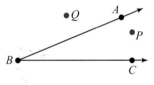

Figure 1.17 Points Interior and Exterior to $\angle ABC$

Two undefined terms that are useful when working with angles are *interior* and *exterior*. Consider $\angle ABC$ and points P and Q as shown in Figure 1.17. We say that P is in the **interior** of $\angle ABC$ and Q is **exterior** to $\angle ABC$.

DEFINITION: Adjacent Angles

Two angles are called **adjacent angles** if they have a common vertex, share a common side, and have no interior points in common.

Note For two angles to have no interior points in common means the angles do not overlap.

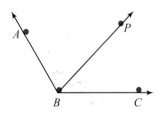

Figure 1.18 Adjacent Angles

For example, in Figure 1.18, $\angle ABP$ and $\angle PBC$ are adjacent, but $\angle ABC$ and $\angle ABP$ are not adjacent because they have some common interior points. Notice that intuitively, $\angle ABP$ and $\angle CBP$ are adjacent angles if P is in the interior of $\angle ABC$.

Figure 1.19 shows examples of nonadjacent angles. In each figure, angles 1 and 2 are nonadjacent angles.

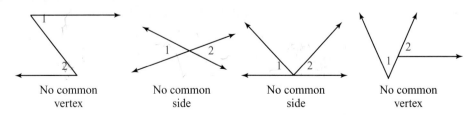

| No common vertex | No common side | No common side | No common vertex |

Figure 1.19

Figure 1.18 can also be used to clarify the next postulate.

> **POSTULATE 1.14   Angle Addition Postulate**
>
> Let A, B, and C be points that determine $\angle ABC$ with P a point in the interior of the angle. Then $m\angle ABC = m\angle ABP + m\angle PBC$, $m\angle PBC = m\angle ABC - m\angle ABP$ and $m\angle ABP = m\angle ABC - m\angle PBC$.

EXAMPLE 6

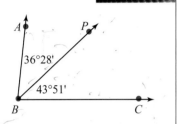

Figure 1.20

Suppose that $\angle ABP$ and $\angle PBC$ are adjacent angles and that $m\angle ABP = 36°28'$ and $m\angle PBC = 43°51'$. Find the measure of $\angle ABC$. Figure 1.20 shows the given angles from which we can see that because P is in the interior of $\angle ABC$, by Postulate 1.14,

$$m\angle ABC = m\angle ABP + m\angle PBC$$
$$= 36°28' + 43°51'$$
$$= 79°79'$$
$$= 80°19' \qquad 79' = 60' + 19' = 1°19'$$

Technology Connection

Geometry software will be needed.

1. Draw an angle and label it $\angle ABC$.
2. Measure $\angle ABC$
3. Locate a point P in the interior of the angle.
4. Draw ray BP ($\overrightarrow{BP}$).
5. Measure $\angle ABP$ and $\angle PBC$.
6. Does $m\angle ABC = m\angle ABP + m\angle PBC$?
7. Does $m\angle PBC = m\angle ABC - m\angle ABP$?
8. Does $m\angle ABP = m\angle ABC - m\angle PBC$?

⌐**Note** Most software packages will round the measurements, so the results may be slightly off.

Do these results confirm Postulate 1.14?

Answers to Practice Exercises

1. $x = 5$ 2. (a) yes   (b) no   (c) yes   (d) only X (e) X and Y (f) no
(g) 8 cm   3. (a) $63°$ (b) $105°33'8''$ 4. $y = 21$

1.3 **Exercises**   FOR EXTRA HELP: 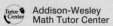 Student's Solutions Manual   Tutor Center Addison-Wesley Math Tutor Center

Consider the line $\overleftrightarrow{AB}$ with point C between A and B and ray $\overrightarrow{CD}$ as shown below. Use this figure and answer *true* or *false* in Exercises 1–20.

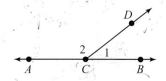

1. Point B is on $\overrightarrow{AC}$.
3. Point C is on $\overline{AB}$.
5. If $AC = 10$ cm and $CB = 13$ cm, then $AB = 23$ cm.
7. $\overrightarrow{CA}$ and $\overrightarrow{AC}$ are the same.
9. $\angle ACD$ is another name for $\angle 2$.
11. A and B are endpoints of $\overleftrightarrow{AB}$.
13. The vertex of $\angle 1$ is C.
15. $\angle ACB$ is a right angle.
17. $\angle 1$ and $\angle 2$ are adjacent angles.
19. $\angle DCB$ is an acute angle.

2. Point A is on $\overrightarrow{CB}$.
4. Point B is on $\overrightarrow{AC}$.
6. If $AB = 30$ cm and $AC = 12$ cm then $CB = 18$ cm.
8. $\angle 1$ is another name for $\angle DCB$.
10. C is the endpoint of $\overrightarrow{BC}$.
12. A and C are endpoints of $\overline{AC}$.
14. The vertex of $\angle DCA$ is A.
16. $\angle BCA$ is a straight angle.
18. $\angle 1$ and $\angle 2$ are complementary angles.
20. $\angle 1$ is the supplement of $\angle DCA$.

Exercises 21–26 refer to the figure below. Give the measure of each angle.

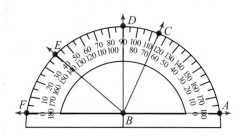

21. $\angle ABC$ **22.** $\angle EBF$ **23.** $\angle ABD$
24. $\angle ABE$ **25.** $\angle FBC$ **26.** $\angle CBE$

Find the complement of each angle in Exercises 27–30.

27. $18°$ **28.** $64°$ **29.** $36° \, 40'$ **30.** $71° \, 45' \, 20''$

Find the supplement of each angle in Exercises 31–34.

31. $74°$ **32.** $136°$ **33.** $57° \, 35'$ **34.** $110° \, 35' \, 40''$
35. What angle has the same measure as its complement?
36. What angle has the same measure as its supplement?
37. What is the complement of the supplement of an angle measuring $130°$?
38. What is the supplement of the complement of an angle measuring $50°$?

39. What is the complement of the complement of an angle measuring 25°?

40. What is the supplement of the supplement of an angle measuring 160°?

41. What is the measure of an angle whose supplement is four times its complement?

42. What is the measure of an angle whose supplement is three times its complement?

State whether each angle given in Exercises 43–51 is straight, right, acute, or obtuse.

43. 65°   **44.** 115°   **45.** 180°   **46.** 90°

47. The complement of an angle measuring 42°.

48. The supplement of an angle measuring 42°.

49. The complement of any acute angle.

50. The supplement of any obtuse angle.

51. The supplement of a right angle.

In Exercises 52–55, $\angle ABP$ and $\angle PBC$ are adjacent angles. Find the measure of $\angle ABC$.

52. $m\angle ABP = 62°20'$ and $m\angle PBC = 31°50'$ **53.** $m\angle ABP = 49°55'$ and $m\angle PBC = 57°15'$

54. $m\angle ABP = 27°25'41''$ and $m\angle PBC = 52°51'35''$ **55.** $m\angle ABP = 120°38'22''$ and $m\angle PBC = 18°41'54''$

For Exercises 56–61, refer to the figure below.

56. On $\overleftrightarrow{AB}$, how many points are located 5 cm from point A?

57. On $\overleftrightarrow{AB}$, how many points are located 5 cm from point B?

58. On $\overrightarrow{AB}$, how many points are located 5 cm from point A?

59. On $\overrightarrow{AB}$, how many points are located 5 cm from point B if $AB = 10$ cm?

60. On $\overline{AB}$, how many points are located 5 cm from point A if $AB = 10$ cm?

61. On $\overrightarrow{AB}$, how many points are located 5 cm from point B?

62. In the given figure of a stained-glass window, if the angles between the panes of glass are equal, what is the measure of each angle?

63. In the given top-down view of an umbrella, if the angles between the spokes are all equal, what is the measure of each angle?

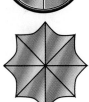

In Exercises 64–65, find the value of x in each figure.

64.

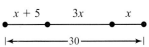

65.

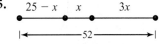

66. If $m\angle A = (5y)°$, $m\angle B = (y + 6)°$, and $\angle A$ and $\angle B$ are complementary, find y.

67. If $m\angle R = (30 - y)°$, $m\angle S = (9y - 10)°$, and $\angle R$ and $\angle S$ are supplementary, find y.

1.4 Introduction to Deductive Proofs

OBJECTIVES

1. Define conditional statements.
2. Define converse statements.
3. Define direct proofs.
4. Define inverse statements and contrapositive statements.

Theorems are important parts of an axiomatic system. We said in Section 1.1 that a theorem is a statement that requires proof. What is a proof? Although mathematicians might disagree on the best proof for a statement, they will agree that a proof involves deductive reasoning rather than inductive reasoning. That is, a theorem is not proved by showing that it is true in a specific number of examples. We must use deductive reasoning to show that a theorem is true based on accepted assumptions and previously proved theorems.

OBJECTIVE 1 **Define conditional statements.** Many theorems we prove in geometry can be stated in the form "If . . . , then . . . ." Such statements are called **conditional statements**, and are often symbolized by using $\longrightarrow$. For example, if we let P represent the statement "the Sun is shining" and Q represent the statement "I can see my shadow," then the conditional statement

"If the Sun is shining, then I can see my shadow,"
has the form "If P, then Q," and can be symbolized by

$$P \longrightarrow Q, \text{ read "If } P \text{ then } Q."$$

The following example offers practice in recognizing and writing conditional statements.

EXAMPLE 1 Let P represent the statement "an animal is a dog," Q represent the statement "it has four legs," and R represent the statement "it barks."

(a) The symbolic form $P \longrightarrow R$ represents the statement "If an animal is a dog, then it barks."

(b) The symbolic form $P \longrightarrow Q$ represents the statement "If an animal is a dog, then it has four legs."

(c) The symbolic form $R \longrightarrow P$ represents the statement "If an animal barks, then it is a dog."

(d) The symbolic form $Q \longrightarrow P$ represents the statement "If an animal has four legs, then it is a dog."

OBJECTIVE 2 **Define converse statements.** A variation of a conditional statement $P \longrightarrow Q$ that is often confused with the given statement is its **converse**, $Q \longrightarrow P$. A common error in reasoning is to assume that the converse of a conditional statement is true. We see from Example 1 that if we accept the statement $P \longrightarrow Q$, "If an animal is a dog, then it has four legs," it would not be correct to assume that $Q \longrightarrow P$, "If an animal has four legs, then it is a dog," follows from it. Clearly $Q \longrightarrow P$ is not true. Thus, a conditional statement and its converse are not the same.

Suppose P represents the statement "we are to win tonight" and Q represents the statement "we must play very hard." Then the conditional statement

"If we are to win tonight, then we must play very hard,"

can be symbolized by $P \longrightarrow Q$. In Example 5 of Section 1.1, we used deductive reasoning to conclude that Q is true; that is, "we played hard" is true, when we assumed $P \longrightarrow Q$ and P to be true.

We can symbolize this form of deductive reasoning as follows:

Premise: If we are to win tonight, then we must play very hard.	$P \longrightarrow Q$
Premise: We won the game tonight.	P
Conclusion: We played very hard.	$\therefore Q$

OBJECTIVE 3 **Define direct proofs.**   The horizontal line is used to separate the premises $P \longrightarrow Q$ and P from the conclusion Q, and the three dots, $\therefore$, symbolize the word *therefore*. This type of reasoning, recognized and used informally in the preceding section, serves as the basis for writing proofs, often called *direct proofs*. The classic format of a direct proof of a theorem $P \longrightarrow Q$ shows a series of statements, starting with the hypothesis P. Each statement follows from the preceding one using the reasoning of the preceding statement, and the final statement is the conclusion Q. The reasons given for the truth of each statement, written to the right of the statement, must be accepted or previously proved statements.

DIRECT PROOF OF $P \longrightarrow Q$

Suppose we already have $P \longrightarrow Q_1, Q_1 \longrightarrow Q_2, Q_2 \longrightarrow Q_3$, and $Q_3 \longrightarrow Q$ as accepted or previously proved statements. The format used to write a **direct proof** of $P \longrightarrow Q$ is:

Given: P (hypothesis)

Prove: Q (conclusion)

Proof

Statements	Reasons
1. P	1. Given
2. Q_1	2. $P \longrightarrow Q_1$
3. Q_2	3. $Q_1 \longrightarrow Q_2$
4. Q_3	4. $Q_2 \longrightarrow Q_3$
5. Q	5. $Q_3 \longrightarrow Q$
$\therefore P \longrightarrow Q$.	

The format of this direct proof shows five statements, but another proof might consist of any number of steps. Notice that P (given in statement 1) together with $P \longrightarrow Q_1$ (assumed true) gives Q_1 (statement 2). Then Q_1 with $Q_1 \longrightarrow Q_2$ gives Q_2 (statement 3), and so on.

EXAMPLE 2   Assume that the following statements are true. Use them to prove the "theorem."

Premise 1: If I have enough money, then I will take a trip.

Premise 2: If I lose my job, I will be unhappy.

Premise 3: If I take a trip, then I will lose my job.

Give a direct proof of the following "theorem."

Theorem: If I have enough money, then I will be unhappy.

Given: I have enough money.

Prove: I will be unhappy.

Proof

Statements	Reasons
1. I have enough money.	1. Given
2. I will take a trip.	2. Premise 1
3. I will lose my job.	3. Premise 3
4. I will be unhappy.	4. Premise 2

∴ If I have enough money, then I will be unhappy.

Notice P (Given) is the statement "I have enough money."

Q_1 is "I will take a trip." Therefore, $P \longrightarrow Q_1$ is Premise 1.

Q_2 is "I will lose my job." Therefore, $Q_1 \longrightarrow Q_2$ is Premise 3.

Q_3 is "I will be unhappy." Therefore, $Q_2 \longrightarrow Q_3$ is Premise 2.

We said earlier that the converse of a conditional $P \longrightarrow Q$ is the statement $Q \longrightarrow P$. The **negation** of P is the statement "not P" and is denoted $\sim P$. For example, if P is the statement

"My car is white,"

then the negation of P ($\sim P$) is

"My car is not white."

OBJECTIVE 4   **Define inverse statements and contrapositive statements.**   If a statement P is true, then $\sim P$ is false; and if P is false, then $\sim P$ is true.

For the statement $P \longrightarrow Q$, we define its **inverse** as $\sim P \longrightarrow \sim Q$ and its **contrapositive** as $\sim Q \longrightarrow \sim P$. Suppose P is the statement "it is a wheel" and Q is the statement "it is round." Then

$P \longrightarrow Q$ "If it is a wheel, then it is round" has the following three variations:

$Q \longrightarrow P$ "If it is round, then it is a wheel"   Converse

$\sim P \longrightarrow \sim Q$ "If it is not a wheel, then it is not round"   Inverse

$\sim Q \longrightarrow \sim P$ "If it is not round, then it is not a wheel"   Contrapositive

In general, the inverse (and converse) of a given conditional need not be true when the conditional is true. However, the contrapositive of a given conditional is always true when the conditional is true. Notice how these facts are supported by the preceding example. A conditional statement can always be replaced with its contrapositive.

Now it is time to develop some guidelines for proving geometric theorems. Recall that many theorems are "If . . . , then, . . ." statements. In other words, a theorem is a statement that can be proven. The word *if* directs us to the hypothesis of the theorem, the information that is given or assumed true, and the word *then* tells us the conclusion of the theorem, what must be shown to be true. Some theorems are not stated in the "if . . . , then, . . ." form. To make it easier to recognize the hypothesis and conclusion, the statement can be reworded.

EXAMPLE 3 Identify the hypothesis and conclusion of each statement.

(a) If x and y are any two quantities with $x = y$, then x can be substituted for y in any expression containing y.

(b) Two right angles are congruent.

Solutions:

(a) The hypothesis is x and y are any two quantities with $x = y$. The conclusion is x can be substituted for y in any expression containing y.

(b) The hypothesis is two angles are right angles. The conclusion is the angles are congruent.

1.4  Exercises

FOR EXTRA HELP: Student's Solutions Manual | Tutor Center — Addison-Wesley Math Tutor Center

Complete the direct proof of the "theorem" in Exercise 1.

1. *Premise 1:* If taxes rise, then the people will be unhappy.
   Premise 2: If people are unhappy, then they will go to the polls.
   Premise 3: If the president gets his budget passed, then taxes will rise.
   Premise 4: If people go to the polls, the president will be voted out of office.
   Theorem: If the president gets his budget passed, then he will be voted out of office.

   Given: _____

   Prove: _____

   Proof _____

Statements	Reasons
1. The president gets his budget passed.	1. _____
2. _____	2. Premise 3
3. The people will be unhappy.	3. _____
4. _____	4. Premise 2
5. _____	5. Premise 4
∴ _____	

Give a direct proof of each "theorem" in Exercises 2 and 3.

 2. *Premise 1:* If the weather report is accurate, then we will get 12 inches of snow.
    Premise 2: If we get 12 inches of snow, then the streets will be treacherous.
    Premise 3: If the streets are treacherous, then school will be canceled.
    Theorem: If the weather report is accurate, then school will be canceled.
 3. *Premise 1:* If I watch TV, then I will not do my homework.
    Premise 2: If I fail geology, then I will lose my scholarship.
    Premise 3: If I do not do my homework, then I will fail geology.
    Premise 4: If I lose my scholarship, then my parents will be upset.
    Theorem: If I watch TV, then my parents will be upset.
 4. In Exercise 2, do we prove that "School will be canceled"?
 5. In Exercise 3, do we prove that "My parents will be upset"?
 6. If Premise 2 is left out in Exercise 2, can the same theorem be proved?
 7. If Premise 3 is left out in Exercise 3, can the same theorem be proved?

Give the converse of each statement in Exercises 8–11, and note that each converse makes a different statement.

 8. If a person is president, then he/she must be a United States citizen.
 9. If it is Sunday, then I will watch football.
10. If a figure is a square, then it is a rectangle.
11. If you fly in an airplane, then you will go to the airport.

Give the negation of each statement in Exercises 12–15.

12. The tree is a pine.
13. The city is large.
14. I received an A in the course.
15. Joe did not run in the race.

Give the inverse of each statement in Exercises 16–19. Note that each inverse makes a different statement.

16. If it rains, then I will stay indoors.
17. If this animal is a bird, then it has two legs.
18. If I have the flu, then I will run a fever.
19. If it's gold, then it glitters.

Give the contrapositive of each statement in Exercises 20–23.

20. If I take a shower, then I will get wet.
21. If a figure is a rectangle, then it is a parallelogram.
22. If I drink orange juice, then I am healthy.
23. If we beat Central State, then we are the conference champions.

State the hypothesis and conclusion for each statement in Exercises 24–28.

24. If a triangle is equilateral, then it is equiangular.
25. Two lines with slopes m_1 and m_2 are parallel if $m_1 = m_2$.
26. If a triangle is isosceles, then the triangle has two congruent sides.
27. Vertical angles are congruent.
28. Two angles are congruent if they are both right angles.

1.5 Formalizing Geometric Proofs

OBJECTIVES

1. Define the format of a formal proof.
2. Describe the thinking process.
3. Use the format of the thinking process.

OBJECTIVE 1 **Define the format of a formal proof.** To formalize the process of writing a two-column geometric proof, we will follow a precise format by listing statements in one column and a reason for each statement in a second column. The following information can be used as reasons for each statement:

- Given (part or all of the hypothesis)
- Postulates
- Definitions
- Theorems (previously proved)

The first step in writing a proof is to draw a figure showing the given information. Next, write *Given:* and list all assumed statements using notation from your drawn figure. Then write *Prove:* and state what is to be proved. Finally, write *Proof:* and head two columns with the words Statements and Reasons. The first statements written are usually taken from the *Given* statements and the final statement will always be the *Prove* statement.

An example of this format would be as follows: Suppose we want to prove the theorem: If B is a point between A and C on segment $\overline{AC}$, Q is a point between P and R on segment $\overline{PR}$, $AB = PQ$, and $BC = QR$, then $AC = PR$.

1.3 postulate

Given:	B is between A and C on $\overline{AC}$	
	Q is between P and R on $\overline{PR}$	Draw figure
	$AB = PQ$	See Figure 1.21 on page 34.
	$BC = QR$	
Prove:	$AC = PR$	

Proof _____

Statements	*Reasons*
1. B is between A and C on $\overline{AC}$	**1.** Given
2. ???	**2.** ???
3. ???	**3.** ???
More statements as needed	More reasons as needed
?. $AC = PR$	**?.** ???

Now look at the completed proof.

THEOREM 1.1 **Addition Theorem for Segments**

If B is a point between A and C on segment $\overline{AC}$, Q is a point between P and R on segment $\overline{PR}$, $AB = PQ$, and $BC = QR$, then $AC = PR$.

Given: B is between A and C on $\overline{AC}$
Q is between P and R on $\overline{PR}$
AB = PQ
BC = QR

Prove: AC = PR

Figure 1.21 Segment Addition

Proof _____

Statements	**Reasons**
1. B is between A and C on $\overline{AC}$	1. Given
2. Q is between P and R on $\overline{PR}$	2. Given
3. AB = PQ	3. Given
4. BC = QR	4. Given
5. AB + BC = PQ + QR	5. Addition-subtraction law (Postulate 1.9)
6. AC = AB + BC and PR = PQ + QR	6. Segment addition postulate (Postulate 1.13)
7. AC = PR	7. Substitution law (Postulate 1.11)

> **THEOREM 1.2 Subtraction Theorem for Segments**
> If B is a point between A and C on segment $\overline{AC}$, Q is a point between P and R on segment $\overline{PR}$, AC = PR, and AB = PQ, then BC = QR.

The proof of Theorem 1.2 is similar to that of Theorem 1.1. It is outlined in the exercises at the end of this section where you are asked to supply reasons for each statement.

> **THEOREM 1.3 Addition Theorem for Angles**
> If D is a point in the interior of ∠ABC, S is a point in the interior of ∠PQR, m∠ABD = m∠PQS, and m∠DBC = m∠SQR, then m∠ABC = m∠PQR.

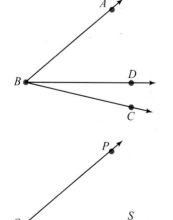

Given: D is interior to ∠ABC
S is interior to ∠PQR
m∠ABD = m∠PQS
m∠DBC = m∠SQR (See Figure 1.22.)

Prove: m∠ABC = m∠PQR

Figure 1.22 Angle Addition

Proof _____

Statements	Reasons
1. D is interior to $\angle ABC$	1. Given
2. S is interior to $\angle PQR$	2. Given
3. $m\angle ABD = m\angle PQS$	3. Given
4. $m\angle DBC = m\angle SQR$	4. Given
5. $m\angle ABD + m\angle DBC =$ $m\angle PQS + m\angle SQR$	5. Addition-subtraction law
6. $m\angle ABC = m\angle ABD + m\angle DBC$ and $m\angle PQR = m\angle PQS + m\angle SQR$	6. Angle addition postulate
7. $m\angle ABC = m\angle PQR$	7. Substitution law

The proof of the next theorem is similar to that of Theorem 1.3 and is left for you to complete as an exercise.

THEOREM 1.4   Subtraction Theorem for Angles

If D is a point in the interior of $\angle ABC$, S is a point in the interior of $\angle PQR$, $m\angle ABC = m\angle PQR$, and $m\angle DBC = m\angle SQR$, then $m\angle ABD = m\angle PQS$.

OBJECTIVE 2   **Describe the thinking process.**   For later work, we will need several theorems that involve supplementary and complementary angles. For these theorems, we will be required to illustrate the given information with a figure.

THEOREM 1.5

Two equal supplementary angles are right angles.

Notice that Theorem 1.5 is not in the form "If . . . , then . . . ." It does state, however, that "If two angles are equal and supplementary, then they are right angles."

First make a sketch of two equal supplementary angles as in Figure 1.23. The statements we make about what is given and what is to be proven refer to the figure. The figure, then, is a part of the proof.

Given: $m\angle ABD = m\angle DBC$
$\angle ABD$ and $\angle DBC$ are supplementary

Prove: $\angle ABD$ and $\angle DBC$ are right angles

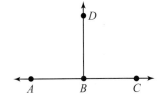

Figure 1.23 Equal Supplementary Angles

In order to develop the proof, a plan is needed. One way to do this is to use a problem-solving technique called "thinking backwards." First draw the figure and mark the given information. Starting with the conclusion, reason backwards to the given information.

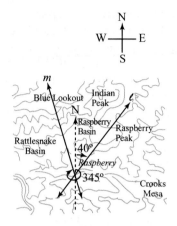

The sport of *orienteering* continues to be popular among wilderness enthusiasts. Using a topographic map and a magnetic compass, you can find your way through a wilderness area by sighting specific landmarks on a map. For example, suppose you look at Raspberry Peak and determine its direction is 40° from your position. You then know that you are somewhere on line ℓ in the figure. If you look toward Blue Lookout and determine its direction is 345° you also know you are somewhere on line *m*. The point of intersection of ℓ and *m* indicates your approximate location in the wilderness area.

In this example of Theorem 1.5, the thinking might go like this: "I want to prove these are right angles so I need to show they measure 90°. The two given angles are supplementary so their sum is 180° by the definition of supplementary angles. Because the angles have equal measures (given) and their sum is 180°, each angle must measure 90° by the division law." The completed proof appears later in the section.

Another way to develop this plan is to think like a computer scientist and make a flowchart to show the step-by-step procedure. It might look like the following:

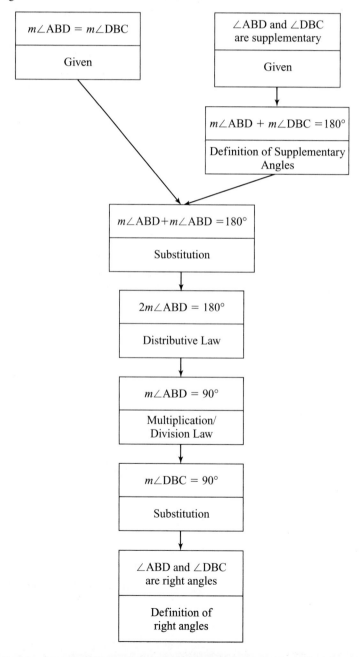

In this example, each piece of information and a reason were enclosed in a rectangle. An oval is another shape often used for the statement, and the reason is written beneath the oval. The process is similar to preparing to write an essay. When writing an essay, you write down the topics and ideas to be used in the paper. This may be done linearly or randomly. Once the ideas are on paper, they are reorganized around the topic in some logical fashion.

When using this technique in geometry, write down what needs to be done and the facts that are known about the problem in separate shapes. List any information that relates to the problem like postulates or previously proven theorems in more shapes. Now go back and rewrite an organized flowchart.

Note Sometimes it's difficult to know what can be assumed from a figure. Betweenness of points and collinear points can be seen and, therefore, assumed from a figure. The vertex of an angle, adjacent angles, vertical angles, and intersecting lines can also be assumed from a figure. All these ideas are about the *location of points* on a plane.

This is the completed proof of Theorem 1.5 written in two-column form. See Figure 1.23.

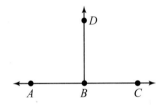

Figure 1.23 (repeated)
Equal Supplementary Angles

Proof

Statements	Reasons
1. $m\angle ABD = m\angle DBC$	1. Given
2. $\angle ABD$ and $\angle DBC$ are supplementary	2. Given
3. $m\angle ABD + m\angle DBC = 180°$	3. Definition of supplementary angles
4. $m\angle ABD + m\angle ABD = 180°$	4. Substitute $\angle ABD$ for $\angle DBC$ in statement 3 using the substitution law
5. $2m\angle ABD = 180°$	5. Distributive law
6. $m\angle ABD = 90°$	6. Divide both sides of statement 5 by 2 using the multiplication-division law
7. $m\angle DBC = 90°$	7. Substitute $\angle DBC$ for $\angle ABD$ in statement 6 using the substitution law
8. $\angle ABD$ and $\angle DBC$ are right angles	8. Definition of right angle

---

THEOREM 1.6

Complements of equal angles are equal in measure.

A point is an undefined term. It can be shown in a figure as a dot and labeled with an italic capital letter.

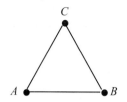

In some figures, the focus is on the angles or segments rather than the points. In this case, the points will be labeled with capital letters but no dots will appear in the figure, as in Figure 1.24.

To begin the proof of this theorem make a sketch as in Figure 1.24.

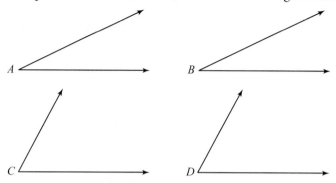

Figure 1.24 Complements of Equal Angles

To illustrate the "thinking backwards" technique for this problem, one might think "I need to prove two angles equal in measure. Since angle A and angle C are complementary, their sum is $90°$ by the definition of complementary angles. The same is true for angles B and D; their sum is $90°$. If both their sums are the same, they must be equal to each other, so I know by the transitive and symmetric laws that $m\angle A + m\angle C = m\angle B + m\angle D$. But since the measure of angle A and the measure of angle B are the same, I can subtract them from the equation I just wrote and the result is the conclusion of the theorem."

OBJECTIVE 3 **Use the format of the thinking process.** This is the completed proof of Theorem 1.6 written in two-column form.

Given: $\angle A$ and $\angle C$ are complementary
$\angle B$ and $\angle D$ are complementary
$m\angle A = m\angle B$ (See Figure 1.24.)

Prove: $m\angle C = m\angle D$

Proof ────────────────────────────

Statements	*Reasons*
1. $\angle A$ and $\angle C$ are complementary	**1.** Given
2. $m\angle A + m\angle C = 90°$	**2.** Definition of complementary angles
3. $\angle B$ and $\angle D$ are complementary	**3.** Given
4. $m\angle B + m\angle D = 90°$	**4.** Definition of complementary angles
5. $m\angle A + m\angle C = m\angle B + m\angle D$	**5.** Transitive law and symmetric law
6. $m\angle A = m\angle B$	**6.** Given
7. $m\angle C = m\angle D$	**7.** Addition-subtraction law

A **corollary** is a theorem that is easy to prove as a direct result of a previously proved theorem. The next corollary follows immediately from Theorem 1.6.

COROLLARY 1.7
Complements of the same angle are equal in measure.

A similar theorem and corollary exist for supplementary angles.

THEOREM 1.8
Supplements of equal angles are equal in measure.

The proof of Theorem 1.8 is left for you to do as an exercise.

COROLLARY 1.9
Supplements of the same angle are equal in measure.

Two adjacent angles whose noncommon sides lie on the same line are supplementary. For example, in Figure 1.25 $\angle 1$ and $\angle 2$ are supplementary adjacent angles. We prove this assertion in the next theorem.

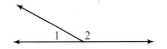

Figure 1.25 Supplementary Adjacent Angles

THEOREM 1.10
If A, B, and C are three points on a line, with B between A and C, and $\angle ABD$ and $\angle DBC$ are adjacent angles, then $\angle ABD$ and $\angle DBC$ are supplementary.

Given: A, B, and C are on the same line
B is between A and C
$\angle ABD$ and $\angle DBC$ are adjacent angles (See Figure 1.26.)

Prove: $\angle ABD$ and $\angle DBC$ are supplementary

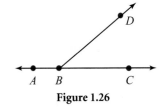

Figure 1.26

Proof _____

Statements

1. A, B, and C are on the same line
2. B is between A and C
3. $\angle ABC$ is a straight angle
4. $m\angle ABC = 180°$
5. $m\angle ABD + m\angle DBC = m\angle ABC$
6. $m\angle ABD + m\angle DBC = 180°$
7. $\angle ABD$ and $\angle DBC$ are supplementary

Reasons

1. Given
2. Given
3. Definition of straight angle
4. Definition of straight angle
5. Angle addition postulate
6. Transitive law
7. Definition of supplementary angles

Sometimes Theorem 1.10 is shortened to "If the exterior sides of two adjacent angles lie in a straight line, the angles are supplementary."

DEFINITION: *Vertical Angles*

Two nonadjacent angles formed by two intersecting lines are called **vertical angles.**

In Figure 1.27a, $\angle 1$ and $\angle 2$ are vertical angles as are $\angle 3$ and $\angle 4$. It appears that $m\angle 1 = m\angle 2$ and $m\angle 3 = m\angle 4$. Let's investigate this statement.

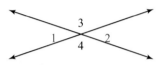

Figure 1.27a Vertical Angles

Technology Connection

Geometry software will be needed.
1. Draw two intersecting lines.
2. Label them $\overleftrightarrow{BC}$ and $\overleftrightarrow{AD}$.
3. Label the intersection point P.
4. Measure $\angle APC$, $\angle CPD$, $\angle DPB$, and $\angle BPA$.
5. Slide point A up or down. Observe what happens to the angle measures in step 4.

Do these results confirm Theorem 1.11? This investigation does not prove the theorem.

THEOREM 1.11

Vertical angles are equal in measure.

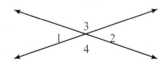

Figure 1.27b Vertical Angles

Given: $\angle 1$ and $\angle 2$ are vertical angles (See Figure 1.27b.)

Prove: $m\angle 1 = m\angle 2$

To illustrate the "thinking backwards" technique for this problem, one might think "I need to prove two angles congruent. What do angles 1 and 2 have in common? They are both next to (adjacent to) angle 3 (or 4). Angles 1 and 3 are supplementary as are angles 2 and 3 because they are adjacent angles whose exterior sides are in a line (Theorem 1.10). We had a theorem that says supplements of the same angle are equal. Since angles 1 and 2 are both supplements of angle 3, they must have the same measure."

Proof _____

Statements	*Reasons*
1. $\angle 1$ and $\angle 2$ are vertical angles	1. Given
2. $\angle 1$ and $\angle 3$ are adjacent angles	2. Def. of adj. $\angle$
3. $\angle 1$ is supplementary to $\angle 3$	3. Adj. $\angle$'s with 2 sides in a line are supp.
4. $\angle 2$ and $\angle 3$ are adjacent angles	4. Def. of adj. $\angle$
5. $\angle 2$ is supplementary to $\angle 3$	5. Adj. $\angle$'s with 2 sides in a line are supp.
6. $m\angle 1 = m\angle 2$	6. Supp. of the same $\angle$ are = in measure.

Note When giving reasons in a proof, we often abbreviate words and use symbols. In the proof of Theorem 1.11, we used *def.* to mean *definition*, *adj.* to mean *adjacent*, and $\angle$ to mean *angle*. It is also appropriate to abbreviate the statement of a theorem or corollary making the proof easier to read without reference to particular numbers. For example, for reason 6 we gave "Supp. of the same $\angle$ are = in measure" instead of "Corollary 1.9" and "Reasons 3 and 5 paraphrase Theorem 1.10."

1.5 Exercises

FOR EXTRA HELP: 📖 Student's Solutions Manual   Tutor Center Addison-Wesley Math Tutor Center

In Exercises 1 and 2, supply reasons for the statements in each proof.

1. Prove Theorem 1.2.

Given: *B* is between *A* and *C* on $\overline{AC}$
        Q is between *P* and *R* on $\overline{PR}$
        $AC = PR$
        $AB = PQ$

Prove: $BC = QR$

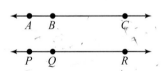

Exercise 1

Proof _____

Statements	*Reasons*
1. *B* is between *A* and *C* on $\overline{AC}$	1. _____
2. *Q* is between *P* and *R* on $\overline{PR}$	2. _____
3. $AC = PR$	3. _____
4. $AB = PQ$	4. _____
5. $AC - AB = PR - PQ$	5. _____
6. $BC = AC - AB$ and $QR = PR - PQ$	6. _____
7. $BC = QR$	7. _____

2. Prove Theorem 1.4.

Given: D is interior to $\angle ABC$
S is interior to $\angle PQR$
$m\angle ABC = m\angle PQR$
$m\angle DBC = m\angle SQR$

Prove: $m\angle ABD = m\angle PQS$

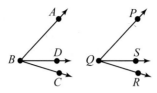

Exercise 2

Proof _____

Statements	*Reasons*
1. D is interior to $\angle ABC$	**1.** Given
2. S is interior to $\angle PQR$	**2.** Given
3. $m\angle ABC = m\angle PQR$ and $m\angle DBC = m\angle SQR$	**3.** Given
4. $m\angle ABC - m\angle DBC = m\angle PQR - m\angle SQR$	**4.** Addition-subtraction law
5. $m\angle ABD = m\angle ABC - m\angle DBC$ and $m\angle PQS = m\angle PQR - m\angle SQR$	**5.** Segment subtraction postulate
6. $m\angle ABD = m\angle PQS$	**6.** Substitution

3. *Given:* $\angle ABC$ is a right angle
$\angle DBC$ and $\angle 1$ are complementary

Prove: $m\angle ABD = m\angle 1$

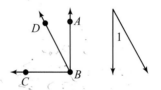

Exercise 3

Proof _____

Statements	*Reasons*
1. $\angle ABC$ is a right angle	**1.** Given
2. $m\angle ABC = 90°$	**2.** Given
3. $m\angle DBC + m\angle ABD = m\angle ABC$	**3.** Addition law
4. $m\angle DBC + m\angle ABD = 90°$	**4.** Addition law
5. $\angle DBC$ and $\angle 1$ are complementary	**5.**
6. $m\angle DBC + m\angle 1 = 90°$	**6.**
7. $m\angle DBC + m\angle ABD = m\angle DBC + m\angle 1$	**7.**
8. _____	**8.** Addition-subtraction law

4. *Given:* Three lines m, n, and ℓ as shown
$m\angle 1 + m\angle 2 = 180°$

Prove: $m\angle 4 = m\angle 3$

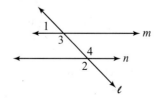

Exercise 4

Proof _____

Statements	*Reasons*
1. _____	1. Given
2. $\angle 1$ and $\angle 3$ are supplementary	2. _____
3. $m\angle 1 + m\angle 3 = 180°$	3. _____
4. $m\angle 1 + m\angle 2 = m\angle 1 + m\angle 3$	4. _____
5. $m\angle 2 = m\angle 3$	5. _____
6. $\angle 4$ and $\angle 2$ are vertical angles	6. _____
7. _____	7. Vert. $\angle$'s are $=$ in measure
8. _____	8. Transitive law

5. Use the figure below and name all pairs of adjacent angles.

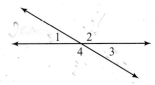

Exercises 5 and 6

6. Use the figure above and name all pairs of vertical angles.

In Exercises 7–10, use the figure below to answer each question.

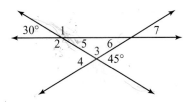

7. What is the measure of $\angle 1$?
8. What is the measure of $\angle 3$?
9. What is the measure of $\angle 2$?
10. What is the measure of $\angle 4$?
11. If $m\angle 3 + m\angle 5 + m\angle 6 = 180°$, what is the measure of $\angle 6$?
12. If $m\angle 3 + m\angle 5 + m\angle 6 = 180°$, what is the measure of $\angle 7$?

For problems 13–26, explain the reasoning in one or two complete sentences.

13. If $\angle A$ and $\angle B$ are vertical angles, must $m\angle A = m\angle B$?
14. If $m\angle A = m\angle B$, must $\angle A$ and $\angle B$ be vertical angles?
15. If $\angle A$ and $\angle B$ are supplementary, can $\angle A$ and $\angle B$ be vertical angles too?
16. If $\angle A$ and $\angle B$ are complementary, can $\angle A$ and $\angle B$ be vertical angles too?
17. Can two supplementary angles both be obtuse?
18. Can two complementary angles both be acute?
19. Can two vertical angles both be obtuse?
20. Can two vertical angles both be acute?

21. Can two right angles be vertical angles?
22. Can two vertical angles both be right angles?
23. If two angles are vertical angles, can one be obtuse and the other acute?
24. If two angles are adjacent angles, can one be obtuse and the other acute?
25. If two angles are adjacent angles, can they be supplementary?
26. If two angles are adjacent angles, can they be complementary?
27. Prove Theorem 1.8.
28. In the figure below, $PQ = RS$. Prove $PR = QS$.

29. In the figure above, $PR = QS$. Prove $PQ = RS$.
30. Prove that the sum of the measures of the complements of complementary angles is 90°.
31. Prove that the sum of the measures of the supplements of complementary angles is 270°.

1.6   Constructions Involving Lines and Angles

OBJECTIVES

1. Use a construction to copy a line segment.
2. Use a construction to copy an angle.
3. Use a construction to bisect a segment.
4. Define perpendicular lines.
5. Construct perpendicular lines.
6. Define an angle bisector.
7. Construct an angle bisector.

Part of geometry involves making accurate drawings of geometric figures. Such drawings are called **geometric constructions**. Two instruments are used to make constructions, a **straightedge** and a **compass**. A straightedge, a ruler with no marks of scale, is used to draw a line between two points. (We often use a standard ruler but do not use it to measure lengths.) A compass is used to draw circles, or portions of circles, called **arcs**. (Circles and arcs will be studied in more detail in Chapter 6.)

OBJECTIVE 1   **Use a construction to copy a line segment.**   The first construction we consider involves duplicating a given line segment.

> **CONSTRUCTION 1.1**
> Construct a line segment with the same length as a given line segment.

Given: Line segment $\overline{AB}$
(See Figure 1.28.)

To Construct: Line segment $\overline{CD}$
with $CD = AB$.

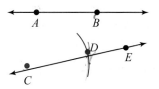

Construction

Figure 1.28 Reconstructing Segment Length

1. Draw a line $\overleftrightarrow{CE}$ containing point C.
2. Place the point of a compass at point A and the pencil point at
3. Without changing the distance between these points on the compass, place
   the point of the compass at point C and with the pencil point draw an arc that
   intersects $\overleftrightarrow{CE}$. The point of intersection of the arc and $\overleftrightarrow{CE}$ determines the
   point D. $\overline{CD}$ is the desired line segment.

OBJECTIVE 2   **Use a construction to copy an angle.**   The next construction involves duplicating a given angle.

CONSTRUCTION 1.2

Construct an angle with the same measure as a given angle.

The bronze sculpture, *Geometria*,
by Renaissance sculptor Antonio
del Pollaiolo (1433–1498) depicts
the study of geometry. It shows
the construction of a geometric
figure using a compass and a
straightedge. The sculpture appears on the base of the tomb of
Pope Sixtus IV and is located in
St. Peter's Cathedral in Rome.

Given: $\angle A$ (See Figure 1.29.)

To Construct: $\angle B$ such that $m\angle A = m\angle B$

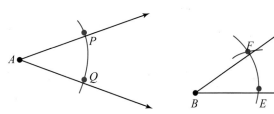

Figure 1.29 Reconstructing Angle Measure

Construction

1. Draw ray $\overrightarrow{BD}$.
2. Choose a convenient setting of the compass, place the point at A and draw
   an arc that intersects the sides of $\angle A$ at two points, for instance P and Q.
3. Without changing the compass setting, place the point at B and draw an
   arc that intersects $\overrightarrow{BD}$ at a point, for instance E.
4. Place the points of the compass so that one is at Q and the other is at P.
5. Without changing the compass setting, place the point at E and draw an arc
   that intercepts the arc drawn in step 3. Label this point F.
6. Draw ray $\overrightarrow{BF}$ to form the second side of the desired angle, $\angle B$.

OBJECTIVE 3   **Use a construction to bisect a segment.**   Next we will divide a line segment into two equal parts.

DEFINITION: *Line Segment Bisection*

Let $\overline{AB}$ be a line segment. To **bisect** $\overline{AB}$ is to identify a point C between A and B such that $AC = CB$. Point C is called the midpoint of $\overline{AB}$. The **midpoint** is a point on a line segment that separates the segment into two equal parts. A line segment that contains the midpoint C but no other point of $\overline{AB}$ is called a **bisector** of $\overline{AB}$.

CONSTRUCTION 1.3

Construct a bisector of a given line segment.

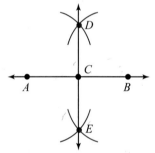

Figure 1.30 Segment Bisector

Given: Line segment $\overline{AB}$ (See Figure 1.30)
To Construct: A bisector $\overleftrightarrow{CD}$ of $\overline{AB}$

Construction _____

1. Set the compass so that the distance between the point and the pencil is greater than $\dfrac{AB}{2}$.
2. Set the point at A and draw two arcs, one above $\overline{AB}$ and the other below $\overline{AB}$. Repeat this procedure using the same compass setting with the point at B.
3. The arcs drawn in Step 2 should intersect at two points we label D and E.
4. Draw the line $\overleftrightarrow{DE}$. Then $\overleftrightarrow{DE}$ is a **bisector** of $\overline{AB}$ and determines **midpoint** C.

Construction 1.3 shows that a given line segment has a midpoint. The next postulate guarantees that this midpoint is unique.

POSTULATE 1.15   *Midpoint Postulate*

Each line segment has exactly one midpoint.

OBJECTIVE 4   **Define perpendicular lines.**   When two lines intersect and form right angles, we call the lines *perpendicular*.

DEFINITION: *Perpendicular Lines*

Two lines are **perpendicular** if they intersect and form equal adjacent angles. The angles formed are right angles. If $\overleftrightarrow{AB}$ is perpendicular to $\overleftrightarrow{CD}$, we write $\overleftrightarrow{AB} \perp \overleftrightarrow{CD}$. Two line segments are perpendicular if they intersect and are contained in perpendicular lines.

In Figure 1.31, lines ℓ and m are perpendicular if and only if $m\angle 1 = m\angle 2$. When $\ell \perp m$, we represent this fact in figures using the symbol ∟ as shown. Also, in Figure 1.32, $\overline{AB} \perp \overline{CD}$ provided $\overleftrightarrow{AB} \perp \overleftrightarrow{CD}$.

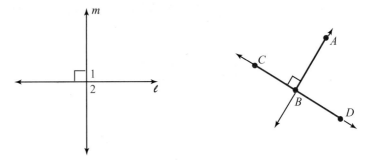

Figure 1.31 Perpendicular Lines
($\ell \perp m$)

Figure 1.32 Perpendicular Segments
($\overline{AB} \perp \overline{CD}$)

Note Relationships like parallel or perpendicular or midpoint may *not* be assumed from the figure. For example, in this figure,

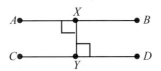

we *may* assume $\angle AXY$ is a right angle because ∟ is at the vertex of the angle but we *may not assume* $\overline{AB}$ is parallel to $\overline{CD}$ and we *may not assume* X and Y are midpoints.

When two lines intersect and form right angles, those right angles are equal in measure. The following theorem proves this statement. Now it can be cited as a reason in future proofs.

THEOREM 1.12

All right angles are equal in measure.

Given: $\angle 1$ is a right angle
$\angle 2$ is a right angle

Prove: $m\angle 1 = m\angle 2$

Proof _____

Statements	**Reasons**
1. $\angle 1$ is a rt $\angle$ $\angle 2$ is a rt $\angle$	1. Given
2. $m\angle 1 = 90°$ $m\angle 2 = 90°$	2. Measure of a right angle is 90° (definition of a right angle)
3. $m\angle 1 = m\angle 2$	3. Substitution

Although a line segment can have many bisectors, the most important one is the *perpendicular bisector*. Construction 1.3 shows this bisector.

> **DEFINITION:**   *Perpendicular Bisector of a Segment*
>
> A line that both bisects and is perpendicular to a given line segment is called a **perpendicular bisector** of the segment.

Construction 1.3 determined not only the midpoint of $\overline{AB}$ but also the perpendicular bisector of $\overline{AB}$. We assume that the perpendicular bisector of a segment is unique.

> **POSTULATE 1.16**   *Perpendicular Bisector Postulate*
>
> Each given line segment has exactly one perpendicular bisector.

OBJECTIVE 5   **Construct perpendicular lines.**   In the next construction, we find a line perpendicular to a given line passing through a given point on the line.

> **CONSTRUCTION 1.4**
>
> Construct a line perpendicular to a given line passing through a given point on the line.

Given: Line $\overleftrightarrow{AB}$ with point C on $\overleftrightarrow{AB}$ (See Figure 1.33.)

To Construct: A line $\overleftrightarrow{DC}$ such that $\overleftrightarrow{DC} \perp \overleftrightarrow{AB}$

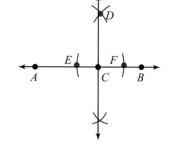

Figure 1.33

Construction _____

1. Choose a convenient setting for the compass, place the point at C, and make two arcs that intersect line $\overleftrightarrow{AB}$. Label these points E and F.
2. Use Construction 1.3 to draw the perpendicular bisector of the segment $\overline{EF}$. The result, $\overleftrightarrow{DC}$ in Figure 1.33, is the desired line perpendicular to $\overleftrightarrow{AB}$ at the point C.

> **POSTULATE 1.17**
>
> There is exactly one line perpendicular to a given line passing through a given point on the line.

Next we consider finding a line perpendicular to a given line passing through a given point *not* on the line.

CONSTRUCTION 1.5

Construct a line perpendicular to a given line passing through a given point not on that line.

Given:	Line $\overleftrightarrow{AB}$ with point C not on $\overleftrightarrow{AB}$ (See Figure 1.34.)
To Construct:	A line $\overleftrightarrow{CD}$ such that $\overleftrightarrow{CD} \perp \overleftrightarrow{AB}$

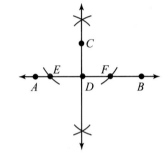

Figure 1.34

Construction _____

1. Choose a convenient setting for the compass, place the point at C, and make two arcs that intersect line $\overleftrightarrow{AB}$. Label these points E and F.

2. Use Construction 1.3 to draw the perpendicular bisector of the segment $\overline{EF}$. The result, $\overleftrightarrow{CD}$ in Figure 1.34, is the desired line perpendicular to $\overleftrightarrow{AB}$ and passing through C.

POSTULATE 1.18

There is exactly one line perpendicular to a given line passing through a given point not on that line.

We said that the **distance** between two points A and B is the length of the segment $\overline{AB}$, denoted by AB. Construction 1.5 and Postulate 1.18 give us a way to find the *distance* from a point to a line. Use Figure 1.34 as a reference for the following definition.

DEFINITION: *Distance from a Point to a Line*

Let $\overleftrightarrow{AB}$ be a line with C a point not on $\overleftrightarrow{AB}$. If D is the point on $\overleftrightarrow{AB}$ such that $\overleftrightarrow{CD} \perp \overleftrightarrow{AB}$, the distance from C to $\overleftrightarrow{AB}$ is CD, the length of $\overline{CD}$.

We now have a way to solve the following applied problem.

EXAMPLE 1   A family is building a recreational cabin at the edge of a wide mountain valley containing a stream that flows north to south. To supply the cabin with water, a pipe must be laid from the cabin to the stream. To minimize construction costs, the family plans to use the least amount of pipe possible. How can they determine the point on the stream bank that is closest to the cabin?

If we think of the stream as a straight line, and the cabin as a point not on the line, the desired point on the bank of the stream, labeled P in Figure 1.35 below, is found by constructing the line ℓ perpendicular to the stream passing through a point, the cabin, not on that line.

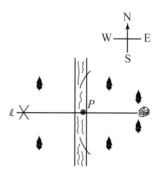

Figure 1.35

OBJECTIVE 6   **Define an angle bisector.**   We have considered the problem of bisecting a line segment, and now we'll try to bisect an angle.

> **DEFINITION: Angle Bisector**
>
> Let $\angle ABC$ be an angle. To bisect $\angle ABC$ is to identify $\overrightarrow{BD}$, where D is in the interior of $\angle ABC$ and $m\angle ABD = m\angle DBC$. $\overrightarrow{BD}$ is called the **angle bisector** of $\angle ABC$. See Figure 1.36.

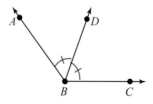

Figure 1.36

Note One way to indicate on a figure that two angles have the same measure, is to draw matching arcs in those angles as shown in Figure 1.36.

OBJECTIVE 7   **Construct an angle bisector.** Now we will use a compass and straightedge to bisect an angle.

> ### CONSTRUCTION 1.6
> Construct a bisector of a given angle.

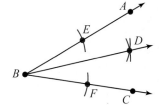

Given: $\angle ABC$
(See Figure 1.37.)

To Construct: $\overrightarrow{BD}$ that bisects $\angle ABC$

Figure 1.37 Angle Bisector

Construction _____

1. Choose a convenient setting for the compass, place the point at B, the vertex of $\angle ABC$, and make two arcs that intersect $\overrightarrow{BA}$ and $\overrightarrow{BC}$. Label these points E and F.

2. Choose a convenient setting for the compass so the distance between the pencil and the point is greater than $\dfrac{EF}{2}$. Set the point at E and draw an arc interior to $\angle ABC$. Then set the point at F and draw a second arc with the same compass setting that intercepts the first arc at a point we label D.

3. Draw $\overrightarrow{BD}$. $\overrightarrow{BD}$ is the desired bisector of $\angle ABC$, thus, $m\angle ABD = m\angle DBC$. We shall assume that the bisector of an angle is unique.

> ### POSTULATE 1.19   *Angle Bisector Postulate*
> Each angle has exactly one bisector.

$m\angle 1 = m\angle 2 = m\angle 3$

For years, mathematicians tried to find a way to trisect (divide into three equal angles) an angle using only a straightedge and compass. Some special angles (such as a straight angle) can be trisected, but the methods used do not carry over to more general angles. In fact, it has now been proved that it is impossible to trisect an arbitrary angle by construction.

1. Draw a line segment $\overline{AB}$ approximately 3 inches in length, and a line ℓ. Construct $\overline{CD}$ on ℓ such that $AB = CD$.
2. Draw a line segment $\overline{AB}$ approximately 2 inches in length, and a line ℓ. Construct $\overline{CD}$ on ℓ such that CD is twice AB.
3. Draw an acute angle and then construct another acute angle with the same measure.
4. Draw an obtuse angle and then construct another obtuse angle with the same measure.
5. Draw an acute angle and then construct another angle whose measure is twice that of the first.
6. Draw an obtuse angle and then construct another angle whose measure is twice that of the first.
7. Draw a line segment $\overline{AB}$ approximately 2 inches in length and locate the midpoint of the segment by bisecting it.
8. Draw a line segment $\overline{AB}$ approximately 4 inches in length and divide the segment into four equal parts by first bisecting $\overline{AB}$ and then bisecting each resulting part.
9. Draw a line segment $\overline{AB}$ approximately 3 inches in length and construct the perpendicular bisector of $\overline{AB}$.
10. Draw a line ℓ and select two points P and Q on ℓ. Construct the perpendicular bisector of $\overline{PQ}$ and locate the midpoint of $\overline{PQ}$.
11. Draw a line ℓ and select a point P on ℓ. Construct the line through P and perpendicular to ℓ.
12. Draw a line ℓ and select a point P not on ℓ. Construct the line through P and perpendicular to ℓ.
13. Draw an acute angle and construct its bisector.
14. Draw an obtuse angle and construct its bisector.
15. Draw an acute angle and use construction techniques to divide the angle into four equal angles.
16. Draw an obtuse angle and use construction techniques to divide the angle into four equal angles.
17. What is the distinction between a ruler and a straightedge?
18. How many midpoints does a line segment have?
19. How many perpendicular bisectors does a line segment have?
20. How many bisectors does a line segment have?
21. Draw a line ℓ with a point P not on ℓ. Construct the line m through P and perpendicular to ℓ and label the point of intersection of ℓ and m as Q. The length of $\overline{PQ}$ is the distance from P to ℓ. Use a ruler to measure $\overline{PQ}$ and approximate this distance.

Assume that every point on the perpendicular bisector of a line segment with endpoints *A* and *B* is equidistant from *A* and *B*. Also, assume that every point on the bisector of an angle is equidistant from the sides of the angle. Use this information and the figure on the right in Exercises 22–24.

22. A ranger wishes to drill a well at the edge of the forest equidistant from the cabin and the ranger station. Explain how she should locate this point at the forest's edge.

23. A ranger wishes to drill a well at the edge of the forest equidistant from the stream and the road. Explain how he should locate this point at the forest's edge.

24. A meteorologist wishes to place a weather station in the meadow equidistant from the ranger station, the bridge, and the cabin. Explain how she should locate this point.

Complete each proof in Exercises 25–28.

25. *Given:* *B* is the midpoint of $\overline{AC}$

   Prove: $AB = \dfrac{AC}{2}$

Exercise 25

Proof _____

Statements	Reasons
1. _____	1. Given
2. $AB = BC$	2. _____
3. $AB + BC = AC$	3. _____
4. $AB + AB = AC$	4. _____
5. _____	5. Distributive law
6. $AB = \dfrac{AC}{2}$	6. _____

26. *Given:* $\overrightarrow{BD}$ bisects $\angle ABC$

   Prove: $m\angle ABD = \dfrac{1}{2}m\angle ABC$

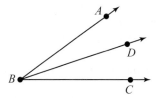

Exercise 26

Proof _____

Statements	Reasons
1. _____	1. Given
2. $m\angle ABD = m\angle DBC$	2. _____
3. $m\angle ABD + m\angle DBC = m\angle ABC$	3. _____
4. $m\angle ABD + m\angle ABD = m\angle ABC$	4. _____
5. _____	5. Distributive law
6. $m\angle ABD = \dfrac{1}{2}m\angle ABC$	6. _____

27. *Given:* *B* is the midpoint of $\overline{AC}$
    Q is the midpoint of $\overline{PR}$
    $AC = PR$
Prove: $AB = PQ$

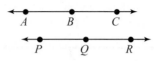

Exercise 27

Proof _____

Statements	*Reasons*
1. _____	1. Given
2. $AB = \dfrac{AC}{2}$	2. Exercise 25
3. *Q* is the midpoint of $\overline{PR}$	3. _____
4. _____	4. Exercise 25
5. _____	5. Given
6. $\dfrac{AC}{2} = \dfrac{PR}{2}$	6. _____
7. $AB = PQ$	7. _____

28. *Given:* $\overrightarrow{BD}$ bisects $\angle ABC$
    $\overrightarrow{QS}$ bisects $\angle PQR$
    $m\angle ABC = m\angle PQR$
Prove: $m\angle ABD = m\angle PQS$

Exercise 28

Proof _____

Statements	*Reasons*
1. _____	1. Given
2. $m\angle ABD = \dfrac{1}{2}m\angle ABC$	2. Exercise 26
3. $\overrightarrow{QS}$ bisects $m\angle PQR$	3. _____
4. _____	4. Exercise 26
5. _____	5. Given
6. $\dfrac{1}{2}m\angle ABC = \dfrac{1}{2}m\angle PQR$	6. _____
7. _____	7. Substitution law

29. You are given two adjacent, acute angles. Their nonshared sides are perpendicular. How are the angle measures related? Explain your reasoning.

Chapter **1** Review

Key Terms and Symbols

1.1 inductive reasoning	**1.3** line segment	negation
axiomatic system	endpoints	inverse
undefined terms	length, (of a segment)	contrapositive
definitions	collinear points	
axioms	ray	**1.5** corollary
postulates	angle ($\angle$)	vertical angles
theorems	vertex	
hypothesis	sides (of an angle)	**1.6** geometric construction
conclusion	degree ($°$)	straightedge
deductive reasoning	straight angle	compass
premises	right angle	arcs
fallacy	acute angle	bisect (a line segment)
	obtuse angle	midpoint
1.2 set	equal angles ($=$)	bisector (of an angle)
point	complementary angles	perpendicular lines ($\perp$)
line	supplementary angles	perpendicular bisector
plane	interior (of an angle)	distance (from a point to a
space	exterior (of an angle)	line)
geometric figure	adjacent angles	
coplanar		
origin	**1.4** conditional statement	
number line	($\longrightarrow$)	
coordinate	converse	
plotting	direct proof	

Chapter 1 Proof Techniques

To Prove:
Two Angles Equal

1. Show they are both right angles. (Theorem 1.12)

2. Show they are complements of the same or equal angles. (Theorem 1.6 or Corollary 1.7)

3. Show they are supplements of the same or equal angles. (Theorem 1.8 or Corollary 1.9)

4. Show they are vertical angles. (Theorem 1.11)

5. Show they can be formed as the sum or difference of equal corresponding angles. (Theorems 1.3 and 1.4)

Two Angles Complementary

1. Show their sum is a right angle that measures 90°. (Definition of complementary angles)

Two Angles Supplementary

1. Show their sum is a straight angle that measures 180°. (Definition of supplementary angles)

2. Show they are adjacent angles whose noncommon sides lie on the same line. (Theorem 1.10)

Two Segments Equal

1. Show that each can be formed by adding or subtracting equal corresponding segments. (Theorems 1.1 and 1.2)

2. Show that they are segments of a third segment formed by the midpoint of the third segment and its endpoints. (Definition of midpoint)

Review Exercises

Section 1.1

1. What are the four parts to any axiomatic system?

2. What role do undefined terms play in an axiomatic system?

Use inductive reasoning in Exercises 3–5 to determine the next element in each list.

3. $-8, -3, 2, 7, 12$ 4. A, C, E, G, I   5. $1, \dfrac{1}{2}, \dfrac{1}{4}, \dfrac{1}{8}, \dfrac{1}{16}$

In Exercises 6–9, determine if each conclusion follows logically from the premises, and state whether the reasoning is inductive or deductive.

6. *Premise:* If I wash my car, then it will rain.
   Premise: I washed my car.
   Conclusion: It will rain.

7. *Premise:* If I wash my car, then it will rain.
   Premise: It is raining.
   Conclusion: I washed my car.

8. *Premise:* Bob Begay owns a VCR.
   Premise: Bob Jones owns a VCR.
   Premise: Bob Santos owns a VCR.
   Conclusion: Everyone named Bob owns a VCR.

9. *Premise:* If it is Sunday, then tomorrow is Monday.
   Premise: If tomorrow is Monday, then it is a holiday.
   Conclusion: If it is Sunday, tomorrow is a holiday.

10. Explain why
    Conclusion: Tomorrow is a holiday.
    does not follow logically from the premises in Exercise 9.

11. If 3 cats kill 3 mice in 3 minutes, how long will it take 100 cats to kill 100 mice?

12. There are 5 white socks and 5 black socks in a drawer. How many socks must be removed without looking to be assured of having a pair that match?

Section 1.2

Answer true or false in Exercises 13–16.

13. Exactly one line can be drawn between two distinct points.

14. Any three distinct points determine a unique plane.

15. All points on a line determined by two distinct points in a plane are also in the plane.

16. No plane contains all points in space.

17. Plot the points associated with 4, $-\frac{3}{4}$, and $\sqrt{6}$ on a number line.

In Exercises 18–22, give the name of the postulate illustrated by the given statement.

18. $3 = 3$

19. If $x = -2$, then $-2 = x$.

20. If $x = 3$ and $3 = w$, then $x = w$.

21. If $\frac{1}{2}x = 6$, then $x = 12$.

22. If $x = 3$ and $x + 2 = 5$, then $3 + 2 = 5$.

Section 1.3

Use the figure below to answer **true** *or* **false** *in Exercises 23–34.*

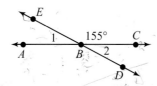

23. C is on $\overrightarrow{BA}$.

24. B is on $\overline{ED}$.

25. $\angle 1$ is another name for $\angle EBA$.

26. B is the vertex of $\angle 2$.

27. $\angle EBD$ is a right angle.

28. $\angle ABD$ and $\angle CBD$ are adjacent angles.

29. $\angle 1$ and $\angle 2$ are complementary.

30. E is an endpoint of $\overleftrightarrow{ED}$.

31. $\angle 2$ and $\angle EBC$ are supplementary.

32. $\overrightarrow{BE}$ and $\overrightarrow{EB}$ are the same.

33. $m\angle 1 = 25°$

34. $\angle 2$ is an obtuse angle.

35. Find the complement of $24°30'45''$.

36. Find the supplement of $136°42'51''$.

37. What is the complement of the supplement of an angle measuring $110°$?

38. Is an angle of measure $145°$ acute?

39. How many points on $\overrightarrow{AB}$ are 3 cm from A?

40. How many points on $\overleftrightarrow{AB}$ are 3 cm from A?

Section 1.4

In Exercises 41 and 42 give a direct proof of the "theorems."

41. *Premise 1:* If it is Gleep, then it is Glop.
Premise 2: If it is Gunk, then it is Grob.
Premise 3: If it is Glop, then it is Gunk.
Theorem: If it is Gleep, then it is Grob.

42. *Premise 1:* If it will burn, it can be destroyed by fire.
Premise 2: If it is a tree, it is made of wood.
Premise 3: If it is made of wood, it will burn.
Theorem: If it is a tree, it can be destroyed by fire.

In Exercises 43 and 44 give the converse of each statement.

43. If it is ice, then it is cold.

44. If you love someone, then you will want the best for that person.

In Exercises 45 and 46 give the negation of each statement.

45. My car is red.

46. The Moon is not made of green cheese.

In Exercises 47 and 48 give the inverse of each statement.

47. If the runner wins the race, then she must be in excellent condition.

48. If the painting is a Picasso, then it is valuable.

In Exercises 49 and 50 give the contrapositive of each statement.

49. If the weather is good, then I will climb the mountain.

50. If I drink, then I do not drive.

Section 1.5

Use the figure below in Exercises 51–54.

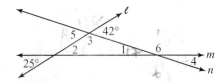

51. What is the measure of $\angle 5$?

52. What is the measure of $\angle 3$?

53. If $m\angle 1 + m\angle 2 + m\angle 3 = 180°$, what is the measure of $\angle 4$?

54. If $m\angle 1 + m\angle 2 + m\angle 3 = 180°$, what is the measure of $\angle 6$?

55. Complete the following proof.

Given: ∠1 and ∠2 are complementary
∠1 and ∠3 are vertical angles
∠2 and ∠4 are vertical angles

Prove: ∠3 and ∠4 are complementary

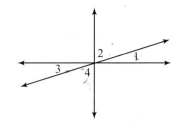

Proof _____

Statements	*Reasons*
1. ∠1 and ∠2 are complementary	1. _____
2. _____	2. Def. of comp. ∠'s
3. _____	3. Given
4. $m∠1 = m∠3$	4. _____
5. _____	5. Given
6. $m∠2 = m∠4$	6. _____
7. $m∠3 + m∠4 = 90°$	7. Substitution law
8. _____	8. Def. of comp. ∠'s

56. Complete the following proof.

Given: The figure of overlapping triangles and $PR = SQ$

Prove: $PS = RQ$

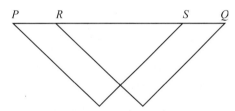

Proof _____

Statements	*Reasons*
1. $PR = SQ$	1. _____
2. $RS = RS$	2. _____
3. $PR + RS = SQ + RS$	3. _____
4. $PS = PR + RS$	4. _____
5. $RQ = SQ + RS$	5. _____
6. $PS = RQ$	6. _____

Section 1.6

57. Draw a line segment $\overline{PQ}$ approximately 2 inches in length, and line *m* containing point *C*. Construct $\overline{CD}$ on *m* such that CD is half PQ.

58. Draw an acute angle and construct its bisector.

59. Draw a line segment $\overline{AB}$ and construct the perpendicular bisector of $\overline{AB}$.

60. How many bisectors does segment $\overline{CD}$ have?

61. How many perpendicular bisectors does segment $\overline{CD}$ have?

62. Draw an acute angle, ∠PQR and ray $\overrightarrow{AB}$. Construct an acute ∠CAB so that $m∠CAB = m∠PQR$.

Chapter (1) Practice Test

1. What is the difference between a postulate and a theorem?
2. Discuss the difference between inductive and deductive reasoning?
3. Which type of reasoning is used in the proof of a theorem in mathematics?
4. Does inductive reasoning always lead to the same conclusion? Explain.
5. Use inductive reasoning to give the next element in the list: $125, -25, 5, -1$.
6. Does the conclusion below follow logically from the premises? What type of reasoning is being used?
   Premise: Michael Jordan eats oatmeal.
   Premise: Larry Bird eats oatmeal.
   Premise: Magic Johnson eats oatmeal.
   Conclusion: All great basketball players eat oatmeal.
7. How many lines can be drawn through one point?
8. Use the addition-subtraction law to complete the following:
   If $x = y$, then $x + 2 = \underline{\quad?\quad}$.

Use the figure below to answer **true** *or* **false** *in Exercises 9–16.*

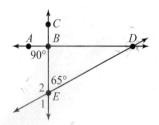

9. $\angle ABC$ is adjacent to $\angle ABE$.
10. $m\angle DBE = 90°$
11. $m\angle 2 = 65°$
12. $m\angle 1 = 65°$
13. A is on $\overleftrightarrow{BD}$
14. A is on $\overrightarrow{BD}$
15. If $m\angle DBE + m\angle BED + m\angle EDB = 180°$, then $m\angle EDB = 25°$.
16. $\angle BED$ and $\angle EDB$ are supplementary.
17. Find the complement of the supplement of an angle measuring $146°20'$.

18. Give a direct proof of the following "theorem."
Premise 1: If I go on trial, then I will be convicted.
Premise 2: If I am arrested, then I will go on trial.
Premise 3: If I am convicted, then I will go to jail.
Premise 4: If I rob a bank, then I will be arrested.
Theorem: If I rob a bank, then I will go to jail.

19. Give the converse of the statement, "If it is milk, then it is white."

20. Give the negation of the statement, "The road to success is difficult."

21. Give the inverse of the statement, "If it is a collie, then it is a dog."

22. Give the contrapositive of the statement, "If the fruit is picked, then it is ripe."

23. *Given:* $m\angle 1 = m\angle 2$
Prove: $m\angle 3 = m\angle 4$

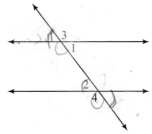

Proof _____

Statements	*Reasons*
1. $m\angle 1 = m\angle 2$	**1.** _____
2. $\angle 1$ and $\angle 3$ are supplementary	**2.** _____
3. $\angle 2$ and $\angle 4$ are supplementary	**3.** _____
4. $m\angle 3 = m\angle 4$	**4.** _____

24. Draw a line segment $\overline{AB}$ and an acute angle $\angle PQR$. Construct an acute angle equal to $\angle PQR$ with vertex at the midpoint C of $\overline{AB}$ and one side $\overrightarrow{CB}$.

25. Draw an obtuse angle and construct its bisector.

26. Complete the following proof.

Given: The figure with $m\angle ABD = m\angle CBE$

Prove: $m\angle 1 = m\angle 3$

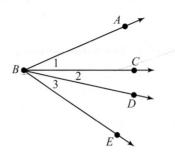

Proof

Statements	Reasons
1. $m\angle ABD = m\angle CBE$	**1.** _____
2. $m\angle 1 + m\angle 2 = m\angle ABD$	**2.** _____
3. $m\angle 2 + m\angle 3 = m\angle CBE$	**3.** _____
4. $m\angle 1 + m\angle 2 = m\angle 2 + m\angle 3$	**4.** _____
5. $m\angle 2 = m\angle 2$	**5.** _____
6. $m\angle 1 = m\angle 3$	**6.** _____

27. Complete the following statements using the given figure where $AD = 2$ cm and $DB = 2$ cm.

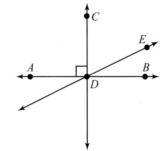

(a) D is a _____ of $\overline{AB}$.

(b) $\overleftrightarrow{DE}$ is a _____ of $\overline{AB}$.

(c) $\overleftrightarrow{CD}$ is a _____ of $\overline{AB}$.

Triangles

The triangle is the most important geometric figure we will study because of its applications in art and engineering as well as its influence in developing other areas of geometry. In this chapter, we classify triangles by their sides and angles and develop three ways to determine if two triangles have the same size and shape.

Properties of triangles give us the necessary tools for solving many applied problems. One of these is given below and solved in Example 2 of Section 2.3.

AN APPLICATION

Mr. Wells owns a cottage on the south shore of Lake Pleasant. Directly north of the cottage on the north shore is a boat dock. To determine the distance from the cottage to the dock, Mr. Wells does the following: He first paces 40 yd due east from the cottage and drives a reference stake into the ground. He then paces another 40 yd due east, at which point he turns due south and walks until he meets the straight line formed by the dock and the stake. The total distance paced off in the southerly direction is 150 yd, and he concludes that the distance from the cottage to the dock is also 150 yd. Explain why this is correct.

2.1 Classifying Triangles

OBJECTIVES

1. Identify parts of a triangle.
2. Classify triangles by their angles.
3. Classify triangles by their sides.
4. Find the perimeter of a triangle.
5. Define interior and exterior angles of a triangle.

Much of geometry involves studying figures such as those in Figure 2.1. The undefined terms *interior* and *exterior* used in conjunction with angles in Chapter 1 are also applicable to other geometric figures. Four additional undefined terms are **sides**, **closed**, **included**, and **opposite**. Figure 2.1(a) shows a closed three-sided figure; Figure 2.1(b) shows a closed six-sided figure with the point *G* in its interior and point *H* in its exterior; Figure 2.1(c) shows a four-sided figure that is not closed.

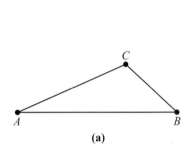 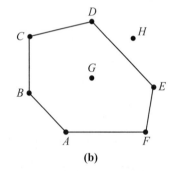 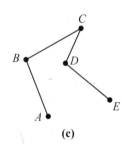

| (a) | (b) | (c) |

Figure 2.1 Geometric Figures

OBJECTIVE 1 **Identify parts of a triangle.** Perhaps the simplest, but most important, geometric figure is a *triangle*, a closed three-sided figure.

How many different triangles can you find in the figure below?

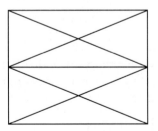

You might be amazed to discover that there are actually eighteen.

> **DEFINITION:** *Triangle*
>
> Let *A*, *B*, and *C* be three points not on the same line. The figure formed by the three segments $\overline{AB}$, $\overline{BC}$, and $\overline{AC}$ is called a **triangle**, denoted by △*ABC*. The three segments $\overline{AB}$, $\overline{BC}$, and $\overline{AC}$ are the **sides** that form △*ABC*, and the three points *A*, *B*, and *C* are the **vertices** (singular—**vertex**) of △*ABC*.

Figure 2.1(a) shows △*ABC*. We say, for example, that side $\overline{BC}$ is opposite ∠*A*, and that ∠*A* is included by the sides $\overline{AC}$ and $\overline{AB}$. Also, side $\overline{BC}$ is included by ∠*B* and ∠*C*, and ∠*A* is opposite side $\overline{BC}$. Notice that every triangle has six basic parts: three sides and three angles.

OBJECTIVE 2 **Classify triangles by their angles.** Triangles are often classified by their angles.

> **USING ANGLES TO CLASSIFY TRIANGLES**
>
> **1.** An **acute triangle** is a triangle in which all angles are acute (measure less than 90°). *continued*

2. A **right triangle** is a triangle in which one angle is a right angle. The side opposite the right angle is the **hypotenuse** of the triangle, and the other two sides are **legs**.
3. An **obtuse triangle** is a triangle in which one angle is obtuse (measures greater than 90°).
4. An **equiangular triangle** is a triangle in which all three angles are equal in measure.

Figure 2.2 shows four triangles classified by their angles.

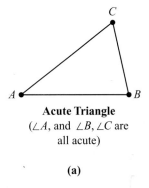
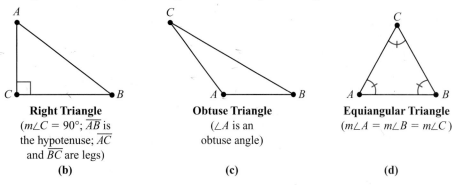

Acute Triangle	**Right Triangle**	**Obtuse Triangle**	**Equiangular Triangle**
($\angle A$, and $\angle B$, $\angle C$ are all acute)	($m\angle C = 90°$; $\overline{AB}$ is the hypotenuse; $\overline{AC}$ and $\overline{BC}$ are legs)	($\angle A$ is an obtuse angle)	($m\angle A = m\angle B = m\angle C$)
(a)	**(b)**	**(c)**	**(d)**

Figure 2.2 Classifying Triangles by Angles

OBJECTIVE 3   Classify triangles by their sides.   Triangles can also be classified by their sides.

USING SIDES TO CLASSIFY TRIANGLES
1. A **scalene triangle** is a triangle in which no two sides are equal in length.
2. An **isosceles triangle** is a triangle in which two sides are equal in length. The third side is its base.
3. An **equilateral triangle** is a triangle in which all three sides are equal in length.

Figure 2.3 shows three triangles classified by their sides. Notice that if a triangle is equilateral, then it is also isosceles. However, the converse is not true because an isosceles triangle need not be equilateral.

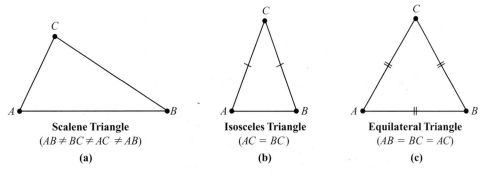

Scalene Triangle	**Isosceles Triangle**	**Equilateral Triangle**
($AB \neq BC \neq AC \neq AB$)	($AC = BC$)	($AB = BC = AC$)
(a)	**(b)**	**(c)**

Figure 2.3 Classifying Triangles by Sides

The marks on Figures 2.3(b) and 2.3(c) show the sides with equal in length.

Technology Connection

Geometry software will be needed.

1. Draw any triangle and label it *ABC*.
2. Measure $\angle A$, $\angle B$, and $\angle C$.
3. Drag point *A*, *B*, or *C* and observe what happens to the three angle measurements.
4. Use the information from step 3 to investigate the following:
   a. How many acute angles can a triangle have?
   b. How many obtuse angles can a triangle have?
   c. How many right angles can a triangle have?
5. Use the software to state whether the triangle is possible or not possible. If it's possible, record the sketch you made on your paper.
   a. isosceles right triangle
   b. acute right triangle
   c. acute scalene triangle
   d. obtuse equilateral triangle

We can also describe triangles with a combination of terms involving both angles and sides as shown in Figure 2.4.

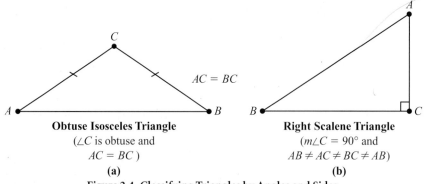

Obtuse Isosceles Triangle
($\angle C$ is obtuse and
$AC = BC$)
(a)

Right Scalene Triangle
($m\angle C = 90°$ and
$AB \neq AC \neq BC \neq AB$)
(b)

Figure 2.4 Classifying Triangles by Angles and Sides

EXAMPLE 1

Refer to $\triangle PQR$ in Figure 2.5, in which $PQ = PR$ and $m\angle P = 40°$. Answer the following:

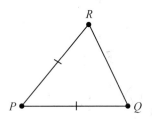

Figure 2.5

(a) What are the vertices of $\triangle PQR$?

(b) What are the sides of $\triangle PQR$?

(c) Classify $\triangle PQR$ using its angles.

(d) Classify $\triangle PQR$ using its sides.

(e) What side is included by $\angle R$ and $\angle Q$?

(f) What side is opposite $\angle Q$?

(g) What angle is included by $\overline{PR}$ and $\overline{PQ}$?

(h) What angle is opposite $\overline{PQ}$?

(i) What is the base of $\triangle PQR$?

(a) *P, Q,* and *R*

(b) $\overline{PQ}, \overline{PR}$, and $\overline{QR}$

(c) **acute triangle**

(d) **isosceles triangle**

(e) $\overline{RQ}$

(f) $\overline{PR}$

(g) $\angle P$

(h) $\angle R$

(i) $\overline{QR}$

PRACTICE EXERCISE **1**

Refer to △*ABC* in which *m*∠*C* = 90° and *AC* ≠ *BC* ≠ *AB* ≠ *AC*. Answer the following:

(a) Classify △*ABC* using its angles.

(b) Classify △*ABC* using its sides.

(c) What angle is included by $\overline{AC}$ and $\overline{BC}$?

(d) What angle is opposite $\overline{AC}$?

(e) What side is included by ∠*A* and ∠*B*?

(f) What side is opposite ∠*A*?

(g) What is the hypotenuse of △*ABC*?

(h) What are the legs of △*ABC*?

ANSWERS ON PAGE 68

OBJECTIVE 4   **Find the perimeter of a triangle.**   A useful number associated with every triangle is its *perimeter*.

> **DEFINITION:** *Perimeter*
>
> **Perimeter** is the sum of the lengths of a shape's sides.

EXAMPLE 2

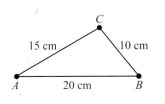

Figure 2.6

Find the perimeter of △*ABC* shown in Figure 2.6. Because the lengths of the sides are

$$AB = 20 \text{ cm}, BC = 10 \text{ cm, and } CA = 15 \text{ cm}$$
$$P = 20 + 10 + 15$$
$$= 45$$

Thus, the perimeter of △*ABC* is 45 cm.

The next example requires the ability to solve a simple equation using techniques learned in beginning algebra.

EXAMPLE 3

Because of their properties of strength as well as aesthetic appeal, triangles are often used in the construction of buildings such as the Cadet Chapel at the Air Force Academy outside Colorado Springs, Colorado.

The base of an isosceles triangle is 12 inches in length and its perimeter is 40 inches. What is the length of each of its equal sides?

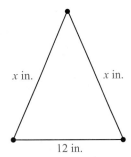

x in.   *x* in.

12 in.

Figure 2.7

First make a sketch of the triangle as shown in Figure 2.7 with the unknown sides labeled x inches. Because the perimeter is 40 inches, we have

$$40 = x + x + 12.$$

$40 = 2x + 12$	Combine terms
$28 = 2x$	Subtract 12 from both sides
$14 = x$	Divide both sides by 2

Thus, each equal side is 14 inches long.

PRACTICE EXERCISE 2

The perimeter of an equilateral triangle is 75 ft. What is the length of each side?

ANSWERS BELOW

OBJECTIVE 5 **Define interior and exterior angles of a triangle.** Several other types of angles can be associated with a given triangle.

> **DEFINITION: Interior and Exterior Angles**
> An **interior angle** of a triangle is an angle formed by two sides of the triangle such that the angle is on the *inside* of the triangle.
> An **exterior angle** of a triangle is an angle formed by a side of the triangle and an *extension* of another side. Both these sides have a common endpoint. The angle lies on the *outside* of the triangle.

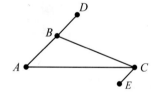

Figure 2.8 Exterior and Interior Angles

Refer to Figure 2.8. Given $\triangle ABC$ and point D on $\overline{AB}$ with B between A and D. Then $\angle DBC$ is an exterior angle of $\triangle ABC$, and $\angle A$, $\angle ABC$, and $\angle BCA$ are interior angles of $\triangle ABC$. Also, $\angle A$ and $\angle BCA$ are **remote interior angles** relative to $\angle DBC$, and $\angle ABC$ is an **adjacent interior angle** relative to $\angle DBC$.

Note In Figure 2.8, $\angle ECA$ is *not* an exterior angle of $\triangle ABC$ because E is not on an extension of $\overline{AC}$ or $\overline{BC}$.

Answers to Practice Exercises

1. **(a)** right triangle **(b)** scalene **(c)** $\angle C$ **(d)** $\angle B$ **(e)** $\overline{AB}$ **(f)** $\overline{BC}$ **(g)** $\overline{AB}$ **(h)** $\overline{AC}$ and $\overline{BC}$ **2.** 25 ft

2.1 **Exercises**   FOR EXTRA HELP: 📖 Student's Solutions Manual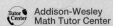

Exercises 1–10 refer to △DEF, shown below, in which *DF = EF* and *m*∠*DFE* = 120°.

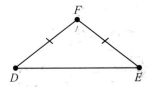

1. What are the vertices of △DEF?
2. What are the sides of △DEF?
3. Classify △DEF using its angles.
4. Classify △DEF using its sides.
5. What side is included by ∠D and ∠E?
6. What side is opposite ∠E?
7. What angle is included by $\overline{DF}$ and $\overline{EF}$?
8. What angle is opposite $\overline{FE}$?
9. What is the base of △DEF?
10. What is the hypotenuse of △DEF?

Exercises 11–20 refer to △ABC, shown below, in which *AB = BC = AC* and *m*∠*A = m*∠*B = m*∠*C*.

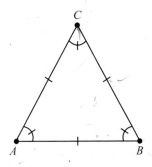

11. What are the sides of △ABC?
12. What are the vertices of △ABC?
13. Classify △ABC using its sides.
14. Classify △ABC using its angles.
15. What side is opposite ∠A?
16. What side is included by ∠A and ∠C?
17. What angle is opposite $\overline{BC}$?
18. What angle is included by $\overline{CA}$ and $\overline{BC}$?
19. What is the hypotenuse of △ABC?
20. Is △ABC an isosceles triangle?
21. Find the perimeter of a triangle with sides 20 cm, 30 cm, and 40 cm.
22. Find the perimeter of an equilateral triangle with sides 18 ft.
23. Find the perimeter of an isosceles triangle with base 8 inches and equal sides 12.5 inches.
24. A triangle with sides 14 cm and 22 cm has a perimeter 66 cm. Find the length of the third side.
25. The perimeter of an equilateral triangle is 69 ft. Find the length of each side.
26. The base of an isosceles triangle is 13 inches and the perimeter is 47 inches. Find the length of its equal sides.
27. The base of an isosceles triangle is one-third the length of each equal side. If the perimeter is 105 cm, find the length of the base and each side.
28. The base of an isosceles triangle is half the length of each equal side. If the perimeter is 30 ft, find the length of the base and each side.

Exercises 29–32 refer to the figure below. Name all triangles that have the given part.

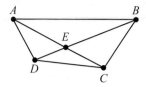

29. $\overline{AE}$ **30.** $\overline{DB}$ **31.** $\angle ECD$ **32.** $\angle ABD$

Exercises 33–40 refer to the figure below. Answer *true* or *false* for each.

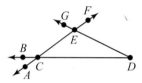

33. $\angle CED$ is an interior angle of $\triangle EDC$.
34. $\angle FED$ is an exterior angle of $\triangle EDC$.
35. $\angle ACB$ is an exterior angle of $\triangle EDC$.
36. $\angle D$ and $\angle CED$ are remote interior angles relative to $\angle ACD$.
37. $\angle CED$ is an interior adjacent angle to $\angle GEF$.
38. $m\angle ACB = m\angle ECD$.
39. Exterior angle $\angle BCE$ is supplementary to interior angle $\angle ECD$.
40. Exterior angle $\angle FED$ is complementary to interior angle $\angle CED$.

For Exercises 41–45, explain each answer. If the triangle is possible, draw a sketch.

41. Can a scalene triangle be an obtuse triangle?
42. Can a scalene triangle be an isosceles triangle?
43. Can a right triangle be an isosceles triangle?
44. Can an isosceles triangle be an acute triangle?
45. Can an equilateral triangle be an obtuse triangle?

2.2 Congruent Triangles

OBJECTIVES

1. Define congruent segments and angles.
2. Define congruent triangles.
3. State the SAS, ASA, and SSS postulates.
4. Use SAS, ASA, and SSS to prove triangles are congruent.

In this section, we compare triangles that are the same size and shape. The term **congruent**, from the Latin words *con* (with) and *gruere* (to agree), is applied to such figures and literally means "in agreement with." Intuitively, congruent triangles can be made to coincide by placing one on top of the other either directly or by *flipping* one of them over.

OBJECTIVE 1   **Define congruent segments and angles.**   In the previous chapter, if two segments had the same measure, the notation $AB = CD$ was used. When two segments have the same measure, they are said to be congruent. The symbol used for congruent is $\cong$.

> **DEFINITION: Congruent Segments**
>
> **Congruent segments** are two segments with the same measure.
> If $AB = CD$ then $\overline{AB} \cong \overline{CD}$ also if $\overline{AB} \cong \overline{CD}$ then $AB = CD$.

EXAMPLE 1   If $\overline{AB} \cong \overline{CD}$ and $AB = 6$ inches then find the measure of CD.

If two segments are congruent then by the definition, their measures are equal. Thus, $CD = 6$ inches.

> **DEFINITION: Congruent Angles**
>
> **Congruent angles** are two angles with the same measure.
> If $m\angle 1 = m\angle 2$ then $\angle 1 \cong \angle 2$.
> Also if $\angle 1 \cong \angle 2$ then $m\angle 1 = m\angle 2$.

EXAMPLE 2   If $\angle 1 \cong \angle 2$ and $m\angle 1 = 27°$, find $m\angle 2$.

Because $\angle 1 \cong \angle 2$ means $m\angle 1 = m\angle 2$, and since $m\angle 1 = 27°$ then $m\angle 2$ must also be $27°$.

EXAMPLE 3   If $\angle 1 \cong \angle 2$ and $m\angle 1 = 3x - 5$ while $m\angle 2 = 2x + 10$, find the measures of both angles.

Because $\angle 1 \cong \angle 2$ means $m\angle 1 = m\angle 2$, then
$3x - 5 = 2x + 10$.
$x = 15$

Therefore,

$$m\angle 1 = 3(15) - 5 = 40$$
$$m\angle 2 = 2(15) + 10 = 40$$

So both angles measure $40°$, which makes sense since they are congruent.

Because every triangle has six **parts**, three angles and three sides, two congruent triangles have equal parts that can be made to match when one is placed on top of the other. In Figure 2.9, $\triangle ABC$ and $\triangle DEF$ are congruent because $\triangle ABC$ can be placed on top of $\triangle DEF$. Also, $\triangle ABC$ and $\triangle HGI$ are congruent because if $\triangle HGI$ were flipped over, it could be made to coincide with $\triangle ABC$. On the other hand, $\triangle ABC$ and $\triangle JKL$ are not

congruent because they cannot be made to coincide either directly or by
flipping over one of them.

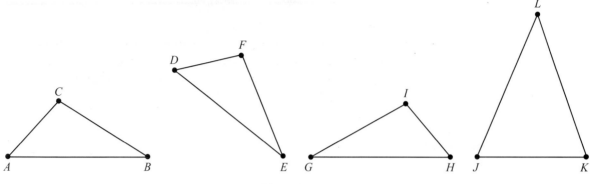

Figure 2.9

OBJECTIVE 2   **Define congruent triangles.**   The parts of congruent triangles that coincide
when one is placed on top of the other are called **corresponding parts**. For
example, congruent triangles $\triangle ABC$ and $\triangle DEF$ in Figure 2.9 have six pairs
of corresponding parts:

$$\angle A \text{ corresponds to } \angle D \qquad \overline{AB} \text{ corresponds to } \overline{DE}$$
$$\angle B \text{ corresponds to } \angle E \qquad \overline{BC} \text{ corresponds to } \overline{EF}$$
$$\angle C \text{ corresponds to } \angle F \qquad \overline{AC} \text{ corresponds to } \overline{DF}$$

Notice that corresponding angles are opposite corresponding sides, and
corresponding sides are opposite corresponding angles. We see that two
triangles are congruent when their corresponding parts are equal.

> **DEFINITION:**   **Congruent Triangles**
>
> If all six parts of one triangle (three angles and three sides) are congru-
> ent to the corresponding parts of another triangle, the two triangles are
> **congruent**. The notation is $\triangle ABC \cong \triangle DEF$.

If $\triangle ABC$ and $\triangle DEF$ are congruent, written $\triangle ABC \cong \triangle DEF$, then
$\angle A \cong \angle D, \angle B \cong \angle E, \angle C \cong \angle F, \overline{AB} \cong \overline{DE}, \overline{BC} \cong \overline{EF}$, and $\overline{AC} \cong \overline{DF}$.

To avoid giving specific lengths and angle measures, we often mark the
parts in working with figures of congruent triangles to indicate those that are
congruent. Figure 2.10 illustrates this and shows that $\angle A \cong \angle D, \angle B \cong \angle E$,
$\angle C \cong \angle F, \overline{AB} \cong \overline{DE}, \overline{BC} \cong \overline{EF}$, and $\overline{AC} \cong \overline{DF}$ in congruent triangles $\triangle ABC$
and $\triangle DEF$.

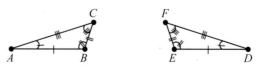

Figure 2.10 Labeling Congruent Triangles

Notice in the statement $\triangle ABC \cong \triangle DEF$, A and D are both in the first positions and $\angle A$ corresponds to $\angle D$ ($\angle A \cong \angle D$). Similarly with B and E as well as C and F. The **positions of the letters indicate the correspondence.**

OBJECTIVE 3   **State the SAS, ASA, and SSS postulates.**   According to our definition, in order to prove that two triangles are congruent, we would need to show that *all* six parts of one triangle are congruent to *all* six parts of the other. Fortunately, it is necessary to show congruence using only *three* pairs of corresponding parts as shown in the next three postulates.

  Technology Connection

Geometry software will be needed.

1. Draw any triangle and label it $\triangle ABC$.
2. Measure $\angle A$ and sides $\overline{AB}$ and $\overline{AC}$.
3. Another place on the screen, copy $\angle A$ and sides $\overline{AB}$ and $\overline{AC}$. Relabel $A \longrightarrow X$, $B \longrightarrow Y$, and $C \longrightarrow Z$.
4. Form $\triangle XYZ$ by drawing $\overline{YZ}$.
5. Measure $\angle Y$ and $\angle Z$. How do these measurements compare to the measurements of $\angle B$ and $\angle C$ respectively? How do the measurements of $\overline{BC}$ and $\overline{YZ}$ compare?

Using the definition of congruent triangles, we see that $\triangle ABC \cong \triangle XYZ$. Notice, all that was needed to form the congruent pair of triangles was two sides and the included angle (SAS).

POSTULATE 2.1   SAS (side-angle-side)

If two sides and the included angle of one triangle are congruent to two corresponding sides and the included angle of a second triangle, then the triangles are congruent.

EXAMPLE 4   Verify that $\triangle ABC \cong \triangle DFE$ in Figure 2.11.

Because AC and DE are both 12 cm, while the included angles $\angle C$ and $\angle E$ both measure $24°$, and BC and FE are both 13 cm, $\triangle ABC \cong \triangle DFE$ by Postulate 2.1, SAS.

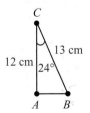

Figure 2.11

POSTULATE 2.2 ASA (angle-side-angle)

If two angles and the included side of one triangle are congruent to the corresponding two angles and the included side of a second triangle, then the triangles are congruent.

EXAMPLE 5

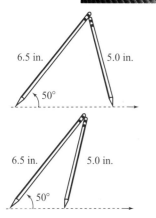

6.5 in.   5.0 in.

50°

6.5 in.   5.0 in.

50°

There will not be a SSA postulate. The figure above shows why we *do not* have a SSA postulate for proving triangles congruent. Two noncongruent triangles can be formed with sides measuring 6.5 inches, 5.0 inches, and a nonincluded angle of 50°.

Use the information given in Figure 2.12 to verify that $\triangle PQR \cong \triangle MNO$.

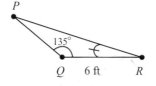
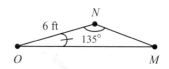

Figure 2.12

Because $QR = 6$ ft and $NO = 6$ ft, $\overline{QR} \cong \overline{NO}$. Also, $\angle Q \cong \angle N$ because both measure 135°, and we are given in the figure that $\angle R \cong \angle O$. Thus, $\triangle PQR \cong \triangle MNO$ by Postulate 2.2, ASA.

POSTULATE 2.3 SSS (side-side-side)

If three sides of one triangle are congruent to the corresponding three sides of a second triangle, then the triangles are congruent.

PRACTICE EXERCISE 1

Use the information given in the figure to verify that $\triangle ABC \cong \triangle STU$.

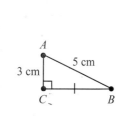
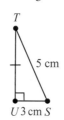

ANSWERS ON PAGE 77

OBJECTIVE 4   Use SAS, ASA, and SSS to prove triangles are congruent.   To prove two triangles congruent, we sometimes need to derive information about equal parts using our knowledge about vertical angles or the fact that two triangles share a common side. The following two examples illustrate.

EXAMPLE 6   Refer to Figure 2.13.

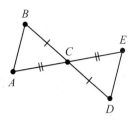

Figure 2.13

Given: $\overline{BC} \cong \overline{DC}$
$\overline{AC} \cong \overline{EC}$

Prove: $\triangle ABC \cong \triangle EDC$

Proof ───────────────────────────────────────

Statements	*Reasons*
1. $\overline{BC} \cong \overline{DC}$	1. Given
2. $\overline{AC} \cong \overline{EC}$	2. Given
3. $\angle BCA \cong \angle DCE$	3. Vert. $\angle$'s are $\cong$
4. $\triangle ABC \cong \triangle EDC$	4. SAS

Because congruent segments (or angles) are related to the equality of their length (or angle measure) by the definition of congruence, here is a list of properties of congruence. They are similar to Postulates 1.6, 1.7, and 1.8 and are true for congruent segments as well as congruent angles.

Reflexive: $\overline{AB} \cong \overline{AB}$ (A segment is congruent to itself.)

Symmetric: If $\overline{AB} \cong \overline{CD}$ then $\overline{CD} \cong \overline{AB}$.

Transitive: If $\overline{AB} \cong \overline{CD}$ and $\overline{CD} \cong \overline{EF}$, then $\overline{AB} \cong \overline{EF}$.

Using similar reasoning, some theorems from Chapter 1 relating angles of equal measure can be restated using congruent angles.

Theorem 1.6: Complements of congruent angles are congruent.

Theorem 1.7: Complements of the same angle are congruent.

Theorem 1.8: Supplements of congruent angles are congruent.

Theorem 1.9: Supplements of the same angle are congruent.

Theorem 1.11: Vertical angles are congruent.

Theorem 1.12: All right angles are congruent.

EXAMPLE 7   Refer to Figure 2.14.

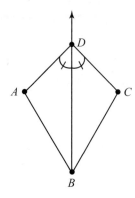

Figure 2.14

Given: $\overrightarrow{BD}$ bisects $\angle ABC$
           $\angle ADB \cong \angle CDB$

Prove: $\triangle ABD \cong \triangle CBD$

Proof _____

Statements	*Reasons*
1. $\overrightarrow{BD}$ bisects $\angle ABC$	1. Given
2. $\angle ABD \cong \angle CBD$	2. Def. of $\angle$ bisector
3. $\angle ADB \cong \angle CDB$	3. Given
4. $\overline{BD} \cong \overline{BD}$	4. Reflexive law
5. $\triangle ABD \cong \triangle CBD$	5. ASA

Example 7 using a flowchart proof:

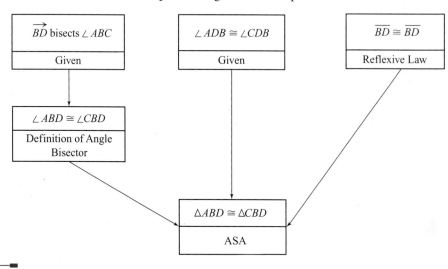

Thales (640–546 B.C.)

Thales of Miletus was a wealthy merchant who became interested in the practical aspects of geometry. In Greece, he taught geometry to many of his friends. His most noteworthy pupil was Pythagoras. Thales has been called the "Father of Greek Mathematics."

┌─ **CAUTION** Although two segments or two angles may *appear* to be congruent,
│ never make this assumption simply because they look the same. When writing
│ proofs, use only given information together with known facts from previously
│ proved theorems, postulates, or definitions.
└─

Student Activity

Explain to a partner what is *wrong* with each picture in Figure 2.15.

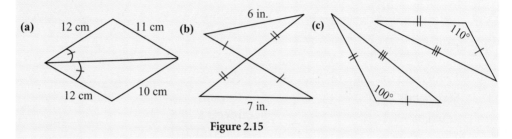

Figure 2.15

Answer to Practice Exercise

1. The desired congruence can be shown in two ways. Because $AC = 3$ cm and
$SU = 3$ cm, $\overline{AC} \cong \overline{SU}$. Also, we are given that $\overline{CB} \cong \overline{UT}$ and $\angle C \cong \angle U$
(both are right angles). Thus, $\triangle ABC \cong \triangle STU$ by SAS. Alternatively, because
$\overline{AC} \cong \overline{SU}$, $\overline{AB} \cong \overline{ST}$, and $\overline{BC} \cong \overline{TU}$, $\triangle ABC \cong \triangle STU$ by SSS.

2.2 Exercises

FOR EXTRA HELP: 📖 Student's Solutions Manual   Tutor Center Addison-Wesley Math Tutor Center

Decide whether the triangles given in Exercises 1–6 are congruent. Explain your reasoning.

1.

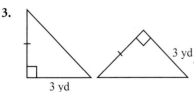

2.

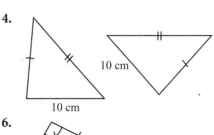

3.

4.

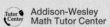

5.

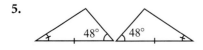

6.

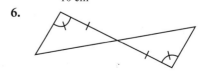

Complete each proof in Exercises 7–10.

7. *Given:* ∠1 ≅ ∠2
   ∠3 ≅ ∠4

Prove: △ABD ≅ △CBD

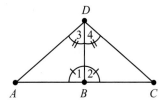

Exercise 7

Proof _____

Statements	**Reasons**
1. ∠1 ≅ ∠2	**1.** _____
2. _____	**2.** Given
3. _____	**3.** Reflexive law
4. △ABD ≅ △CBD	**4.** _____

8. *Given:* C is the midpoint of $\overline{AE}$
   ∠E ≅ ∠A

Prove: △ABC ≅ △EDC

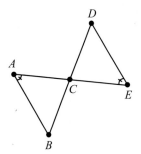

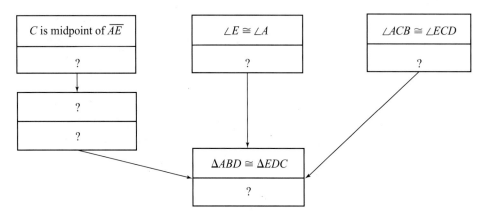

Exercise 8

9. *Given:* $\overline{DB} \perp \overline{AB}$
     $\overline{DB} \perp \overline{DC}$
     $\overline{AB} \cong \overline{DC}$

   Prove: $\triangle ABD \cong \triangle CDB$

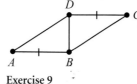

Exercise 9

Proof _____

Statements	**Reasons**
1. $\overline{DB} \perp \overline{AB}$ and $\overline{DB} \perp \overline{DC}$	**1.** _____
2. $\angle ABD$ and $\angle BDC$ are right angles	**2.** _____
3. _____	**3.** Rt $\angle$'s are $\cong$
4. _____	**4.** Reflexive law
5. $\overline{AB} \cong \overline{DC}$	**5.** _____
6. $\triangle ABD \cong \triangle CDB$	**6.** _____

10. *Given:* $\overline{AB} \cong \overline{CD}$
      $\overline{AC} \cong \overline{BD}$

   Prove: $\triangle ABC \cong \triangle DCB$

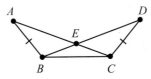

Exercise 10

Proof _____

Statements	**Reasons**
1. $\overline{AB} \cong \overline{CD}$	**1.** _____
2. _____	**2.** Given
3. _____	**3.** Reflexive law
4. $\triangle ABC \cong \triangle DCB$	**4.** _____

Write a two-column or flowchart proof in Exercises 11–14.

11. *Given:* $\overline{AD}$ bisects $\overline{BE}$
      $\overline{BE}$ bisects $\overline{AD}$

   Prove: $\triangle ABC \cong \triangle DEC$

12. *Given:* $\angle B$ and $\angle E$ are right angles
      $\overline{AD}$ bisects $\overline{BE}$

   Prove: $\triangle ABC \cong \triangle DEC$

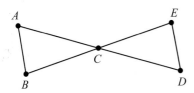

Exercise 11

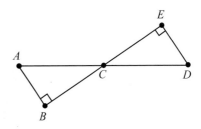

Exercise 12

13. *Given:* $\overline{AD} \cong \overline{BD}$
$\overline{AE} \cong \overline{BC}$

    Prove: $\triangle ACD \cong \triangle BED$

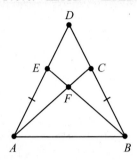

Exercise 13

14. *Given:* $\overline{DB} \perp \overline{AC}$
$\overline{DB}$ bisects $\overline{AC}$

    Prove: $\triangle ABD \cong \triangle CBD$

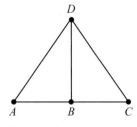

Exercise 14

Student Activity

Answer the questions, justifying each answer.

15. Using the given figure, determine which of the statements are correct and which are not. Justify each answer.

    (a) $\triangle BAC \cong \triangle EDF$
    (b) $\triangle BAC \cong \triangle FED$
    (c) $\triangle EDF \cong \triangle ABC$
    (d) The two triangles are not congruent.

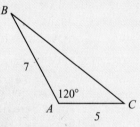
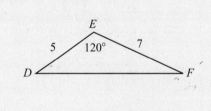

Exercise 15

16. State which triangles (if any) are congruent and justify the answer. *Do not* judge by the drawings but by the markings on the figures.

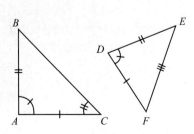
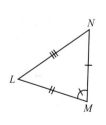
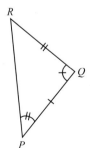

Exercise 16

Student Activity

17. Investigating the SSS theorem. A compass, protractor, and straight-edge will be needed.

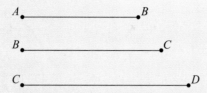

(a) Using the given segments, *construct* a triangle on a sheet of paper.
(b) Compare that triangle with a partner's triangle. Is it different or is it the same? Did you have to flip one over so the two triangles match?
(c) Compare your triangle to a third person's triangle. Do they match?
(d) Draw three different line segments on the paper. You and a partner use these line segments to construct another triangle. Do they match? Will any three lengths always make a triangle?
(e) Does it seem reasonable that if three sides of one triangle are congruent to three sides of another triangle, the triangles will be congruent?
(f) Now using a protractor, draw a triangle given three angles that measure $30°, 60°$, and $90°$. Compare your triangle with at least two other people. Do they match? Do you think there will be an AAA theorem of congruence? Explain.

18. Investigating the SAS theorem. A compass and straightedge are needed.

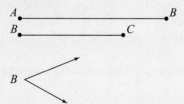

(a) *Construct* a triangle with the given segments and included angle where angle B is between the two segments.
(b) Form a triangle by connecting A and C.
(c) Compare your triangle with a partner's triangle. Do they match? Did you have to flip one over so the two triangles match?
(d) Draw two different line segments and a different angle. You and a partner use these sides and included angle to construct another triangle. Do they match?
(e) Does it seem reasonable that if two sides and an included angle are congruent to two sides and an included angle of another triangle the triangles are congruent?

Use the figure below in Exercises 19–30. You will need to recall what you learned in beginning algebra to solve several equations. Assume that $\triangle ABC \cong \triangle PQR$, $AC = x + 1$, $PR = 3x - 5$, $m\angle B = (100 + y)°$, and $m\angle Q = (5y + 20)°$.

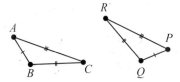

19. Find the value of x.
20. Find the value of y.
21. Find AC.
22. Find PR.
23. Find the measure of $\angle B$.
24. Find the measure of $\angle Q$.
25. $\overline{AC}$ corresponds to which side in $\triangle PQR$?
26. $\overline{QP}$ corresponds to which side in $\triangle ABC$?
27. $\overline{BC}$ corresponds to which side in $\triangle PQR$?
28. $\angle Q$ corresponds to which angle in $\triangle ABC$?
29. $\angle A$ corresponds to which angle in $\triangle PQR$?
30. $\angle R$ corresponds to which angle in $\triangle ABC$?

(2.3) Proofs Involving Congruence

OBJECTIVES

1. Define CPCTC.
2. Use CPCTC.
3. Prove the segment bisector theorem.
4. Prove the angle bisector theorem.

OBJECTIVE 1 Define CPCTC. In the previous section, congruent triangles were defined as two triangles in which the six corresponding parts are congruent to each other. Recall all definitions are reversible. Thus, if two triangles are congruent, their six corresponding parts are also congruent. If we proved $\triangle ABC \cong \triangle XYZ$ by SAS (using the congruent parts marked in Figure 2.16 below), then we know the remaining corresponding parts are also congruent to each other by the definition of congruent triangles. This is abbreviated CPCTC (corresponding parts of congruent triangles are congruent). We can conclude $\angle A \cong \angle X$, $\angle C \cong \angle Z$, and $\overline{AC} \cong \overline{XZ}$.

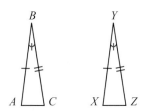

Figure 2.16

DEFINITION: CPCTC: Corresponding Parts of Congruent Triangles Are Congruent

If two triangles are congruent, the six pairs of corresponding parts are congruent.

EXAMPLE 1   Refer to Figure 2.17.

Given: $\angle A \cong \angle E$
    $\overline{AC} \cong \overline{EC}$

Prove: $\angle B \cong \angle D$

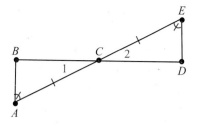

Figure 2.17

Proof

Statements	*Reasons*
1. $\angle A \cong \angle E$	1. Given
2. $\overline{AC} \cong \overline{EC}$	2. Given
3. $\angle 1 \cong \angle 2$	3. Vert. $\angle$'s are $\cong$
4. $\triangle ABC \cong \triangle EDC$	4. ASA
5. $\angle B \cong \angle D$	5. CPCTC

OBJECTIVE 2   **Use CPCTC.**   The next example solves the applied problem given in the chapter introduction.

EXAMPLE 2   Mr. Wells owns a cottage on the south shore of Lake Pleasant. Directly north of the cottage on the north shore is a boat dock. To determine the distance from the cottage to the dock, Mr. Wells does the following: He first paces 40 yd due east from the cottage and drives a reference stake into the ground. He then paces another 40 yd due east, at which point he turns due south and walks until he meets the straight line formed by the dock and the stake. The total distance paced off in the southerly direction is 150 yd, and he concludes that the distance from the cottage to the dock is also 150 yd. Explain why this is correct.

We'll make a sketch (at left) with point D corresponding to the boat dock, point C corresponding to the cottage, and point S corresponding to the stake as in Figure 2.18. Because $\angle DCS$ and $\angle BAS$ are right angles, $\angle DCS \cong \angle BAS$. Because $\angle DSC$ and $\angle BSA$ are vertical angles, $\angle DSC \cong \angle BSA$. And because $CS = 40$ yd and $SA = 40$ yd, $\overline{CS} \cong \overline{AS}$. Thus, $\triangle CSD \cong \triangle ASB$ by ASA. Because $\overline{DC}$ and $\overline{BA}$ are corresponding parts of congruent triangles, $\overline{DC} \cong \overline{BA}$. Thus, CD, the distance from the cottage to the boat dock is the same as BA, which is 150 yd.

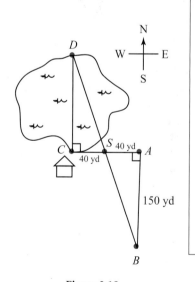

Figure 2.18

> **THEOREM 2.1   *Transitive Law for Congruent Triangles***
> If $\triangle ABC \cong \triangle DEF$ and $\triangle DEF \cong \triangle GHI$, then $\triangle ABC \cong \triangle GHI$.

Simply stated, Theorem 2.1 says that two triangles congruent to the same triangle are congruent to each other. Refer to Figure 2.19 for the proof of this theorem.

Given: $\triangle ABC \cong \triangle DEF$, $\triangle DEF \cong \triangle GHI$
Prove: $\triangle ABC \cong \triangle GHI$

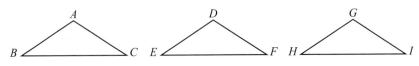

Figure 2.19

Proof

Statements	Reasons
1. $\triangle ABC \cong \triangle DEF$	1. Given
2. $\overline{AB} \cong \overline{DE}$, $\overline{AC} \cong \overline{DF}$, $\overline{BC} \cong \overline{EF}$	2. CPCTC
3. $\triangle DEF \cong \triangle GHI$	3. Given
4. $\overline{DE} \cong \overline{GH}$, $\overline{DF} \cong \overline{GI}$, $\overline{EF} \cong \overline{HI}$	4. CPCTC
5. $\overline{AB} \cong \overline{GH}$, $\overline{AC} \cong \overline{GI}$, $\overline{BC} \cong \overline{HI}$	5. Using statements 2, 4, and trans. law
6. $\triangle ABC \cong \triangle GHI$	6. SSS

OBJECTIVE 3 **Prove the segment bisector theorem.** In Chapter 1, we showed how to construct the bisector of a line segment. Congruent triangles can illustrate that our construction gives the perpendicular bisector of the segment.

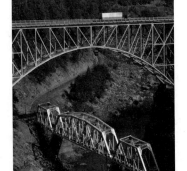

When constructing a roof, bridge, or other type of structure that must support a heavy load, a triangular-shaped frame called a truss is used. The SSS postulate guarantees that a triangle with given sides cannot be distorted in shape. It is this fact that gives a truss its strength.

> **THEOREM 2.2** *SEGMENT BISECTOR THEOREM*
> Construction 1.3 gives the perpendicular bisector of a given line segment.

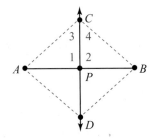

Figure 2.20 Perpendicular Bisector

Given: $\overline{AC} \cong \overline{BC}$ (by construction)
$\overline{AD} \cong \overline{BD}$ (by construction)
(See Figure 2.20.)
Prove: $\overleftrightarrow{CD}$ is the perpendicular bisector of $\overline{AB}$

Proof _____

Statements	Reasons
1. $\overline{AC} \cong \overline{BC}$ and $\overline{AD} \cong \overline{BD}$	1. Given
2. $\overline{CD} \cong \overline{CD}$	2. Reflexive law
3. $\triangle ACD \cong \triangle BCD$	3. SSS
4. $\angle 3 \cong \angle 4$	4. CPCTC
5. $\overline{CP} \cong \overline{CP}$	5. Reflexive law
6. $\triangle ACP \cong \triangle BCP$	6. SAS
7. $\overline{AP} \cong \overline{BP}$	7. CPCTC
8. P is the midpoint of $\overline{AB}$	8. Def. of midpoint
9. $\overleftrightarrow{CD}$ bisects $\overline{AB}$	9. Def. of bisector
10. $\angle 1$ and $\angle 2$ are supplementary	10. Adj. $\angle$'s whose noncommon sides are in line are supp. (Thm l.10)
11. $\angle 1 \cong \angle 2$	11. CPCTC
12. $\overleftrightarrow{CD} \perp \overline{AB}$	12. Def. of $\perp$
13. $\overleftrightarrow{CD}$ is the perpendicular bisector of $\overline{AB}$	13. Statements 9 and 12 and def. of $\perp$ bisector

The next theorem is closely related to Theorem 2.2, and its proof is requested in the exercises.

> **THEOREM 2.3**
> Every point on the perpendicular bisector of a segment is equidistant from the two endpoints.

OBJECTIVE 4   **Prove the angle bisector theorem.**   Now that we can prove two triangles are congruent, we can show that the method used to construct an angle bisector (Construction 1.6) actually produces the angle bisector.

> **THEOREM 2.4**   *ANGLE BISECTOR THEOREM*
> Construction 1.6 gives the bisector of a given angle.

Refer to Figure 2.21 for the proof of Theorem 2.4.

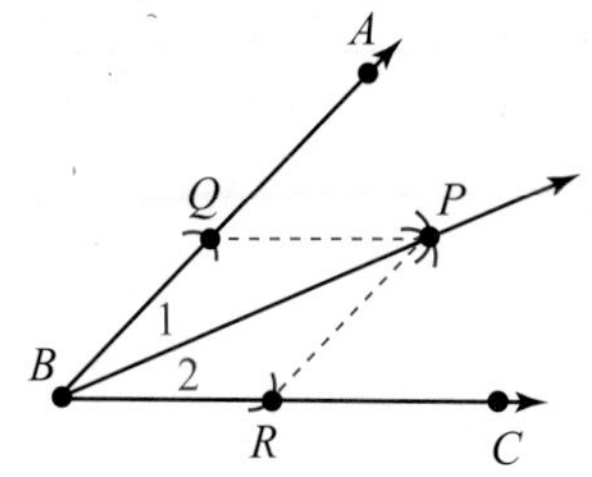

Figure 2.21 Angle Bisector

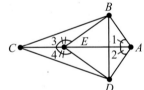

Why does the frame of a bicycle maintain its shape?

Given: $\overline{BQ} \cong \overline{BR}$ (by construction)
$\overline{QP} \cong \overline{RP}$ (by construction)
Prove: $\overrightarrow{BP}$ bisects $\angle ABC$

Proof _____

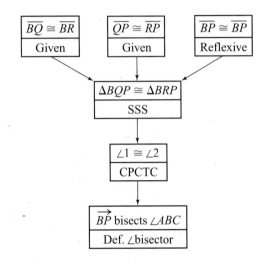

Sometimes it is necessary to prove that several pairs of triangles are congruent to show that two segments or angles are equal. The following example illustrates.

EXAMPLE 3   Refer to Figure 2.22.
Given: $\angle 1 \cong \angle 2$
$\angle 3 \cong \angle 4$

Prove: $\overline{CB} \cong \overline{CD}$

Figure 2.22

Proof _____

Statements	**Reasons**
1. $\angle 1 \cong \angle 2$	1. Given
2. $\angle 3 \cong \angle 4$	2. Given
3. $\angle BEA$ is supplementary to $\angle 3$ and $\angle DEA$ is supplementary to $\angle 4$	3. Adj. $\angle$'s whose noncommon sides are in line are supp.
4. $\angle BEA \cong \angle DEA$	4. Supp. of $\cong$ $\angle$'s are $\cong$
5. $\overline{AE} \cong \overline{AE}$	5. Reflexive law
6. $\triangle BEA \cong \triangle DEA$	6. ASA
7. $\overline{BE} \cong \overline{DE}$	7. CPCTC
8. $\overline{CE} \cong \overline{CE}$	8. Reflexive law
9. $\triangle CEB \cong \triangle CED$	9. SAS
10. $\overline{CB} \cong \overline{CD}$	10. CPCTC

Note Theorem 1.8 states "supplements of equal angles are equal in measure." In this chapter, angles equal in measure are defined as congruent angles. So now supplements of congruent angles are congruent.

 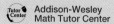
Exercises 1–4 refer to the congruent triangles given below in which $\angle A$ and $\angle F$ correspond as do $\angle C$ and $\angle E$.

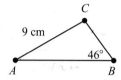

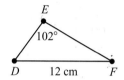

1. Find EF.
3. Find the measure of $\angle D$.

2. Find AB.
4. Find the measure of $\angle C$.

Complete each proof in Exercises 5–8.

5. *Given:* $\overline{BG} \cong \overline{CE}$
$\angle 1 \cong \angle 2$
Prove: $\angle G \cong \angle E$

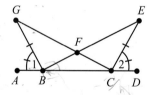

Exercise 5

Proof

Statements	**Reasons**
1. $\overline{BG} \cong \overline{CE}$	1. _____
2. _____	2. Given
3. $\angle 1$ and $\angle GBC$ are supplementary	3. _____
4. _____	4. Adj. $\angle$'s whose noncommon sides are in a line are supp.
5. $\angle GBC \cong \angle ECB$	5. _____
6. $\overline{BC} \cong \overline{BC}$	6. _____
7. $\triangle GBC \cong \triangle ECB$	7. _____
8. _____	8. CPCTC

6. *Given:* $\overline{BC} \cong \overline{CD}$
$\angle 1 \cong \angle 2$
Prove: $\overline{GC} \cong \overline{FC}$

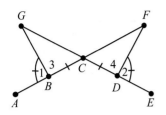

Exercise 6

Proof _____

Statements	Reasons
1. $\overline{BC} \cong \overline{CD}$	**1.** _____
2. _____	**2.** Given
3. $\angle 1$ and $\angle 3$ are supplementary	**3.** _____
4. _____	**4.** Adj. $\angle$'s whose noncommon sides are in a line are supp.
5. _____	**5.** Supp. of $\cong$ $\angle$'s are $\cong$
6. $\angle GCB$ and $\angle FCD$ are vertical angles	**6.** _____
7. _____	**7.** Vert. $\angle$'s are $\cong$
8. _____	**8.** ASA
9. $\overline{GC} \cong \overline{FC}$	**9.** _____

7. *Given:* $\overline{AB} \cong \overline{CD}$
$\angle A \cong \angle D$
$\angle 1 \cong \angle 2$
Prove: $\overline{AF} \cong \overline{DE}$

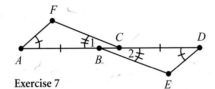

Exercise 7

Proof _____

Statements	Reasons
1. $\angle A \cong \angle D$	**1.** _____
2. $\angle 1 \cong \angle 2$	**2.** _____
3. _____	**3.** Given and def. $\cong$ seg.
4. $AC = AB + BC$	**4.** _____
5. _____	**5.** Seg. add. post.
6. $AB + BC = CD + BC$	**6.** Addition law using statement 3
7. $AC = BD;\ \overline{AC} \cong \overline{BD}$	**7.** _____
8. _____	**8.** ASA
9. $\overline{AF} \cong \overline{DE}$	**9.** _____

8. *Given:* $\overline{BD} \perp \overline{AC}$
$\overline{BC} \cong \overline{BE}$
$\angle 1 \cong \angle C$

Prove: $\angle A \cong \angle D$

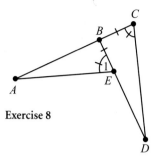

Exercise 8

Proof _____

Statements	**Reasons**
1. $\angle 1 \cong \angle C$	**1.** _____
2. $\overline{BD} \perp \overline{AC}$	**2.** _____
3. _____	**3.** Def. $\perp$ lines
4. _____	**4.** Given
5. $\triangle ABE \cong \triangle BCD$	**5.** _____
6. _____	**6.** CPCTC

Write a two-column or flowchart proof in Exercises 9–16. [Note: For the remainder of the book, the key will only show a two-column proof. This is done for efficiency.]

9. *Given:* $\overline{AC} \cong \overline{CE}$
$\overline{DC} \cong \overline{CB}$

Prove: $\angle A \cong \angle E$

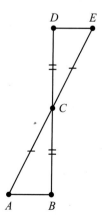

Exercise 9

10. *Given:* $\overrightarrow{AC}$ bisects $\angle BAD$
$\overrightarrow{CA}$ bisects $\angle BCD$

Prove: $\angle B \cong \angle D$

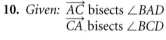
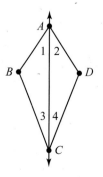

Exercise 10

11. *Given:* $\overline{BC} \cong \overline{CD}$
$\qquad \angle 1 \cong \angle 2$
$\quad$ *Prove:* $\overline{AB} \cong \overline{AD}$

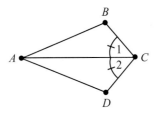

Exercise 11

12. *Given:* $\overrightarrow{DB}$ bisects $\angle ADC$
$\qquad \overline{AD} \cong \overline{CD}$
$\quad$ *Prove:* $\overline{DB} \perp \overline{AC}$

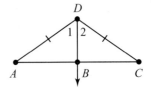

Exercise 12

13. *Given:* $\angle 1 \cong \angle 2$
$\qquad \angle 3 \cong \angle 4$
$\quad$ *Prove:* $\angle A \cong \angle C$

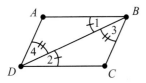

Exercise 13

14. *Given:* $\overline{BC} \perp \overline{AB}$
$\qquad \overline{AD} \perp \overline{DC}$
$\qquad \angle 1 \cong \angle 2$
$\quad$ *Prove:* $\overline{AB} \cong \overline{CD}$

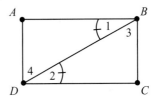

Exercise 14

15. *Given:* $\overline{GB} \perp \overline{AF}$
$\qquad \overline{FD} \perp \overline{GE}$
$\qquad \overline{GD} \cong \overline{FB}$
$\qquad \overline{GB} \cong \overline{FD}$
$\quad$ *Prove:* $\overline{BC} \cong \overline{DC}$
$\qquad$ [Hint: Several pairs of triangles
$\qquad$ must be shown congruent.]

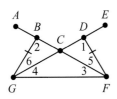

Exercise 15

16. *Given:* $\angle A \cong \angle C$
$\qquad \angle 1 \cong \angle 2$
$\qquad B$ is the midpoint of $\overline{AC}$
$\quad$ *Prove:* $\overline{GF} \cong \overline{DE}$
$\qquad$ [Hint: Several pairs of triangles
$\qquad$ must be shown congruent.]

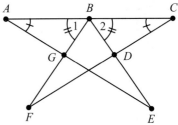

Exercise 16

17. To measure the distance between two points A and B on opposite sides of a lake, a ranger first sets a stake at point C as shown in the Exercise 17 figure. He then sets a stake at point D in line with B and C so that $BC = CD$. Next he sets a stake at point E in line with A and C so that $AC = CE$. He measures the distance between D and E, 105 yd, and concludes that the distance between A and B is also 105 yd. Why is this true?

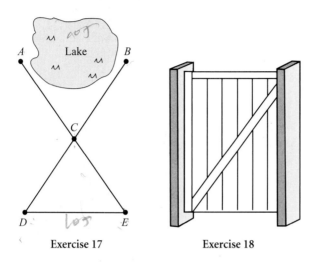

Exercise 17   Exercise 18

18. Why do we put a diagonal brace on a gate as shown in the figure above? [Hint: Your answer will involve one of the postulates for congruent triangles.]

19. The figure below shows a bridge. Why is it composed of triangles? Can the bridge change in shape without breaking?

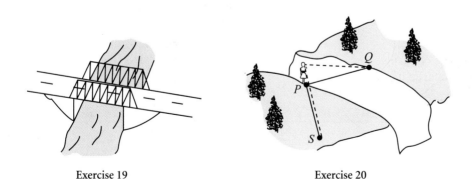

Exercise 19   Exercise 20

20. To measure the distance across a canyon, Diana stood at the edge of the canyon at point P and sighted a point Q on the far rim. Without raising or lowering her eyes, she turned around and sighted a point S on the ground. She then measured the distance between P and S and found it to be 45 ft. She concluded that the width of the canyon is also 45 ft. Why is this true?

In Exercises 21–24, use the given information to find the value of x. Justify all answers.

21.

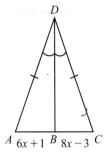

Exercise 21

22.

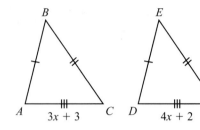

Exercise 22

23.

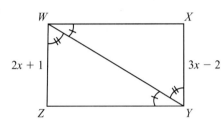

Exercise 23

24.

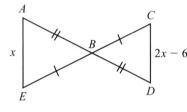

Exercise 24

 25. Write a persuasive argument to prove that $\overline{AB} \cong \overline{CD}$.

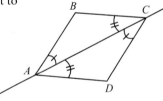

Exercise 25

2.4 Isosceles Triangles, Medians, Altitudes, and Concurrent Lines

OBJECTIVES

1. Prove some properties of isosceles triangles.
2. Identify the converse of a statement.
3. Define concurrent lines.
4. Define median, altitude, and angle bisectors of a triangle.
5. Investigate some properties of concurrent lines in a triangle.

Recall that an isosceles triangle is a triangle having two equal sides with the third side called the base of the triangle. The angle included between the equal sides of an isosceles triangle is the **vertex angle**, and the remaining angles are the **base angles**.

OBJECTIVE 1   Prove some properties of isosceles triangles.   The next theorem shows that the base angles of an isosceles triangle are congruent. Refering to Figure 2.23, $\angle BAC$ is the vertex angle, $\angle B$ and $\angle C$ are the base angles.

> **THEOREM 2.5**
>
> If two sides of a triangle are congruent, then the angles opposite these sides are also congruent.

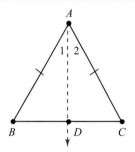

Figure 2.23

Given: $\triangle ABC$
$\overline{AB} \cong \overline{AC}$ (See Figure 2.23.)

Prove: $\angle C \cong \angle B$

Construction: Construct the bisector
$\overrightarrow{AD}$ of $\angle BAC$

Proof _____

Statements	**Reasons**
1. $\triangle ABC$ with $\overline{AB} \cong \overline{AC}$	1. Given
2. $\overrightarrow{AD}$ bisects $\angle BAC$	2. An $\angle$ can be bisected
3. $\angle 1 \cong \angle 2$	3. Def. of $\angle$ bisector
4. $\overline{AD} \cong \overline{AD}$	4. Reflexive law
5. $\triangle ABD \cong \triangle ACD$	5. SAS
6. $\angle C \cong \angle B$	6. CPCTC

This theorem is sometimes stated "The base angles of an isosceles triangle are congruent."

Technology Connection

Geometry software will be needed.

1. Draw two congruent segments, $\overline{AB}$ and $\overline{AC}$, with a common endpoint A, as shown.
2. Connect B and C. What kind of triangle is formed?
3. Measure $\angle B$ and $\angle C$, the angles opposite the congruent sides. They are called base angles. Is there a relationship between these angles?
4. Repeat steps 1–3. Label the new common endpoint X and the congruent segments $\overline{XY}$ and $\overline{XZ}$ where $XY \neq AB$ and $m\angle X \neq m\angle A$.
5. Is the relationship of the base angles in the new triangle the same as in $\triangle ABC$? Are the results consistent with Theorem 2.5?

The next corollary follows directly from Theorem 2.5.

COROLLARY 2.6

If a triangle is equilateral, then it is equiangular.

In Section 1.1, the *hypothesis* was defined as the information that is given in a theorem. The hypothesis is the phrase that follows the word *if* in the statement. The phrase following the word *then* includes the statement to be verified and is the *conclusion*. For example, in the statement "If two sides of a triangle are congruent, then the angles opposite these sides are also congruent." The hypothesis is "two sides of a triangle are congruent." The conclusion follows the word *then*. The conclusion is "the angles opposite these sides are also congruent."

OBJECTIVE 2 **Identify the converse of a statement.** The **converse** of a statement interchanges the hypothesis and conclusion. The converse of any statement must be proven to be true before it can be used as a reason in any proof. The converse to the above statement is "If two angles of a triangle are congruent, then the sides opposite these angles are also congruent." The converse of Theorem 2.5 is Theorem 2.7.

THEOREM 2.7

If two angles of a triangle are congruent, then the sides opposite these angles are also congruent.

Figure 2.24

Given: △ABC
$\angle B \cong \angle C$ (See Figure 2.24.)

Prove: $\overline{AB} \cong \overline{AC}$

Proof _____

Statements	**Reasons**
1. $\angle B \cong \angle C$	1. Given
2. $\angle C \cong \angle B$	2. Symmetric law
3. $\overline{BC} \cong \overline{CB}$	3. Reflexive law
4. $\triangle ABC \cong \triangle ACB$	4. ASA
5. $\overline{AB} \cong \overline{AC}$	5. CPCTC

In the proof of Theorem 2.7 we showed that a triangle is congruent to itself, and as a result, we found congruent sides. This may seem strange at first because in all previous cases we had two distinct triangles in a congruence proof. Actually, we are thinking of $\triangle ABC$ and $\triangle ACB$ as overlapping triangles much like the example in which two triangles shared a common side or common angle. Another way to state this theorem is "angles in a triangle opposite congruent sides are congruent."

The next corollary follows directly from Theorem 2.7 and is the converse of Corollary 2.6.

> **COROLLARY 2.8**
> If a triangle is equiangular, then it is equilateral.

PRACTICE EXERCISE 1

Given: $\angle A \cong \angle D$
$\qquad \angle 1 \cong \angle 2$

Prove: $\triangle BCE$ is isosceles

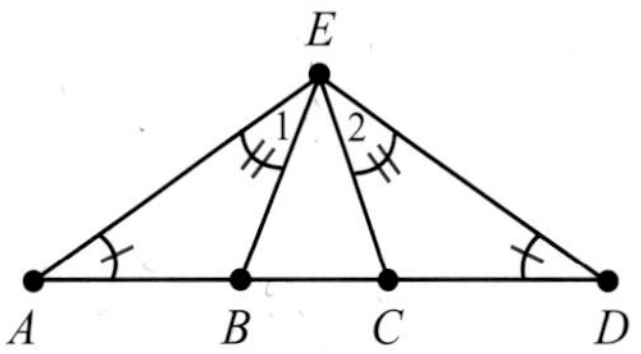

Proof _____

Statements	**Reasons**
1. $\angle A \cong \angle D$	1. _____
2. _____	2. If 2 $\angle$'s of $\triangle$ are $\cong$, sides opp. are $\cong$
3. _____	3. Given
4. $\triangle AEB \cong \triangle DEC$	4. _____
5. $\overline{BE} \cong \overline{CE}$	5. _____
6. _____	6. Def. of isosceles $\triangle$ ANSWERS ON PAGE 105

Some proofs require the use of **auxiliary lines** or **segments**, which we illustrated in the proof of Theorem 2.5 when we constructed the bisector of an angle in a triangle. Often these lines are drawn in a figure using a dashed line because they are not part of the original problem. The following example illustrates.

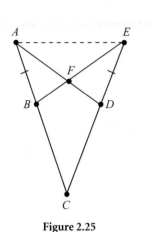

Figure 2.25

EXAMPLE 1

Refer to Figure 2.25.

Given: $\overline{AB} \cong \overline{ED}$
$\overline{AD} \cong \overline{BE}$

Prove: $\overline{AC} \cong \overline{EC}$

Proof _____

Statements	*Reasons*
1. Construct auxiliary segment $\overline{AE}$	1. A line can be drawn between two points
2. $\overline{AB} \cong \overline{ED}$ and $\overline{AD} \cong \overline{BE}$	2. Given
3. $\overline{AE} \cong \overline{AE}$	3. Reflexive law
4. $\triangle AEB \cong \triangle EAD$	4. SSS
5. $\angle BAE \cong \angle DEA$	5. CPCTC
6. $\overline{AC} \cong \overline{EC}$	6. If 2 ∠'s of △ ≅, sides opp. ≅ (In $\triangle ACE$, AC and EC are sides opposite congruent ∠'s)

Four concurrency theorems will be informally investigated by paper folding and geometric software.

OBJECTIVE 3 Define concurrent lines.

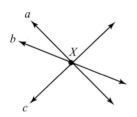

Figure 2.26

> **DEFINITION:** **CONCURRENT LINES**
>
> Two or more lines are **concurrent** if they intersect in one and only one point.

Concurrent lines share a common point. In Figure 2.26, lines *a*, *b*, and *c* are concurrent at point *X*.

In Chapter 1, we learned by combining Constructions 1.3 and 1.4, the perpendicular bisector of a line segment is constructed. A perpendicular bisector is a line that both bisects and is perpendicular to a given segment. In Figure 2.27, if *M* is the midpoint of $\overline{AC}$ and $\overline{MN} \perp \overline{AC}$ then $\overline{MN}$ is the perpendicular bisector of $\triangle ABC$.

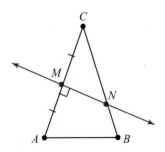

Figure 2.27

> **THEOREM 2.9**
>
> The perpendicular bisectors of the sides of a triangle are concurrent.

Inductive reasoning through paper folding and geometric software will be used to investigate Theorem 2.9. These demonstrations are not proofs of the theorem. The theorem is stated without a formal proof.

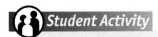

Student Activity

For this activity, the following equipment will be needed: a 6-inch square of unlined, lightweight, white paper like tracing paper and a ruler.

In groups of three, draw a large triangle on the paper. One member of the group draws an acute triangle and the other members draw either a right or obtuse triangle. Be sure to use a straightedge when drawing.

1. Select one of the sides of the triangle. Fold the paper so two endpoints of the side meet. Crease the paper. The crease is a perpendicular bisector of the side of the triangle.
2. Reopen the paper and fold the other two perpendicular bisectors in the same manner.

Are these three lines concurrent for all three types of triangles as Theorem 2.9 states?

Extension

Measure the distances from the point of concurrency to the vertices of the triangle. What do you observe?

A compass and straightedge could also be used to investigate the concurrency of the perpendicular bisectors of a triangle by using Construction 1.3. After constructing perpendicular bisectors, note they are concurrent.

> **DEFINITION: Circumcenter**
>
> The **circumcenter** of a triangle is the point of intersection of the perpendicular bisectors of a triangle.

Technology Connection

Geometry software will be needed.

1. Draw any triangle and label it $\triangle ABC$. It can be an acute, obtuse, or right triangle.
2. Locate the midpoints on sides $\overline{AB}$, $\overline{BC}$, and $\overline{AC}$ and label them X, Y, and Z, respectively.
3. Construct the perpendicular bisector of each side of the triangle. Color the perpendicular bisectors red. What do you notice about these lines?
4. Drag vertex A and observe what happens to the perpendicular bisectors. Do all three red lines meet no matter where point A is located? The point of intersection is called the circumcenter of the triangle.

Extension

5. Label the circumcenter P. Measure the distance from P to the vertices of the triangle. Is $AP = PB = PC$? It can be proven that the circumcenter is equidistant from the vertices of the triangle.

OBJECTIVE 4 **Define median, altitude, and angle bisectors of a triangle.** In addition to its three sides, every triangle has nine segments associated with it: three *medians*, three *altitudes*, and three *angle bisectors*.

> **DEFINITION: *Median of a Triangle***
>
> A **median** of a triangle is the segment joining a vertex to the midpoint of the side opposite that vertex.

In △ABC shown in Figure 2.28, D is the midpoint of $\overline{BC}$, and $\overline{AD}$ is a median of △ABC. △ABC has three medians, one from each of the three vertices.

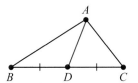

Figure 2.28 Median of a Triangle

OBJECTIVE 5 **Investigate some properties of concurrent lines in a triangle.** Will the three medians of any triangle be concurrent? Investigate this possibility for all triangles using paper folding or the geometric software experiment below.

 **Student Activity**

For this activity, the following equipment will be needed: a 6-inch square of unlined, lightweight, white paper like tracing paper and a ruler with centimeter scale.

Using groups of three, draw a large triangle on the paper. One member of the group draws an acute triangle and the other two members draw either a right triangle or an obtuse triangle. Be sure to use a straightedge when drawing it.

1. Select one of the sides of the triangle and locate its midpoint. To do this, fold the paper so the two endpoints of a particular side meet. Mark the point with a *small* crease. Open the paper and make a fold from the midpoint just located to the opposite vertex (the one you didn't use). Crease the paper so a segment is constructed between the vertex and the midpoint. This is a median of the triangle.
2. Reopen the paper and fold the other two medians in the same manner. Do the three medians meet at one point for all three types of triangles as Theorem 2.10 states? The point of intersection is called the *centroid*.

continued

3. Using a ruler, measure the distance from a vertex to the midpoint of the opposite side (length of the median), in centimeters. Calculate two-thirds of this distance.
4. Measure the distance from the vertex used in Step 3 to the centroid, in centimeters. Is the distance from the vertex to the centroid two-thirds the length of the median? Compare these results with Theorem 2.10.

DEFINITION: Centroid

The **centroid** of a triangle is the point of intersection of the medians of a triangle.

 Technology Connection

Geometry software will be needed.

1. Draw any triangle and label it $\triangle ABC$. It may be an acute, obtuse, or right triangle.
2. Locate the midpoints of each side of the triangle. Label X the midpoint of $\overline{AB}$, Y as the midpoint of $\overline{BC}$, and Z as the midpoint of $\overline{AC}$.
3. Draw $\overline{CX}$, $\overline{BZ}$, and $\overline{AY}$ and color them red. The red segments are the medians of the triangle.

Are all three medians concurrent? The point of intersection is called the centroid of the triangle. Label it P.

4. Drag vertex C to another location. Do the medians still intersect at point P?
5. Measure CX, BZ, and AY (the distances from a vertex to the midpoint).
6. Measure CP, BP, and AP (the distances from a vertex to the centroid).
7. Calculate two-thirds of CX, BZ, and AY. How do these compare to CP, BP, and AP respectively?

Compare these results to Theorem 2.10.

Cut a triangle from a piece of cardboard, draw the medians of the triangle. If you are careful, the medians will intersect at a point. Insert the point of a sharp pencil through this point. Does the triangle balance without tilting? Why do you suppose the intersection of the medians is called the *centroid* or *balance point* of the triangle?

Theorem 2.10 is stated without a formal proof.

THEOREM 2.10

The medians of a triangle are concurrent and meet at a point that is two-thirds the distance from the vertex to the midpoint of the opposite side.

In $\triangle ABC$ shown in Figure 2.29, D is the midpoint of $\overline{AC}$, F is the midpoint of $\overline{BC}$, and E is the midpoint of $\overline{AB}$. The three medians are $\overline{AF}$, $\overline{BD}$, and $\overline{CE}$. From Theorem 2.10, $AP = \frac{2}{3}AF$, $BP = \frac{2}{3}BD$, and $CP = \frac{2}{3}CE$.

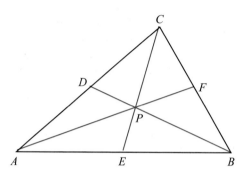

Figure 2.29

EXAMPLE 2

Given $\triangle RST$, medians $\overline{AT}$, $\overline{BR}$, and $\overline{CS}$ intersecting at centroid D in Figure 2.30.

(a) Find RD, if $RB = 9$ in.

(b) Find DT and AD if $AT = 3.6$ cm

(c) Find SD if $CS = \dfrac{12}{5}$m

(d) Find AT if $DT = 10$ ft

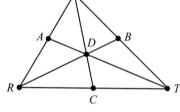

Figure 2.30

Solution:

(a) $RD = \frac{2}{3}(RB) = \frac{2}{3}(9) = 6$ in.

(b) $DT = \frac{2}{3}(AT) = \frac{2}{3}(3.6) = 2.4$ cm

$AD = AT - DT = 3.6 - 2.4 = 1.2$ cm

(c) $SD = \frac{2}{3}(SC) = \frac{2}{3}\left(\dfrac{12}{5}\right) = \dfrac{8}{5}$ m

(d) $\frac{2}{3}(AT) = DT$; let $x = AT$

$\frac{2}{3}x = 10; x = 10\left(\dfrac{3}{2}\right) = 15$ ft

> **DEFINITION:** *Altitude of a Triangle*
>
> An **altitude** of a triangle is a line segment from a vertex *perpendicular* to the side opposite that vertex (possibly extended).

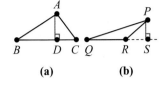

Figure 2.31 Altitudes of a Triangle

In $\triangle ABC$ shown in Figure 2.31(a), $\overline{AD}$ is perpendicular to $\overline{BC}$, and $\overline{AD}$ is the altitude of $\triangle ABC$ from vertex A. In $\triangle PQR$ shown in Figure 2.31(b), $\overline{PS}$ is perpendicular to the *extension* of $\overline{QR}$, that is $\overline{QS}$, and $\overline{PS}$ is the altitude of $\triangle PQR$ from vertex P. An altitude may be outside the triangle. Notice that any triangle has three altitudes, one from each of the three vertices.

The length of the altitude is called the height of the triangle.

EXAMPLE 3   Sketch the three altitudes of $\triangle ABC$ in Figure 2.32.

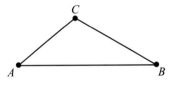

Figure 2.32

Solution:

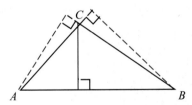

> **DEFINITION:** *Orthocenter*
>
> The **orthocenter** of a triangle is the point of intersection of the three altitudes of a triangle.

Are the three altitudes of any triangle concurrent? Investigate this possibility for all types of triangles using the directions in the Student Activity on page 102, with paper folding or the Technology Connection with geometric software.

Leonard Euler (1707–1783)

Swiss mathematician Leonard Euler (pronounced "Oy'-ler") proved that the orthocenter, the circumcenter, and the centroid of a triangle all lie on the same line. It is called the *Euler line*.

Student Activity

For this activity, the following equipment will be needed: a 6-inch square of unlined, lightweight, white paper like tracing paper and a ruler.

Using groups of three, draw a large triangle on the paper. One member of the group draws an acute triangle and the other two members draw either a right triangle or an obtuse triangle. Be sure to use a straightedge.

1. Select one of the vertices of the triangle and the side opposite it. Darken the entire side and the vertex with a pencil. Bring the endpoints of the darkened side together but don't crease the paper yet. Slide the paper keeping the darkened segment on top of itself until the fold goes through the darkened vertex. Crease the paper. This is one altitude of the triangle.

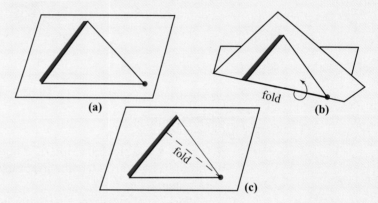

(a) fold (b)

fold (c)

2. Reopen the paper and fold the other two altitudes in the same manner.

Are all three altitudes concurrent? Compare these results with Theorem 2.11.

Technology Connection

Geometry software will be needed.

1. Draw any triangle and label it $\triangle ABC$. It can be an acute, right, or obtuse triangle.
2. Construct a line perpendicular to $\overline{AB}$ that goes through C and color it red. [Hint: Side $\overline{AB}$ may need to be extended.]
3. Repeat step 2 for side $\overline{BC}$ and point A as well as side $\overline{AC}$ and point B.

Are the three altitudes concurrent?

4. What happens to the altitudes if you drag vertex A?

Compare these results to Theorem 2.11.

Theorem 2.11 is stated without a formal proof.

THEOREM 2.11
The altitudes of a triangle are concurrent.

DEFINITION: *Angle Bisector of a Triangle*
An **angle bisector** of a triangle is a line segment (or a ray) that separates the given angle into two congruent adjacent angles.

In $\triangle ABC$ shown in Figure 2.33, $\overline{AD}$ is the bisector of $\angle CAB$ because $\angle CAD \cong \angle DAB$. Notice that $\triangle ABC$ has three angle bisectors, one for each angle in the triangle.

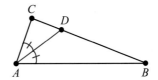

Figure 2.33 Angle Bisector of a Triangle

Student Activity

For this activity, the following equipment will be needed: a 6-inch square of unlined, lightweight, white paper like tracing paper and a ruler with centimeter scale.

Working in groups of three, draw a large triangle on the paper. One member of the group draws an acute triangle and the other two members draw either a right triangle or an obtuse triangle. Be sure to use a straightedge.

1. Select one of the angles of the triangle and fold an angle bisector by folding one side of the angle on top of the other side. Crease the paper.
2. Reopen the paper and fold the other two angle bisectors in the same manner.

Are all three lines concurrent as Theorem 2.12 states?

3. Measure the distances from the incenter (angle bisectors intersection point) to the sides of the triangle. Remember the distance from a point to a line segment is the length of the perpendicular segment from the point to the segment (not the distance along the fold). What do you notice about the three distances? From Theorem 2.12, the incenter is equidistant from the sides of the triangle.

DEFINITION: *Incenter*
The **incenter** of a triangle is the point of intersection of the three angle bisectors of the triangle.

Technology Connection

Geometry software will be needed.

1. Draw any triangle and label it $\triangle ABC$. It can be an acute, right, or obtuse triangle.
2. Draw the angle bisector of each angle and color them red.

What do you observe about the angle bisectors?

3. Drag vertex A to change the angle measures in the triangle.

Are all three angle bisectors still concurrent? Does this agree with Theorem 2.12?

4. Label the incenter point P.
5. Measure the distances from P (incenter) to the sides of the triangle. Remember the distance from a point to a line segment is the length of the perpendicular segment from the point to the segment.

What do you observe about these measurements? Does this observation agree with Theorem 2.12?

Theorem 2.12 is stated without a formal proof.

THEOREM 2.12

The bisectors of the angles of a triangle are concurrent and meet at a point equidistant from the sides of the triangle.

SUMMARY

Put in journal

Triangle Concurrency	Definition	Intersection Point
Perpendicular bisector	Line through the midpoint and perpendicular to one side of the triangle.	Circumcenter
Median	Segment joining a vertex with the midpoint of the opposite side of triangle.	Centroid
Altitude	Segment from a vertex, perpendicular to side opposite the vertex (or extension of the side).	Orthocenter
Angle bisector	Segment or ray that separates the given angle into two congruent adjacent angles.	Incenter

PRACTICE EXERCISE 2

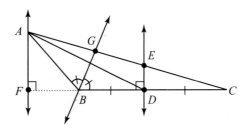

In $\triangle ABC$,

1. Which line or line segment will contain the circumcenter?

2. Which line or line segment will contain the centroid?

3. Which line or line segment will contain the orthocenter?

4. Which line or line segment will contain the incenter?

ANSWERS BELOW

Answers to Practice Exercises

1. 1. Given 2. $\overline{EA} \cong \overline{ED}$ 3. $\angle 1 \cong \angle 2$ 4. ASA 5. CPCTC 6. $\triangle BCE$ is isosceles

2. 1. $\overleftrightarrow{ED}$ 2. $\overline{AD}$ 3. $\overleftrightarrow{AF}$ 4. $\overleftrightarrow{BG}$

2.4   Exercises

FOR EXTRA HELP: 📖 Student's Solutions Manual   Tutor Center   Addison-Wesley Math Tutor Center

From the given information on each figure, decide which angles or sides are congruent using Theorems 2.5 or 2.7.

1.

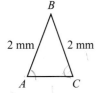

2.

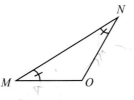

3.

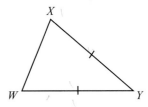

4.

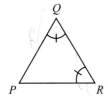

5.

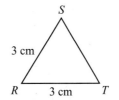

6.

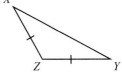

Complete each proof in Exercises 7–10.

7. *Given:* $\overline{AB} \cong \overline{AD}$
$\overline{AC}$ bisects $\angle BAD$

Prove: $\triangle BDE$ is isosceles

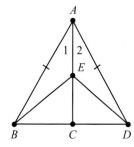

Exercise 7

Proof _____

Statements	**Reasons**
1. $\overline{AB} \cong \overline{AD}$	1. _____
2. _____	2. Given
3. $\angle 1 \cong \angle 2$	3. _____
4. $\overline{AE} \cong \overline{AE}$	4. _____
5. _____	5. SAS
6. $\overline{BE} \cong \overline{DE}$	6. _____
7. _____	7. Def. of isos. $\triangle$

8. *Given:* $\angle 1 \cong \angle 2$
$\overline{AB} \cong \overline{AE}$

Prove: $\angle 3 \cong \angle 4$

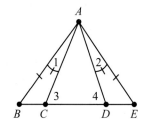

Exercise 8

Proof _____

Statements	**Reasons**
1. $\angle 1 \cong \angle 2$	1. _____
2. _____	2. Given
3. $\angle B \cong \angle E$	3. _____
4. $\triangle ABC \cong \triangle AED$	4. _____
5. $\overline{AC} \cong \overline{AD}$	5. _____
6. _____	6. $\angle$'s opp. $\cong$ sides are $\cong$

9. *Given:* Isosceles $\triangle ACD$ with base $\overline{CD}$
   B is the midpoint of $\overline{AC}$
   E is the midpoint of $\overline{AD}$
Prove: $\angle 1 \cong \angle 2$

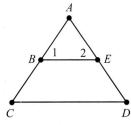

Exercise 9

Proof

Statements	*Reasons*
1. $\triangle ACD$ is isosceles with base $\overline{CD}$	1. _____
2. _____	2. Def. of isos. $\triangle$
3. B is the midpoint of $\overline{AC}$	3. _____
4. $\overline{AB} \cong \overline{BC}$	4. _____
5. _____	5. Given
6. $\overline{AE} \cong \overline{ED}$ so $AE = ED$	6. _____
7. $AC = AB + BC$ and $AD = AE + ED$	7. _____
8. $AC = AB + AB$ and $AD = AE + AE$	8. _____
9. $AC = 2AB$ and $AD = 2AE$	9. Distributive law
10. $AB = \dfrac{AC}{2}$ and $AE = \dfrac{AD}{2}$	10. _____
11. $AB = \dfrac{AC}{2}$ and $AE = \dfrac{AC}{2}$	11. Substitution law
12. $AB = AE$	12. Sym. and trans. laws
13. _____	13. $\angle$'s opp. $\cong$ sides are $\cong$

10. *Given:* $\overline{AC} \cong \overline{AD}$
   $\overline{BD} \cong \overline{CE}$
Prove: $\overline{AB} \cong \overline{AE}$

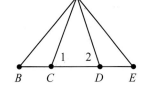

Exercise 10

Proof

Statements	*Reasons*
1. $\overline{AC} \cong \overline{AD}$	1. _____
2. _____	2. $\angle$'s opp. $\cong$ sides are $\cong$
3. _____	3. Given
4. _____	4. SAS
5. $\overline{AB} \cong \overline{AE}$	5. _____

Write a two-column proof in Exercises 11 and 12.

11. *Given:* $\angle 1 \cong \angle 2$
$\qquad \angle 3 \cong \angle 4$

Prove: $\angle A \cong \angle D$

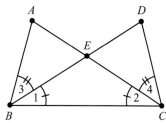

Exercise 11

12. *Given:* $\overline{AB} \cong \overline{AE}$
$\qquad \overline{BC} \cong \overline{DE}$

Prove: $\angle 1 \cong \angle 2$

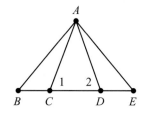

Exercise 12

13. Every triangle has three altitudes, angle bisectors, and medians. Describe the differences among these line segments.

Exercises 14–19 refer to the figure below in which $\angle 1 \cong \angle 2$, $\overline{BD} \cong \overline{DC}$, and $\overline{CF} \perp \overline{BF}$. Answer *true* or *false*.

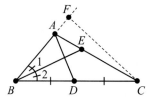

14. $\overline{AD}$ is a median of $\triangle ABC$.

15. $\overline{CF}$ is an altitude of $\triangle ABC$.

16. $\overline{BE}$ is an altitude of $\triangle ABC$.

17. $\overline{BE}$ is an angle bisector of $\triangle ABC$.

18. $\overline{AD}$ is an angle bisector of $\triangle ABC$.

19. $\overline{CF}$ is a median of $\triangle ABC$.

In the figure below, *E* is the incenter, *G* is the circumcenter, and *F* is the centroid of $\triangle ABC$. Use this figure for Exercises 20–27.

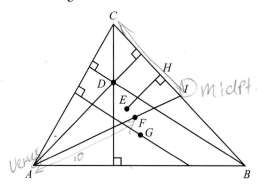

20. If $FI = 3$ units, what is the measure of AF?

21. What is the orthocenter of $\triangle ABC$?

22. If $\triangle ABC$ was equilateral, what can be said about the incenter, circumcenter, orthocenter, and centroid?

23. If $AI = 10.5$ in., what is the measure of FI?

24. If $AF = 10$ mm, what is the measure of AI?

25. If $FI = 6.8$ m, what is the measure of AI?

26. If $CI = 2x + 3$ and $IB = x + 5$, find x.

27. Which point is equidistant from the sides of $\triangle ABC$?

28. A police officer is patrolling three residential areas that form a triangle. At break time, she wants to park in a location that is equidistant from the three areas. Which point of concurrency does she need to locate?

29. Do you think the centroid of a triangle can ever be outside a triangle? Explain.

30. Do you think the orthocenter of a triangle can ever be outside the triangle? Explain.

(2.5) Proving Right Triangles Congruent

OBJECTIVES

1. Define the LA theorem.
2. Define the LL theorem.

Recall that a right triangle is a triangle that contains a right angle. The hypotenuse of a right triangle is the side opposite the right angle, and the remaining two sides are called legs. In this section, we prove congruence theorems that apply to right triangles. We'll present three theorems using the terminology of right triangles with A representing *acute angle*, L representing *leg*, and H representing *hypotenuse*.

OBJECTIVE 1   **Define the LA theorem.**   The first theorem is a special case of Postulate 2.2, ASA.

> **THEOREM 2.13   LA (LEG-ANGLE)**
>
> If a leg and acute angle of one right triangle are congruent, to a leg and the corresponding acute angle of another right triangle, then the two right triangles are congruent.

Given: Right triangles $\triangle ABC$ and $\triangle DEF$ with $\angle A$ and $\angle D$ are right angles, $\overline{AB} \cong \overline{DE}$ and $\angle B \cong \angle E$. (See Figure 2.34.)

Prove: $\triangle ABC \cong \triangle DEF$

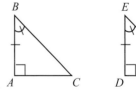

Figure 2.34

Proof _____

Statements	Reasons
1. $\triangle ABC$ and $\triangle DEF$ are right triangles with $\angle A$ and $\angle D$ right angles.	**1.** Given
2. $\angle A \cong \angle D$	**2.** All right $\angle$'s are $=$ in measure (Thm 1.12); def. $\cong \angle$'s.
3. $\overline{AB} \cong \overline{DE}$ and $\angle B \cong \angle E$	**3.** Given
4. $\triangle ABC \cong \triangle DEF$	**4.** ASA

OBJECTIVE 2   **Define the LL theorem.**   The next theorem is a special case of Postulate 2.1, SAS.

> **THEOREM 2.14   LL (LEG-LEG)**
> If the two legs of one right triangle are congruent to the two legs of another right triangle, then the two right triangles are congruent.

Given: Right triangles $\triangle ABC$ and $\triangle DEF$ with $\angle A$ and $\angle D$ are right angles, $\overline{AB} \cong \overline{DE}$ and $\overline{AC} \cong \overline{DF}$. (See Figure 2.35.)

Prove: $\triangle ABC \cong \triangle DEF$

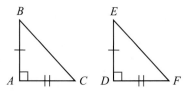

Figure 2.35

Proof

Statements	Reasons
1. $\triangle ABC$ and $\triangle DEF$ are right triangles with $\angle A$ and $\angle D$ right angles.	1. Given
2. $\angle A \cong \angle D$	2. All right $\angle$'s are = in measure (Thm 1.12); def. $\cong \angle$'s.
3. $\overline{AB} \cong \overline{DE}$ and $\overline{AC} \cong \overline{DF}$	3. Given
4. $\triangle ABC \cong \triangle DEF$	4. SAS

EXAMPLE 1   Refer to Figure 2.36.

Given: Right triangles $\triangle ABC$ and $\triangle ABD$ where $\angle ABC$ and $\angle ABD$ are right angles, and $\angle 1 \cong \angle 2$.

Prove: $\triangle ABC \cong \triangle ABD$

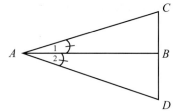

Figure 2.36

Proof

Statements	Reasons
1. $\triangle ABC$ and $\triangle ABD$ are right $\triangle$'s with $\angle ABC$ and $\angle ABD$ right angles.	1. Given
2. $\overline{AB} \cong \overline{AB}$	2. Reflexive law
3. $\angle 1 \cong \angle 2$	3. Given
4. $\triangle ABC \cong \triangle ABD$	4. *LA*

PRACTICE EXERCISE 1

Given: $\overline{AB} \perp \overline{BE}, \overline{AC} \perp \overline{CD}, \overline{BE} \cong \overline{CD}$,
  B is midpoint of $\overline{AC}$

Prove: $\triangle ABE \cong \triangle BCD$

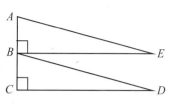

Proof

Statements	Reasons
1. $\overline{AB} \perp \overline{BE}, \overline{AC} \perp \overline{CD}$	1. Given
2. $\angle ABE$ and $\angle BCD$ are right angles	2. Def. $\perp$ lines
3. $\triangle ABE \cong \triangle BCD$ are right triangles	3. _____ SAS _____
4. $\overline{BE} \cong \overline{CD}$, B is midpoint of $\overline{AC}$	4. _____
5. $\overline{AB} \cong \overline{BC}$	5. _____
6. $\triangle ABE \cong \triangle BCD$	6. _____

ANSWERS BELOW

Answers to Practice Exercises

1. 3. Def. right triangle 4. Given 5. Def. midpoint 6. LL

2.5 Exercises

FOR EXTRA HELP: 📖 Student's Solutions Manual   📞 Tutor Center Addison-Wesley Math Tutor Center

Exercises 1–3 refer to $\triangle ABC$ and $\triangle DEF$ given below. State why $\triangle ABC \cong \triangle DEF$ under the given conditions.

Exercises 1–3

1. $BC = 5$ cm, $EF = 5$ cm, $m\angle A = 35°, m\angle D = 35°$
2. $BC = 7$ yd, $EF = 7$ yd, $AC = 10$ yd, $DF = 10$ yd
3. $AC = 9.2$ ft, $m\angle A = 32°, m\angle D = 32°, DF = 9.2$ ft

In the figure shown, $\angle ABC, \angle BCD, \angle CDA, \angle DAB$, and $\angle ATB$ are right angles.

4. Name each right triangle that has $\overline{AB}$ as one of its legs.
5. Name each right triangle that has $\overline{AB}$ as its hypotenuse.
6. Name each right triangle that has $\overline{AT}$ as a leg.

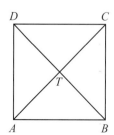

Exercises 4–6

In the figure shown, $\overline{CB} \perp \overline{CA}$, $\overline{DB} \perp \overline{AD}$, $\overline{AC} \cong \overline{AD}$, and $\overline{CB} \cong \overline{DB}$. Two different students wrote different proofs to prove $\triangle ACB \cong \triangle ADB$. Write the reasons for each statement.

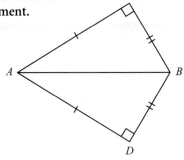

STUDENT 1

Statements	**Reasons**
7. $\overline{AC} \cong \overline{AD}$	7. _____
8. $\overline{CB} \cong \overline{DB}$	8. _____
9. $\overline{CB} \perp \overline{CA}, \overline{DB} \perp \overline{AD}$	9. _____
10. $\angle C$ and $\angle D$ are rt. angles	10. _____
11. $\triangle ACB$ and $\triangle ADB$ are rt. $\triangle$'s	11. _____
12. $\triangle ACB \cong \triangle ADB$	12. _____

STUDENT 2

Statements	**Reasons**
13. $\overline{AC} \cong \overline{AD}$	13. _____
14. $\overline{CB} \cong \overline{DB}$	14. _____
15. $\overline{AB} \cong \overline{AB}$	15. _____
16. $\triangle ACB \cong \triangle ADB$	16. _____

17. *Given:* $\overline{WY} \perp \overline{XZ}, \overline{XY} \cong \overline{YZ}$
    Prove: $\triangle XYW \cong \triangle ZYW$

Exercises 18 and 19 refer to the figure to the right.

18. *Given:* C is the midpoint of $\overline{AE}$ and
    of $\overline{BD}$, $\overline{AB} \perp \overline{BD}$, and $\overline{DE} \perp \overline{BD}$

    Prove: $\triangle ABC \cong \triangle EDC$

19. *Given:* $\overline{AB} \cong \overline{DE}, \overline{AB} \perp \overline{BD}$, and
    $\overline{DE} \perp \overline{BD}$

    Prove: $\overline{AC} \cong \overline{EC}$

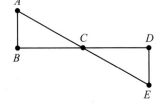

20. *Given:* $\overline{BD} \perp \overline{AC}, \overline{CE} \perp \overline{AB}$, and $\overline{BD} \cong \overline{CE}$
    Prove: $\overline{AE} \cong \overline{AD}$

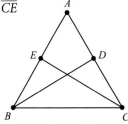

2.6 Constructions Involving Triangles

OBJECTIVES

1. Construct triangles with various given parts.
2. Construct the altitude of a triangle.
3. Construct the median of a triangle.

OBJECTIVE 1   Construct triangles with various given parts.   The first construction we'll consider involves duplicating a given triangle.

> **CONSTRUCTION 2.1**
>
> Construct a triangle that is congruent to a given triangle.

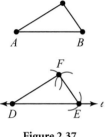

Figure 2.37

Given: △ABC (See Figure 2.37.)

To Construct: △DEF such that △DEF ≅ △ABC

Construction _____

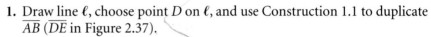

1. Draw line ℓ, choose point D on ℓ, and use Construction 1.1 to duplicate $\overline{AB}$ ($\overline{DE}$ in Figure 2.37).
2. Set the compass at the length $\overline{AC}$, place the point at D and draw an arc. Then set the compass at the length $\overline{BC}$, place the point at E, and draw an arc that intersects the first arc in point F.
3. Use a straightedge to draw $\overline{DF}$ and $\overline{EF}$. The resulting △DEF is congruent to △ABC by SSS because $\overline{AB} \cong \overline{DE}$, $\overline{AC} \cong \overline{DF}$, and $\overline{BC} \cong \overline{EF}$.

One way to view Construction 2.1 is to construct a triangle given the three sides. Next we construct a triangle when two sides and the included angle are given.

> **CONSTRUCTION 2.2**
>
> Construct a triangle with two sides and the included angle given.

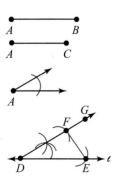

Figure 2.38

Given: Sides $\overline{AB}$ and $\overline{AC}$ and ∠A (See Figure 2.38.)

To Construct: The triangle with these parts

Construction _____

1. Draw line ℓ, choose point D on ℓ, and use Construction 1.1 to duplicate $\overline{AB}$ ($\overline{DE}$ in Figure 2.38).
2. Use Construction 1.2 to duplicate ∠A with vertex at D (∠GDE in Figure 2.38).
3. On $\overrightarrow{DG}$ use Construction 1.1 to duplicate $\overline{AC}$ ($\overline{DF}$ in Figure 2.38).
4. Use a straightedge to draw $\overline{EF}$ forming △DEF with the required properties. By SAS, △DEF is the triangle that satisfies the given conditions.

Next we show how to construct a triangle when two angles and the included side are given.

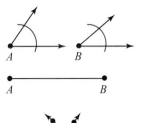

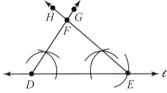

Figure 2.39

> **CONSTRUCTION 2.3**
> Construct a triangle with two angles and the included side given.

Given: $\angle A$, $\angle B$, and side $\overline{AB}$ (See Figure 2.39.)

To Construct: The triangle with these parts

Construction _____

1. Draw line ℓ, choose point D on ℓ, and use Construction 1.1 to duplicate $\overline{AB}$ ($\overline{DE}$ in Figure 2.39).
2. Use Construction 1.2 to duplicate $\angle A$ with vertex at D ($\angle GDE$) and $\angle B$ with vertex at E ($\angle HED$).
3. Locate point F at the intersection of rays $\overrightarrow{DG}$ and $\overrightarrow{EH}$ forming sides $\overline{DF}$ and $\overline{EF}$ of the desired triangle $\triangle DEF$. By ASA, $\triangle DEF$ is the triangle that satisfies the given conditions.

EXAMPLE 1 Construct an isosceles triangle with base $\overline{AB}$ and sides CD, given in Figure 2.40.

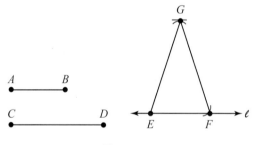

Figure 2.40

The construction is similar to Construction 2.1. Draw line ℓ and duplicate $\overline{AB}$ on ℓ, call it $\overline{EF}$. Set the compass at length CD, place the point at E and draw an arc, then at F and draw another arc intersecting the first one. This determines point G. Use a straightedge and draw $\overline{EG}$ and $\overline{FG}$ to form the desired isosceles triangle, $\triangle EFG$.

Plato (430–349 B.C.)

The Greek philosopher Plato also studied geometry. Much of the mathematical work done in the fourth century B.C. was done by his students and friends. Plato is credited with the concept that geometric constructions should be formed with only a compass and straightedge. Over the door to his school, the Academy, was the statement "Let no one ignorant of geometry enter here."

> **PRACTICE EXERCISE 1**
>
> Construct an isosceles triangle with vertex angle $\angle A$ and sides $\overline{BC}$, given in the figure below.
>
>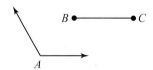
>
> ANSWERS ON PAGE 116

Equilateral triangles are easy to construct.

> **CONSTRUCTION 2.4**
>
> Construct an equilateral triangle when given a single side.

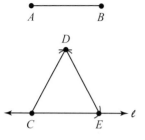

Figure 2.41

Given: Side $\overline{AB}$ (See Figure 2.41.)

To Construct: An equilateral triangle with sides of length AB

Construction _____

1. Draw line ℓ and duplicate $\overline{AB}$ on $\ell\,(\overline{CE})$.
2. Using the compass set at length AB, make two arcs by placing the point at C and at E. The point of intersection of the arcs is D.
3. Use a straightedge and draw $\overline{CD}$ and $\overline{ED}$. Then $\triangle CDE$ is the desired equilateral triangle.

OBJECTIVE 2   **Construct the altitude of a triangle.**   Constructing an altitude of a triangle is a direct application of Construction 1.5, constructing a line perpendicular to a given line from a point not on that line.

> **CONSTRUCTION 2.5**
>
> Construct an altitude of a given triangle.

Given: $\triangle ABC$ (See Figure 2.42.)

To Construct: The altitude of $\triangle ABC$ from vertex A

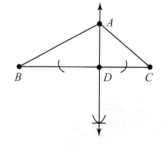

Figure 2.42

Construction _____

1. Use Construction 1.5 to construct the line $\overleftrightarrow{AD}$ such that $\overrightarrow{AD} \perp \overline{BC}$. To do this, we sometimes need to extend $\overline{BC}$ and consider $\overleftrightarrow{BC}$.
2. Then $\overline{AD}$ is the desired altitude from vertex A.

⌐**Note** In Construction 2.5, we said that $\overline{AD}$ is the desired altitude from vertex A implying that it is also the *only* such altitude from A. This is true because there is exactly one line perpendicular to a given line passing through a given point not on the line by Postulate 1.18.

OBJECTIVE 3   **Construct the median of a triangle.**   Constructing a median of a triangle uses Construction 1.3 to find the midpoint of a side.

> ### CONSTRUCTION 2.6
> Construct a median of a given triangle.

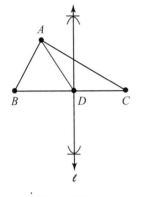

Figure 2.43

Given:	$\triangle ABC$ (See Figure 2.43.)
To Construct:	The median of $\triangle ABC$ from vertex A

Construction

1. Use Construction 1.3 to construct line ℓ, the (perpendicular) bisector of $\overline{BC}$, to determine the midpoint D of $\overline{BC}$.
2. Use a straightedge to draw $\overline{AD}$, the desired median from vertex A. We say *the* median because by the midpoint postulate, $\overline{BC}$ has exactly one midpoint D, and by Postulate 1.1, the two points A and D determine exactly one line.

Note Some students might wonder why we do not specifically give the construction of an angle bisector of a triangle similar to that for medians and altitudes. This is because we already know how to bisect an angle by Construction 1.6, and this procedure works even when the given angle is part of a triangle.

Answer to Practice Exercise

1. Construct the desired triangle using Construction 2.2 with sides $\overline{BC}$ and included angle $\angle A$.

2.6 Exercises

FOR EXTRA HELP: Student's Solutions Manual   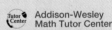 Addison-Wesley Math Tutor Center

1. Draw an acute scalene triangle and construct a triangle that is congruent to it.
2. Draw an obtuse scalene triangle and construct a triangle that is congruent to it.
3. Draw two segments, one about 2 inches long and the other about 3 inches long. Draw an obtuse angle, and construct the triangle with these sides and included angle.
4. Draw two segments, one about 2 inches long and the other about 3 inches long. Draw an acute angle, and construct the triangle with these sides and included angle.
5. Draw a segment about 2 inches long. Draw two angles, both acute, and construct the triangle with these angles and included side.
6. Draw a segment about 3 inches long. Draw an angle that is about 90° and another acute angle. Construct the triangle with these angles and included side.

7. Give an example to show that two obtuse angles and a segment cannot be used in Construction 2.3.

8. Give examples to show that an obtuse angle, an acute angle, and a segment may or may not yield a triangle using Construction 2.3.

9. Draw a segment about 2 inches long and an obtuse angle. Construct the isosceles triangle with these as sides and the vertex angle.

10. Draw a segment about 2 inches long and a second segment 5 inches long. Construct an isosceles triangle with the base as the 2-inch segment and each side as the 5-inch segment.

11. What happens in Exercise 10 if the base is a 5-inch segment and each side is a 2-inch segment?

12. Construct a right triangle with legs about 2 inches and 3 inches in length, respectively.

13. Construct a right isosceles triangle with legs that are about 2 inches in length.

14. Construct a right angle and bisect it to obtain an acute angle, $\angle A$.

15. Use $\angle A$ in Exercise 14, a segment about 3 inches in length, and a segment about 2.5 inches in length. Try to construct a triangle using $\angle A$, the 3-inch segment as a side of $\angle A$, and the 2.5-inch segment as the side opposite $\angle A$. What happens?

16. Repeat Exercise 15 using a 1-inch segment instead of the 2.5-inch segment. What happens?

17. Draw a scalene acute triangle and construct its three altitudes. Label the orthocenter.

18. Draw a scalene obtuse triangle and construct its three altitudes. Label the orthocenter.

19. Draw a scalene obtuse triangle and construct its three medians. Label the centroid.

20. Draw a scalene acute triangle and construct its three medians. Label the centroid.

21. Draw a scalene acute triangle and construct its three angle bisectors. Label the incenter.

22. Draw a scalene obtuse triangle and construct its three angle bisectors. Label the incenter.

23. Will the orthocenter of a triangle always be inside the triangle? Explain.

24. Will the centroid of a triangle always be inside the triangle? Explain.

25. Will the incenter of a triangle of a triangle always be inside the triangle? Explain.

26. What is the point of concurrency called?

27. Construct the perpendicular bisectors of the sides of an obtuse triangle. What common relationships do these lines seem to have?

28. Starting with three given line segments, can you always construct a triangle (using the method in Construction 2.1) with these segments as its sides? Explain.

29. Construct an equilateral triangle with sides about 3 inches in length.

30. Use the triangle constructed in Exercise 29 and construct its three medians, three altitudes, and three angle bisectors. What happens?

Chapter (2) Review

Key Terms and Symbols

2.1   sides
closed
included
opposite
triangle
vertex
acute triangle
right triangle
hypotenuse
legs
obtuse triangle
equiangular triangle
scalene triangle

isosceles triangle
base
equilateral triangle
perimeter
exterior angle
interior angle
remote interior angles

2.2   congruent
congruent segments
congruent triangles
congruent angles
corresponding parts

2.4   vertex angles (of an isosceles triangle)
base angles (of an isosceles triangle)
auxiliary lines or segments
concurrent lines
median
centroid
circumcenter
altitude
orthocenter
angle bisector (of a triangle)
incenter

Chapter 2 Proof Techniques

To Prove:

Two Triangles Congruent

1. Show two triangles are congruent to a third triangle. (Theorem 2.1)

2. (SAS) Show two sides and the included angle of one triangle are congruent respectively to two sides and the included angle of the other triangle. (Postulate 2.1)

3. (ASA) Show two angles and the included side of one triangle are congruent respectively to two angles and the included side of the other triangle. (Postulate 2.2)

4. (SSS) Show three sides of one triangle are congruent respectively to three sides of the other triangle. (Postulate 2.3)

5. (LA) In two right triangles show a leg and acute angle are congruent to a leg and the corresponding acute angle of the other triangle. (Theorem 2.13)

6. (LL) In two right triangles show the two legs are congruent respectively to the two legs of the other triangle. (Theorem 2.14)

Two Segments Congruent

1. Show they are corresponding parts of congruent triangles. (Def. of CPCTC)

2. Show they are drawn from any point on the perpendicular bisector of a segment to the endpoints of that segment. (Theorem 2.3)

3. Show they are sides opposite congruent angles in a triangle. (Theorem 2.7)

Two Angles Congruent

1. Show they are corresponding parts of congruent triangles. (Def. of CPCTC)

2. Show they are angles formed by the bisector of an angle. (Theorem 2.4)

3. Show they are angles opposite the equal sides of a triangle. (Theorem 2.5)

SUMMARY

TRIANGLE CONCURRENCY	DEFINITION	INTERSECTION POINT
Perpendicular bisector	Line through the midpoint and perpendicular to one side of the triangle.	Circumcenter
Median	Segment joining a vertex with the midpoint of the opposite side of triangle.	Centroid
Altitude	Segment from a vertex, perpendicular to side opposite the vertex (or extension of the side).	Orthocenter
Angle bisector	Ray that separates the given angle into two congruent adjacent angles.	Incenter

Review Exercises

Section 2.1

Exercises 1–8 refer to △ABC given below in which all angles are unequal and all sides are unequal.

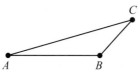

1. What are the vertices of △ABC?

2. What are the sides of △ABC?

3. Classify △ABC using its sides.

4. Classify △ABC using its angles.

5. What side is included by ∠A and ∠C?

6. What angle is included by $\overline{AB}$ and $\overline{BC}$?

7. What side is opposite ∠A?

8. What angle is opposite $\overline{AB}$?

9. Find the perimeter of an isosceles triangle with base 12 cm and equal sides 16 cm.

10. The base of an isosceles triangle is half the length of equal side. If the perimeter is 65 ft, find the length of the base and each side.

Exercises 11–14 refer to the figure below. Answer true *or* false *in each.*

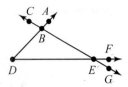

11. ∠*D* is an interior angle of △*BDE.*   **12.** ∠*BEF* is an exterior angle of △*BDE.*

13. ∠*FEG* is an exterior angle of △*BDE.*

14. ∠*D* and ∠*BED* are remote interior angles relative to ∠*ABE.*

15. Can an isosceles triangle be an obtuse triangle?

Section 2.2

In Exercises 16 and 17, state why the given pair of triangles are congruent.

16. 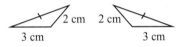   **17.**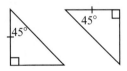

Complete each proof in Exercises 18 and 19.

18. *Given:* ∠*C* ≅ ∠*E*
$\overline{AC} ≅ \overline{AE}$

Prove: △*ACF* ≅ △*AEB*

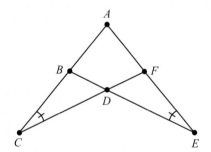

Exercise 18

Proof _____

Statements	*Reasons*
1. ∠*C* ≅ ∠*E*	**1.** _____
2. _____	**2.** Given
3. _____	**3.** Reflexive law
4. _____	**4.** ASA

19. *Given:* $\overline{AB} \cong \overline{AE}$
$\overline{AC}$ bisects $\angle BAD$

Prove: $\triangle ABC \cong \triangle AEC$

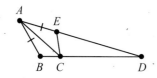

Exercise 19

Proof _____

Statements	*Reasons*
1. $\overline{AB} \cong \overline{AE}$	1. _____
2. _____	2. Given
3. _____	3. Def. of $\angle$ bisector
4. $\overline{AC} \cong \overline{AC}$	4. _____
5. _____	5. SAS

Write a two-column proof or a flowchart proof for Exercises 20 and 21.

20. *Given:* $\overline{AD} \cong \overline{CB}$
$\overline{AB} \cong \overline{CD}$

Prove: $\triangle ADB \cong \triangle CBD$

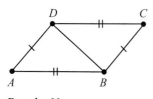

Exercise 20

21. *Given:* $\overline{AC} \perp \overline{BD}$
$\angle 1 \cong \angle 2$

Prove: $\overline{AB} \cong \overline{AD}$

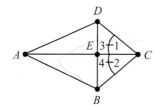

Exercise 21

Use the figure below in Exercises 22–27. Assume that $\triangle ABC \cong \triangle EDF$,
$AC = x + 2$, $EF = 4x - 4$, $m\angle E = (y + 10^\circ)$, and $m\angle A = (2y - 15)^\circ$.

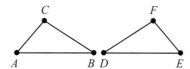

22. Find the value of x.

23. Find the value of y.

24. Find AC.

25. Find the measure of $\angle A$.

26. $\overline{DF}$ corresponds to which side in $\triangle ABC$?

27. $\angle B$ corresponds to which angle in $\triangle EDF$?

Section 2.3

28. Complete the following proof.

Given: $\angle DAB \cong \angle DBA$
$\overline{AC}$ bisects $\angle DAB$
$\overline{BE}$ bisects $\angle DBA$

Prove: $\overline{AC} \cong \overline{BE}$

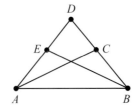

Exercise 28

Proof

Statements	*Reasons*
1. $\angle DAB \cong \angle DBA$ so $m\angle DAB = m\angle DBA$	**1.** _____
2. _____	**2.** Given
3. $\angle CAB \cong \angle DAC$ so $m\angle CAB = m\angle DAC$	**3.** _____
4. $\overline{BE}$ bisects $\angle DBA$	**4.** _____
5. _____	**5.** Def. of $\angle$ bisector; Def. $\cong \angle$'s
6. $m\angle CAB + m\angle DAC = m\angle DAB$ and $m\angle EBA + m\angle DBE = m\angle DBA$	**6.** _____
7. $m\angle CAB + m\angle CAB = m\angle DAB$ and $m\angle EBA + m\angle EBA = m\angle DBA$	**7.** _____
8. $2m\angle CAB = m\angle DAB$ and $2m\angle EBA = m\angle DBA$	**8.** Distributive law
9. $2m\angle CAB = 2m\angle EBA$	**9.** _____
10. $m\angle CAB = m\angle EBA$ so $\angle CAB \cong \angle EBA$	**10.** _____
11. $\overline{AB} \cong \overline{AB}$	**11.** _____
12. $\triangle ACB \cong \triangle BEA$	**12.** _____
13. _____	**13.** CPCTC

Write a two-column proof or a flowchart proof in Exercises 29–32.

29. *Given:* E is the midpoint of $\overline{AC}$
E is the midpoint of $\overline{BD}$

Prove: $\overline{AB} \cong \overline{CD}$

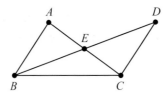

Exercise 29

Section 2.4

30. *Given:* $\overline{AB} \cong \overline{CD}$
$\overline{BC} \cong \overline{DE}$
$\angle CAE \cong \angle CEA$

Prove: $\angle B \cong \angle D$

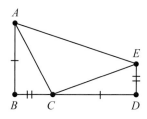

Exercise 30

31. *Given:* $\triangle ABD$ is isosceles with base $\overline{BD}$
$\triangle BDE$ is isosceles with base $\overline{BD}$

Prove: $\angle 1 \cong \angle 2$

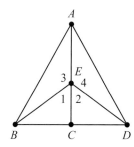

Exercise 31

32. *Given:* $\overline{AB} \cong \overline{CB}$
$\overline{AD} \cong \overline{CD}$

Prove: $\overline{BD}$ is bisector of $\angle ABC$

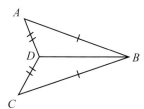

Exercise 32

For Exercises 33–39, determine if each statement is **true** *or* **false.**

33. A median of a triangle is a segment joining a vertex to the midpoint of the side opposite that vertex.

34. The orthocenter of a triangle is the point of concurrency where the angle bisectors intersect.

35. The centroid will always be located inside the triangle.

36. An altitude of a triangle is a segment from a vertex perpendicular to the side opposite that vertex.

37. The altitudes of a triangle will always be in the interior of the triangle.

38. The segment from the vertex that bisects the angle determined by the vertex contains the circumcenter.

39. If $\overline{BD}$ and $\overline{CF}$ are medians of $\triangle ABC$ where $CE = 17$ m and $AD = 12$ m, find EF and AC.

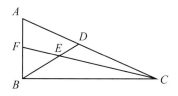

Exercise 39

Section 2.5

Write a two-column or flowchart proof in Exercises 40 and 41.

40. *Given:* $\angle C \cong \angle E$
$\overline{AD} \perp \overline{BC}$; $\overline{AD} \perp \overline{DE}$
B is midpoint $\overline{AD}$

Prove: $\triangle ABC \cong \triangle BDE$

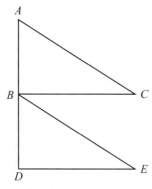

Exercise 40

41. *Given:* $\overline{WY} \perp \overline{XZ}$
$\triangle WXY$ is isosceles $\triangle$ where $\overline{WX} \cong \overline{YX}$

Prove: $\triangle WXZ \cong \triangle YXZ$

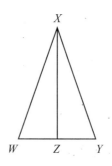

Exercise 41

Section 2.6

42. Draw an obtuse triangle and construct a triangle that is congruent to it.

43. Draw a segment 3 inches long and an acute angle. Construct a right triangle with the 3-inch segment as the leg included between the right angle and the acute angle.

44. Construct the median from the vertex at the right angle in the triangle you constructed in Exercise 43.

45. Draw a segment about 2 inches long and an obtuse angle. Construct an isosceles triangle with the two sides as the 2-inch segment and vertex angle as the obtuse angle.

46. Construct the altitude from the vertex at the obtuse angle in the triangle you constructed in Exercise 45.

Chapter **2** **Practice Test**

Refer to the figure below in Problems 1–10.

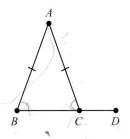

1. Classify △ABC using its angles.
2. Classify △ABC using its sides.
3. What is the vertex angle?
4. What can be said about ∠B and ∠ACB?
5. What angle is opposite $\overline{AC}$?
6. What side is opposite ∠A?
7. Is ∠B an interior angle of △ABC?
8. Is ∠ACD an exterior angle of △ABC?
9. ∠A and ∠B are remote interior angles to what angle?
10. If AB = 5 cm and BC = 3 cm, what is the perimeter of △ABC?
11. What is the perimeter of an equilateral triangle with one side measuring 11 inches?
12. Suppose $\overline{AB}$ and $\overline{EF}$ are corresponding sides of congruent triangles with ∠C opposite $\overline{AB}$ and ∠D opposite $\overline{EF}$. What is the value of x if m∠C = 48° and m∠D = (x + 20)?

Write a two-column proof or a flowchart proof in Problems 13 and 14.

13. *Given:* $\overline{AC} \cong \overline{AE}$
    D is the midpoint of $\overline{CE}$
    ∠1 ≅ ∠2
    Prove: $\overline{BD} \cong \overline{FD}$

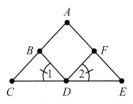

Exercise 13

14. *Given:* ∠3 ≅ ∠4
    $\overline{BC} \cong \overline{DE}$
    Prove: ∠5 ≅ ∠6

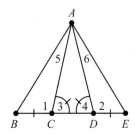

Exercise 14

15. *Given:* $\overline{AB}$ bisects $\overline{CD}$
  $\angle C$ and $\angle D$ are
  right angles

  Prove: $\triangle ACE \cong \triangle BDE$

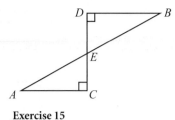

Exercise 15

16. Given $\angle A$ and segment $\overline{BC}$. Construct an isosceles triangle with vertex angle equal to $\angle A$ and both congruent sides of length BC. Then construct the altitude from one of the base angles and the median from the vertex angle.

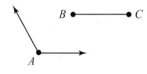

Exercise 16

For Problems 17–19, state whether the bold segment is an altitude, perpendicular bisector, both, or neither.

17.

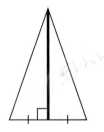

18.

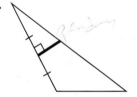

19.

20. In $\triangle MNP$, $\overline{MY}$, $\overline{NZ}$, and $\overline{PX}$ are medians.
  (a) What is the special name for point W?
  (b) If $\overline{PW} = 29$ m, what is $\overline{WX}$?
  (c) If $\overline{XN} = 9$ m, what is $\overline{MX}$?

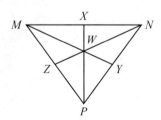

Exercise 20

3 Parallel Lines and Polygons

*T*his chapter begins with a discussion of another type of proof, an indirect proof. Such proofs are *especially* important when we want to show that two lines in a plane never intersect, that is, when the two lines are parallel. The study of parallel lines played a major role in the development of geometry and centers on one of the more controversial postulates, the parallel postulate.

The word *polygon* means a plane figure with "many sides" bounded by straight lines. Consider the different examples of parallel lines and polygons in your environment particularly in construction and architecture. Look at the different shapes of traffic signs, bridges, buildings, and product logos.

AN APPLICATION

You may have seen chicken-wire fencing around people's gardens—even if they don't have chickens. Why do you think chicken wire is a hexagonal shape? What are the benefits of this shape? Some of the answers concern tensile strength and economy of design. If the shape is assumed to be a regular hexagon, what is the measure of each angle inside the hexagon? Does this have anything to do with economy of design?

 ## (3.1) Indirect Proof and the Parallel Postulate

OBJECTIVES

1. Define indirect proof.
2. State the parallel postulate.

OBJECTIVE 1 Define indirect proof. In a direct proof of a geometry theorem, the reason given for each statement is either a definition, a postulate, or a theorem that has already been proven. Although direct proofs are the most common type of proof, some theorems are more easily proved using the format of an *indirect proof*. If $P \longrightarrow Q$ is a theorem we are to prove and we start by assuming P is true, then we must show that Q is true. Now we know that there are two possibilities for Q: either Q is true or Q is false. If we can show that the assumption that Q is false leads to a contradiction of a known fact, such as P, then we are forced to conclude that Q *cannot* be false, or that Q *is* true (the only remaining possibility). This is the basis of an indirect proof. The following example illustrates this type of reasoning by solving a classic puzzle involving truth-telling and lying.

EXAMPLE 1 Inaz loves to explore in the woods and pick samples of different plants. While exploring in the woods, Inaz finds a plant she suspects is poison ivy. Inaz is very concerned because she knows she can get a terrible rash if she touches poison ivy, so she carries a plant identification book with her when she explores. Give an indirect proof to show that the plant is not poison ivy.

First, Inaz assumes the plant is poison ivy. If she can show this assumption is wrong, then she is forced to conclude that the plant is not poison ivy, the desired result. From the picture in the book, she sees that poison ivy leaves grow in groups of three. The plant she found in the woods does not have leaves in groups of three. This is a contradiction to the assumption the plant is poison ivy. Thus, the statement, the plant is not poison ivy, is true.

Indirect proofs require the ability to form the negation of a statement. Recall that if Q is the statement,

"Line m is perpendicular to line n,"

then the negation of Q, $\sim Q$, is

"Line m is *not* perpendicular to line n."

In an indirect proof of the theorem $P \longrightarrow Q$, if we were to assume that the theorem is not true, we would be assuming that P is true and that Q is false.

This is the same as assuming that P is true and that $\sim Q$ (**negation** of Q) is also true. With this notation we can outline the format of an indirect proof.

INDIRECT PROOF OF $P \longrightarrow Q$

Suppose we already know that $\sim Q \longrightarrow Q_1, Q_1 \longrightarrow Q_2,$ $Q_2 \longrightarrow Q_3,$ and $Q_3 \longrightarrow \sim P$ are accepted or previously proved statements, then the format used to write an **indirect proof** of $P \longrightarrow Q$ is:

Given: P

Prove: Q

Proof

Statements	Reasons
1. P	1. Given
2. Assume $\sim Q$ is true	2. Assumption
3. Q_1	3. $\sim Q \longrightarrow Q_1$
4. Q_2	4. $Q_1 \longrightarrow Q_2$
5. Q_3	5. $Q_2 \longrightarrow Q_3$
6. $\sim P$	6. $Q_3 \longrightarrow \sim P$

But this is a contradiction because we were given that P is true and now we have $\sim P$ also true. Thus, our assumption that $\sim Q$ was true is incorrect; so Q must be true.

$\therefore P \longrightarrow Q.$

Note The format for an indirect proof can consist of any number of steps. Notice that the progression of statements uses the same deductive reasoning (for example, $\sim Q$ and $\sim Q \longrightarrow Q_1$, together give Q_1) as was used in direct proofs. An indirect proof can also be written in paragraph form. Each statement must still be justified whether it is in two-column or paragraph form.

EXAMPLE 2

Consider the following information taken from a newspaper article. "If the governor has committed a crime, the House of Representatives will impeach him. If impeached by the House, he will stand trial in the Senate. If tried in the Senate, the governor will be found guilty. If guilty, he will be permanently removed from office by May."

Give an indirect proof of the statement "If the governor is still in office during the summer, then he did not commit a crime."

Given: The governor is still in office during the summer.

Prove: The governor did not commit a crime.

Proof

Statements	Reasons
1. The governor is still in office during the summer.	1. Given
2. Assume that the governor committed a crime.	2. Assumption
3. The House of Representatives will impeach the governor.	3. First sentence in the article
4. The governor will be tried in the Senate.	4. Second sentence in the article
5. The governor will be found guilty.	5. Third sentence in the article
6. The governor will be permanently removed from office by May.	6. Fourth sentence in the article

But statement 6 contradicts the given statement 1 "The governor is still in office during the summer." Thus, our assumption that "The governor committed a crime" was incorrect; so we must conclude that he did not commit a crime. ∴ If the governor is still in office during the summer, then he did not commit a crime. ∎

Both the direct and indirect methods of proof use a chain of conditional statements. In a direct proof of $P \longrightarrow Q$, we start with the hypothesis P and end with the conclusion Q. In an indirect proof, we start with P and $\sim Q$ and end when we reach a contradiction (often $\sim P$). The contradiction tells us that if P is true, then Q cannot be false, which means that $P \longrightarrow Q$.

OBJECTIVE 2   **State the parallel postulate.**   Up to now, we've worked mostly with lines that intersect and the angles formed by these lines. Using indirect proof, we will work with lines that do not intersect.

> **DEFINITION: *Parallel Lines***
> Two lines in the same plane that do not intersect are called **parallel lines**.

If two lines ℓ and m are parallel, we write $\ell \parallel m$. We also say that two segments, two rays, or a segment and a ray are parallel when the lines containing them are parallel. For example in Figure 3.1, if $\ell \parallel m$, we also have $\overline{AB} \parallel \overline{CD}$, $\overline{AB} \parallel \overrightarrow{CD}$, and $\overrightarrow{AB} \parallel \overrightarrow{DC}$. Also, ℓ is not parallel to n, written $l \nparallel n$, because ℓ and n intersect at point B. Similarly, $\overline{CB} \nparallel \overline{CD}$ and $\overrightarrow{AB} \nparallel \overrightarrow{BC}$.

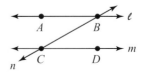

Figure 3.1 Parallel and Nonparallel Lines

The next postulate is perhaps the most famous and controversial of all postulates in geometry. Euclid tried to prove it, as did many mathematicians who followed him. All were unsuccessful. One of the major discoveries in mathematics history involved showing that this postulate *was* independent from the other postulates and *could not* be proved using them.

POSTULATE 3.1 *Parallel Postulate*

For a given line ℓ and a point P not on ℓ, one and only one line through P is parallel to ℓ.

EXAMPLE 3

In the nineteenth century, the Russian mathematician Nicholas Lobachevsky (1793–1856) and a Hungarian colleague, John Bolyai (1802–1860), working independently, replaced Euclid's parallel postulate with the assumption that there are *infinitely many* lines through a point, not on a given line, parallel to the line. Somewhat later, Bernard Riemann (1826–1866), a German mathematician, assumed that there are *no* lines through a point *P*, not on a given line, parallel to the line. These assumptions led to new geometries, called non-Euclidean geometries, which have played a major role in the evolution of mathematics. Riemann's work was of extreme importance in Einstein's theory of relativity. In this text, we follow the approach taken by Euclid.

In Figure 3.2, $\ell \parallel m$ and P is a point on m. How many lines through P are parallel to ℓ? By the parallel postulate, there is one and only one line through P parallel to ℓ; that line must be m.

Figure 3.2

We need to be able to determine when two lines are parallel. The definition of parallel lines might be called a *negative definition*; it tells us that lines are parallel when they *do not* intersect. As a result, many proofs involving parallels are indirect proofs. To give an indirect proof of a theorem $P \longrightarrow Q$ we assume that P is true and that Q is false. Thus, when Q is the statement "two lines are parallel," we often assume that the lines are not parallel, that is, the lines intersect, and then see where the assumption leads. If we arrive at a contradictory fact, we can conclude that the lines do not intersect and therefore must be parallel. The proof of the next theorem illustrates and provides us with a more positive method for proving that two lines are parallel.

THEOREM 3.1

If two lines in a plane are both perpendicular to a third line, then they are parallel.

Given: Lines ℓ, m, and n with $m \perp \ell$ and $n \perp \ell$. (See Figure 3.3.)

Prove: $m \parallel n$

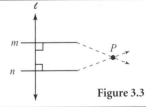

Figure 3.3

Proof _____

Statements	Reasons
1. Assume $m \nparallel n$	**1.** Assumption we wish to show is incorrect
2. m and n intersect at some point, say P	**2.** Def. of $\parallel$ lines
3. $m \perp \ell$ and $n \perp \ell$	**3.** Given

But this is a contradiction of Postulate 1.18 that says that there is only one line perpendicular to a given line passing through a point not on that line. Thus, the assumption $m \nparallel n$ is incorrect; so we must conclude that $m \parallel n$.

3.1 Exercises

FOR EXTRA HELP: 📖 Student's Solutions Manual Tutor Center Addison-Wesley Math Tutor Center

Complete the indirect proof of each "theorem" in Exercises 1 and 2.

1. *Premise 1:* If Bob arrives on time for work, then he woke up on schedule.
   Premise 2: If he wakes up on time, then his alarm rang.
   Premise 3: If his alarm rings, then the power did not fail.
   Theorem: If the power fails, then Bob will be late for work.

   Given: The power fails.

   Prove: Bob will be late for work.

   Proof _____

Statements	Reasons
1. The power fails.	1. Given
2. Assume Bob arrives on time for work.	2. _Assumption_
3. He woke up on schedule.	3. _1st statement_
4. His alarm rang.	4. _2nd statement_
5. The power did not fail.	5. _3rd statement_

   But not having a power failure contradicts the given statement 1 "The power fails." Thus, our assumption in statement 2 was incorrect; so we must conclude that Bob will be late for work. ∴ If the power fails, then Bob will be late for work.

2. *Premise 1:* If I gamble, then I will lose.
   Premise 2: If I am unhappy, then I will get a divorce.
   Premise 3: If I go to Las Vegas, then I will gamble.
   Premise 4: If I lose at gambling, then I will be unhappy.
   Theorem: If I stay married, then I did not go to Las Vegas.

   Given: I stay married.

   Prove: I did not go to Las Vegas.

   Proof _____

Statements	Reasons
1. I stay married.	1. Given
2. _I did go to Vegas_	2. Assumption
3. I will gamble.	3. _____
4. I will lose.	4. _____
5. _____	5. Premise 4
6. I will get a divorce.	6. _____

   But this is a contradiction of Statement 1. Thus, we must conclude that our assumption in statement 2 was incorrect. ∴ _____

Give an indirect proof of each "theorem" in Exercises 3 and 4.

3. *Premise 1:* If I don't exercise, then I'm not playing tennis regularly.
Premise 2: If I'm unhealthy, then I don't exercise.
Premise 3: If I don't play tennis regularly, then the weather is bad.
Theorem: If the weather is nice, then I am healthy.

4. *Premise 1:* If it's not a parallelogram, then it's not a rectangle.
Premise 2: If it's not a four-sided figure, then it's not a quadrilateral.
Premise 3: If it's not a rectangle, then it's not a square.
Premise 4: If it's not a quadrilateral, then it's not a parallelogram.
Theorem: If it's a square, then it's a four-sided figure.

Use the figure below to answer Exercises 5–14.

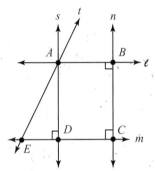

5. Is $\ell \parallel m$?

6. Is $s \parallel n$?

7. Is $\overleftrightarrow{AB} \parallel \overleftrightarrow{ED}$?

8. Is $\overleftrightarrow{AD} \parallel \overleftrightarrow{EC}$?

9. Is $\overleftrightarrow{EC} \parallel \overleftrightarrow{DA}$?

10. Is $\overleftrightarrow{AB} \parallel \overleftrightarrow{CE}$?

11. Can t be parallel to n?

12. Does there exist a line through E parallel to n?

13. How many lines through B are parallel to m?

14. Do $\overleftrightarrow{AE}$ and $\overleftrightarrow{BC}$ intersect?

Give an indirect proof of each theorem in Exercises 15 and 16.

15. If two lines are parallel to a third line then they are parallel to each other.
[Hint: Use Postulate 3.1.] Another way to say this is if two lines are parallel to the same line, they are parallel to each other.

16. If a line intersects one of two parallel lines, then it intersects the other.
[Hint: Use Postulate 3.1.]

In Exercises 17–22, assume that ℓ, m, and n are three distinct lines in a plane and P is a point in the plane.

17. If $\ell \parallel m$ and $\ell \parallel n$, is $m \parallel n$?

18. If $\ell \parallel m$ and P is a point on both m and n, is $n \parallel \ell$?

19. If $\ell \perp m$ and $\ell \perp n$, is $m \perp n$?

20. If $\ell \perp m$ and $\ell \perp n$, is $m \parallel n$?

21. If $\ell \parallel m$, $\ell \parallel n$, and P is on m, is P on n?

22. If $\ell \perp m$, $m \perp n$, and P is on ℓ, is P on n?

23. Give examples of parallel lines found in a classroom.
24. Give examples of perpendicular lines found in a classroom.
25. Give examples of perpendicular lines found on a football field.
26. Give examples of parallel lines found on a football field.
27. Can a triangle have two right angles?
28. Can a triangle have two sides that are parallel?
29. A suspect claimed to be deaf. Sherlock Holmes did not believe the suspect and wanted to prove he was not deaf. Holmes stood behind the suspect and shouted, "Surprise!" The startled suspect jumped just a little. Use a paragraph form indirect proof to prove Holmes was correct. Start by assuming the suspect is deaf.
30. The definition of parallel lines in this section indicated the lines must be in the same plane. This is a pivotal assumption for Euclidean geometry, the geometry in this textbook. There are other non-Euclidean geometries. These geometries are characterized by at least one contradiction of a Euclidean geometry postulate. How many non-Euclidean geometries can you find? Do a little research on how they are different from Euclidean geometry.

(3.2) Transversals and Angles

OBJECTIVES

1. Define angles formed by parallel lines and a transversal.
2. Describe ways to prove lines parallel.
3. Prove theorems about parallel lines.

One of the best ways to study two parallel lines is to consider the various angles that are formed when a third line intersects them.

> **DEFINITION:** *Transversal*
>
> A **transversal** is a line that intersects two or more distinct lines in different points.

In Figure 3.4(a) ℓ is a transversal that **cuts** (intersects) lines m and n in the two points A and B, respectively. However, in Figure 3.4(b), s is not a transversal because it intersects t and u in only one point P.

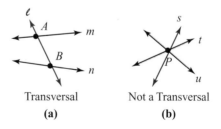

Transversal   Not a Transversal
(a)   (b)

Figure 3.4

When a transversal cuts two lines, several pairs of angles are formed.

OBJECTIVE 1   Define angles formed by parallel lines and a transversal.

> **DEFINITION: *Angles Formed by a Transversal***
> Suppose two lines are cut by a transversal.
> 1. The nonadjacent angles on opposite sides of the transversal but on the interior of the two lines are called **alternate interior angles**.
> 2. The nonadjacent angles on the same side of the transversal and in the same corresponding positions with respect to the two lines are called **corresponding angles**.
> 3. The nonadjacent angles on opposite sides of the transversal and on the exterior of the two lines are called **alternate exterior angles**.

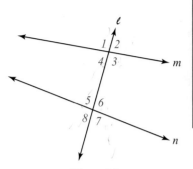

Figure 3.5

In Figure 3.5, ℓ is a transversal that cuts m and n. There are two pairs of alternate interior angles: $\angle 4$ and $\angle 6$; and $\angle 3$ and $\angle 5$. There are four pairs of corresponding angles: $\angle 1$ and $\angle 5$; $\angle 2$ and $\angle 6$; $\angle 3$ and $\angle 7$; and $\angle 4$ and $\angle 8$. There are two pairs of alternate exterior angles: $\angle 1$ and $\angle 7$; and $\angle 2$ and $\angle 8$.

OBJECTIVE 2   Describe ways to prove lines parallel.   A pair of alternate interior angles can be used to show that two lines are parallel.

> **THEOREM 3.2**
> If two lines are cut by a transversal and a pair of alternate interior angles are congruent, then the lines are parallel.

Given: Lines m and n cut by transversal ℓ
  $\angle 1 \cong \angle 2$ (See Figure 3.6.)
Prove: $m \parallel n$
Construction: Construct the midpoint of $\overline{AB}$, C.
  Construct $\overleftrightarrow{CD} \perp m$ through C.

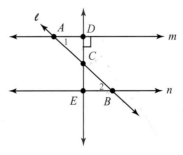

Figure 3.6

Proof

Statements	Reasons
1. m and n are lines with transversal ℓ	1. Given
2. C is the midpoint of $\overline{AB}$	2. Construction 1.3
3. $\overline{AC} \cong \overline{CB}$	3. Def. of midpoint
4. $\overline{CD} \perp m$	4. Construction 1.5
5. $\angle ADC$ is a right angle	5. $\perp$ lines form right angles (def. of $\perp$ lines)
6. $m\angle ADC = 90°$	6. Def. of right angle
7. $\angle DCA \cong \angle BCE$	7. Vertical angles are $\cong$
8. $\angle 1 \cong \angle 2$	8. Given
9. $\triangle ACD \cong \triangle BCE$	9. ASA
10. $\angle ADC \cong \angle BEC$ so $m\angle ADC = m\angle BEC$	10. CPCTC and def. $\cong$ angles
11. $m\angle CEB = 90°$	11. Substitution in statements 6 and 10
12. $\angle CEB$ is a right angle	12. Def. of right angle
13. $\overleftrightarrow{CD} \perp n$	13. If 2 lines form right angles they are $\perp$ (def. $\perp$ lines)
14. $m \parallel n$	14. Two lines $\perp$ to third line are $\parallel$ (Thm 3.1)

Corresponding angles can also be used to show that lines are parallel.

 Student Activity

For this activity, the following equipment will be needed: two 6-inch squares of tracing paper, a ruler, and a protractor.
1. Draw a line and label it line m.
2. Draw a new line that intersects line m and label this line t.
3. Label one of the vertical angles $\angle 1$.
4. Place another piece of tracing paper over the original and copy these lines onto the new piece of tracing paper. Label them as in step 1. Be sure to use a straightedge.
5. Keeping line t exactly over each other, slide the top sheet of paper about an inch. Line m and $\angle 1$ will move compared to the bottom piece of paper.
6. Measure the distance between the two lines labeled m. Do this in several places.
What is special about these distances? Are they the same? Measure the two angles labeled $\angle 1$. What is their special name? What conjecture might you make? Compare these results to Theorem 3.3.

THEOREM 3.3

If two lines are cut by a transversal and a pair of corresponding angles are congruent, then the lines are parallel.

Given: Lines m and n cut by
transversal ℓ
$\angle 1 \cong \angle 2$ (See Figure 3.7.)

Prove: $m \parallel n$

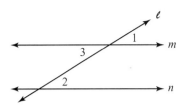

Figure 3.7

Proof _____

Statements	Reasons
1. $\angle 1 \cong \angle 2$	1. Given
2. $\angle 3 \cong \angle 1$	2. Vert. $\angle$'s are $\cong$
3. $\angle 3 \cong \angle 2$	3. Trans. law
4. $m \parallel n$	4. If alt. int. $\angle$'s are $\cong$ lines are $\parallel$

THEOREM 3.4

If two lines are cut by a transversal and a pair of alternate exterior angles are congruent, then the lines are parallel.

The proof of this theorem is similar to that for Theorem 3.3 and is left for you to do as an exercise.

THEOREM 3.5

If two lines are cut by a transversal and two interior angles on the same side of the transversal are supplementary, then the lines are parallel.

Sometimes our eyes play tricks on us. The two horizontal lines in the figure above, cut by many transversals, are actually parallel. Do they appear this way to you?

PRACTICE EXERCISE 1

Complete the proof of Theorem 3.5.

Given: Lines m and n cut by
transversal ℓ
$\angle 1$ and $\angle 2$ are supplementary

Prove: $m \parallel n$

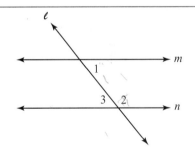

Proof _____

Statements	Reasons
1. $\angle 1$ and $\angle 2$ are supplementary	1. <u>Given</u>
2. $\angle 3$ and $\angle 2$ are supplementary	2. <u>Adj $\angle$'s whose noncommon sides are in a line are supplementary</u>
3. <u>$\angle 1 \cong \angle 3$</u>	3. Supp. of same $\angle$ are $\cong$
4. $m \parallel n$	4. <u>If alt. int $\angle$'s are $\cong$ the lines are $\parallel$</u>

ANSWERS ON PAGE 141

> **SUMMARY: WAYS TO PROVE LINES PARALLEL**
>
> 1. Two lines are both perpendicular to a third line. (Theorem 3.1)
> 2. A pair of alternate interior angles are congruent. (Theorem 3.2)
> 3. A pair of corresponding angles are congruent. (Theorem 3.3)
> 4. A pair of alternate exterior angles are congruent. (Theorem 3.4)
> 5. Interior angles on the same side of a transversal are supplementary. (Theorem 3.5)

OBJECTIVE 3 **Prove theorems about parallel lines.**   Remember to form the converse of a statement, interchange the hypothesis and the conclusion. The converse of any statement must be proven true before it can be used in a proof.

⌐**Note** The symbol $\not\cong$ means *not congruent*.

The walls and corners in the rooms of a house are usually not perfectly vertical. A paperhanger uses a plumb line, a string with a weight attached to it, to form a vertical line that is then used to align the first piece of wallpaper. Properties of parallel lines guarantee that the remaining strips of wallpaper will be vertical when seams are aligned.

> **THEOREM 3.6**   **(CONVERSE OF THEOREM 3.2)**
> If two parallel lines are cut by a transversal, then all pairs of alternate interior angles are congruent.

Given: $m \parallel n$ with ℓ a transversal (See Figure 3.8.)

Prove: $\angle 1 \cong \angle 2$

We give an indirect proof by showing that the assumption $\angle 1 \not\cong \angle 2$ leads to a contradiction.

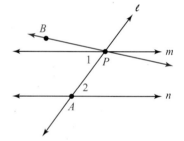

Figure 3.8

Proof _____

Statements	**Reasons**
1. Assume $\angle 1 \not\cong \angle 2$	1. Assumption
2. Construct $\angle APB$ such that $\angle APB \cong \angle 2$	2. Construction 1.2
3. $\overline{BP} \parallel n$	3. If alt. int. $\angle$'s are $\cong$, lines are $\parallel$.
4. $m \parallel n$	4. Given

Because $\angle APB \cong \angle 2$ and $\angle 2 \not\cong \angle 1$, $\angle APB \not\cong \angle 1$, so $\overline{BP}$ and m are two distinct lines through P parallel to n, a contradiction of the parallel postulate. Thus, $\angle 1 \not\cong \angle 2$ leads to a contradiction forcing us to conclude that $\angle 1 \cong \angle 2$.

■

We could show that the other pair of alternate interior angles are also equal using a similar proof.

Technology Connection

Geometry software will be needed.
1. Construct a pair of parallel lines.
2. Draw a transversal that cuts through the parallel lines.
3. Measure both pairs of alternate interior angles.

What do you notice? Does this support Theorem 3.6?

THEOREM 3.7   (*Converse of Theorem 3.1*)

If two lines are parallel and a third line is perpendicular to one of them, then it is also perpendicular to the other.

Given: $m \parallel n$
$\ell \perp m$ (See Figure 3.9.)

Prove: $\ell \perp n$

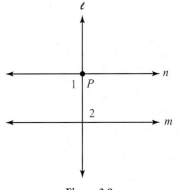

Figure 3.9

Proof

Statements	Reasons
1. $m \parallel n$	1. Given
2. $\angle 1 \cong \angle 2$ therefore $m\angle 1 = m\angle 2$	2. If lines are $\parallel$ then alternate interior angles are $\cong$ (Thm 3.6) and def. $\cong$ angles
3. $\ell \perp m$	3. Given
4. $\angle 2$ is right angle	4. Def. $\perp$ lines
5. $m\angle 2 = 90°$	5. Def. right angle
6. $m\angle 1 = 90°$	6. Substitution
7. $\ell \perp n$	7. Def. $\perp$ lines

THEOREM 3.8   (*Converse of Theorem 3.3*)

If two parallel lines are cut by a transversal, then all pairs of corresponding angles are congruent.

PRACTICE EXERCISE 2

Complete the proof of Theorem 3.8.

Given: $m \parallel n$ and m and n are cut
by transversal ℓ

Prove: $\angle 1 \cong \angle 2$

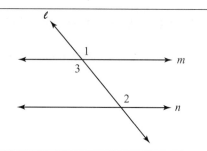

Proof _____

Statements	*Reasons*
1. $m \parallel n$	1. <u>Given</u>
2. $\angle 2 \cong \angle 3$	2. <u>If ‖ lines are cut by a transversal the alt. int ∠'s are ≅</u>
3. $\angle 3 \cong \angle 1$	3. <u>Vert ∠'s are ≅</u>
4. <u>$\angle 1 \cong \angle 2$</u>	4. Trans. and sym. laws

The remaining corresponding angles are proved congruent in a similar way.

ANSWERS ON PAGE 141

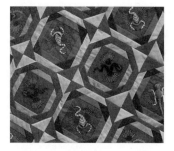

When the pieces in a quilt are sewn together, much care must be taken to maintain congruent angles so that the appropriate sides of each quilt block remain parallel. Which theorems involving lines cut by a transversal are suggested by the quilt shown?

Note In Practice Exercise 2, the statement: "$\angle 1$ and $\angle 3$ are vertical angles" was omitted. For the remainder of the textbook, vertical angles will be identified by their position. Thus, in a proof, we will directly state vertical angles are congruent.

Also, in the same proof, you may have thought statement 4 should read $\angle 2 \cong \angle 1$ by the Transitive Law and $\angle 1 \cong \angle 2$ by the Symmetric Law. As you see in the key, the last statement will be written to match the "prove" statement even if the previous line does not have the correct order. This same thinking will be used when proving lines perpendicular or parallel. The order is not important. We will do this in all future proofs.

THEOREM 3.9   (CONVERSE OF THEOREM 3.4)

If two parallel lines are cut by a transversal, then all pairs of alternate exterior angles are congruent.

The proof of Theorem 3.9 is similar to that of Theorem 3.8 and is left for you to do as an exercise.

THEOREM 3.10   (CONVERSE OF THEOREM 3.5)

If two parallel lines are cut by a transversal, then all pairs of interior angles on the same side of the transversal are supplementary.

The proof of Theorem 3.10 is left for you to do as an exercise.

EXAMPLE 1

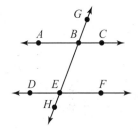

Figure 3.10

In Figure 3.10, $m \parallel n$, $\overline{AB} \cong \overline{BC}$, $m\angle BAC = 40°$. What is the measure of $\angle 1$?
Because $\overline{AB} \cong \overline{BC}$, $m\angle BCA = 40°$. Because $\angle BCA$ and $\angle ACD$ are supplementary, $m\angle ACD = 140°$. Because $\angle 1$ and $\angle ACD$ are alternate interior angles formed by transversal $\overleftrightarrow{BD}$ cutting parallel lines m and n, $m\angle 1 = 140°$.

Technology Connection

Geometry software will be needed.

1. Construct two parallel lines.
2. Construct a transversal that intersects both lines and label the figure as shown at the right.
3. Measure all eight angles in the figure.
4. Name all the pairs of corresponding angles. Are they all congruent in the figure?
5. Name all the pairs of alternate interior angles. Are they all congruent in the figure?
6. Name all the pairs of alternate exterior angles. Are they all congruent in the figure?
7. Name all the pairs of same side interior angles. Are they supplementary in the figure?
8. Drag point B (left or right) and see which angles remain congruent.
9. In a new sketch, draw two lines that are not parallel. Construct a transversal.
10. Label point similar to the figure above.
11. Measure one set of corresponding angles.
12. Move the lines until the corresponding angles are congruent.

Did you find the corresponding angles did not become congruent until the lines were parallel? Thus to have corresponding angles congruent, parallel lines are needed.

ANSWERS ON PAGE A-8

Answers to Practice Exercises

1. 1. Given 2. Adj. $\angle$'s whose noncommon sides are in a line are supp. 3. $\angle 1 \cong \angle 3$ 4. If alt. int. $\angle$'s are $\cong$ the lines are $\parallel$. **2.** 1. Given 2. If $\parallel$ lines are cut by a transv. the alt. int. $\angle$'s are $\cong$ **3.** Vert. $\angle$'s are $\cong$ **4.** $\angle 1 \cong \angle 2$

Exercises 1–12 refer to the figure below in which $m \parallel n$ and ℓ is a transversal.

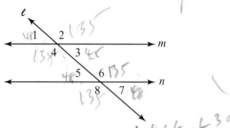

1. List two pairs of alternate interior angles.
2. List two pairs of alternate exterior angles.
3. List four pairs of corresponding angles.
4. List four angles that are supplementary to $\angle 1$.
5. List four angles that are supplementary to $\angle 2$.
6. List three angles that are congruent to $\angle 1$.
7. List three angles that are congruent to $\angle 2$.
8. If $m\angle 1 = 48°$, find the measure of every other angle.
9. If $m\angle 6 = 135°$, find the measure of every other angle.
10. Find the measure of each angle if $m\angle 4$ is 30° more than twice $m\angle 5$.
11. Find the measure of each angle if $m\angle 6$ is 15° more than twice $m\angle 1$.
12. Find the measure of each angle if $m\angle 2$ is 10° more than $m\angle 8$.

For Exercises 13–17, refer to the figure below where $a \parallel b \parallel c$. Each number is a separate problem. Do not use information from a prior problem to solve the current problem.

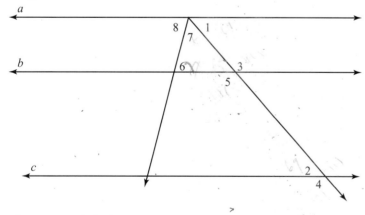

13. If $m\angle 1 = 69°$, find $m\angle 2$.
14. If $m\angle 3 = 125°$, find $m\angle 4$.
15. If $m\angle 2 = 52°$, find $m\angle 5$.
16. If $m\angle 3 = 140°$, find $m\angle 2$.
17. If $m\angle 6 = 70°$, $m\angle 3 = 115°$, find $m\angle 7$.

Exercises 18–25 refer to the figure at the right. Explain your reasoning.

18. If $\angle 2 \cong \angle 4$, can $\overleftrightarrow{BC}$ and $\overleftrightarrow{DE}$ intersect?

19. If $\angle 1 \cong \angle 4$, can $\overleftrightarrow{BC}$ and $\overleftrightarrow{DE}$ intersect?

20. If $\angle 2 \cong \angle 5$, can $\overleftrightarrow{BC}$ and $\overleftrightarrow{DE}$ intersect?

21. If $\angle 3$ and $\angle 5$ are supplementary, can $\overleftrightarrow{BC}$ and $\overleftrightarrow{DE}$ intersect?

22. If $AB = AC$ and $\angle 5 \cong \angle 6$, can $\overleftrightarrow{BC}$ and $\overleftrightarrow{DE}$ intersect?

23. If $AB = AC$ and $\angle 4 \cong \angle 6$, can $\overleftrightarrow{BC}$ and $\overleftrightarrow{DE}$ intersect?

24. If $AD = AE$ and $\angle 5$ and $\angle 3$ are supplementary, are $\overleftrightarrow{BC}$ and $\overleftrightarrow{DE}$ parallel?

25. If $AD = AE$ and $\angle 4 \cong \angle 6$, are $\overleftrightarrow{BC}$ and $\overleftrightarrow{DE}$ parallel?

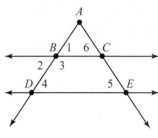

Exercises 18–25

Use the figure below in Exercises 26–33.

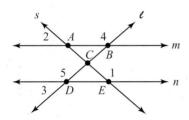

Exercises 26–33

Note Proofs are not unique. Remember, your proof may differ slightly from the key or someone else's proof. This does not mean your proof is necessarily incorrect. Consult your instructor.

26. *Given:* $\angle 1$ is supplementary to $\angle 2$
Prove: $m \parallel n$

27. *Given:* $\angle 3$ is supplementary to $\angle 4$
Prove: $m \parallel n$

28. *Given:* $m \parallel n$
$\overline{AB} \cong \overline{DE}$
Prove: $\triangle ABC \cong \triangle CDE$

29. *Given:* $m \parallel n$
$\angle 1 \cong \angle 5$
Prove: $\triangle ABC$ is isosceles

30. *Given:* C is the midpoint of $\overline{AE}$ and $\overline{BD}$
Prove: $m \parallel n$

31. *Given:* $m \parallel n$
$\overline{CD} \cong \overline{CE}$
Prove: $\overline{AC} \cong \overline{BC}$

32. *Given:* $\overline{AC} \cong \overline{CE}$
$m \parallel n$
Prove: $\overline{DC} \cong \overline{CB}$

33. *Given:* $\overline{AB} \cong \overline{DE}$
$\overline{AD} \cong \overline{BE}$ (Draw auxiliary lines.)
Prove: $m \parallel n$

34. Prove Theorem 3.4.

35. Prove Theorem 3.9.

36. Prove Theorem 3.10.

37. Use the figure below to find the values of x and y that make $\overline{AB} \parallel \overline{CD}$ and $\overline{AD} \parallel \overline{BC}$.

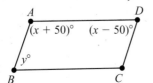

Exercise 37

38. Use the figure below to find the values of x and y that make $\overline{AB} \parallel \overline{CD}$ and $\overline{BC} \parallel \overline{DE}$.

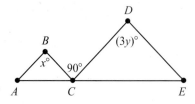

Exercise 38

39. Use the figure below to find the values of x that make $m \parallel n$ if $m\angle 1 = (8x)°$ and $m\angle 2 = (x^2)°$

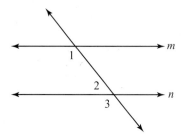

Exercises 39 and 40

40. Use the figure above to find the value of x that makes $m \parallel n$ if $m\angle 1 = (x^2 - 15)°$ and $m\angle 3 = x(x - 1)°$

41. John knows several things about the figure to the right: $\overline{AC} \cong \overline{BD}$, $\overline{AX} \cong \overline{BZ}$, and $\overline{AX} \parallel \overline{BZ}$. His friend Jose asks if $\overline{CX} \parallel \overline{DZ}$? John says, "They look parallel so I think they are." Jose reminds John not to go by appearances and asks to be convinced. John knows there are two congruent triangles in the figure. Write a paragraph convincing Jose that $\overline{CX} \parallel \overline{DZ}$.

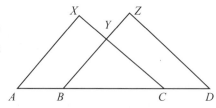

Exercise 41

42. The snow has collected on the corners of these windowpanes. If it is assumed the window framing is parallel, where has the snow collected (name the type of angles). Make a conjecture about the reason the snow collected there.

3.3 Polygons and Angles

OBJECTIVES

1. Define a polygon.
2. Construct parallel lines.
3. Establish that the sum of angles of triangle is 180°.
4. Find the sum of measures of angles in any polygon.

What could be more geometrically impressive than the honeycomb of a bee? Note that each opening is in the shape of a convex polygon with six sides.

A triangle is a special case of a general class of geometric figures called *polygons*. Recall that a triangle is composed of three distinct segments no two of which are on the same line.

OBJECTIVE 1   Define a polygon.

> **DEFINITION: Polygon**
>
> A **polygon** is a closed figure in a plane. It has n segments (where $n \geq 3$) called **sides** that intersect only at their endpoints. Each endpoint is called the **vertex** of the polygon. No two consecutive sides are on the same line.

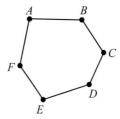

Figure 3.11

The polygon in Figure 3.11 is an example of a convex polygon. The angles of a **convex** polygon measure between 0° and 180°. Figure 3.12 shows a polygon that is not convex. It is called concave. A polygon is **concave** if a line segment joining two points in the polygon may include points not in the interior to the polygon. When we use the term *polygon* from now on, we will mean *convex polygon*.

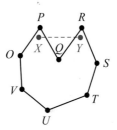

Figure 3.12

> **DEFINITION: Regular Polygon**
>
> A polygon is a **regular polygon** if all its sides are congruent and all its angles are congruent.

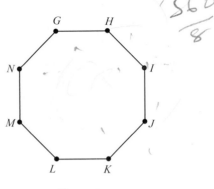

$$\frac{360}{8}$$

$$\frac{(n-2)180}{n}$$

$$\frac{(8-2)180}{8}$$

Figure 3.13

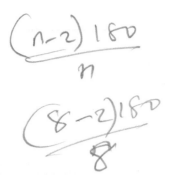

The most well-known polygonal building in the world is the Pentagon in Washington, DC.

The polygon in Figure 3.13 is a regular polygon because $\overline{GH} \cong \overline{HI} \cong \overline{IJ} \cong \overline{JK} \cong \overline{KL} \cong \overline{LM} \cong \overline{MN} \cong \overline{NG}$ and $\angle GHI \cong \angle HIJ \cong \angle IJK \cong \angle JKL \cong \angle KLM \cong \angle LMN \cong \angle MNG \cong \angle NGH$. Another example of a regular polygon is an equilateral (equiangular) triangle.

Polygons are given special names according to their number of sides as shown in the following table and Figure 3.14.

NUMBER OF SIDES	POLYGON
3	triangle
4	quadrilateral
5	pentagon
6	hexagon
7	heptagon
8	octagon
9	nonagon
10	decagon
n	n-gon

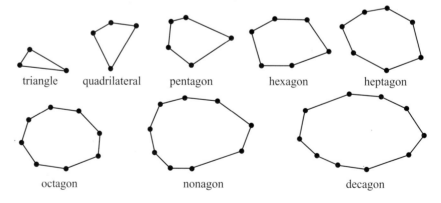

triangle   quadrilateral   pentagon   hexagon   heptagon

octagon   nonagon   decagon

Figure 3.14

The polygon in Figure 3.11 is a hexagon and the polygon in Figure 3.13 is a regular octagon.

The next postulate gives the relationship between the number of sides and the number of angles of a polygon.

POSTULATE 3.2

A polygon has the same number of angles as sides.

DEFINITION: *Diagonal of a Polygon*

A **diagonal** of a polygon is a segment that joins two nonadjacent vertices.

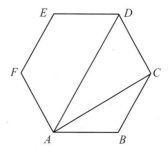

Figure 3.15

In Figure 3.15 for example, if we were to construct segment $\overline{AD}$, then $\overline{AD}$ would be a diagonal of the hexagon. Notice in the definition of a diagonal that the term *nonadjacent* is important. Segments joining adjacent vertices are sides, not diagonals. $\overline{AC}$ is another example of a diagonal.

We have previously defined the perimeter of a triangle. This definition can be extended to all polygons.

DEFINITION: *Perimeter of a Polygon*

The **perimeter** of a polygon is the sum of the lengths of its sides.

EXAMPLE 1   Consider the polygon in Figure 3.16.

(a) What kind of polygon is this?
Because it has 5 sides, it is a pentagon.

(b) How many diagonals does it have?
Because the diagonals are $\overline{AC}, \overline{AD}, \overline{BD}, \overline{BE}$, and $\overline{CE}$, there are 5 diagonals.

(c) What is the perimeter of the polygon?
The perimeter is

$$P = AB + BC + CD + DE + EA.$$
$$= 2 + 3 + 4 + 3 + 1.5 = 13.5 \text{ cm}$$

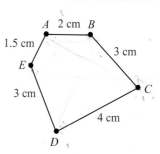

Figure 3.16 Perimeter of a Polygon

EXAMPLE 2   If each side of a 12-sided regular polygon measures 56 mm, what is the perimeter of the polygon?

Because this is a regular polygon, all sides are congruent.

$$\text{Perimeter} = \text{Sum of sides}$$
$$P = 12(56)$$
$$P = 672 \text{ mm}$$

OBJECTIVE 2   **Construct parallel lines.**   To determine properties of the angles of any polygon, we will first prove several theorems about the angles of a triangle and use the results in more general cases. To do this, we need to be able to construct the line parallel to a given line and passing through a point not on the given line.

CONSTRUCTION 3.1

Construct the line parallel to a given line that passes through a point not on the given line.

Given: Line ℓ and point P (See Figure 3.17.)

To Construct: Line m through P such that $m \parallel \ell$

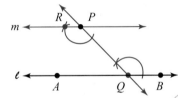

Figure 3.17

Construction

1. Draw any line through P that intersects ℓ at a point we label Q.
2. At P, use Construction 1.2 to construct $\angle RPQ$ such that $m\angle RPQ = m\angle PQB$.
3. Line $\overleftrightarrow{RP}$ (which is the same as m) is the desired parallel line because $m \parallel \ell$ since $\angle RPQ$ and $\angle PQB$ are congruent alternate interior angles. By the parallel postulate, m is *the* line satisfying the given conditions.

Do the following exploration before moving to the next theorem.

 Technology Connection

Geometry software will be needed.

1. Construct any triangle. It may be acute, obtuse, or right.
2. Measure the three angles.
3. What is the sum of the measures of the three angles?
4. Drag one vertex of the triangle to a different position and observe the angle sum. Repeat several times.

[Note: Because the computer may round angle measures, the sum may not be exactly 180° but it will be close.] These results will be proven in Theorem 3.11.

OBJECTIVE 3 Establish that the sum of angles of triangle is 180°. The followers of Pythagoras (about 584–495 B.C.) are thought to have been the first to prove the next important theorem.

THEOREM 3.11

The sum of the measures of the angles of a triangle is 180°.

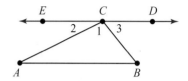

Figure 3.18

Given: △ABC (See Figure 3.18.)
Prove: $m\angle A + m\angle 1 + m\angle B = 180°$

Construction: Construct line $\overleftrightarrow{ED}$ through C parallel to $\overline{AB}$.

Proof

Statements	Reasons
1. $\overline{ED} \parallel \overline{AB}$ and C is on $\overline{ED}$	1. Construction 3.1
2. $\angle ECD$ is a straight angle and $m\angle ECD = 180°$	2. Def. of straight $\angle$
3. $m\angle 2 + m\angle 1 + m\angle 3 = 180°$	3. Angle add. post.
4. $\angle A \cong \angle 2$ and $\angle B \cong \angle 3$	4. If lines $\parallel$, alt. int. $\angle$'s $\cong$.
5. $m\angle A = m\angle 2; m\angle B = m\angle 3$	5. Def. $\cong$ $\angle$'s
6. $m\angle A + m\angle 1 + m\angle B = 180°$	6. Substitution law

Theorem 3.11 gives rise to a series of corollaries.

COROLLARY 3.12

Any triangle can have at most one right angle or at most one obtuse angle.

Given: △ABC (See Figure 3.19.)
Prove: △ABC can have at most one right angle.

Proof

Statements	Reasons
1. Assume $\angle A$ and $\angle B$ are both right angles	1. Assumption we wish to show incorrect.
2. $m\angle A = 90°, m\angle B = 90°$	2. Def. of right angles
3. $m\angle A + m\angle B + m\angle C = 180°$	3. Sum of angles of △ = 180°
4. $90° + 90° + m\angle C = 180°$	4. Substitution
5. $m\angle C = 0°$	5. Add.-subt. law

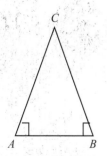

Figure 3.19

Many floor coverings are made by joining together various polygons without leaving gaps or overlapping the edges. Such coverings are called *tessellations*. The tessellation shown above is made up of squares and regular octagons. Can you find other examples of tessellations?

A contradiction has been reached. There must be three angles in a triangle by definition and this proof says there are only two since the third angle measures 0°. Therefore, the assumption is incorrect and a triangle may have at most, one right angle. ∎

A similar proof may be used to show there is at most one obtuse angle in a triangle.

> **COROLLARY 3.13**
>
> If two angles of one triangle are congruent, respectively, to two angles of another triangle, then the third angles are also congruent.

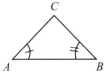

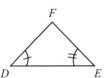

Figure 3.20

Given: $\triangle ABC$ and $\triangle DEF$ with $\angle A \cong \angle D$, $\angle B \cong \angle E$ (See Figure 3.20.)

Prove: $\angle C \cong \angle F$

Proof _____

Statements	Reasons
1. $\triangle ABC$ and $\triangle DEF$ have $\angle A \cong \angle D$, $\angle B \cong \angle E$, so $m\angle A = m\angle D$, $m\angle B = m\angle E$	1. Given and def. $\cong$ angles
2. $m\angle A + m\angle B + m\angle C = 180°$ $m\angle D + m\angle E + m\angle F = 180°$	2. Sum of angles of $\triangle = 180°$
3. $m\angle A + m\angle B + m\angle C = m\angle D + m\angle E + m\angle F$	3. Substitution
4. $m\angle C = m\angle F$	4. Add.-subt. law
5. $\angle C \cong \angle F$	5. Def. $\cong$ angles

∎

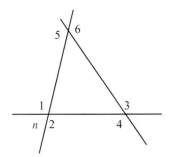

Figure 3.21

The next corollary involves exterior angles of a triangle, which was defined in Chapter 2. In Figure 3.21, $\angle 1$ is an exterior angle of the triangle as are angles 2, 3, 4, 5, and 6. [Note: Angle n is *not* an exterior angle because it's formed by two extensions of sides of the triangle.]

> **COROLLARY 3.14**
>
> The measure of an exterior angle of a triangle is equal to the sum of the measures of the nonadjacent interior angles.

Given: △*ABC* with exterior ∠1
(See Figure 3.22.)

Prove: $m\angle 1 = m\angle A + m\angle B$

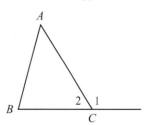

Figure 3.22

Proof

Statements	*Reasons*
1. △*ABC* with exterior ∠1	**1.** Given
2. $m\angle A + m\angle B + m\angle 2 = 180°$	**2.** Sum of angles of △ = 180°
3. ∠1, ∠2 are supplementary	**3.** If the exterior sides of two adjacent angles lie in a straight line, the angles are supplementary (Thm 1.10).
4. $m\angle 1 + m\angle 2 = 180°$	**4.** Def. of supp. angles
5. $m\angle A + m\angle B + m\angle 2 = m\angle 1 + m\angle 2$	**5.** Substitution
6. $m\angle A + m\angle B = m\angle 1$	**6.** Add.-subt. law

OBJECTIVE 4   **Find the sum of measures of angles in any polygon.**   We now turn our attention to the angles of polygons in general.

Student Activity

For this activity, you will need one pair of scissors, one protractor, and one straightedge per group, and several pieces of regular paper. Form groups of 2 to 4 students according to your instructor's directions.

1. One member of each group draws and cuts out any size triangle. (Make it large so it's easy to work with.) Be sure to use a straightedge when drawing the triangle. Number the angles 1, 2, and 3. Put the numbers *very* close to the vertex of the angle. It might help to put a circle around the number. Cut or tear off the angles. Align the angles so they have a common vertex (the ones you numbered) and the sides are adjacent. See the figure at the right.

   What do you observe about the sum of the angles of a triangle?
   In Theorem 3.11, the sum was proved to be 180°, thus we can put 180° under 3 in the chart on page 152.

For the next part of the activity, each group of students "counts off" starting at 4, 5, 6, and so on.

continued

2. Draw any polygon with the number of sides given to your group by the "count off." (Make it large so it's easy to work with.) Be sure to use a straightedge. Number the angles 1, 2, 3, 4, and so on. Put the numbers *very* close to the vertex of the angle. With a protractor measure all the angles in the polygon. What do you observe about the sum of the angles of your polygon? Fill in the table below.

3. This exercise is about looking for a pattern. All groups put their data in a table for the entire class to see and make a conjecture about the sum of the measures of any polygon with n sides. You are using inductive reasoning to make the conjecture.

Number of sides	3	4	5	6	7	8	9	10	n
Sum of angles									

Compare your conjecture with Theorem 3.15.

PRACTICE EXERCISE 1

If 1440° is the sum of the interior angles of a polygon, find the number of sides of the polygon. Use the table in the Student Activity.

ANSWER ON PAGE 154

THEOREM 3.15

The sum of the measures of the angles of a polygon with n sides is given by the formula $S = (n - 2)180°$.

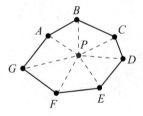

Figure 3.23

It is difficult to write a formal two-column proof of this theorem. Instead, we will present a paragraph-style proof using the polygon in Figure 3.23.

PROOF: Select a point P in the interior of the polygon and construct segments joining P to each vertex of the polygon. Because the polygon has n sides, we will obtain n triangles. Because the sum of the angle measures of a triangle is 180°, the sum of the angle measures of these n triangles is $n(180°)$. However, the angles of the triangles with their vertex at P are not parts of the angles of the polygon. Because the sum of these angle measures is 360°, we must subtract this amount from $n(180°)$. Then,

$$n(180°) - 360° = n(180°) - 2(180°)$$
$$= (n - 2)180°.$$

Although the polygon in Figure 3.23 has seven sides, we can see that this polygon was used for illustrative purposes only and that the formula

$$S = (n - 2)180°$$

works for any number of sides n. Notice that when $n = 3$ (a triangle), we obtain $S = (3 - 2)180° = (1)180° = 180°$, the same result we obtained earlier.

EXAMPLE 3   Find the sum of the angle measures of an octagon.
Because an octagon has 8 sides, set $n = 8$ in the formula $S = (n - 2)180°$. Thus,

$$S = (8 - 2)180° = (6)180° = 1080°.$$

COROLLARY 3.16

The measure of each angle of a *regular* polygon with n sides is given by the formula $a = \dfrac{(n - 2)180°}{n}$.

PROOF: Because there are n equal angles in a regular polygon with n sides, if a is the measure of one of these angles, na is the sum of all these measures. By Theorem 3.15,

$$na = (n - 2)180°$$

from which the desired formula results by dividing both sides by n.

EXAMPLE 4   At the beginning of this chapter, there is a picture of a chicken-wire fence. Why do you think chicken wire is a hexagonal shape? What are the benefits of this shape? Some of the answers concern tensile strength and economy of design. If the shape is assumed to be a regular hexagon, what is the measure of each angle inside the hexagon? Does this have anything to do with economy of design?

To find the measure of each interior angle of a hexagon use Corollary 3.16. Because a hexagon has 6 sides, $n = 6$. Thus,

$$a = \frac{(6 - 2)180°}{6} = \frac{720°}{6} = 120°$$

Concerning economy of design, if each interior angle is 120°, at the junction of three hexagons, the sum of all three angles is $3(120°) = 360°$, the number of degrees around a circle.

> **THEOREM 3.17**
>
> The sum of the measures of the exterior angles of a polygon, one at each vertex, is 360°.

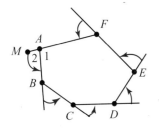

Figure 3.24

PROOF: Consider the polygon in Figure 3.24. Through each vertex we have a straight angle each of which is formed by one interior angle and one exterior angle. For example, straight angle $\angle FAM$ at vertex A is formed by interior angle $\angle 1$ and exterior angle $\angle 2$ because $m\angle FAM = m\angle 1 + m\angle 2$. The total measure of these n straight angles is $(n)180°$, and the sum of the n interior angles of the polygon is $(n-2)180°$. If we let T be the sum of the n exterior angles, then

$$T + (n-2)180° = (n)180°.$$

$T + (n)(180°) - (2)(180°) = (n)180°$	Distributive law
$T - (2)(180°) = 0$	Subtract $(n)180°$ from both sides
$T - 360° = 0$	$(2)(180°) = 360°$
$T = 360°$	Add 360° to both sides

Thus, the sum of the exterior angles is 360°.

> **COROLLARY 3.18**
>
> The measure of each exterior angle of a regular polygon with n sides is determined with the formula $e = \dfrac{360°}{n}$.

The proof of Corollary 3.18 is left for you to do as an exercise.

EXAMPLE 5 What is the measure of each exterior angle of a regular hexagon? Because a hexagon has 6 sides, substitute 6 for n in $e = \dfrac{360°}{n}$.

$$e = \frac{360°}{6} = 60°$$

> **PRACTICE EXERCISE 2**
>
> Find the measure of each exterior angle in a 30-sided regular polygon.
>
> **ANSWER BELOW**

Answers to Practice Exercises

1. 10   2. 12°

3.3 Exercises

FOR EXTRA HELP: Student's Solutions Manual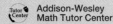

In Exercises 1–8, answer each of the following questions for a regular polygon with the given number of sides.

(a) What is the name of the polygon?
(b) What is the sum of the angles of the polygon?
(c) What is the measure of each angle of the polygon?
(d) What is the sum of the measures of the exterior angles of the polygon?
(e) What is the measure of each exterior angle of the polygon?
(f) If each side is 5 cm long, what is the perimeter of the polygon?

1. 3	**2.** 4	**3.** 5	**4.** 6
5. 7	**6.** 8	**7.** 9	**8.** 10

In Exercises 9–12, solve the equation $S = (n - 2)\,180°$ for n when S is a given value. Find the number of sides of each polygon (if possible) if the given value corresponds to the number of degrees in the sum of the interior angles of a polygon. Remember that n must be a whole number greater than 2, or no such polygon can exist.

9. 1620°   **10.** 2700°   **11.** 2000°   **12.** 3200°

In Exercises 13–16, solve the equation $a = \dfrac{(n - 2)\,180°}{n}$ for n when a is a given value. Find the number of sides of each polygon (if possible) if the given value is the measure of one interior angle of a regular polygon.

13. 157.5°   **14.** 162°   **15.** 145°   **16.** 105°

17. Two exterior angles of a triangle sum to 200°. What is the measure of the third exterior angle?

18. Three exterior angles of a quadrilateral sum to 300°. What is the measure of the fourth exterior angle?

19. As the number of sides of a regular polygon increases, does each exterior angle increase or decrease?

20. As the number of sides of a regular polygon increases, does an interior angle increase or decrease?

21. What is the smallest angle that any regular polygon can have?

22. What is the largest exterior angle that any regular polygon can have?

23. Find the number of sides of a polygon if the sum of its angles is twice the sum of its exterior angles.

24. If the number of sides of a polygon were doubled, the sum of the angles of the polygon would be increased by 900°. How many sides does the original polygon have?

25. If the sum of the angles of a polygon is equal to the sum of the exterior angles of the polygon, how many sides does the polygon have?

26. By how many degrees is the sum of the angles of a polygon increased when the number of sides is increased by 4?

27. Given an isosceles triangle where the base measures 3 inches longer than the other two sides and the perimeter of the triangle is 24 inches, find the length of all three sides.

In Exercise 28a and 28b, find the measures of the numbered angles. Justify each answer.

28. (a)

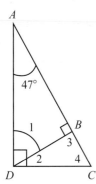

(b)

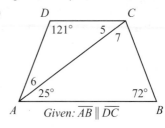

Given: $\overline{AB} \parallel \overline{DC}$

29. In the given figure, find the value of x then find the measure of each angle. In a paragraph, explain if this is a regular figure.

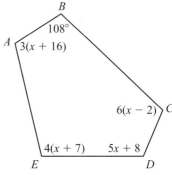

Exercise 29

30. Prove that the exterior angles of a regular polygon are equal.

31. Prove Corollary 3.18.

32. Prove that every point on the bisector of an angle is equidistant from the sides of the angle.

Use the figure below in Exercises 33 and 34.

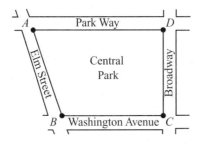

33. A city planner wishes to locate the point in Central Park that is equidistant from Elm Street, Washington Avenue, and Park Way. Explain how she can determine this point. [Hint: Use Exercise 32.]

34. Explain how a city planner can find the point in Central Park that is equidistant from a library at point A, a monument at point B, and a fountain at point C.

35. Copy this table on your paper, fill in the information and then look for a pattern. What conjecture can you make about the number of diagonals, one from each vertex, in any polygon?

POLYGON	DRAWING	NUMBER OF SIDES	NUMBER OF DIAGONALS FROM ONE VERTEX
Triangle		3	
Quadrilateral		4	
Pentagon			
Hexagon			
Octagon			
Decagon			
Dodecagon			
n-gon			

36. Using the given figure, *ABCDE* is a regular pentagon. Find $m\angle F$ and write an explanation of your answer.

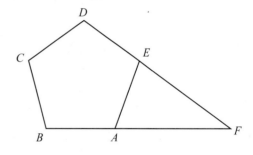

Exercise 36

3.4 More Congruent Triangles

OBJECTIVES

1. Prove the AAS theorem.
2. Prove the HA theorem.
3. Prove the HL theorem.

In Chapter 2, several methods were presented to prove triangles congruent. They included SAS, ASA, SSS, LA, and LL. The last two are only used to prove right triangles congruent. In this section, we will prove three more theorems concerning the congruency of triangles.

OBJECTIVE 1   **Prove the AAS theorem.**   The next theorem is another way to prove two triangles congruent.

> **THEOREM 3.19   AAS (ANGLE-ANGLE-SIDE)**
> If two angles and any side of one triangle are congruent to the corresponding two angles and side of another triangle, then the triangles are congruent.

Notice this is different from ASA. In ASA, the side must be included between the two congruent angles. For AAS, the side is *not* included between the two angles.

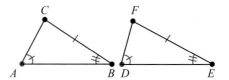

Given: $\triangle ABC$ and $\triangle DEF$
$\angle A \cong \angle D$ and
$\angle B \cong \angle E$
$CB \cong FE$
(See Figure 3.25.)

Prove: $\triangle ABC \cong \triangle DEF$

Figure 3.25

Proof _____

Statements	*Reasons*
1. $\angle A \cong \angle D$ and $\angle B \cong \angle E$	1. Given
2. $\angle C \cong \angle F$	2. If 2 $\angle$'s of one $\triangle \cong$ 2 $\angle$'s of 2nd $\triangle$, the 3rd $\angle$'s are $\cong$
3. $\overline{CB} \cong \overline{FE}$	3. Given
4. $\triangle ABC \cong \triangle DEF$	4. ASA

PRACTICE EXERCISE 1

Complete the following proof.
Given: $\overline{AB} \cong \overline{EB}$; $\overline{AC} \parallel \overline{ED}$
(See Figure 3.26.)

Prove: $\triangle ABC \cong \triangle EBD$

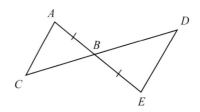

Figure 3.26

Proof _____

Statements	*Reasons*
1. $\overline{AB} \cong \overline{EB}$; $\overline{AC} \parallel \overline{ED}$	1. _____
2. $\angle C \cong \angle D$	2. _____
3. _____	3. _____
4. $\triangle ABC \cong \triangle EBD$	4. _____

ANSWERS ON PAGE 161

OBJECTIVE 2   **Prove the HA theorem.**   The next theorem is a special case of Theorem 3.19, AAS.

> **THEOREM 3.20   HA (HYPOTENUSE ANGLE)**
>
> If the hypotenuse and an acute angle of one right triangle are congruent to the hypotenuse and an acute angle of another triangle, then the two right triangles are congruent.

Given: Right triangles △*ABC*, △*DEF*
with ∠*A*, ∠*D* right angles;
$\overline{BC} \cong \overline{EF}$; ∠*C* ≅ ∠*F*
(See Figure 3.27.)

Prove: △*ABC* ≅ △*DEF*

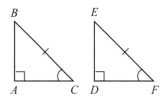

Figure 3.27

Proof _____

Statements	**Reasons**
1. Right triangles △*ABC*, △*DEF*	1. Given
2. ∠*A* ≅ ∠*D*	2. All right angles are congruent
3. $\overline{BC} \cong \overline{EF}$; ∠*C* ≅ ∠*F*	3. Given
4. △*ABC* ≅ △*DEF*	4. AAS

OBJECTIVE 3   **Prove the HL theorem.**   You may have noticed that we had no theorem of the form SSA for arbitrary triangles. For example, the triangles in Figure 3.28 have this property but they are clearly not congruent. The next theorem, however, provides the counterpart for right triangles giving us a totally new congruence theorem.

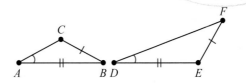

Figure 3.28

> **THEOREM 3.21**   **HL (HYPOTENUSE LEG)**
>
> If the hypotenuse and a leg of one right triangle are congruent to the hypotenuse and a leg of another right triangle, then the two right triangles are congruent.

Given: Right triangles △*ABC* and
△*DEF* with ∠*C* and
∠*F* right angles, $\overline{AB} \cong \overline{DE}$,
and $\overline{AC} \cong \overline{DF}$ (See Figure 3.29.)

Prove: △*ABC* ≅ △*DEF*

Auxiliary lines: Extend $\overleftrightarrow{FE}$, construct $\overline{FG}$
such that $\overline{FG} \cong \overline{BC}$, and
construct $\overline{DG}$

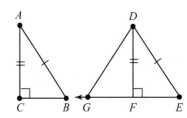

Figure 3.29

Proof _____

Statements	Reasons
1. $\triangle ABC$ and $\triangle DEF$ are right triangles with $\angle C$ and $\angle F$ right angles	1. Given
2. $\overline{AB} \cong \overline{DE}$ and $\overline{AC} \cong \overline{DF}$	2. Given
3. $\angle DFG$ is a right angle making $\triangle DFG$ a right triangle	3. Supp. of rt. $\angle$ is also a rt. $\angle$
4. $\overline{BC} \cong \overline{FG}$	4. By construction
5. $\triangle ABC \cong \triangle DGF$	5. LL
6. $\overline{AB} \cong \overline{DG}$	6. CPCTC
7. $\overline{DE} \cong \overline{DG}$	7. Sym. and trans. laws (statements 2 and 6)
8. $\angle DGF \cong \angle DEF$	8. $\angle$'s opp $\cong$ sides of $\triangle$ are $\cong$
9. $\triangle DGF \cong \triangle DEF$	9. HA
10. $\triangle ABC \cong \triangle DEF$	10. Transitive law for $\cong$ (Thm 2.1)

EXAMPLE 1   Refer to Figure 3.30.

Given: Isosceles triangle $\triangle ABC$ with base $\overline{BC}$ and $\overline{AD} \perp \overline{BC}$

Prove: $\overline{BD} \cong \overline{DC}$

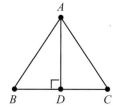

Figure 3.30

Proof _____

Statements	Reasons
1. $\triangle ABC$ is isosceles with base $\overline{BC}$	1. Given
2. $\overline{AB} \cong \overline{AC}$	2. Def. of isos. $\triangle$
3. $\overline{AD} \perp \overline{BC}$	3. Given
4. $\angle ADB$ and $\angle ADC$ are right angles	4. $\perp$ lines form rt. $\angle$'s (def. $\perp$ lines)
5. $\triangle ADB$ and $\triangle ADC$ are right triangles	5. Def. of rt. $\triangle$
6. $\overline{AD} \cong \overline{AD}$	6. Reflexive law
7. $\triangle ADB \cong \triangle ADC$	7. HL
8. $\overline{BD} \cong \overline{DC}$	8. CPCTC

> ### SUMMARY: METHODS TO PROVE TRIANGLES CONGRUENT
>
> SAS, ASA, SSS, AAS
>
> Only for right triangles: LA, LL, HA, HL

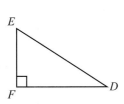 **Student Activity**

For this activity, you will need a protractor and a ruler.
1. What is the measure of each angle in an equiangular triangle? Using your protractor, make an angle with this measure.
2. Since an equiangular triangle is also equilateral, using the angle in step 1, make an equilateral triangle with sides measuring 2 inches.
3. Using your protractor and ruler, make a second equilateral triangle where the sides measure 3 inches.
4. Are the triangles congruent?
5. There are SSS, ASA, SAS, and AAS patterns of congruence of triangles. Will there be an AAA pattern of congruence? Show another example that does not involve equilateral triangles.

Answers to Practice Exercises

1. Given   **2.** If lines are parallel, alternate interior angles $\cong$.   **3.** $\angle ABC \cong \angle EBD$, vertical angles are $\cong$.   **4.** AAS

3.4 Exercises

FOR EXTRA HELP: 📖 Student's Solutions Manual   Tutor Center   Addison-Wesley Math Tutor Center

Exercises 1–6 refer to right $\triangle ABC$ and right $\triangle DEF$. State why $\triangle ABC \cong \triangle DEF$ under the given conditions.

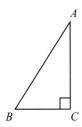

1. $BC = 5$ cm, $EF = 5$ cm, $m\angle A = 35°$, $m\angle D = 35°$
2. $AB = 15$ ft, $ED = 15$ ft, $m\angle B = 40°$, $m\angle E = 40°$
3. $BC = 7$ yd, $EF = 7$ yd, $AC = 10$ yd, $DF = 10$ yd
4. $AB = 12$ cm, $BC = 5$ cm, $ED = 12$ cm, $EF = 5$ cm
5. $ED = 4.3$ yd, $FE = 3.1$ yd, $AB = 4.3$ yd, $BC = 3.1$ yd
6. $AC = 9.2$ ft, $m\angle A = 32°$, $FD = 9.2$ ft, $m\angle D = 32°$

In Exercises 7–10, state the congruence theorem and other reasons needed to justify why these triangles are congruent.

7.
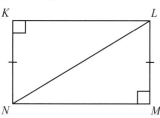

8.
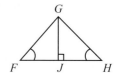

9.
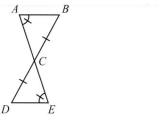

10.
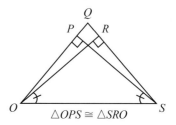

$\triangle OPS \cong \triangle SRO$

11. Complete the following flowchart proof.
Given: Isosceles $\triangle ABC$ with $\overline{BD} \perp \overline{AC}$

Prove: $\triangle ADB \cong \triangle CDB$

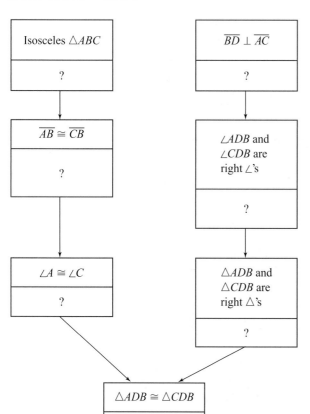

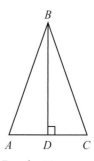
Exercise 11

12. *Given:* ∠*BDA* ≅ ∠*BDC*; ∠*A* ≅ ∠*C*

    Prove: △*ADB* ≅ △*CDB*

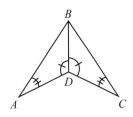

Exercise 12

13. *Given:* $\overline{AB} \perp \overline{CD}$ at *E*, $\overline{AE} \cong \overline{BE}$; $\overline{AC} \cong \overline{BD}$

    Prove: △*AEC* ≅ △*BED*

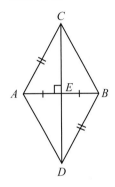

Exercise 13

14. *Given:* Isosceles △*QRS* and ∠*RUT* ≅ ∠*RTU*

    Prove: △*QRT* ≅ △*SRU*

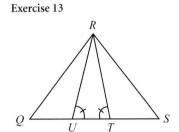

Exercise 14

15. *Given:* ∠*V* and ∠*Y* are right angles; $\overline{WX} \cong \overline{ZX}$

    Prove: △*YXZ* ≅ △*VXW*

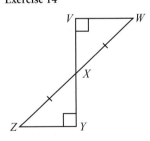

Exercise 15

16. *Given:* Right angles ∠*B* and ∠*A* with ∠1 ≅ ∠2

    Prove: △*ACD* ≅ △*BCD*

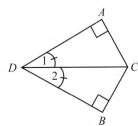

Exercise 16

17. *Given:* Isosceles $\triangle VWX$ with base $\overline{VX}$,
$\overline{WY}$ is perpendicular bisector of $\overline{VX}$

  Prove: $\triangle VWY \cong \triangle XWY$

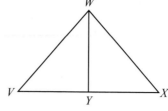

Exercise 17

18. *Given:* $\overline{AD} \parallel \overline{BC}$; $\angle A \cong \angle C$

  Prove: $\triangle ABD \cong \triangle CDB$

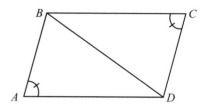

Exercise 18

19. *Given:* $\angle ACB$ and $\angle ACD$ are
right angles with $\overline{AB} \cong \overline{AD}$

  Prove: $\overline{AC}$ bisects $\angle BAD$

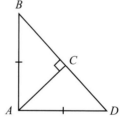

Exercise 19

20. A classmate is confused about three of the congruence theorems. She says the letters all look alike. In a paragraph, explain to the classmate the difference between AAS, ASA, and SAS.

21. There will *not* be a SSA pattern of congruence of triangles. Investigate this with geometric software or by drawing figures. Show an example of two *noncongruent* triangles with two pairs of congruent sides and one pair of congruent *nonincluded* angles.

Chapter ③ Review

Key Terms and Symbols

3.1 negation ($\sim$)
indirect proof
parallel lines ($\parallel$)

3.2 transversal
alternate interior angles
corresponding angles
alternate exterior angles

3.3 polygon
sides

vertex
convex polygon
concave polygon
regular polygon
triangle
quadrilateral
pentagon
hexagon
heptagon
octagon

nonagon
decagon
n-gon
diagonal
perimeter

3.4 AAS
HA
HL

Chapter 3 Proof Techniques

To Prove:
Two Lines Parallel

1. Show they are both perpendicular to a third line. (Theorem 3.1)

2. Show a pair of alternate interior angles are congruent when the lines are cut by a transversal. (Theorem 3.2)

3. Show a pair of corresponding angles are congruent when the lines are cut by a transversal. (Theorem 3.3)

4. Show a pair of alternate exterior angles are congruent when the lines are cut by a transversal. (Theorem 3.4)

5. Show two interior angles on the same side of a transversal are supplementary. (Theorem 3.5)

Two Lines Perpendicular

1. Show one is parallel to a third line that is perpendicular to the second. (Theorem 3.1)

Two Angles Congruent

1. Show they are alternate interior angles formed when parallel lines are cut by a transversal. (Theorem 3.6)

2. Show they are corresponding angles formed when parallel lines are cut by a transversal. (Theorem 3.8)

3. Show they are alternate exterior angles formed when parallel lines are cut by a transversal. (Theorem 3.9)

Two Angles Supplementary

1. Show they are interior angles on the same side of a transversal that cuts parallel lines. (Theorem 3.10)

Two Triangles Congruent

1. (AAS) Show two angles and any side of one are congruent, respectively, to two angles and the corresponding side of the other. (Theorem 3.19)

2. (HA) In a right triangle, show the hypotenuse and an acute angle of one triangle are congruent respectively to the hypotenuse and an acute angle of the other triangle. (Theorem 3.20)

3. (HL) In a right triangle show the hypotenuse and a leg of one triangle are congruent respectively to the hypotenuse and leg of the other triangle. (Theorem 3.21)

SUMMARY: METHODS TO PROVE TRIANGLES CONGRUENT

SAS, ASA, SSS, AAS

Only for right triangles: LA, LL, HA, HL

Review Exercises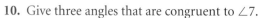

Section 3.1

1. Discuss the major differences between giving a direct proof and an indirect proof of $P \longrightarrow Q$.

2. Give an indirect proof of the following "theorem."
   Premise 1: If I don't have clean shirts, then I can't go to work.
   Premise 2: If I can't work, then I won't have any money.
   Premise 3: If I don't buy my wife a present, then she'll be unhappy.
   Premise 4: If my wife is unhappy, then she won't wash my shirts.
   Theorem: If I have the money, then I'll buy my wife a present.

3. State the parallel postulate.

4. If $\ell \perp m$, $\ell \perp n$, and m and n are distinct lines, is $m \parallel n$? Explain.

5. If $\ell \parallel m$, ℓ, m, and n are distinct lines, P is a point on m, and n is a line through P, is $n \parallel \ell$? Explain.

Section 3.2

Exercises 6–12 refer to the figure to the right in which $\ell \parallel m$.
In Exercises 6–9, explain the reasoning.

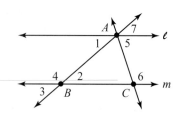

6. Is $\angle 1 \cong \angle 2$?

7. Is $\angle 5 \cong \angle 1$?

8. Is $\angle 1 \cong \angle 3$?

9. Is $\angle 5$ supplementary to $\angle 6$?

10. Give three angles that are congruent to $\angle 7$.

11. If $m\angle 1 = (x + 20)°$ and $m\angle 2 = (3x - 40)°$, find x.

12. If $m\angle 4 = (y + 30)°$ and $m\angle 7 = (2y - 90)°$, find y.

13. *Given:* $\ell \parallel m$ and $\angle 1 \cong \angle 2$
    Prove: $\triangle ABC$ is isosceles

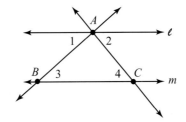

Exercise 13

14. Use the figure to the right and find the value of x and y that make $\ell \parallel m$.

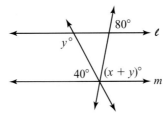

Exercise 14

Section 3.3

15. What is the sum of the angles of a hexagon?

16. What is the measure of each angle of a regular hexagon?

17. What is the sum of the measures of the exterior angles of a hexagon?

18. What is the measure of each exterior angle of a regular hexagon?

19. Give the number of sides of a regular polygon if each interior angle measures $156°$.

20. Give the number of sides of a polygon if the sum of its interior angles is $3600°$.

21. If three exterior angles of a quadrilateral sum to $325°$, what is the measure of the fourth exterior angle?

22. Is an interior angle of a polygon always greater than its corresponding exterior angle? Explain.

23. Can a triangle have more than one right angle? Explain.

24. *Given:* $\overline{AE} \perp \overline{ED}, \overline{DB} \perp \overline{AC}, m\angle C = 30°$
    Find: the measures of the numbered angles.

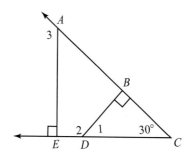

Exercise 24

25. *Given:* ∠C is a right angle,
    $m\angle 1 = 2x + 1, m\angle 2 = 3(x - 2)$
    Find: x and the exact measures of the
    three numbered angles.

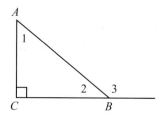

Exercise 25

Section 3.4

For Exercises 26–28 determine what additional "given" information is missing to prove the triangles congruent using only the theorem stated.

26. *Given:* $\overline{BC} \parallel \overline{AD}$
    Prove: △ADC ≅ △CBA by AAS

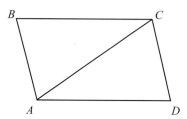

Exercise 26

27. *Given:* $\overline{BD}$ is the perpendicular bisector of $\overline{AC}$
    Prove: △ADB ≅ △CDB by HL

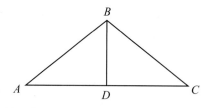

Exercise 27

28. *Given:* $\overline{AB} \perp \overline{BC}, \overline{CD} \perp \overline{BC}$
    Prove: △ABC ≅ △DCB by LL

29. Use the figure from Exercise 28.
    Given: $\overline{AB} \perp \overline{BC}, \overline{CD} \perp \overline{BC}, \overline{DB} \cong \overline{AC}$
    Prove: △ABC ≅ △DCB

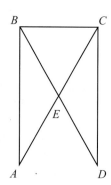

Exercises 28–29

1. If P is a point not on line ℓ, how many lines through P are parallel to ℓ? Explain.

For Exercises 2–5 refer to the figure to the right. Answer **true** *or* **false** *assuming* $\ell \parallel m$ *and* $m \parallel n$. *Explain each answer.*

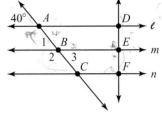

2. $m\angle 1 = 40°.$

3. $m\angle 2 = 120°.$

4. $m\angle 3 = 40°.$

5. $\angle DEB$ is supplementary to $\angle EFC.$

Exercises 2–5

6. What is the sum of the measures of the angles of an octagon?

7. What is the sum of the measures of the exterior angles of an octagon?

8. What is the measure of each interior angle of a regular octagon?

9. If $a \parallel b$, $m\angle 1 = (6x + 5)°$ and $m\angle 2 = (9x - 16)°$ find the degree measure of $\angle 1$.

10. In the given figure where $a \parallel b$, find the measure of $\angle 3$ if it is four times the measure of $\angle 1$.

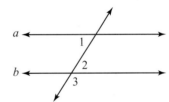

Exercises 9 and 10

11. *Given:* $\overline{AB} \cong \overline{CB}$, $\angle A \cong \angle C$, $\angle B$ is a right angle

   Prove: $\triangle ABE \cong \triangle CBD$

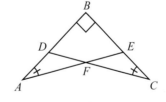

Exercise 11

12. Given an isosceles triangle where the base measures 12 centimeters and the perimeter of the triangle is 46 centimeters, find the length of the other two sides of the triangle.

13. Find the number of sides of a regular polygon if the measure of one interior angle is $162°$.

14. *Given:* $\angle B \cong \angle D$, $\overline{AB} \parallel \overline{CD}$

   Prove: $\triangle ABC \cong \triangle CDA$

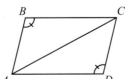

15. Given a right triangle, what is the sum of the two acute angles? Explain your answer.

Exercise 14

Chapters 1-3   Cumulative Review

Chapter 1

For Exercises 1–14, use the given figure to answer the questions as either true or false.

1. $\angle 1$ and $\angle 2$ are adjacent angles.

2. If $m\angle 1 = 60°$ then $m\angle 2 = 60°$.

3. $\overleftrightarrow{WY} \perp \overleftrightarrow{ZV}$.

4. W is on $\overrightarrow{XY}$.

5. $\angle ZXY$ is a right angle.

6. $\angle UVY$ is supplementary to $\angle YVX$.

7. W is on $\overleftrightarrow{XY}$.

8. V is an endpoint of $\overleftrightarrow{VY}$.

9. $\angle 1$ is an acute angle.

10. W is on $\overline{XY}$.

11. $m\angle 1 = m\angle 3$.

12. $\angle 2$ is an obtuse angle.

13. $\angle UVY$ is a straight angle.

14. $\angle WXV$ and $\angle VXY$ are complementary.

15. Find the complement of $27°$.

16. With compass and straightedge, draw $\overline{XY}$ and construct its perpendicular bisector.

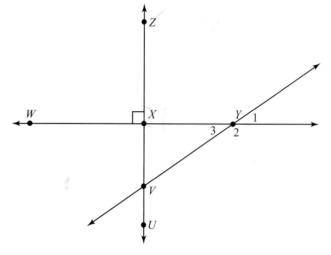

Exercises 1–14

Chapter 2

In Exercises 17–21, use the given figure where all sides are unequal and all angles are unequal.

17. Classify the triangle according to its sides.

18. Classify the triangle according to its angles.

19. What angle is opposite $\overline{BC}$?

20. Name an exterior angle of $\triangle ABC$.

21. What side is included by $\angle A$ and $\angle B$?

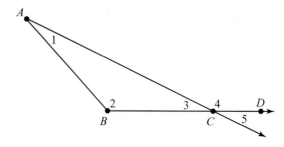

Exercises 17–21

22. What is the perimeter of an equilateral triangle with one side measuring 7.5 inches?

23. *Given:* $\overline{AB} \cong \overline{BC}; \overline{DE} \cong \overline{EC}$

    Prove: $\angle A \cong \angle D$

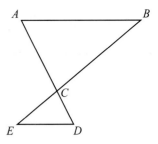

Exercise 23

24. State which segment in the figure is an altitude, which is a perpendicular bisector, and which is a median.

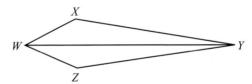

Exercise 24

25. *Given:* $\overline{WY}$ bisects $\angle XYZ$, $\overline{YW}$ bisects $\angle XWZ$

    Prove: $\triangle WXY \cong \triangle WZY$

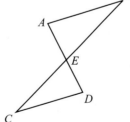

Exercise 25

26. *Given:* $\overline{AD}$ and $\overline{BC}$ bisect each other

    Prove: $\overline{AB} \cong \overline{DC}$

Exercise 26

27. If point X is the centroid of the $\triangle ABC$ in the figure, find XE if $AE = 162$ mm.

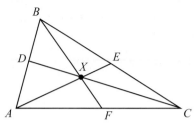

Exercise 27

28. Using definitions and theorems learned in Chapters 1 and 2, deduce the measures of the lettered angles.

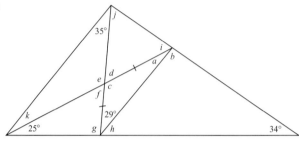

Exercise 28

Chapter 3

29. Find the following angle measures using the given figure where $a \parallel b$, $m\angle 3 = (6x + 5)°$, $m\angle 6 = (9x - 16)°$

Find $m\angle 1, m\angle 2, m\angle 4, m\angle 5, m\angle 7, m\angle 8$

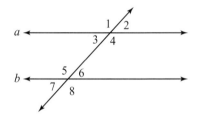

Exercise 29

30. What is the sum of the interior angles of an octagon?

31. What is the sum of the measure of the exterior angles, one at each vertex, of an octagon?

32. *Given:* $\overline{AB} \perp \overline{BE}, \overline{DE} \perp \overline{BE}, \angle A \cong \angle D$
    Prove: $\triangle ABE \cong \triangle DEB$

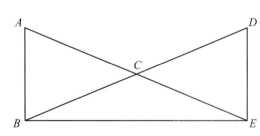

Exercise 32

In Exercises 33–40, use the information given in the diagram to decide if the two triangles are congruent. If they are congruent, complete the congruence statement and state the reason. If the triangles cannot be shown to be congruent based on the information given, state not enough information given.

33.

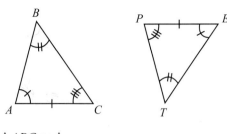

$\triangle ABC \cong \triangle$ _____

Reason _____

34.

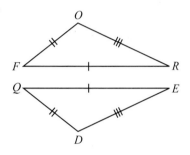

$\triangle FOR \cong \triangle$ _____

Reason _____

35.

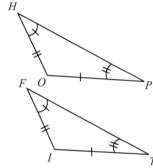

$\triangle HOP \cong \triangle$ _____

Reason _____

36.

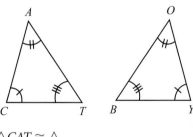

$\triangle CAT \cong \triangle$ _____

Reason _____

37.

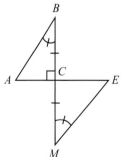

$\triangle ABC \cong \triangle$ _____

Reason _____

38.

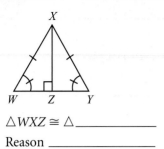

$\triangle WXZ \cong \triangle$ _____

Reason _____

39.

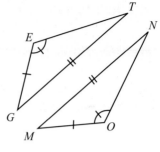

$\triangle GET \cong \triangle$ _____

Reason _____

40.

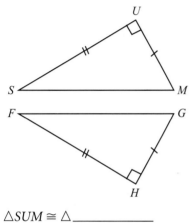

$\triangle SUM \cong \triangle$ _____

Reason _____

Quadrilaterals

*I*n this chapter, we will study the properties of several special types of quadrilaterals including the parallelogram, rhombus, kite, rectangle, square, and trapezoid. We will discover that geometric figures involving parallel line segments have numerous properties and applications in such diverse areas as navigation, physics, geology, construction, and architecture. One of these applications, presented below, is solved later in the chapter in Example 1 of Section 4.1.

AN APPLICATION

The wind is blowing due north at 50 mph. The pilot of an airplane wants to fly due west. Because the wind will take the airplane off course toward the north, the pilot must set a course to the south of due west to maintain a true westerly resultant direction. If the velocity of the airplane is 500 mph, how can a parallelogram be used to assist the pilot in determining the correct course?

4.1   Parallelograms

OBJECTIVES

1. Define a parallelogram.
2. State the properties of a parallelogram.
3. State theorems to prove a quadrilateral is a parallelogram.

In Chapter 3, we investigated some polygons. Recall a polygon is a closed figure in a plane. If it has four sides it is called a **quadrilateral**. See Figure 4.1 for an example. Quadrilateral *ABCD* has four vertices, points *A, B, C,* and *D. Consecutive sides* are two sides of a quadrilateral that intersect at the endpoint, such as $\overline{AB}$ and $\overline{BC}$. Pairs of sides of a quadrilateral that do not intersect are called *opposite sides*, such as $\overline{AB}$ and $\overline{DC}$. Similarly, angles of a quadrilateral are consecutive if they are next to each other like $\angle A$ and $\angle B$. *Opposite angles* are across from one another like $\angle A$ and $\angle C$ and do not share a common side.

OBJECTIVE 1   **Define a parallelogram.**   The first quadrilateral to be investigated is the *parallelogram.*

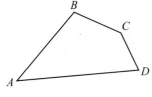

Figure 4.1

> **DEFINITION: Parallelogram**
> A **parallelogram** is a quadrilateral whose opposite sides are parallel.

The symbol $\square$ represents the word *parallelogram*. Figure 4.2 shows $\square ABCD$ with $\overline{AB} \parallel \overline{CD}$ and $\overline{AD} \parallel \overline{BC}$.

OBJECTIVE 2   State the properties of a parallelogram.

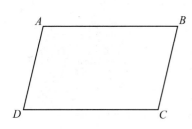

Figure 4.2 $\square ABCD$

> **THEOREM 4.1**
> Each diagonal divides a parallelogram into two congruent triangles.

Given: $\square ABCD$ with diagonal $\overline{BD}$
(See Figure 4.3.)

Prove: $\triangle ABD \cong \triangle CDB$

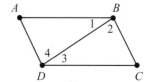

Figure 4.3

Proof _____

Statements	**Reasons**
1. $\square ABCD$ with diagonal $\overline{BD}$	1. Given
2. $\overline{AB} \parallel \overline{DC}$ and $\overline{AD} \parallel \overline{BC}$	2. Def. of $\square$
3. $\angle 1 \cong \angle 3$ and $\angle 2 \cong \angle 4$	3. If $\parallel$, alt. int. $\angle$'s are $\cong$
4. $\overline{BD} \cong \overline{BD}$	4. Reflexive law
5. $\triangle ABD \cong \triangle CDB$	5. ASA

In a similar manner, we could show that diagonal $\overline{AC}$ divides the parallelogram into congruent triangles $\triangle ACD$ and $\triangle CAB$.

COROLLARY 4.2

The opposite sides and opposite angles of a parallelogram are congruent.

Given: $\square ABCD$ (See Figure 4.4.)

Prove: $\overline{AB} \cong \overline{CD}$, $\angle BAD \cong \angle DCB$,
$\overline{AD} \cong \overline{CB}$, $\angle ABC \cong \angle CDA$

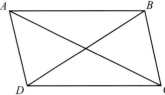

Figure 4.4

Proof

Statements	Reasons
1. $\square ABCD$	1. Given
2. $\triangle ABD \cong \triangle CDB$	2. Diagonal of $\square$ forms $2 \cong \triangle$'s
3. $\overline{AB} \cong \overline{CD}$, $\angle BAD \cong \angle DCB$	3. CPCTC
4. $\triangle ACD \cong \triangle CAB$	4. Diagonal of $\square$ forms $2 \cong \triangle$'s
5. $\overline{AD} \cong \overline{CB}$, $\angle ABC \cong \angle CDA$	5. CPCTC

The next theorem proves another property of parallelograms.

THEOREM 4.3

Consecutive angles of a parallelogram are supplementary. *sum to 180°*

consecutively doing in

Given: $\square ABCD$ with consecutive angles $\angle A$ and $\angle B$ (See Figure 4.5.)

Prove: $\angle A$ and $\angle B$ are supplementary

$\angle 1 \mathscr{e} \angle 2 = 180°$ $\angle 4$ and $\angle 1 = 180°$
$\angle 2 \mathscr{e} \angle 3 = 180$
$\angle 3 \mathscr{e} \angle 4 = 180$

Figure 4.5

Proof

Statements	Reasons
1. $\square ABCD$ with consecutive angles A and B	1. Given
2. $\overline{AD} \parallel \overline{BC}$	2. Opposite sides of $\square$ are $\parallel$
3. AB is a transversal between parallel lines $\overline{AD}$ and $\overline{BC}$	3. Definition of transversal
4. $\angle A$ and $\angle B$ are supplementary	4. Interior angles on the same side of transversal are supplementary

In a similar way, it can be shown $\angle A$ is supplementary to $\angle D$, $\angle B$ is supplementary to $\angle C$, and $\angle C$ is supplementary to $\angle D$.

The next theorem is another property of parallelograms.

THEOREM 4.4

The diagonals of a parallelogram bisect each other.

Given: $\square ABCD$ with diagonals $\overline{AC}$ and $\overline{BD}$
(See Figure 4.6.)

Prove: $\overline{AC}$ and $\overline{BD}$ bisect each other

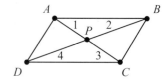

Figure 4.6

Proof _____

Statements	Reasons
1. $\overline{AC}$ and $\overline{BD}$ are diagonals of $\square ABCD$	1. Given
2. $\angle 1 \cong \angle 3$ and $\angle 2 \cong \angle 4$	2. If $\parallel$, alt. int. $\angle$'s are $\cong$
3. $\overline{AB} \cong \overline{CD}$	3. Opp. sides of $\square$ are $\cong$
4. $\triangle APB \cong \triangle CPD$	4. ASA
5. $\overline{AP} \cong \overline{CP}$ and $\overline{PB} \cong \overline{PD}$	5. CPCTC
6. $\overline{AC}$ and $\overline{BD}$ bisect each other	6. Def. of seg. bisector

---

SUMMARY: PROPERTIES OF A PARALLELOGRAM

1. Opposite sides parallel. Def ◇
2. Diagonal divides it into two $\cong$ $\triangle$'s 4.1
3. Opposite sides $\cong$ 4.2
4. Opposite angles $\cong$ 4.2
5. Consecutive angles are supplementary 4.2
6. Diagonals bisect each other 4.4

OBJECTIVE 3   **State theorems to prove a quadrilateral is a parallelogram.** Now we will look at ways to prove a quadrilateral is a parallelogram. The first method is to use the definition of a parallelogram and prove opposite sides are parallel. The next way is with Theorem 4.5.

THEOREM 4.5

If both pairs of opposite sides of a quadrilateral are congruent, then the quadrilateral is a parallelogram.

PRACTICE EXERCISE 1

Complete the proof of Theorem 4.5.

Given: $\overline{AB} \cong \overline{CD}$ and $\overline{AD} \cong \overline{CB}$

Prove: ABCD is a $\square$

Construction: Construct diagonal $\overline{BD}$

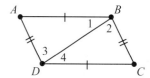

Proof _____

Statements	Reasons
1. $\overline{AB} \cong \overline{CD}$ and $\overline{AD} \cong \overline{CB}$	1. _given_
2. $\overline{BD} \cong \overline{BD}$	2. Reflexive law
3. $\triangle ABD \cong \triangle CDB$	3. SSS
4. $\angle 1 \cong \angle 4$	4. _CPCTC_
5. $\overline{AB} \parallel \overline{CD}$	5. _If alt. int. ∠'s are ≅, then ∥._
6. $\angle 2 \cong \angle 3$	6. CPCTC
7. $\overline{BC} \parallel \overline{AD}$	7. If alt. int. ∠'s ≅, then lines ∥.
8. ABCD is a $\square$	8. _Def of ⊠_

ANSWERS ON PAGE 182

THEOREM 4.6

If both pairs of opposite angles of a quadrilateral are congruent, then the quadrilateral is a parallelogram.

The proof of Theorem 4.6 is requested in the exercises.

THEOREM 4.7

If two opposite sides of a quadrilateral are congruent and parallel, then the quadrilateral is a parallelogram.

The proof of Theorem 4.7 is requested in the exercises.

The converse of Theorem 4.4 is also true and gives us another way to prove that a quadrilateral is a parallelogram.

THEOREM 4.8

If the diagonals of a quadrilateral bisect each other, then the quadrilateral is a parallelogram.

PRACTICE EXERCISE 2

Complete the proof of Theorem 4.8.

Given: $\overline{AC}$ and $\overline{BD}$ bisect each other

Prove: $ABCD$ is a $\square$

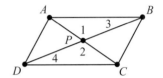

Proof ───

Statements	**Reasons**
1. $\overline{AC}$ and $\overline{BD}$ bisect each other	1. ___given___
2. $\overline{AP} \cong \overline{CP}$ and $\overline{PB} \cong \overline{PD}$	2. ___Def. of Bisector___
3. ___$\angle 1 \cong \angle 2$___	3. Vert. $\angle$'s are $\cong$
4. ___$\triangle ADB \cong \triangle CPD$___	4. SAS
5. $\angle 3 \cong \angle 4$	5. ___CPCTC___
6. $\overline{AB} \parallel \overline{DC}$	6. ___If alt. int $\angle$'s $\cong$ then $\parallel$.___
7. $\overline{AB} \cong \overline{CD}$	7. ___CPCTC___
8. ___ABCD is parallelogram___	8. If two opp. sides of quadrilateral are $\cong$ and $\parallel$, then quadrilateral is $\square$.

ANSWERS ON PAGE 182

SUMMARY: METHODS TO PROVE A QUADRILATERAL IS A PARALLELOGRAM

1. Show both pairs opposite sides parallel
2. Show both pairs of opposite sides $\cong$
3. Show both pairs of opposite angles $\cong$
4. Show one pair of opposite sides $\cong$ and parallel
5. Show diagonals bisect each other

Parallelograms have many applications. For example, in physics, when a quantity has both magnitude and direction in a plane, a **vector** is used to describe the quantity. A vector is represented by a line segment, the length of which corresponds to its magnitude. An arrowhead on the segment shows its direction. For example, the velocity of an airplane flying at 400 mph due east can be represented by the vector in Figure 4.7, in which each inch of length corresponds to 100 mph, and the arrow points east in the direction of travel. If the wind is blowing in a northeasterly direction at 50 mph, the vector corresponding to this velocity (of length $\frac{1}{2}$ inch) could be placed with the above vector as shown in Figure 4.8. When the two velocities are combined, the **resultant vector** is found with the **parallelogram law**—by using the diagonal of the parallelogram with the given sides. From the figure we can see that the airplane will fly in the direction of the resultant, and its speed will be a little more than 400 mph, about 425 mph, because the length of the vector is about 4.25 inches.

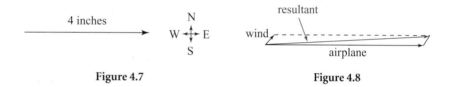

Figure 4.7   **Figure 4.8**

Using this information, we can solve the applied problem given in the chapter introduction.

EXAMPLE 1

The wind is blowing due north at 50 mph. The pilot of an airplane wants to fly due west. Because the wind will take the airplane off course toward the north, the pilot must set a course somewhat to the south of due west to maintain a true westerly resultant direction. If the velocity of the plane is 500 mph, how can a parallelogram be used to assist the pilot in determining the correct course?

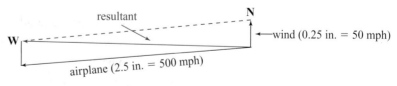

Figure 4.9

According to the parallelogram law, the pilot must set his course in the direction shown in Figure 4.9 so that the resultant of the two forces will be due west. Notice in this case that the actual speed of the plane will be diminished somewhat from the rate of 500 mph because the length of the resultant is a little less than 2.5 inches.

▌ *Answers to Practice Exercises*

1. 1. Given 2. $\overline{BD} \cong \overline{BD}$ 3. $\triangle ABD \cong \triangle CDB$ 4. CPCTC 5. If alt. int. $\angle$'s are $\cong$, then $\parallel$. 6. $\angle 2 \cong \angle 3$ 7. $\overline{BC} \parallel \overline{AD}$ 8. Def. of $\square$ **2.** 1. Given 2. Def. of seg. bisector 3. $\angle 1 \cong \angle 2$ 4. $\triangle APB \cong \triangle CPD$ 5. CPCTC 6. If alt. int. $\angle$'s $\cong$ then $\parallel$. 7. CPCTC 8. *ABCD* is a parallelogram

---

4.1   **Exercises**

FOR EXTRA HELP: 📖   Student's Solutions Manual   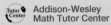 Addison-Wesley Math Tutor Center

In Exercises 1–20, refer to the figure below. Answer *true* or *false*.

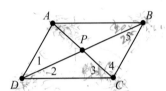

1. If *ABCD* is a parallelogram, then $\triangle ABC \cong \triangle CDA$.
2. If *ABCD* is a parallelogram, then $\triangle ABD \cong \triangle CDB$.
3. If *ABCD* is a parallelogram, then $\overline{AB} \cong \overline{CD}$.
4. If *ABCD* is a parallelogram, then $\overline{CD} \cong \overline{DB}$.
5. If *ABCD* is a parallelogram, then $\angle ACD \cong \angle BAC$.
6. If *ABCD* is a parallelogram, then $\angle CAB \cong \angle ACD$.
7. If $\overline{AB} \cong \overline{CD}$, then *ABCD* is a parallelogram.
8. If $\overline{AB} \cong \overline{CD}$ and $\overline{AD} \cong \overline{BC}$, then *ABCD* is a parallelogram.
9. If $\angle CAB \cong \angle BDC$ and $\angle ACD \cong \angle ABD$, then *ABCD* is a parallelogram.
10. If $\angle ACD \cong \angle ABD$, then *ABCD* is a parallelogram.
11. If *ABCD* is a parallelogram, then $\angle BAD$ and $\angle ADC$ are supplementary.
12. If *ABCD* is a parallelogram, then $\angle CAB$ and $\angle CDB$ are supplementary.
13. If $\overline{AD} \cong \overline{CB}$ and $\overline{AD} \parallel \overline{CB}$, then *ABCD* is a parallelogram.
14. If $\overline{AD} \cong \overline{CB}$ and $\overline{AB} \parallel \overline{CD}$, then *ABCD* is a parallelogram.
15. If *P* is the midpoint of $\overline{AC}$ and $\overline{BD}$, then *ABCD* is a parallelogram.
16. If *ABCD* is a parallelogram, then $\angle 1 \cong \angle 2$.
17. If *ABCD* is a parallelogram and $m\angle ACD = 55°$, then $m\angle ABD = 55°$.
18. If *ABCD* is a parallelogram and $m\angle BAC = 125°$, then $m\angle ACD = 125°$.
19. If $m\angle 1 = 25°$ and $m\angle 2 = 20°$ in $\square ABCD$, then $m\angle 3 + m\angle 4 = 135°$.
20. If $m\angle 1 = 25°$ in $\square ABCD$, then $m\angle 5 = 25°$.

21. Complete the proof of Theorem 4.6.
Given: quadrilateral *ABCD* with ∠*A* ≅ ∠*C* and ∠*B* ≅ ∠*D*.
Prove: *ABCD* is a parallelogram

Proof _____

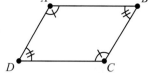

Exercise 21

Statements	Reasons
1. $m\angle A + m\angle B + m\angle C + m\angle D = 360°$	**1.** _____
2. ∠*A* ≅ ∠*C* and ∠*B* ≅ ∠*D* thus, $m\angle A = m\angle C$ and $m\angle B = m\angle D$	**2.** _____
3. $m\angle A + m\angle B + m\angle A + m\angle B = 360°$	**3.** _____
4. $2m\angle A + 2m\angle B = 360°$	**4.** Distributive law
5. $m\angle A + m\angle B = 180°$	**5.** _____
6. _____	**6.** Def. of supp. ∠'s
7. _____	**7.** If int. ∠'s same side of transv. are supp, then lines ∥.
8. $m\angle A + m\angle D = 180°$	**8.** Substitution
9. ∠*A* and ∠*D* are supplementary	**9.** _____
10. _____	**10.** If int. ∠'s same side of transv. are supp, then lines ∥.
11. _____	**11.** Def. of ▱

22. Complete the proof of Theorem 4.7.
Given: *ABCD* with $\overline{AB} \cong \overline{CD}$ and $\overline{AB} \parallel \overline{CD}$
Prove: *ABCD* is a parallelogram

Proof _____

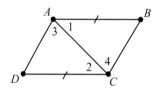

Exercise 22

Statements	Reasons
1. Construct diagonal $\overline{AC}$	**1.** A line is determined by two points
2. $\overline{AB} \cong \overline{CD}$	**2.** _____
3. _____	**3.** Given
4. _____	**4.** If ∥, alt. int. ∠'s are ≅
5. $\overline{AC} \cong \overline{AC}$	**5.** _____
6. _____	**6.** SAS
7. _____	**7.** CPCTC
8. $\overline{AD} \parallel \overline{CB}$	**8.** _____
9. _____	**9.** Def. of ▱

23. Do you suppose there could be a congruency theorem similar to SSS but for parallelograms (SSSS)? Explain your reasoning.

For Problems 24–26, use quadrilateral *VWXZ*.

24. *Given:* $\overline{VZ} \cong \overline{YW}$, $\angle 2 \cong \angle 3$
    Prove: *VWYZ* is a parallelogram
25. *Given:* $\overline{VZ} \parallel \overline{WY}$, $\overline{VZ} \cong \overline{WX}$, $\overline{WY} \cong \overline{WX}$
    Prove: *VWYZ* is a parallelogram
26. *Given:* $\angle 1$ is supplementary to $\angle 2$, $\overline{VW} \cong \overline{ZY}$
    Prove: *VWYZ* is a parallelogram
27. If one angle of a parallelogram is twice the measure of another angle, what is the measure of each angle?
28. If one angle of a parallelogram is $(2x + 20)°$ and a consecutive angle is $(x - 50)°$, find the value of x.

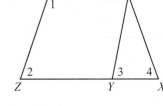

Exercises 24–26

29. *Given:* $\square ABCD$
    $\overline{AP} \cong \overline{QC}$
    Prove: *PBQD* is a parallelogram

30. *Given:* $\square ABCD$
    $\overline{AE} \cong \overline{CG}$ and $\overline{BF} \cong \overline{DH}$
    Prove: *EFGH* is a parallelogram

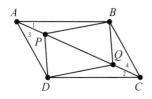

Exercise 29

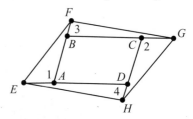

Exercise 30

Student Activity

31. The following equipment will be needed for this activity: cardboard tube, scissors, ruler, and protractor. Examine a cardboard tube from a paper product such as paper towels, wrapping paper, or toilet paper. If the tube was cut on the spiral glue line you see around it, make a conjecture about the shape of the cardboard when it lies flat. Now take scissors and actually cut the tube on the spiral line to confirm your conjecture. Describe the shape. Measure its sides with a ruler and its angles with a protractor. Compare your measurements with another group. Explain the differences and similarities in measurements. What shape is it?

32. An airplane is flying a course due north at a rate of 400 mph, and the wind is blowing from the west at 75 mph. Make a sketch that illustrates this information and use the parallelogram law to find the resultant of these two velocities. Measure the length of the resultant and estimate the ground speed of the airplane.

33. Two children are dragging a heavy crate by pulling ropes attached to the crate. One is pulling due east with a force of 40 lb, and the other is pulling due south with a force of 60 lb. Use the parallelogram law to find the resultant of these two forces. Measure the length of the resultant to estimate the force required to move the crate in the same manner in the direction of the resultant.

4.2 Rhombus and Kite

OBJECTIVES

1. Define a rhombus.
2. Properties of a rhombus.
3. Prove a quadrilateral is a rhombus.
4. Define a kite.
5. Properties of a kite.

OBJECTIVE 1   Define a rhombus.   A *rhombus* is a special kind of parallelogram.

> **DEFINITION: Rhombus**
> A **rhombus** is a parallelogram that has two congruent adjacent sides.

OBJECTIVE 2   Properties of a rhombus.   Because a rhombus is a parallelogram, it has all the properties of a parallelogram listed in the previous section. See Figure 4.10. A rhombus has:

1. Opposite sides parallel
2. Diagonal divides it into two $\cong$ $\triangle$'s
3. Opposite sides $\cong$
4. Opposite angles $\cong$
5. Consecutive angles are supplementary
6. Diagonals bisect each other

In addition to these, the rhombus has several other properties.

> **THEOREM 4.9**
> All four sides of a rhombus are congruent.

The proof of Theorem 4.9 follows directly from Corollary 4.2. Using a paragraph style proof, Figure 4.10 is given as a rhombus. To prove all four sides are congruent, begin with $\overline{AD} \cong \overline{AB}$ from the definition of a rhombus. Since opposite sides of a parallelogram are congruent $\overline{AD} \cong \overline{BC}$ and $\overline{AB} \cong \overline{DC}$. By the transitive law, $\overline{AB} \cong \overline{DC} \cong \overline{AD} \cong \overline{BC}$. Hence, all four sides are congruent.

A rhombus is an equilateral parallelogram because all the sides are congruent.

> **THEOREM 4.10**
> The diagonals of a rhombus are perpendicular.

Given: $\square ABCD$ is a rhombus (See Figure 4.11.)
Prove: $\overline{AC} \perp \overline{BD}$

Figure 4.10

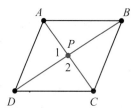

Figure 4.11

Proof _____

Statements	Reasons
1. $\square ABCD$ is a rhombus	1. Given
2. $\overline{AD} \cong \overline{CD}$	2. The sides of a rhombus are $\cong$
3. $\overline{AP} \cong \overline{CP}$	3. The diag. of $\square$ bisect each other
4. $\overline{DP} \cong \overline{DP}$	4. Reflexive law
5. $\triangle APD \cong \triangle CPD$	5. SSS
6. $\angle 1 \cong \angle 2$	6. CPCTC
7. $\angle 1$ and $\angle 2$ are adjacent angles	7. Def. adj. $\angle$'s
8. $\overline{AC} \perp \overline{DB}$	8. Def. of $\perp$ lines

OBJECTIVE 3 **Prove a quadrilateral is a rhombus.** The converse of Theorem 4.10 is also true, and it provides us with a way to prove that a parallelogram is a rhombus.

THEOREM 4.11

If the diagonals of a parallelogram are perpendicular, then the parallelogram is a rhombus.

PRACTICE EXERCISE 1

Complete the proof of Theorem 4.11.

Given: $\square ABCD$
$\overline{AC} \perp \overline{BD}$

Prove: $\square ABCD$ is a rhombus

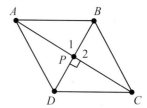

Proof _____

Statements	Reasons
1. $\overline{AC} \perp \overline{BD}$	1. Given
2. $\angle 1 \cong \angle 2$	2. Def $\perp$ lines
3. $\overline{BP} \cong \overline{BP}$	3. Reflexive law
4. $ABCD$ is a parallelogram	4. given
5. $\overline{AP} \cong \overline{CP}$	5. Diag. of $\square$ bisect each other
6. $\triangle ABP \cong \triangle CBP$	6. SAS
7. $\overline{AB} \cong \overline{CB}$	7. CPCTC
8. $\square ABCD$ is a rhombus	8. Def. of rhombus

ANSWERS ON PAGE 190

> **THEOREM 4.12**
> The diagonals of a rhombus bisect the angles of the rhombus.

PRACTICE EXERCISE 2

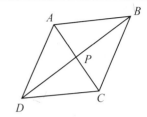

Complete the proof of Theorem 4.12.

Given: ▱$ABCD$ is a rhombus with diagonals $\overline{AC}$ and $\overline{BD}$

Prove: $\overline{AC}$ bisects $\angle BAD$ and $\angle DCB$
$\overline{BD}$ bisects $\angle ABC$ and $\angle ADC$

Proof _____

Statements	Reasons
1. ▱$ABCD$ is a rhombus with diagonals $\overline{AC}$ and $\overline{BD}$	1. Given
2. $\overline{AD} \cong \overline{AB}$	2. Def. of Rhombus
3. $\overline{AP} \cong \overline{AP}$	3. Reflexive law
4. $\overline{DP} \cong \overline{BP}$	4. Diag. of parallelogram bisect each other
5. $\triangle APD \cong \triangle APB$	5. SSS
6. $\angle DAP \cong \angle BAP$	6. CPCTC
7. $\overline{AC}$ bisects $\angle BAD$	7. Def. angle bisector

The other three parts of the prove statement are proved similarly.   **ANSWERS ON PAGE 190**

In addition to the six properties of a rhombus given earlier in this section, there are three more properties:

7. All sides are $\cong$
8. Diagonals are $\perp$
9. Diagonals bisect angles of the rhombus

Student Activity

For this activity, work in groups of three. You will need a worksheet from your instructor, a ruler, and a protractor.

(a) One member of the group draws a rectangle with diagonals on regular dot paper. Another member of the group draws a square with diagonals on regular dot paper. The third member of the group draws a rhombus with diagonals on isometric dot paper. All group members drawing a rectangle or square should label their figure just like the one pictured. Those drawing a rhombus, copy the labels on the worksheet.

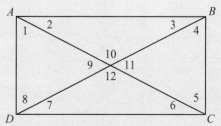

continued

(b) Each person measures $\overline{AC}$ and $\overline{BD}$ in their figure. Record the measurement on the table.

(c) With a protractor, each person measure angles 1–12 on their figure and record the measurements on the worksheet table.

	AC	BD	m∠1	m∠2	m∠3	m∠4	m∠5	m∠6	m∠7	m∠8	m∠9	m∠10	m∠11	m∠12
Rectangle														
Square														
Rhombus														

(d) All group members share data and complete the table.

(e) By looking at the measures of angles 9–12, for which figures are the diagonals perpendicular?

(f) By looking at the measures of angles 1–8, for which figures do the diagonals bisect a pair of opposite angles?

(g) As a prelude to the next section, for which figures are the diagonals congruent?

Figure 4.12

OBJECTIVE 4 Define a kite.   The next quadrilateral is a *kite*.

> **DEFINITION: Kite**
> A **kite** is a quadrilateral with exactly two distinct pairs of congruent consecutive sides.

OBJECTIVE 5 Properties of a kite.   The kite in Figure 4.12 looks just like the child's toy. Do you see two isosceles triangles in the figure? Notice the kite is *not* a parallelogram. It has some interesting properties.

> **THEOREM 4.13**
> If a quadrilateral is a kite, one pair of opposite angles is congruent.

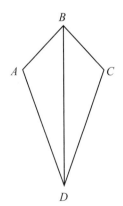

Figure 4.13

Given: Kite *ABCD* with diagonal $\overline{BD}$ (See Figure 4.13.)

Prove: $\angle A \cong \angle C$

Proof _____

Statements	**Reasons**
1. Kite *ABCD* with diagonal $\overline{BD}$	1. Given
2. $\overline{AD} \cong \overline{CD}, \overline{AB} \cong \overline{CB}$	2. Def. kite
3. $\overline{BD} \cong \overline{BD}$	3. Reflexive law
4. $\triangle ABD \cong \triangle CBD$	4. SSS
5. $\angle A \cong \angle C$	5. CPCTC

> **THEOREM 4.14**
> If a quadrilateral is a kite, one diagonal is the perpendicular bisector of the other diagonal.

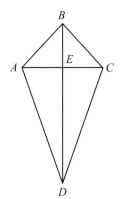

Figure 4.14

Given: Kite *ABCD* with diagonals $\overline{AC}$ and $\overline{BD}$ (See Figure 4.14.)
Prove: $\overline{BD}$ is the $\perp$ bisector of $\overline{AC}$

Proof _____

Statements	Reasons
1. Kite *ABCD* with diagonals $\overline{AC}$ and $\overline{BD}$	1. Given
2. $\overline{AB} \cong \overline{CB}$; $\overline{AD} \cong \overline{CD}$	2. Def. kite
3. $\overline{BD} \cong \overline{BD}$	3. Reflexive law
4. $\triangle ABD \cong \triangle CBD$	4. SSS
5. $\angle ABD \cong \angle CBD$	5. CPCTC
6. $\overline{BE} \cong \overline{BE}$	6. Reflexive law
7. $\triangle ABE \cong \triangle CBE$	7. SAS
8. $\overline{AE} \cong \overline{CE}$	8. CPCTC
9. $\angle BEA \cong \angle BEC$	9. CPCTC
10. $\overline{BD} \perp \overline{AC}$	10. Def. $\perp$ lines
11. $\overline{BD} \perp$ bisector of $\overline{AC}$	11. Def. segment bisector and statements 8 and 10

PRACTICE EXERCISE 3

Complete the proof of the statement: One diagonal of a kite bisects two of its angles.

Given: Kite *ABCD* with diagonal $\overline{BD}$
Prove: $\overline{BD}$ bisects $\angle ABC$ and $\angle ADC$

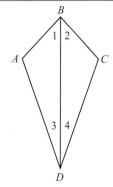

Proof _____

Statements	Reasons
1. Kite *ABCD* with diagonal $\overline{BD}$	1. _____
2. $\overline{AD} \cong \overline{CD}, \overline{AB} \cong \overline{CB}$	2. _____
3. _____	3. _____
4. _____	4. SSS
5. _____	5. CPCTC
6. $\overline{BD}$ bisects $\angle ABC$ and $\angle ADC$	6. _____

ANSWERS ON PAGE 190

SUMMARY: PROPERTIES OF A KITE

1. It's a quadrilateral but *not* a parallelogram with two pair of ≅, consecutive sides.
2. *One* pair of opposite angles ≅
3. *One* diagonal is ⊥ bisector of the other diagonal.
4. *One* diagonal bisects two of the kite's angles.

Answers to Practice Exercises

1. 1. $\overline{AC} \perp \overline{BD}$ 2. Def. ⊥ lines 3. Reflexive law 4. Given 5. $\overline{AP} \cong \overline{CP}$ 6. $\triangle ABP \cong \triangle CBP$ 7. CPCTC 8. $\square ABCD$ is a rhombus   **2.** 2. Def. rhombus 3. Reflexive law 4. Diagonals of parallelogram bisect each other 5. SSS 6. CPCTC 7. Def. angle bisector   **3.** 1. Given 2. Def. kite 3. $\overline{BD} \cong \overline{BD}$, reflexive law 4. $\triangle ABD \cong \triangle CBD$ 5. $\angle 1 \cong \angle 2, \angle 3 \cong \angle 4$ 6. Def. angle bisector

---

4.2 Exercises

FOR EXTRA HELP: 📖 Student's Solutions Manual   ☎ Tutor Center Addison-Wesley Math Tutor Center

1. Refer to the figure of a rhombus.
   a. What kind of triangle is $\triangle ABC$? Why?
   b. What kind of triangle is $\triangle ADE$? Why?
   c. Is $\triangle ABE \cong \triangle ADE$? Why or why not?

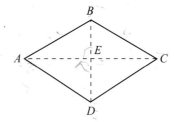

Exercise 1

2. Refer to the figure of a kite.
   a. What kind of triangle is $\triangle ABC$? Why?
   b. What kind of triangle is $\triangle ADE$? Why?
   c. Is $\triangle ABE \cong \triangle ADE$? Why or why not?

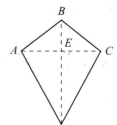

Exercise 2

3. Describe the similarities and differences between a rhombus and a kite.

4. If one side of a rhombus measures 14 centimeters, find the perimeter of the rhombus.

5. If a rhombus measures 12.6 inches on one side, find the perimeter of the rhombus.

6. Find the perimeter of the kite in the figure.

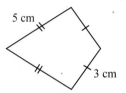

5 cm

3 cm

Exercise 6

7. Find the perimeter of the kite in the figure.

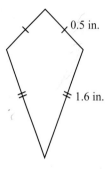

0.5 in.

1.6 in.

Exercise 7

8. Find the measures of ∠1 and ∠2 in the kite below.

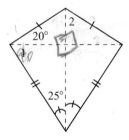

20°

2

25°

Exercise 8

9. Find the measures of ∠1 and ∠2 in the rhombus below.

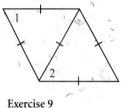

1

2

Exercise 9

10. Find the measures of ∠1 and ∠2 in the kite below.

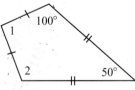

100°

1

2   50°

Exercise 10

11. Find the measures of ∠1 and ∠2 in the rhombus.

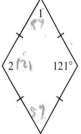

1

2   121°

Exercise 11

12. *Given:* $\square ABCD$ *is a rhombus* $\overline{BE} \perp \overline{AD}$ *and* $\overline{DF} \perp \overline{BC}$

Prove: $\overline{BE} \cong \overline{DF}$

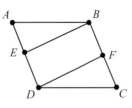

Exercise 12

13. *Given:* $\square ABCD$ *is a rhombus*

Prove: $\angle 1$ and $\angle 2$ are complementary

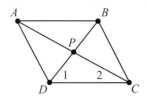

Exercise 13

14. Use the properties of a kite to construct a kite given sides $\overline{WX}$ and $\overline{XY}$ and diagonal $\overline{WY}$.

(**4.3**) **Rectangles and Squares**

OBJECTIVES

1. Define a rectangle.
2. Properties of a rectangle.
3. Define a square.
4. Properties of a square.
5. Discuss the construction of a rectangle.
6. Prove the theorem about the segment joining midpoints of a triangle.

OBJECTIVE 1   Define a rectangle.   A *rectangle* is a special type of parallelogram that has many useful applications.

> **DEFINITION: Rectangle**
> A **rectangle** is a parallelogram with one right angle.

The symbol $\square$ represents the word *rectangle*.

OBJECTIVE 2   Properties of a rectangle.   Because a rectangle is also a parallelogram, all of the properties of parallelograms are also properties of rectangles. For example, the opposite sides and opposite angles of a rectangle are congruent.

> **THEOREM 4.15**
> All angles of a rectangle are right angles.

Given: *ABCD* is a rectangle where ∠*A* is a right angle (See Figure 4.15.)

Prove: ∠*B*, ∠*C*, ∠*D* are right angles

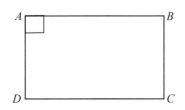

Figure 4.15

Proof

Statements	Reasons
1. *ABCD* is a rectangle where ∠*A* is a right angle	1. Given
2. ∠*A* ≅ ∠*C*; ∠*B* ≅ ∠*D*	2. Opposite angles of ▱ are ≅
3. *m*∠*A* = *m*∠*C*; *m*∠*B* = *m*∠*D*	3. Def. ≅ ∠'s
4. *m*∠*A* = 90°	4. Def. right ∠
5. *m*∠*C* = 90°	5. Substitution
6. ∠*A* and ∠*B* are supplementary	6. Consecutive angles of ▱ are supplementary
7. *m*∠*A* + *m*∠*B* = 180°	7. Def. supplementary ∠'s
8. 90° + *m*∠*B* = 180°	8. Substitution
9. *m*∠*B* = 90°	9. Subtraction
10. *m*∠*D* = 90°	10. Substitution
11. ∠*B*, ∠*C*, ∠*D* are rt. ∠'s	11. Def. rt. ∠

Thus all the angles of the rectangle are right angles.

THEOREM 4.16

The diagonals of a rectangle are congruent.

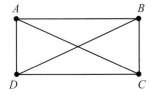

Figure 4.16

Given: ▱*ABCD* (See Figure 4.16.)

Prove: $\overline{AC} \cong \overline{BD}$

Proof

Statements	Reasons
1. ▱*ABCD*	1. Given
2. $\overline{AD} \cong \overline{BC}$	2. Opp. sides of a ▱ are ≅
3. $\overline{DC} \cong \overline{DC}$	3. Reflexive law
4. ∠*ADC* and ∠*BCD* are right angles	4. All ∠'s of ▱ are rt. ∠'s
5. ∠*ADC* ≅ ∠*BCD*	5. All rt. ∠'s are ≅
6. △*ADC* ≅ △*BCD*	6. SAS
7. $\overline{AC} \cong \overline{BD}$	7. CPCTC

Theorem 4.17 has many practical applications. To show that a quadrilateral is a rectangle, you must do more than show that the opposite sides are equal (the figure might simply be a parallelogram). Also, it might not be possible to show that the figure has a right angle, but it might be possible to measure the diagonals. For example, when a construction worker is framing the wall of a post-and-beam house, he might measure the diagonals of the wall framing to be sure the wall is rectangular and not simply a parallelogram with equal opposite sides.

The converse of Theorem 4.16 is also true and provides us with another method in addition to the definition for proving that a parallelogram is a rectangle.

> **THEOREM 4.17**
> If the diagonals of a parallelogram are congruent, then the parallelogram is a rectangle.

This theorem was investigated in the dot paper Student Activity on pages 187–188. The proof of this theorem will be done in the exercises.

OBJECTIVE 3 Define a square. A *square* is another special type of parallelogram.

> **DEFINITION: Square**
> A **square** is a rhombus with one right angle.

The symbol □ represents the word *square*. Because a rhombus is a parallelogram, a square is a parallelogram with a right angle, making it a rectangle.

OBJECTIVE 4 Properties of a square. Because all sides of a rhombus are equal, a square is a rectangle with all sides equal. Thus, a square has all the properties of both a rhombus and a rectangle. For example, the diagonals of a square are perpendicular, they bisect each other, and they bisect the angles of the square. Notice that every square is a rhombus, but not every rhombus is a square. Similarly, every square is a rectangle, but not every rectangle is a square.

PRACTICE EXERCISE 1

Classify the following statements as true or false.
1. Every square is a parallelogram.
2. Every rhombus is a square.
3. Every rectangle is equiangular.
4. Every square is a rectangle.
5. The diagonals of every parallelogram are congruent.

ANSWERS ON PAGE 197

Because rectangles and squares are perhaps the most important quadrilaterals, we should be able to construct them.

OBJECTIVE 5   Discuss the construction of a rectangle.

> **CONSTRUCTION 4.1**
> Construct a rectangle when two adjacent sides are given.

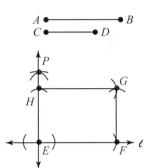

Figure 4.17

Given: Segments $\overline{AB}$ and $\overline{CD}$ (See Figure 4.17.)

To Construct: $\square EFGH$ with $\overline{EF} \cong \overline{AB}$ and $\overline{EH} \cong \overline{CD}$

Construction _____

1. Draw line ℓ, identify point E on ℓ, use Construction 1.4 to construct $\overleftrightarrow{EP} \perp \ell$.
2. Use Construction 1.1 to copy $\overline{AB}$ onto line ℓ obtaining $\overline{EF}$, and to copy $\overline{CD}$ onto $\overleftrightarrow{EP}$ obtaining $\overline{EH}$.
3. Set the compass for length AB, place the tip at H, and make an arc. Now set the compass for length CD, place the tip at F, and make an arc that intersects the first arc in point G.
4. Use a straightedge to draw segments $\overline{HG}$ and $\overline{FG}$. Then $EFGH$ is the desired rectangle because it is a parallelogram ($\overline{EH} \cong \overline{FG}$ and $\overline{EF} \cong \overline{HG}$) with right angle $\angle HEF$.

Constructing a square with a given side consists of the same steps as in Construction 4.1 except that $\overline{AB} \cong \overline{CD}$. Construction 4.2 is left for you to do as an exercise.

> **CONSTRUCTION 4.2**
> Construct a square when a side is given.

Recall from Section 1.6 that the distance from a point to a line is the length of a perpendicular segment drawn from the point to the line. We use this fact to prove the following theorem.

> **THEOREM 4.18**
> Two parallel lines are always the same distance apart.

Given: $\ell \parallel m$ with A and B arbitrary points on ℓ, and C and D on m such that $\overrightarrow{AC} \perp m$ and $\overline{BD} \perp m$
AC is the distance from A to m
BD is the distance from B to m
(See Figure 4.18.)

Prove: $AC = BD$ (to ℓ and m are always the same distance apart)

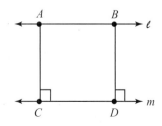

Figure 4.18

Proof _____

Statements	Reasons
1. $\ell \parallel m$, $\overline{AC} \perp m$, and $\overline{BD} \perp m$	1. Given
2. $\overline{AC} \parallel \overline{BD}$	2. Lines $\perp$ to third line are $\parallel$
3. $ABCD$ is a parallelogram	3. Opp. sides are $\parallel$
4. $\overline{AC} \cong \overline{BD}$	4. Opp. sides of $\square$ are $\cong$
5. $AC = BD$	5. Def. $\cong$ segments, thus all points on ℓ are the same distance from m because A and B were arbitrary points on ℓ.

OBJECTIVE 6   **Prove the theorem about the segment joining midpoints of a triangle.**

Theorems about parallelograms will be used to prove a theorem about triangles.

> **THEOREM 4.19**
>
> The segment joining the midpoints of two sides of a triangle is parallel to the third side and its length is one-half the length of the third side.

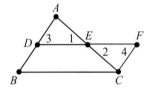

Figure 4.19

Given:	$\triangle ABC$ with D the midpoint of $\overline{AB}$ and E the midpoint of $\overline{AC}$ (See Figure 4.19.)
Prove:	$\overline{DE} \parallel \overline{BC}$ and $DE = \frac{1}{2}BC$
Auxiliary lines:	Extend $\overline{DE}$ to point F so that $\overline{DE} \cong \overline{EF}$ by Construction 1.1 Draw segment $\overline{FC}$

Proof _____

Statements	Reasons
1. D is the midpoint of $\overline{AB}$, and E is the midpoint of $\overline{AC}$ in $\triangle ABC$	1. Given
2. $\overline{AE} \cong \overline{EC}$ and $\overline{AD} \cong \overline{DB}$	2. Def. of midpt.
3. $\angle 1 \cong \angle 2$	3. Vert. $\angle$'s are $\cong$
4. $\overline{DE} \cong \overline{EF}$	4. By construction
5. $\triangle ADE \cong \triangle CFE$	5. SAS
6. $\overline{AD} \cong \overline{CF}$	6. CPCTC
7. $\overline{DB} \cong \overline{CF}$	7. Transitive law
8. $\angle 3 \cong \angle 4$	8. CPCTC
9. $\overline{CF} \parallel \overline{DB}$	9. If alt. int. $\angle$'s are $\cong$, lines $\parallel$
10. $BCFD$ is a parallelogram	10. Opp. sides are $\cong$ and $\parallel$

continued

11. $\overline{DE} \parallel \overline{BC}$	**11.** Opp. sides of $\square$ are $\parallel$
12. $\overline{DF} \cong \overline{BC}$ and $DF = BC$	**12.** Opp. sides of $\square$ are $\cong$; def. $\cong$ segments
13. $DF = DE + EF$	**13.** Seg. add. post.
14. $DF = DE + DE$	**14.** Substitution law
15. $DF = 2DE$	**15.** Distributive law
16. $2DE = BC$	**16.** Substitution law
17. $DE = \frac{1}{2}BC$	**17.** Mult.-div. law

 Technology Connection

Geometry software will be needed.

1. Draw any $\triangle XYZ$.
2. Construct the midpoints of $\overline{XY}$ and $\overline{YZ}$.
3. Label the midpoint of $\overline{XY}$, M, and the midpoint of $\overline{YZ}$, N.
4. Draw $\overline{MN}$ and measure its length.
5. Measure the length of $\overline{XZ}$.
6. Measure $\angle YMN$ and $\angle YXZ$.
7. Measure $\angle YNM$ and $\angle YZX$.
8. Compare these results to Theorem 4.19.

EXAMPLE 1 Refer to $\triangle PQR$ in Figure 4.20. If M is the midpoint of $\overline{PR}$, N is the midpoint of $\overline{PQ}$, and $MN = 5$ cm, find the length of $\overline{QR}$.

By Theorem 4.19,

$$MN = \frac{1}{2}QR, \text{ so } QR = 10 \text{ cm}$$

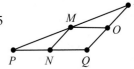

Figure 4.20

PRACTICE EXERCISE 2

Refer to $\triangle PQR$ in Figure 4.20. Assume $\triangle PQR$ is isosceles with base $\overline{PR}$ and $RQ = 30$ inches, and M and O are the midpoints of $\overline{PR}$ and $\overline{QR}$, respectively. Find the length of $\overline{MO}$.

ANSWER BELOW

Answers to Practice Exercises

1. 1. True 2. False 3. True 4. True 5. False
2. 15 inches

4.3   Exercises

For Exercises 1–13, check the box for each correct property of the given quadrilateral.

Quadrilaterals

Properties ↓ / Definitions →	PARALLELOGRAM ▱ Quad. in which both pairs of opposite sides are parallel.	RHOMBUS ◇ Parallelogram with two congruent adjacent sides.	KITE ◇ Quad. with exactly two distinct pairs of ≅ adjacent sides.	RECTANGLE ▭ Parallelogram that has a right angle.	SQUARE ▢ Rhombus with a right angle.
1. Both pairs of opposite sides ∥	yes	yes		yes	yes
2. Diagonals form two ≅ △'s	yes	yes	yes	yes	yes
3. Opposite ∠'s ≅	yes	yes		yes	
4. Opposite sides ≅	yes	yes	yes	yes	
5. Diagonals bisect each other					
6. Consecutive ∠'s supplementary					
7. All ∠'s are right ∠'s				✓	
8. Diagonals are ≅	✓		✓	✓	✓
9. Diagonals are ⊥					
10. All sides ≅	✓	✓		✓	
11. Two ≅ adjacent sides		✓	✓		✓
12. One diagonal ⊥ bisector of other	✓	✓	✓	✓	✓
13. Diagonals bisect ∠'s	✓		✓	✓	

In Exercises 14–17, refer to the square shown.

14. How many right angles are in the figure?
15. How many angles in the figure measure 45°?
16. How many right triangles are in the figure?
17. How many of the triangles in the figure are isosceles triangles?

Exercises 14–17

For Exercises 18–21, refer to rectangle *ABCD* shown where *AB* = 24 inches, *BC* = 10 inches, and *AE* = 13 inches.

18. Find the perimeter of rectangle *ABCD*.
19. Find the perimeter of △*BCD*.
20. Find the perimeter of △*BEC*.
21. Find the perimeter of △*DEC*.
22. Construct a rectangle with sides measuring 3 inches and 2 inches.
23. Construct a square with each side measuring 2 inches.
24. Jeanne wants to enclose a garden with a fence in the shape of a rectangle 15 ft by 20 ft. To be certain she has formed a rectangle, she measures the diagonals and finds they are equal. Does this make the garden rectangular in shape?
25. Galen wishes to find the distance between two points *A* and *B* on opposite sides of a lake. He places a stake at point *C* and determines the midpoints of $\overline{AC}$ and $\overline{BC}$ to be *D* and *E*, respectively (see the figure below). He measures the length of $\overline{DE}$ and finds it to be 56 ft. What is the length of $\overline{AB}$?

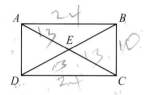

Exercises 18–21

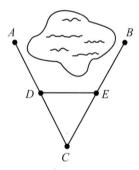

Exercise 25

26. In △*ABC*, *X* and *Y* are midpoints of their respective sides. If *XY* = 12.86 centimeters, how long is *AB*?

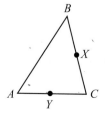

Exercise 26

27. Given △ABC with X, Y, Z midpoints of the respective sides with
AB = 10 cm, BC = 14 cm, and AC = 18 cm find the following:
   a. XY
   b. YZ
   c. XZ
   d. perimeter △ABC
   e. perimeter △XYZ
   f. Do you see a relationship between the two perimeters? Do you think
   this relationship will be true regardless of the lengths of the sides of the
   triangle? Why or why not?

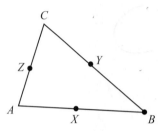

Exercise 27

28. Many people have heard of a family tree. It
is a diagram showing genealogical (family)
relationships. Fill in the "quadrilateral tree"
to the right. The arrows mean the answer
in the oval is a special member of the "family" of the oval it is pointing to. Think carefully and select from the quadrilaterals
studied thus far: rhombus, kite, rectangle,
and square.

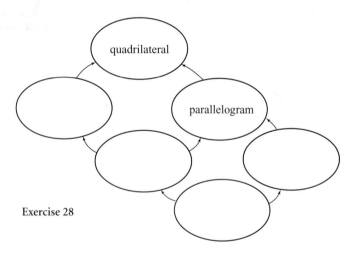

Exercise 28

The statements in Exercises 29–32 are about any quadrilateral ABCD. If
the statement is true explain why; if it is false, give a counterexample and
explanation as needed.

29. If $\overline{AB} \cong \overline{CD}$, then ABCD is a rectangle.
30. If $\overline{AC} \cong \overline{BD}$, then ABCD is a rectangle.
31. If ABCD is a rectangle, then $\angle A \cong \angle B \cong \angle D$.
32. If $\angle A \cong \angle B \cong \angle C$, then ABCD is a square.

33. As you play checkers, you begin to wonder how many squares are actually on the board. Use the figure to help solve the problem. Use a systematic counting method. Think about how many small squares, how many next larger squares, and so on. Don't forget about overlapping squares.

34. Complete the proof of Theorem 4.17.

Given: □ABCD
$\overline{AC} \cong \overline{BD}$

Prove: ABCD is a rectangle

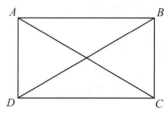

Exercise 34

Proof _____

Statements	Reasons
1. ABCD is a parallelogram	**1.** _____
2. _____	**2.** Opp. sides of □ are ≅
3. $\overline{DC} \cong \overline{DC}$	**3.** _____
4. _____	**4.** Given
5. $\triangle ADC \cong \triangle BCD$	**5.** _____
6. $\angle ADC \cong \angle BCD$, thus, $m\angle ADC = m\angle BCD$	**6.** _____
7. $\angle ADC$ and $\angle BCD$ are supplementary	**7.** _____
8. $m\angle ADC + m\angle BCD = 180°$	**8.** _____
9. $m\angle ADC + m\angle ADC = 180°$ or $2m\angle ADC = 180°$	**9.** Substitution
10. $m\angle ADC = 90°$	**10.** _____
11. ABCD is a rectangle	**11.** _____

35. To solve this quadrilateral puzzle, record the given information on the diagram. Using your knowledge of quadrilaterals and triangles, find the measures of angles 1 through 20. Do not use a protractor or ruler or go by how the figure "looks." A hint in working this type of puzzle is to record the measures of every angle you know even if it is not asked for. It may help you find a missing angle. It is usually best to start with a relationship you see immediately instead of starting at angle 1.

Given: *ABCD* is a parallelogram; *DEFG* is a rectangle;

$$\overline{AD} \cong \overline{DC};\ m\angle DAC = 59°;\ m\angle EBF = 70°$$

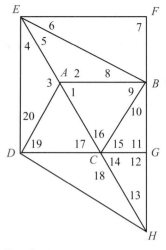

Exercise 35

36. **Technology Connection**

Geometry software will be needed.

1. Construct quadrilateral *WXYZ*. Make the sides nonparallel and of different lengths.
2. Construct the midpoints of the sides.
3. Connect the midpoints and label the new quadrilateral.
4. Measure the lengths of the sides and the angles of the new quadrilateral.
5. What kind of quadrilateral is the new figure? Use the theorems studied to support your answer.
6. Drag any point on the quadrilateral.

Do the results you found in step 5 change? Use the theorems studied to support your answer.

(4.4) Trapezoids

OBJECTIVES

1. Define a trapezoid and an isosceles trapezoid.
2. Properties of trapezoids.
3. Divide segment into *n* equal parts.

To conclude our study of quadrilaterals, we will learn about the *trapezoid*.

OBJECTIVE 1   Define a trapezoid and an isosceles trapezoid.

> **DEFINITION: Trapezoid**
> A **trapezoid** is a quadrilateral with exactly one pair of parallel sides. The parallel sides are called **bases** and the nonparallel sides are called **legs**. If the legs of a trapezoid are congruent, the trapezoid is an **isosceles trapezoid**. A pair of angles of a trapezoid are called **base angles** if they include the same base.

A trapezoid is a figure with properties similar to those of a triangle and a parallelogram. Figure 4.21 shows a trapezoid with bases $\overline{AB}$ and $\overline{CD}$ and sides $\overline{AD}$ and $\overline{BC}$. One pair of base angles is $\angle D$ and $\angle C$, and the other pair is $\angle A$ and $\angle B$. If $\overline{AD}$ and $\overline{BC}$ were congruent, the trapezoid would be isosceles.

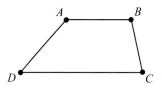

Figure 4.21 Trapezoid

OBJECTIVE 2   **Properties of trapezoids.** The first two properties are about isosceles trapezoids.

> **THEOREM 4.20**
> The base angles of an isosceles trapezoid are congruent.

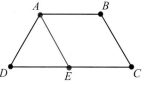

Given:	Isosceles trapezoid $ABCD$ with $\overline{AB} \parallel \overline{DC}$ (See Figure 4.22.)
Prove:	$\angle D \cong \angle C$
Auxiliary lines:	Construct $\overline{AE}$ through A parallel to $\overline{BC}$ using Construction 3.1.

Figure 4.22

Proof

Statements	*Reasons*
1. Isosceles trapezoid $ABCD$ with $\overline{AB} \parallel \overline{DC}$.	1. Given
2. $\overline{AE} \parallel \overline{BC}$	2. By construction
3. $AECB$ is a parallelogram	3. Opp. sides are $\parallel$
4. $\overline{BC} \cong \overline{AE}$	4. Opp. sides of $\square$ are $\cong$
5. $\overline{AD} \cong \overline{BC}$	5. Def. isos. trapezoid
6. $\overline{AD} \cong \overline{AE}$	6. Transitive law
7. $\angle D \cong \angle AED$	7. $\angle$'s opp $\cong$ sides are $\cong$
8. $\angle AED \cong \angle C$	8. Corr. $\angle$'s are $\cong$ if lines $\parallel$
9. $\angle D \cong \angle C$	9. Transitive law

THEOREM 4.21

The diagonals of an isosceles trapezoid are congruent.

The proof of Theorem 4.21 is left for you to do as an exercise. When doing the proof, consider Figure 4.23. This is an isosceles trapezoid so the legs are congruent and the base angles are congruent. Can you show $\triangle ADC \cong \triangle BCD$? How does this help prove the diagonals are congruent?

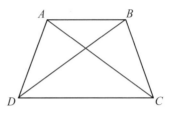

Figure 4.23

DEFINITION: *Median of a Trapezoid*

The segment joining the midpoints of the legs of a trapezoid is the **median of the trapezoid**.

When the string on a guitar is plucked, the vibration creates the sound we hear. Pythagoras knew that the length of the string is related to the pitch of the note. The string in the top figure produces a certain note. If you press the center of the string and pluck it, the note heard is one octave above the first. In general, the most pleasing notes to our ear are formed by dividing the string into an equal number of congruent segments.

In Figure 4.24, $ABCD$ is a trapezoid where E is the midpoint of $\overline{AD}$ and F is the midpoint of $\overline{BC}$. $\overline{EF}$ is the median of trapezoid $ABCD$.

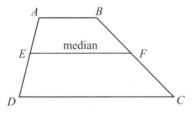

Figure 4.24

THEOREM 4.22

The median of a trapezoid is parallel to the bases and equal to one-half their sum.

The proof of Theorem 4.22 is requested in the exercises.

OBJECTIVE 3 **Divide segment into *n* equal parts.** The next theorem provides a way to divide a given line segment into any number of equal parts.

> **THEOREM 4.23**
> If three or more parallel lines intercept congruent segments on one transversal, then they intercept congruent segments on all transversals.

The proof of Theorem 4.23 is requested in the exercises.

> **CONSTRUCTION 4.3**
> Divide a given segment into a given number of congruent segments.

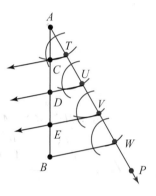

Figure 4.25

A student wants to cut a rectangular piece of plastic into seven congruent rectangles all with length equal to the width of the plastic rectangle. Assume that a measuring device is not available. Explain how the student can discover where to make the cuts by placing the plastic rectangle on a piece of ruled paper. What theorem is being used?

Given: Segment $\overline{AB}$ (See Figure 4.25.)

To Construct: Divide $\overline{AB}$ into *n* congruent parts. For purposes of illustration, let $n = 4$. The technique we will use works for any *n*.

Construction _____

1. Choose point P not on $\overleftrightarrow{AB}$ and draw ray $\overrightarrow{AP}$.
2. Set the compass at any length and mark off equal segments $\overline{AT}, \overline{TU}, \overline{UV},$ and $\overline{VW}$ on $\overrightarrow{AP}$, and draw segment $\overline{BW}$.
3. Use Construction 1.2 to construct $\angle ATC, \angle TUD, \angle UVE$ (all of which are equal in measure to $\angle VWB$). The points $C, D,$ and E divide $\overline{AB}$ into congruent segments $\overline{AC}, \overline{CD}, \overline{DE},$ and $\overline{EB}$ by Theorem 4.23 because $\overline{AB}$ and $\overrightarrow{AP}$ are transversals intercepting parallel lines $\overline{TC}, \overline{UD}, \overline{VE},$ and $\overline{WB}$.

4.4 Exercises

FOR EXTRA HELP: Student's Solutions Manual | Tutor Center | Addison-Wesley Math Tutor Center

Exercises 1–20 refer to the figure below in which $ABCD$ is a trapezoid with bases $\overline{AB}$ and $\overline{CD}$, $\overline{EG} \perp \overline{DC}$ and $\overline{FH} \perp \overline{DC}$, E the midpoint of $\overline{AD}$, and F the midpoint of $\overline{BC}$. Answer *true* or *false*.

1. $\overline{AB} \parallel \overline{DC}$.

2. $\overline{AD} \parallel \overline{BC}$.

3. $\overline{EF} \parallel \overline{AB}$.

4. $\overline{EF} \parallel \overline{DC}$.

5. $\overline{EG} \cong \overline{FH}$.

6. $\overline{AC} \cong \overline{BD}$.

7. $EFHG$ is a parallelogram.

8. $EFHG$ is a rhombus.

9. $EFHG$ is a rectangle.

10. $EFHG$ is a square.

11. $EFHG$ is a trapezoid.

12. $\overline{EH} \cong \overline{FG}$.

13. $\overline{EH}$ and $\overline{FG}$ bisect each other.

14. $\overline{EH} \perp \overline{FG}$.

15. If $AB = 4$ cm and $DC = 6$ cm, then $EF = 5$ cm.

16. If $EF = 10$ inches, then $AB + CD = 30$ inches.

17. If $m\angle ADC = 75°$, then $m\angle BAD = 105°$.

18. If $m\angle ADC = 75°$ and $\overline{AD} \cong \overline{BC}$, then $m\angle BCD = 75°$.

19. If $m\angle BAD = 110°$ and $\overline{AD} \cong \overline{BC}$, then $m\angle ADC = 110°$.

20. If $\overline{AD} \cong \overline{BC}$, then $\overline{AC} \cong \overline{DB}$.

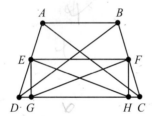

Exercises 1–20

Exercises 21–28 refer to the figure where $k \parallel \ell$, $\ell \parallel m$, $m \parallel n$. Find the value of each of the following:

21. PA

22. AB

23. $m\angle BCF$

24. $m\angle PAD$

25. $m\angle DEB$

26. $m\angle ADE$

27. $m\angle DAB$

28. $m\angle APD$

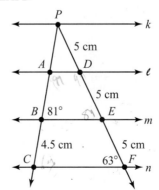

Exercises 21–28

In Exercises 29–32, use the trapezoid shown where $\overline{WX} \parallel \overline{YZ}$ and points A and B are midpoints of their respective legs.

29. If $WX = 4.8$ and $ZY = 13.2$, find AB.
30. If $WX = 3.4$ and $AB = 7.05$, find YZ.
31. If $WX = 2x + 5$ and $ZY = 6x - 1$, find AB (in terms of x).
32. If $WX = 5x + 3$ and $ZY = 13x - 1$ and $AB = 6x + 7$, find x.
33. Complete the proof of Theorem 4.22.

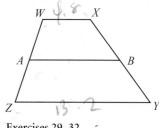

Exercises 29–32

> *Given:* Trapezoid $ABCD$ with $\overline{AB} \parallel \overline{CD}$ and median $\overline{EF}$
>
> *Prove:* $\overline{EF} \parallel \overline{AB}$, $\overline{EF} \parallel \overline{CD}$, and $EF = \dfrac{1}{2}(AB + CD)$
>
> *Auxiliary lines:* Construct $\overleftrightarrow{AF}$. By the parallel postulate $\overleftrightarrow{AF}$ must intersect $\overleftrightarrow{DC}$ at a point, call it G.

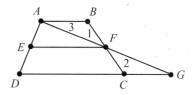

Exercise 33

Proof _____

Statements	Reasons
1. $\overline{AF}$ intersects $\overline{DC}$ at G	1. _____
2. _____	2. Given
3. E is the midpoint of $\overline{AD}$ and F is the midpoint of $\overline{BC}$	3. _____
4. _____	4. Def. of midpoint
5. $\angle 1 \cong \angle 2$	5. _____
6. _____	6. Given
7. $\angle 3 \cong \angle G$	7. _____
8. $\triangle ABF \cong \triangle GCF$	8. _____
9. $\overline{AF} \cong \overline{GF}$ and $\overline{AB} \cong \overline{GC}$	9. _____
10. $EF = \dfrac{1}{2}DG$	10. Theorem 4.19
11. $DG = GC + DC$	11. _____
12. $DG = AB + DC$	12. _____
13. $EF = \dfrac{1}{2}(AB + DC)$	13. Substitution law
14. $\overline{EF} \parallel \overline{DC}$	14. Theorem 4.19
15. $\overline{EF} \parallel \overline{AB}$	15. _____

34. Complete the proof of Theorem 4.23.

> *Given:* $\ell \parallel m$ and $m \parallel n$
> $\overline{AB} \cong \overline{BC}$
>
> *Prove:* $\overline{DE} \cong \overline{EF}$
>
> *Auxiliary lines:* Construct $\overline{AG} \parallel \overline{DE}$ and $\overline{BH} \parallel \overline{EF}$ using Construction 3.1.

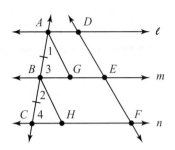

Exercise 34

Proof _____

Statements	**Reasons**
1. $\overline{AG} \parallel \overline{DE}$ and $\overline{BH} \parallel \overline{EF}$	1. _____
2. $\overline{AG} \parallel \overline{BH}$	2. _____
3. $\angle 1 \cong \angle 2$	3. _____
4. _____	4. Given
5. $\angle 3 \cong \angle 4$	5. _____
6. _____	6. ASA
7. _____	7. CPCTC
8. *ADEG* and *BEFH* are parallelograms	8. _____
9. $\overline{DE} \cong \overline{AG}$ and $\overline{EF} \cong \overline{BH}$	9. _____
10. _____	10. Sym. and trans. laws

35. Prove Theorem 4.21

36. Prove that the segments joining the midpoints of the consecutive sides of any quadrilateral form a parallelogram.

> *Given:* Quadrilateral *ABCD* with midpoints *P, Q, R, S*
>
> *Prove:* *PQRS* is parallelogram
>
> *Auxiliary line:* Draw diagonal $\overline{BD}$

37. Prove that the segment joining the midpoints of the adjacent sides of a rectangle form a rhombus.

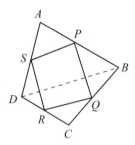

Exercise 35

Student Activity

38. In Exercises 36 and 37, you worked with midpoints of sides of quadrilaterals. Now consider the bisectors of angles of quadrilaterals.

1. The group assigns the following quadrilaterals to individual members: parallelogram, rhombus, kite, rectangle, trapezoid, isosceles trapezoid, and a general quadrilateral that is not any of the previously listed quadrilaterals.

2. Draw the assigned quadrilateral as accurately as possible and construct the four angle bisectors using Construction 1.6.

3. Extend the bisectors so they intersect forming a new quadrilateral. List as many properties as you can for these four points justifying each of them.

4. Are there any properties of the new quadrilaterals your group has in common? Remember this is a conjecture, not a proof.

Chapter (4) Review

Key Terms and Symbols

4.1 parallelogram ($\square$)
vector
resultant vector
parallelogram law

4.2 rhombus
kite

4.3 rectangle
square

4.4 trapezoid
base
legs
isosceles trapezoid
base angles
median

Chapter 4 Proof Techniques

To Prove:

A Quadrilateral Is a Parallelogram

1. Show both pairs of opposite sides are parallel. (definition)
2. Show both pairs of opposite sides are congruent. (Theorem 4.5)
3. Show both pairs of opposite angles are congruent. (Theorem 4.6)
4. Show two opposite sides are congruent and parallel. (Theorem 4.7)
5. Show the diagonals bisect each other. (Theorem 4.8)

A Quadrilateral Is a Rhombus

1. Show it's a parallelogram with two congruent adjacent sides. (definition)
2. Show it's a parallelogram with diagonals perpendicular. (Theorem 4.11)

A Quadrilateral Is a Rectangle

1. Show it's a parallelogram with one right angle. (definition)
2. Show it's a parallelogram with diagonals congruent. (Theorem 4.17)

A Quadrilateral Is a Square

1. Show it's a rhombus with one right angle. (definition)

Properties of Some Quadrilaterals

Parallelogram

1. Opposite sides parallel
2. Opposite sides congruent
3. Opposite angles congruent
4. Consecutive angles supplementary
5. Diagonals divide it into two congruent triangles
6. Diagonals bisect each other

Rhombus

1. Opposite sides parallel (It is a parallelogram.)
2. Opposite sides congruent
3. Opposite angles congruent
4. Consecutive angles supplementary
5. Diagonals divide it into two congruent triangles
6. Diagonals bisect each other
7. All sides congruent
8. Diagonals are perpendicular
9. Diagonals bisect angles
10. One diagonal is perpendicular bisector of the other

Kite

1. It is *not* a parallelogram

2. Exactly two distinct pairs of congruent consecutive sides

3. One pair of opposite angles are congruent

4. One diagonal is the perpendicular bisector of the other diagonal

5. One diagonal bisects two of the kite's angles

Rectangle

1. Opposite sides parallel (It is a parallelogram.)
2. Opposite sides congruent
3. Opposite angles congruent
4. Consecutive angles supplementary
5. Diagonals divide it into two congruent triangles
6. Diagonals bisect each other
7. All angles are right angles
8. Diagonals are congruent

Square

1. Opposite sides parallel
2. Opposite sides congruent
3. Opposite angles congruent
4. Consecutive angles supplementary
5. Diagonals divide it into two congruent triangles
6. Diagonals bisect each other
7. All angles are right angles
8. Diagonals are congruent
9. Diagonals bisect the angles
10. Diagonals are perpendicular
11. All sides congruent

Review Exercises

Sections 4.1 and 4.2

Exercises 1–10 refer to the figure to the right in which ABCD is a parallelogram. Answer **true** *or* **false.**

1. $\triangle ABD \cong \triangle CDB$.

2. $\overline{AB} \cong \overline{DC}$.

3. $\overline{AD} \cong \overline{AB}$.

4. $\angle 1 \cong \angle 2$.

5. $\angle BAD \cong \angle DCB$

6. $\angle BAD$ and $\angle ADC$ are complementary.

7. $\overline{AE} \cong \overline{EB}$.

Exercises 1–10

8. If $\overline{AB} \cong \overline{BC}$ then $\square ABCD$ is a rhombus.

9. If $\overline{AB} \cong \overline{BC}$ then $\overline{AC} \perp \overline{BD}$.

10. If $\square ABCD$ is a rhombus, then $\angle 1 \cong \angle 3$.

11. *Given:* Quadrilateral *MNPQ*
    $\angle 1 \cong \angle 2$; $\overline{MQ} \cong \overline{NO}$; $\overline{MN} \cong \overline{QP}$
    Prove: *MNPQ* is a parallelogram

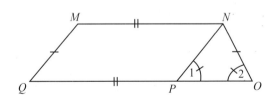

Exercise 11

12. *Given:* $\overline{AB} \cong \overline{CD}$; $\angle 1 \cong \angle 2$

    Prove: ABCD is a parallelogram

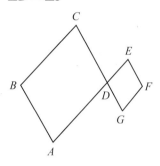

Exercise 12

13. *Given:* $\angle 1 \cong \angle 2$; $\angle 3 \cong \angle 4$

    Prove: LMNO is a kite

Exercise 13

14. *Given:* $\square ABCD$; $\square DEFG$

    Prove: $\angle B \cong \angle F$

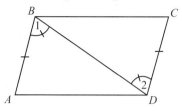

Exercise 14

15. In the given kite, find the measures of $\angle 1$ and $\angle 2$.

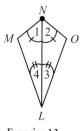

Exercise 15

16. *Given:* $\triangle EBF \cong \triangle GDH$; $\triangle AEH \cong \triangle CGF$

    Prove: EFGH is a parallelogram

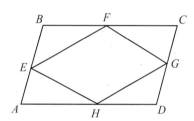

Exercise 16

For Exercises 17–23, answer **true** *or* **false.**

17. The diagonals of a parallelogram are never perpendicular.

18. Two consecutive angles of a parallelogram are always supplementary.

19. Two consecutive angles of a parallelogram may be congruent.

20. If two consecutive sides of a parallelogram are congruent, the parallelogram must be a rhombus.

21. Opposite angles of a rhombus are congruent.

22. The four sides of a kite are congruent.

23. The diagonals of a kite are perpendicular bisectors of each other.

24. Find the perimeter of a rhombus if one side is 12.7 centimeters.

Section 4.3

For Exercises 25–32, answer **true** *or* **false.**

25. Every square is a rhombus.

26. All parallelograms are rectangles.

27. A square is always a rectangle.

28. A rectangle is always a square.

29. The diagonals of a rectangle are congruent.

30. Sometimes the diagonals of a parallelogram are congruent.

31. Every rhombus is also a square.

32. Every square is also a rhombus.

33. *Given:* Rectangle *ABCD* with *AD* = 6*x* − 1;
          AB = 2*x* + 3; *BC* = 5*x* + 2

   Find: *x* and *CD*

Exercise 33

34. *Given:* △*ABC* with midpoints *E*
          and *D* of the respective
          sides and *DE* = 9*x* − 4;
          BC = 6*x* + 4

   Find: *x*, *DE*, and *BC*

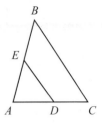

Exercise 34

35. *Given:* Rectangle *ABCD* and diagonals $\overline{AC}$ and $\overline{BD}$

   Prove: ∠1 ≅ ∠2

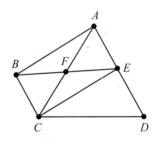

Exercise 35

36. Construct a square with diagonals of 2 inches.

37. *Given:* ▱*ABCE*
          ▱*BCDE*

   Prove: △*ACD* is isosceles

Section 4.4

For Exercises 38–42, answer true or false.

38. The diagonals of a trapezoid bisect each other.

39. All four angles of a trapezoid can have different measures.

40. Opposite angles of an isosceles trapezoid are supplementary.

41. Every trapezoid is a parallelogram.

42. The base angles of an isosceles trapezoid are always congruent.

43. Is the median of a trapezoid perpendicular to the segment joining midpoints of the parallel sides?

44. Draw a segment about 4 inches long and show how to divide it into three equal segments.

45. If a trapezoid has bases of 32 feet and 40 feet, what is the length of its median?

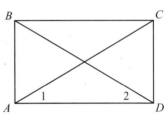

Exercise 37

46. Given trapezoid *ABCD* where $\overline{BC} \parallel \overline{AD}$ and *X* and *Y* are midpoints of the legs. If *XY* = 12.6 and *BC* = 3, find *AD*.

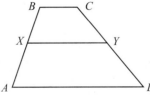

Exercise 46

47. Given isosceles trapezoid *ABCD* with $\overline{BC} \parallel \overline{AD}$ and *m*∠*A* = 67°. Find the measures of the other angles.

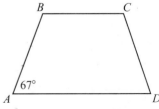

Exercise 47

For Problems 1–4, answer true *or* false.

1. The diagonals of a parallelogram are sometimes perpendicular. *True*

2. Every square is a rectangle.

3. The base angles of a trapezoid are always congruent.

4. The opposite angles of a rhombus are congruent.

5. Given parallelogram *ABCD* with the given measures, find *m∠ADC*

Exercise 5

Use the figure below in Problems 6 and 7. Assume that ABCD is a rectangle and $\overline{AE} \cong \overline{FC}$.

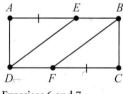

Exercises 6 and 7

6. Prove that $\overline{DE} \cong \overline{BF}$.

7. Prove that *DEBF* is a parallelogram.

8. Draw an acute angle and a segment about 2 inches in length. Construct a rhombus with this angle and sides.

9. The line segment joining the midpoints of two sides of a triangle measures 44 cm. What is the length of the third side?

10. For the given triangle with *M* and *N* the midpoint of sides $\overline{AB}$ and $\overline{BC}$ and *AC* = 4*x* + 6 and *MN* = *x* + 9, find the length of *AC*.

11. In kite *ABCD*, $\overline{AC}$ is the perpendicular bisector of $\overline{BD}$. If *AB* = 12.4 inches and the perimeter of the kite is 54.6 inches, find *DC*.

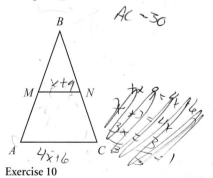

Exercise 10

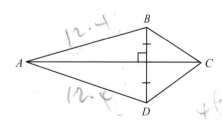

Exercise 11

12. *Given:* Trapezoid *ABCD* with $\overline{AD} \parallel \overline{BC}$, $\overline{AB} \cong \overline{CD}$ and $m\angle A = 67°$

    Find: $m\angle B$ and $m\angle C$

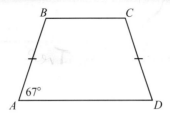

Exercise 12

For Problems 13 and 14, use trapezoid WXYZ, where M and N are the midpoints of the respective legs. And $\overline{WX} \parallel \overline{YZ}$.

13. If $WX = 17$ cm and $YZ = 31$ cm, find MN.

14. If $WX = 4x - 7$, $MN = 2x + 11$, and
    $ZY = 2x + 1$, find x.

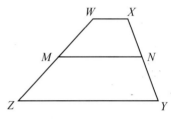

Exercises 13 and 14

15. Suppose the diagonals of a parallelogram are congruent. Must the quadrilateral be a rectangle? Explain.

16. *Given:* *ABCD* is a parallelogram with
    $\overline{AD} \cong \overline{CN}$; $\angle 1 \cong \angle 2$

    Prove: *ABCD* is a rhombus

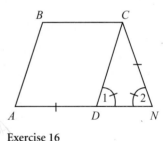

Exercise 16

Name the figure(s) that satisfies each set of properties.

17. It has four sides
    All sides are congruent.
    Opposite sides are parallel.
    All angles are right angles.

18. It has four sides.
    Opposite angles are congruent.
    Opposite sides are parallel.
    All sides are congruent.

19. It has four sides.
    Only two sides are parallel.

20. It has four sides.
    It has two pairs of congruent adjacent sides.

5

Similar Polygons and the Pythagorean Theorem

We begin this chapter by reviewing two concepts learned in beginning algebra: ratio and proportion. We'll use these concepts to define the notion of similar polygons and consider applications related to similarity. One application of similar polygons is given below and solved in Example 5 of Section 5.2.

AN APPLICATION

While standing near the Washington Monument in Washington, DC, Scott observed that the monument (shown here in a top view) cast a shadow measuring 185 ft at the same time his shadow was 2 ft long. If Scott is 6 ft tall, how can he determine the height of the monument?

(5.1) Ratio and Proportion

OBJECTIVES

1. Define ratio and proportion.
2. Define and state theorems about proportions and geometric mean.
3. Solve problems involving proportions.
4. Define proportional segments.

The topics presented in this section review many of the concepts you learned in beginning algebra. You should also be familiar with basic arithmetic operations on fractions.

OBJECTIVE 1   Define ratio and proportion.

> **DEFINITION: Ratio**
> The **ratio** of one number a to another number b, $b \neq 0$, is the fraction $\dfrac{a}{b}$. The ratio of a to b is sometimes written as $a{:}b$ and read "a is to b."

Ratios compare numbers by using division. They are used in many applications such as rate of speed, gas mileage, and unit cost.

APPLICATION	RATIO
32 students for 2 teachers	$\dfrac{32}{2} = 16$ students per teacher
300 miles in 6 hours	$\dfrac{300 \text{ mi}}{6 \text{ hr}} = 50$ mph (miles per hour)
100 miles on 5 gallons of gas	$\dfrac{100 \text{ mi}}{5 \text{ gal}} = 20$ mpg (miles per gallon)
$8 for 4 pounds of meat	$\dfrac{\$8}{4 \text{ lb}} = \$2$ per lb

EXAMPLE 1   Consider the triangle in Figure 5.1. What is the ratio of the length of the shortest side to the length of the longest side?

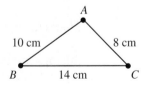

Figure 5.1

Because the shortest side is 8 cm and the longest side is 14 cm, the ratio of the lengths is $\dfrac{8}{14}$, which reduces to $\dfrac{4}{7}$, or 4:7.

PRACTICE EXERCISE 1

Find the ratio of the length of the longest side to the length of the next longest side in the triangle in Figure 5.1.

ANSWER ON PAGE 222

Sometimes two pairs of numbers have the same ratio.

DEFINITION: Proportion

An equation showing that two ratios are equal is called a **proportion**.

The following are several examples of proportions:

$$\frac{1}{2} = \frac{2}{4}, \quad \frac{5}{7} = \frac{15}{21}, \quad \text{and} \quad \frac{10}{3} = \frac{40}{12}.$$

The statement "a is to b as c is to d" translates to the proportion

$$\frac{a}{b} = \frac{c}{d},$$

assuming $b \neq 0$ and $d \neq 0$. (Remember that division by zero is not defined.) We shall assume from now on that all denominators in ratios are not zero.

DEFINITION: Terms of a Proportion

Consider the proportion $\frac{a}{b} = \frac{c}{d}$. The numbers a and d are called **extremes**, and b and c are called **means** of the proportion.

EXAMPLE 2 Name the means and extremes in the proportion

$$\frac{9}{20} = \frac{18}{40}.$$

The means are 20 and 18, and the extremes are 9 and 40. Notice that by the symmetric law

$$\frac{18}{40} = \frac{9}{40}.$$

In this form, the means and extremes are interchanged.

A proportion can be extended to contain several ratios. For example, we might say that a, b, and c are proportional to p, q, and r. This means

$$\frac{a}{p} = \frac{b}{q} = \frac{c}{r}.$$

PRACTICE EXERCISE 2

Are 1, 2, 3, and 4 proportional to 5, 10, 15, and 20?

ANSWER ON PAGE 222

OBJECTIVE 2   Define and state theorems about proportions and geometric mean.

> **DEFINITION:**   *Geometric Mean or Mean Proportional*
>
> In the proportion $\dfrac{a}{b} = \dfrac{b}{c}$, b is called the **geometric mean** or **mean proportional** between a and c.

Because $\dfrac{3}{6} = \dfrac{6}{12}$, 6 is a mean proportional between 3 and 12. We know that $\dfrac{3}{6}$ and $\dfrac{6}{12}$ are equal because they both reduce to $\dfrac{1}{2}$. Also, we might observe that the product of the means, $6 \cdot 6 = 36$, is equal to the product of the extremes, $3 \cdot 12 = 36$.

> **THEOREM 5.1**   *Means-Extremes Property*
>
> In any proportion, the product of the means is equal to the product of the extremes. That is,
>
> $$\text{if } \frac{a}{b} = \frac{c}{d}, \text{ then } ad = bc.$$

Given: $\dfrac{a}{b} = \dfrac{c}{d}$

Prove: $ad = bc$

Proof _____

Statements	Reasons
1. $\dfrac{a}{b} = \dfrac{c}{d}$	1. Given
2. $bd = bd$	2. Reflexive law
3. $bd\left(\dfrac{a}{b}\right) = bd\left(\dfrac{c}{d}\right)$	3. Mult.-div. law
4. $ad = bc$	4. Simplify each side

There are many ways to derive different proportions from a given one. The next six theorems show some of these ways. They are stated without proof.

> **THEOREM 5.2**   *Reciprocal Property of Proportions*
>
> The reciprocals of both sides of a proportion are also proportional. That is,
>
> $$\text{if } \frac{a}{b} = \frac{c}{d}, \text{ then } \frac{b}{a} = \frac{d}{c}.$$

Notice that because $\frac{2}{4} = \frac{5}{10}$, by Theorem 5.2 we also have $\frac{4}{2} = \frac{10}{5}$. If we interchange the means in $\frac{2}{4} = \frac{5}{10}$, we get $\frac{2}{5} = \frac{4}{10}$, which is also a proportion. Similarly, by interchanging the extremes, $\frac{10}{4} = \frac{5}{2}$ is also a proportion. This example introduces the next two theorems.

THEOREM 5.3   MEANS PROPERTY OF PROPORTIONS

If the means are interchanged in a proportion, a new proportion is formed. That is,

$$\text{if } \frac{a}{b} = \frac{c}{d}, \text{ then } \frac{a}{c} = \frac{b}{d}.$$

THEOREM 5.4   EXTREMES PROPERTY OF PROPORTIONS

If the extremes are interchanged in a proportion, a new proportion is formed. That is,

$$\text{if } \frac{a}{b} = \frac{c}{d}, \text{ then } \frac{d}{b} = \frac{c}{a}.$$

Consider the proportion $\frac{2}{8} = \frac{5}{20}$. Suppose we add the denominator to the numerator in each ratio.

$$\frac{2 + 8}{8} = \frac{5 + 20}{20} \text{ or } \frac{10}{8} = \frac{25}{20}$$

The results are still proportional. This example introduces the next two theorems, which are presented without proof.

THEOREM 5.5   ADDITION PROPERTY OF PROPORTIONS

If the denominators in a proportion are added to their respective numerators, a new proportion is formed. That is,

$$\text{if } \frac{a}{b} = \frac{c}{d}, \text{ then } \frac{a + b}{b} = \frac{c + d}{d}.$$

A similar result follows when the denominators are subtracted from the numerators. For example, $\frac{2}{8} = \frac{5}{20}$, and $\frac{2 - 8}{8} = \frac{5 - 20}{20}$ because $\frac{-6}{8} = \frac{-15}{20}$.

THEOREM 5.6   SUBTRACTION PROPERTY OF PROPORTIONS

If the denominators in a proportion are subtracted from their respective numerators, a new proportion is formed. That is,

$$\text{if } \frac{a}{b} = \frac{c}{d}, \text{ then } \frac{a - b}{b} = \frac{c - d}{d}.$$

A calendar designer wishes to enlarge a picture for the month of July to fill a space 5 inches high by 8 inches wide. Before the enlargement process, she must crop the picture so that its dimensions are in the same ratio as the dimensions of the space to be filled. How would you crop the picture for the project?

Consider the continued proportion

$$\frac{2}{3} = \frac{4}{6} = \frac{6}{9}.$$

Then consider the following:

$$\frac{2 + 4 + 6}{3 + 6 + 9} = \frac{2}{3}.$$

Because $2 + 4 + 6 = 12$ and $3 + 6 + 9 = 18$, $\frac{12}{18} = \frac{2}{3}$. This example illustrates the following theorem.

THEOREM 5.7

If a, b, c, d, e, and f are numbers satisfying $\frac{a}{b} = \frac{c}{d} = \frac{e}{f}$, then

$$\frac{a + c + e}{b + d + f} = \frac{a}{b}.$$

PRACTICE EXERCISE 3

Complete the proof of Theorem 5.7.

Given: $\frac{a}{b} = \frac{c}{d} = \frac{e}{f}$

Prove: $\frac{a + c + e}{b + d + f} = \frac{a}{b}$

Proof _____

Statements	*Reasons*
1. Let $x = \frac{a}{b} = \frac{c}{d} = \frac{e}{f}$	1. _____
2. $a = bx$, $c = dx$, and $e = fx$	2. _____
3. $a + c + e = bx + dx + fx$	3. _____
4. _____	4. Distributive law
5. $\frac{a + c + e}{b + d + f} = x$	5. _____
6. _____	6. Substitution law

ANSWERS ON PAGE 222

OBJECTIVE 3   **Solve problems involving proportions.**   The process of finding unknown terms in a given proportion, sometimes called **solving the proportion**, applies the means-extremes property. The following examples illustrate.

EXAMPLE 3   Solve each proportion.

(a) $\dfrac{x}{5} = \dfrac{8}{20}$

$20x = 5 \cdot 8$ Means-extremes property

$20x = 40$ Simplify

$x = 2$ Divide both sides by 20

(b) $\dfrac{4}{y} = \dfrac{y}{9}$

$4 \cdot 9 = y \cdot y$ Means-extremes property

$36 = y^2$ Simplify

Because $6^2 = 36$ and $(-6)^2 = 36$, $y = 6$ or $y = -6$.

(c) $\dfrac{z + 5}{z} = \dfrac{9}{4}$

$4(z + 5) = 9z$ Means-extremes property

$4z + 20 = 9z$ Distributive law

$20 = 5z$ Subtract $4z$ from both sides

$4 = z$ Divide both sides by 5

EXAMPLE 4   In an election, the winning candidate won by a ratio of 5:4. If she received 600 votes, how many votes did the losing candidate receive?

If a represents the number of votes for the losing candidate, the following proportion describes the problem.

Winning ratio $\longrightarrow$ $\dfrac{5}{4} = \dfrac{600}{a}$ $\longleftarrow$ Votes for winner
$\longleftarrow$ Votes for loser

$5a = 4 \cdot 600$ Means-extremes property

$5a = 2400$ Simplify

$a = 480$ Divide both sides by 5

Thus, the loser received 480 votes.

[Note: The proportion could have been written as:

$\dfrac{4}{5} = \dfrac{a}{600}$ or $\dfrac{5}{600} = \dfrac{4}{a}$ or $\dfrac{600}{5} = \dfrac{a}{4}$.]

An **extended ratio** compares more than two quantities and is written as $a:b:c$. If the sides of a triangle are 8, 10, and 12 inches long, then the sides are in the ratio of 4 to 5 to 6 (4:5:6).

EXAMPLE 5   If the sides of a triangle are in the ratio 4:5:6 and the perimeter is 60 inches, find the measures of each side.

Represent the lengths of the sides as $4x$, $5x$, and $6x$.

$$4x + 5x + 6x = 60$$
$$15x = 60$$
$$x = 4$$

Thus, the sides of the triangle are 16, 20, and 24 inches.

OBJECTIVE 4 **Define proportional segments.** Some of the work in this chapter involves segments that are proportional.

> **DEFINITION: Proportional Segments**
> If the lengths of segments are proportional, the segments are called **proportional segments.** That is, segments $\overline{AB}$ and $\overline{CD}$ are proportional to segments $\overline{EF}$ and $\overline{GH}$ when
> $$\frac{AB}{CD} = \frac{EF}{GH}.$$

Consider the segments in Figure 5.2. If $AB = 2$ cm, $CD = 3$ cm, $EF = 4$ cm, and $GH = 6$ cm, then because $\frac{2}{3} = \frac{4}{6}$, segments $\overline{AB}$ and $\overline{CD}$ are proportional to segments $\overline{EF}$ and $\overline{GH}$.

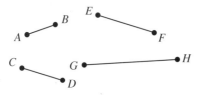

Figure 5.2

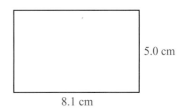

5.0 cm

8.1 cm

Figure 5.3 Golden Rectangle

EXAMPLE 6 Consider the rectangle in Figure 5.3. The length ℓ and width w approximately satisfy the proportion $\frac{l}{w} = \frac{l + w}{l}$.

Architects and artists have determined that a rectangle with this property, called a *golden rectangle,* is especially pleasing to the eye. Before the top of the Parthenon was destroyed, the front of the building could be placed into a golden rectangle. The ratio $\frac{l}{w}$ is called the *golden ratio* and is approximated by $\frac{8.1}{5.0} = 1.62$.

The Parthenon

Answers to Practice Exercises

1. $\frac{7}{5}$ or 7:5 **2.** Yes, because $\frac{1}{5} = \frac{2}{10} = \frac{3}{15} = \frac{4}{20}$. **3.** 1. Given 2. Mult.-div. law
3. Add.-subt. law **4.** $a + c + e = (b + d + f)x$ **5.** Mult.-div. law **6.** $\frac{a + c + e}{b + d + f} = \frac{a}{b}$

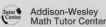

Write each ratio in Exercises 1–10 as a fraction and simplify.

1. 20 to 35   **2.** 18 to 54   **3.** 8 cm to 32 cm   **4.** 12 ft to 72 ft

5. 200 mi in 4 hr   **6.** 150 mi in 5 hr   **7.** 4 in. to 2 ft   **8.** 5 cm to 1 m

9. $\frac{1}{2}$ in. to $\frac{3}{8}$ in.   **10.** $\frac{1}{4}$ ft to $\frac{3}{2}$ ft

Solve each proportion in Exercises 11–20.

11. $\dfrac{a}{3} = \dfrac{14}{21}$ **12.** $\dfrac{4}{7} = \dfrac{x}{28}$ **13.** $\dfrac{40}{35} = \dfrac{2}{y}$ **14.** $\dfrac{2}{1} = \dfrac{1}{a}$

15. $\dfrac{25}{x} = \dfrac{x}{1}$ **16.** $\dfrac{4}{y} = \dfrac{y}{100}$ **17.** $\dfrac{a+2}{a} = \dfrac{7}{5}$ **18.** $\dfrac{x}{x-3} = \dfrac{11}{8}$

19. $\dfrac{y+2}{12} = \dfrac{y-2}{4}$ **20.** $\dfrac{6}{z-3} = \dfrac{15}{z}$

21. Find the geometric mean between 4 and 36.

22. Find the geometric mean between 36 and 4.

23. Find the geometric mean between 9 and 64.

24. Find the geometric mean between 25 and 144.

25. Representative Wettaw won an election by a ratio of 8 to 5. If he received 10,400 votes, how many votes did his opponent receive?

26. If a wire is 70 ft long and weighs 84 lb, how much will 110 ft of the same wire weigh?

27. If $\frac{1}{2}$ inch on a map represents 10 miles, how many miles are represented by $6\frac{1}{2}$ inches?

28. A ranger wants to estimate the number of antelope in a preserve. He catches 58 antelope, tags their ears, and returns them to the preserve. Some time later, he catches 29 antelope and discovers that 7 of them are tagged. Estimate the number of antelope in the preserve.

29. It has been estimated that a family of four produces 115 lb of garbage in one week. Estimate the number of pounds of garbage produced by 7 such families in one week.

30. A baseball pitcher gave up 60 earned runs in 180 innings. Estimate the number of earned runs he will give up every 9 innings. This number is called his *earned run average*.

31. The geometric mean occurs frequently in nature. For example, a starfish has the shape of a *pentagram* shown to the right. AB is the mean proportional between BC and AC. That is, $\dfrac{BC}{AB} = \dfrac{AB}{AC}$. Also, the ratios $\dfrac{AD}{AC}, \dfrac{AC}{AB}$, and $\dfrac{AB}{BC}$ all equal the golden ratio. Measure the lengths in the figure and try to confirm these results.

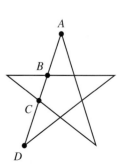

Exercise 31

32. Proportions are also used to determine actual lengths from scale drawings. For example, shown below is the scale drawing of a family room in which $\frac{1}{8}$ inch corresponds to 1 ft. Use proportions to find the following.

  a. What is the length of the room?
  b. What is the width of the room?
  c. What is the width of the window?
  d. What is the width of the fireplace?

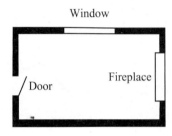

Exercise 32

33. The sides of a triangle are in the ratio of 3:4:5. If the perimeter is 90 centimeters, find the lengths of each side.

34. If the sides of a quadrilateral are in the ratio of 3:5:7:9 and the perimeter is 240 inches, find the length of each side.

35. The measures of two supplementary angles are in the ratio of 2:7. Find the measures of each angle. $2x + 7x = 180$

36. The measures of two complementary angles are in the ratio of 2:3, find the measures of each angle. $2x + 3x = 90$

 37. Explain the difference between a ratio and a proportion. Use words not just a numeric example.

5.2 Similar Polygons

OBJECTIVES

1. Define similar polygons.
2. State the AAA postulate.
3. State the AA theorem.
4. Discuss the transitive property of similar triangles.
5. Divide a triangle proportionally.
6. Construct proportional segments.
7. Construct a polygon similar to a given polygon.

OBJECTIVE 1 Define similar polygons. When an architect draws a blueprint for a house or an engineer makes a drawing of a machine part, the result is drawn to scale showing the same objects in reduced sizes. Similarly, a biologist looking through a microscope sees shapes enlarged in size. Two figures with the same shape are called *similar*. In geometry, we study *similar polygons*.

> **DEFINITION: Similar Polygons**
>
> Two polygons are **similar** if their vertices can be paired in such a way that corresponding angles are congruent and corresponding sides are proportional.

In designing a new automobile, an engineer first makes a scale model. The model and the finished automobile are similar to each other. If the wheelbase on the actual car will be 100 inches in length, and the initial model is to be $\frac{1}{20}$ the actual size, what should be the length of the wheelbase on the model?

When we refer to similar polygons, we list the corresponding vertices in the same order. For example, if polygon $ABCDE$ is similar to polygon $PQRST$ in Figure 5.4, we write $ABCDE \sim PQRST$. From the definition of similar polygons,

$$\angle A \cong \angle P, \angle B \cong \angle Q, \angle C \cong \angle R, \angle D \cong \angle S, \text{ and } \angle E \cong \angle T,$$

and

$$\frac{AB}{PQ} = \frac{BC}{QR} = \frac{CD}{RS} = \frac{DE}{ST} = \frac{EA}{TP}.$$

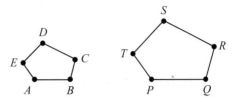

Figure 5.4 Similar Polygons

When working with similar polygons, it is helpful to label corresponding vertices as A and A' (read A prime), B and B', and so on. The following example illustrates.

EXAMPLE 1 Assume that $ABCD \sim A'B'C'D'$ in Figure 5.5.

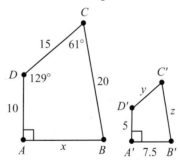

Figure 5.5

(a) Find x. Because $\dfrac{AB}{A'B'} = \dfrac{AD}{A'D'}$, substituting we have

$$\frac{x}{7.5} = \frac{10}{5}$$

$$5x = (10)(7.5) \quad \text{Means-extremes property}$$

$$5x = 75$$

$$x = 15$$

(b) Find z. Because $\dfrac{BC}{B'C'} = \dfrac{AD}{A'D'}$, substituting we have

$$\frac{20}{z} = \frac{10}{5}$$

$$(20)(5) = 10z$$

$$10 = z.$$

(c) Find $m\angle B'$. By Theorem 3.15 the sum of the angles of a quadrilateral is
$(4 - 2)(180°) = (2)(180°) = 360°$,

$$m\angle A + m\angle B + m\angle C + m\angle D = 90° + m\angle B + 61° + 129°$$

$$= 280° + m\angle B,$$

so

$$360° = 280° + m\angle B$$

$$80° = m\angle B.$$

Because $m\angle B' = m\angle B$, $m\angle B' = 80°$.

PRACTICE EXERCISE 1

Refer to Figure 5.5 in which $ABCD \sim A'B'C'D'$. Find y and $m\angle D'$.

ANSWERS ON PAGE 233

OBJECTIVE 2   **State the AAA postulate.**   Similar triangles, such as $\triangle ABC$ and $\triangle A'B'C'$ in Figure 5.6, have many applications in geometry, and it is appropriate to investigate some of their properties. We can use the definition of similar polygons to prove that two triangles are similar, but the next postulate provides a more useful method.

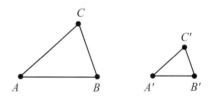

Figure 5.6 Similar Triangles

POSTULATE 5.1   AAA

Two triangles are similar if three angles of one triangle are congruent to the corresponding three angles of the other triangle.

Thus, to determine similarity of triangles, we do not need to verify proportionality of corresponding sides, but only equality of corresponding angles.

OBJECTIVE 3   **State the AA theorem.**   By Corollary 3.13, we know that if two angles of one triangle are congruent, respectively, to two angles of another triangle, then the third angles are congruent. So we can show similarity using only two pairs of angles. We have just given an informal proof of the next useful theorem.

> **THEOREM 5.8   AA**
> Two triangles are similar if two angles of one triangle are congruent to the corresponding two angles of the other triangle.

EXAMPLE 2   Refer to Figure 5.7.
Given: $\overline{AB} \perp \overline{BD}$ and $\overline{ED} \perp \overline{BD}$
Prove: $\triangle ABC \sim \triangle EDC$

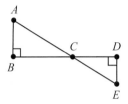

Figure 5.7

Proof

Statements	Reasons
1. $\overline{AB} \perp \overline{BD}$ and $\overline{ED} \perp \overline{BD}$	1. Given
2. $\angle ABC$ and $\angle EDC$ are right angles	2. $\perp$ lines form rt. $\angle$'s
3. $\angle ABC \cong \angle EDC$	3. Rt. $\angle$'s are $\cong$
4. $\angle ACB$ and $\angle ECD$ are vertical angles	4. Def. of vert. $\angle$'s
5. $\angle ACB \cong \angle ECD$	5. Vert. $\angle$'s are $\cong$
6. $\triangle ABC \sim \triangle EDC$	6. AA

PRACTICE EXERCISE 2

In Figure 5.7, if $AB = 9$ ft, $BC = 12$ ft, $CE = 10$ ft, and $DE = 6$ ft, find AC and CD.

ANSWERS ON PAGE 233

Figure 5.6 on the opposite page clearly shows that similar triangles do not have to be congruent. However, congruent triangles are similar.

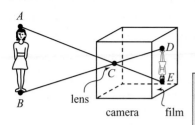

When you take a photograph of a person, the image is captured on the film inside the camera as shown. The height of the image on the film will vary in proportion with the distance of the person from the lens. By using the fact that $\triangle ABC \sim \triangle EDC$, and by knowing the height of the person, a photographer can determine the best location of the camera to give any desired image size on the film.

THEOREM 5.9

If $\triangle ABC \cong \triangle DEF$, then $\triangle ABC \sim \triangle DEF$

Given: $\triangle ABC \cong \triangle DEF$
(See Figure 5.8.)

Prove: $\triangle ABC \sim \triangle DEF$

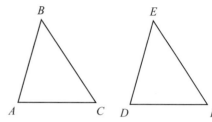

Figure 5.8

Proof

Statements	Reasons
1. $\triangle ABC \cong \triangle DEF$	1. Given
2. $\angle A \cong \angle D, \angle B \cong \angle E$	2. CPCTC
3. $\triangle ABC \sim \triangle DEF$	3. AA

OBJECTIVE 4 **Discuss the transitive property of similar triangles.** The next theorem shows that triangles similar to the same triangle are themselves similar.

THEOREM 5.10 *TRANSITIVE LAW FOR SIMILAR TRIANGLES*

If $\triangle ABC \sim \triangle DEF$ and $\triangle DEF \sim \triangle GHI$, then $\triangle ABC \sim \triangle GHI$.

Given: $\triangle ABC \sim \triangle DEF$,
$\triangle DEF \sim \triangle GHI$
(See Figure 5.9.)

Prove: $\triangle ABC \sim \triangle GHI$

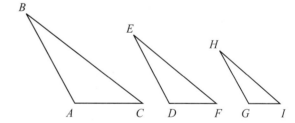

Figure 5.9

Proof

Statements	Reasons
1. $\triangle ABC \sim \triangle DEF$	1. Given
2. $\angle A \cong \angle D, \angle B \cong \angle E$	2. Def. of similar triangles
3. $\triangle DEF \sim \triangle GHI$	3. Given
4. $\angle D \cong \angle G, \angle E \cong \angle H$	4. Def. of similar triangles
5. $\angle A \cong \angle G, \angle B \cong \angle H$	5. Transitive law
6. $\triangle ABC \sim \triangle GHI$	6. Def. of similar triangles

 **Student Activity**

The following equipment is needed for this activity: Two 6-inch square sheets of tracing paper and a ruler.
1. Draw a large triangle on one piece of tracing paper and label it *ABC*.
2. Choose any point of $\overline{AB}$ and call it *D*.
3. On a second piece of tracing paper, make a copy of angle *A* using a ruler.
4. Place the paper with the triangle on top of the paper with the angle matching the vertex of the angle with point *D* on the triangle and matching side $\overline{AB}$.
5. Draw a line through point *D* parallel to $\overline{AC}$. Label the point where the parallel line intersects $\overline{BC}$ as *E*. Now $\overline{DE} \parallel \overline{AC}$. Why are they parallel?
6. Explain why $\triangle ABC \sim \triangle DBE$
7. With a ruler find, $\dfrac{AD}{DB}$ and $\dfrac{CE}{EB}$
8. Compare your results with other groups and read Theorem 5.11.

OBJECTIVE 5 **Divide a triangle proportionally.** The next theorem illustrates how similar triangles are often used to determine certain properties of a given triangle and proves the results from the Student Activity above.

THEOREM 5.11 *TRIANGLE PROPORTIONALITY THEOREM*

A line parallel to one side of a triangle that intersects the other two sides divides the two sides into proportional segments.

Given: $\triangle ABC$ with $\ell \parallel \overline{BC}$
(See Figure 5.10.)

Prove: $\dfrac{DB}{AD} = \dfrac{EC}{AE}$

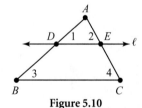

Figure 5.10

Proof

Statements	Reasons
1. $\triangle ABC$ with $\ell \parallel \overline{BC}$	1. Given
2. $\angle 1 \cong \angle 3$ and $\angle 2 \cong \angle 4$	2. If $\parallel$, corr. $\angle$'s are $\cong$
3. $\triangle ADE \sim \triangle ABC$	3. AA
4. $\dfrac{AB}{AD} = \dfrac{AC}{AE}$	4. Corr. sides of $\sim \triangle$'s are proportional
5. $\dfrac{AB - AD}{AD} = \dfrac{AC - AE}{AE}$	5. Subt. prop. of proportions
6. $AB - AD = DB$ and $AC - AE = EC$	6. Seg. add. prop.
7. $\dfrac{DB}{AD} = \dfrac{EC}{AE}$	7. Substitution law

EXAMPLE 3 In Figure 5.11, $\overline{DE} \parallel \overline{BC}$, $AC = 8.8$ yd, $AD = 1.3$ yd, and $DB = 3.1$ yd. Find AE and EC.

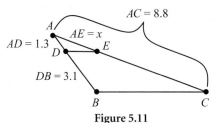

Figure 5.11

Let $x = AE$, then $EC = AC - AE = 8.8 - x$. Using Theorem 5.11,

$$\frac{1.3}{3.1} = \frac{x}{8.8 - x}.$$

$$1.3(8.8 - x) = 3.1x \qquad \text{Means-extremes prop.}$$

$$11.44 - 1.3x = 3.1x \qquad \text{Distributive law}$$

$$11.44 = 4.4x \qquad \text{Add } 1.3x \text{ to both sides}$$

$$2.6 = x \qquad \text{Divide both sides by 4.4}$$

Thus, $AE = 2.6$ yd and $EC = 8.8 - 2.6 = 6.2$ yd.

Theorem 5.11 offers a way to construct a segment proportional to three given line segments.

OBJECTIVE 6 Construct proportional segments.

CONSTRUCTION 5.1
Construct a segment proportional to three given line segments.

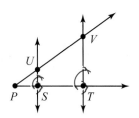

Figure 5.12

Given: Segments $\overline{AB}$, $\overline{CD}$, and $\overline{EF}$ (See Figure 5.12.)

To Construct: $\overline{UV}$ so that $\dfrac{AB}{CD} = \dfrac{EF}{UV}$

Construction _____

1. Construct an arbitrary angle with vertex P. On one ray of $\angle P$, construct $\overline{PS}$ such that $PS = AB$ and $\overline{ST}$ such that $ST = CD$.

2. On the other ray of $\angle P$, construct $\overline{PU}$ such that $PU = EF$. Then construct $\overleftrightarrow{SU}$.

3. Use Construction 3.1 to construct the line through T parallel to $\overleftrightarrow{SU}$ intersecting $\overleftrightarrow{PU}$ at point V. Then $\overline{UV}$ is the desired segment by Theorem 5.11.

Technology Connection

Geometry software will be needed.

1. Draw any triangle *XYZ*.
2. Draw the bisector of ∠*X* and label the point where the bisector intersects $\overline{YZ}$ as *W*.
3. Find *XY, XZ, YW,* and *ZW*.
4. Compare the ratios of $\dfrac{XY}{XZ}$ and $\dfrac{YW}{ZW}$ [Note: The computer may approximate but the ratios should be close.]
5. Write about your observations and try them out on another type of triangle. Compare these results with Theorem 5.12.

THEOREM 5.12   TRIANGLE ANGLE-BISECTOR THEOREM
The bisector of one angle of a triangle divides the opposite side into segments that are proportional to the other two sides.

PRACTICE EXERCISE 3

Complete the proof of Theorem 5.12.

Given: △*ABC* and $\overline{AD}$ bisects ∠*A*

Prove: $\dfrac{BD}{DC} = \dfrac{AB}{AC}$

Auxiliary lines: Construct the line segment through *B* parallel to $\overline{AD}$ intersecting the extension of $\overline{AC}$ at point *E*

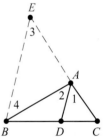

Proof _____

Statements	Reasons
1. $\overline{AD}$ bisects ∠*A*	1. _____
2. _____	2. Def. of ∠ bisector
3. $\overline{BE} \parallel \overline{AD}$	3. _____
4. _____	4. If ∥, corr. ∠'s are ≅
5. ∠2 ≅ ∠4	5. _____
6. ∠3 ≅ ∠4	6. _____
7. _____	7. Sides opp. ≅ ∠'s are ≅; def. ≅ seg.
8. $\dfrac{BD}{DC} = \dfrac{AE}{AC}$	8. _____
9. $\dfrac{BD}{DC} = \dfrac{AB}{AC}$	9. _____

ANSWERS ON PAGE 233

EXAMPLE 4

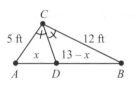

Figure 5.13

Refer to Figure 5.13 showing right $\triangle ABC$ with legs 5 ft and 12 ft and hypotenuse 13 ft. If $\overline{CD}$ bisects right $\angle C$, approximate the lengths of $\overline{AD}$ and $\overline{DB}$, correct to the nearest tenth of a foot.

Let $x = AD$, then $DB = AB - AD = 13 - x$. By the triangle angle-bisector theorem,

$$\frac{x}{13 - x} = \frac{5}{12}.$$

$$12x = 5(13 - x) \qquad \text{Means-extremes prop.}$$
$$12x = 65 - 5x \qquad \text{Distributive law}$$
$$17x = 65 \qquad \text{Add } 5x \text{ to both sides}$$
$$x = \frac{65}{17}$$

Using a calculator, we get $x \approx 3.8$, to the nearest tenth. Then $13 - x \approx 9.2$. Thus, $AD \approx 3.8$ ft and $DC \approx 9.2$ ft, correct to the nearest tenth of a foot.

[Note: In example 4, we used the symbol $\approx$ which stands for the phrase "is **approximately** equal to." Another symbol for approximately is $\doteq$. These symbols are often used in applied problems to indicate approximate or rounded values.]

The next example solves the applied problem given in the chapter introduction.

EXAMPLE 5

While standing near the Washington Monument in Washington, DC, Scott discovered that the length of the shadow of the monument was 185 ft at the same time his shadow was 2 ft long. If Scott is 6 ft tall, how did he determine the height of the monument?

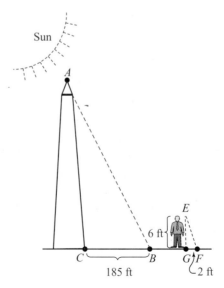

Figure 5.14

Figure 5.14 shows the information given (not drawn to scale). Because the Sun is so far away, we will assume that the rays from the Sun are parallel, making $\angle BAC \cong \angle FEG$. Also, we can assume that $\angle ACB$ and $\angle EGF$ are right angles and, hence, are congruent. Thus, $\triangle ABC \sim \triangle EFG$ by AA. If x is the height of the monument (side $\overline{AC}$), then

$$\frac{x}{6} = \frac{185}{2}$$

$$2x = (6)(185)$$

$$x = 555$$

Thus, Scott concluded that the monument is 555 ft high.

The Washington Monument is officially 555 ft., 5⅛ in. tall. Scott's estimate was very accurate.

OBJECTIVE 7   **Construct a polygon similar to a given polygon.**   We can use Theorem 5.8 (AA) in another way to show how to construct a polygon similar to a given one. We illustrate this with a pentagon, but the process applies to any polygon.

CONSTRUCTION 5.2

Construct a polygon similar to a given polygon.

Given: Pentagon *ABCDE* and segment $\overline{GH}$ (See Figure 5.15.)

To Construct: Pentagon *GHIJK* such that *ABCDE ~ GHIJK*

Construction _____

1. Construct diagonal $\overline{AC}$ forming $\angle 1$. Copy $\angle 1$ at G and $\angle B$ at H. Label their point of intersection I. Then $\triangle ABC \sim \triangle GHI$ by AA.
2. Now draw diagonal $\overline{AD}$ and repeat the construction in step 1 obtaining $\triangle GIJ \sim \triangle ACD$.
3. Now construct $\triangle GJK$ in the same manner, so that $\triangle ADE \sim \triangle GJK$.
4. Then *ABCDE ~ GHIJK*.

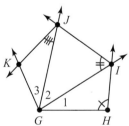

Figure 5.15 Construction of Similar Polygons

Answers to Practice Exercises

1. 7.5; 129°   **2.** *AC* = 15 ft, *CD* = 8 ft   **3.** 1. Given 2. $\angle 1 \cong \angle 2$ 3. By construction 4. $\angle 3 \cong \angle 1$ 5. If lines ∥, alt. int. $\angle$'s are $\cong$ 6. Trans. law 7. $\overline{AE} \cong \overline{AB}$ so *AE* = *AB* 8. $\triangle$ proportionality thm. 9. Substitution law

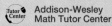

Exercises 1–6 refer to the hexagons in the figure below. Assume that
$ABCDEF \sim A' B' C' D' E' F'$.

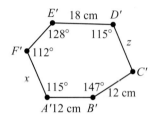

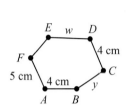

Exercises 1–6

1. Find the value of x.

2. Find the value of y.

3. Find the value of z.

4. Find the value of w.

5. Find $m\angle C'$.

6. Find $m\angle C$.

In Exercises 7–18, state whether the two polygons are *always, sometimes,*
or *never* similar.

7. Two squares.

8. Two rectangles.

9. Two right triangles.

10. Two rhombuses.

11. Two equilateral triangles.

12. Two equiangular triangles.

13. Two isosceles triangles.

14. Two scalene triangles.

15. A right triangle and an acute triangle.

16. An acute triangle and an obtuse triangle.

17. A right triangle and a scalene triangle.

18. A right triangle and an isosceles triangle.

Exercises 19–24 refer to the figure on the right in which $\overline{DE} \parallel \overline{BC}$.

19. If $AD = 12$ ft, $DB = 6$ ft, and $AE = 20$ ft, find EC.

20. If $BD = 5$ cm, $CE = 9$ cm, and $EA = 27$ cm, find AD.

21. If $AB = 22$ yd, $AE = 8$ yd, and $EC = 3$ yd, find AD and BD.

22. If $AC = 40$ in., $AD = 12$ in., and $DB = 8$ in., find AE and EC.

23. If $AC = 18$ ft, $AE = 12$ ft, and $AB = 15$ ft, find AD and BD.

24. If $AB = 30$ cm, $BD = 5$ cm, and $AC = 42$ cm, find AE and CE.

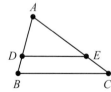

Exercises 19–24

Exercises 25–28 refer to the figure on the right in which $\overline{AD}$ bisects $\angle A$.

25. If $AB = 6$ ft, $AC = 9$ ft, and $BD = 2$ ft, find DC.

26. If $AB = 20$ cm, $BD = 8$ cm, and $DC = 14$ cm, find AC.

27. If $BC = 45$ in., $AB = 24$ in., and $AC = 36$ in., find BD and DC.

28. If $BC = 28$ yd, $BA = 18$ yd, and $CA = 24$ yd, find CD and DB.

29. Find the height of a tree that casts an 80-foot shadow at the same time that a telephone pole 18 ft tall casts a 12-foot shadow.

30. To measure the distance between points, A and B, on two islands in the figure below, a man takes the following measurements on one of the islands: $AC = 5$ yd, $DE = 6$ yd, and $AE = 15$ yd. How far apart are the islands?

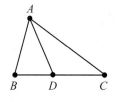

Exercises 25–28

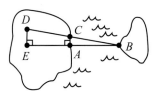

Exercise 30

31. *Given:* $\triangle ABC$ and $\angle 1 \cong \angle 2$

Prove: $\dfrac{AD}{AB} = \dfrac{DE}{BC}$

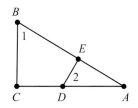

Exercise 31

32. *Given:* $\triangle ABC$ and $\square CDEF$

Prove: $\dfrac{AD}{DC} = \dfrac{CF}{FB}$

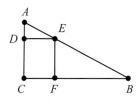

Exercise 32

33. *Given:* $\triangle ABC$ and $\overline{DE} \parallel \overline{BC}$

Prove: $\dfrac{DP}{BF} = \dfrac{PE}{FC}$

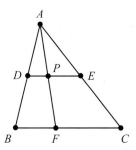

Exercise 33

34. *Given:* $\triangle ABC$, $\overline{AD}$ bisects $\angle BAC$, and $\overline{AE} \cong \overline{ED}$

Prove: $\dfrac{AE}{AC} = \dfrac{BD}{BC}$

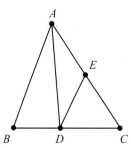

Exercise 34

35. Construct a segment proportional to $\overline{AB}$, $\overline{CD}$, and $\overline{EF}$.

Exercise 35

36. *Given:* $\triangle ABC$ with $\overline{DE} \parallel \overline{FG} \parallel \overline{AC}$ where $BE = 24$,
$BD = 18$, $EB = 16$, and $FA = 15$
Find: DF and GC

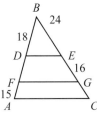

Exercise 36

37. *Given:* Trapezoid $WXYZ$ with $\overline{XY} \parallel \overline{WZ}$
Prove: $\triangle XAY \sim \triangle ZAW$

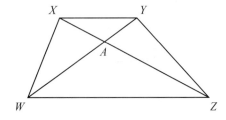

Exercise 37

38. Answer the following questions *true* or *false* and explain your reasoning.
  a. If each of two isosceles triangles has an angle that measures 120 degrees, then the two isosceles triangles must be similar.
  b. If each of two isosceles triangles has an angle that measures 40 degrees, then the two isosceles triangles must be similar.

39. Describe some real-world examples of similar figures. Some ideas might include certain business logos and seashells enlarging proportionally to accommodate the growth of the shellfish. The examples mentioned at the beginning of this section involved scale drawings.

40. Prove that the ratio of the perimeters of two similar triangles equals the ratio of the lengths of any two corresponding sides.

41. Prove that the ratio of the lengths of the altitudes from corresponding angles in similar triangles equals the ratio of the lengths of any two corresponding sides.

42. Draw segments $\overline{AB}$, $\overline{CD}$, and $\overline{EF}$ such that AB is about 2 inches, CD is about 3 inches, and EF is about 4 inches. Construct $\overline{UV}$ so that $\dfrac{AB}{CD} = \dfrac{EF}{UV}$.

43. Construct a quadrilateral similar to a given quadrilateral, where $\overline{EF}$ (on the new quadrilateral) corresponds to $\overline{AB}$.

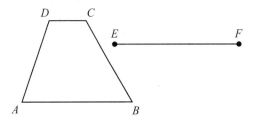

Exercise 43

44. Wires are stretched from the top of each of two poles to the bottom of the other as shown in the following figure. If one pole is 4 ft tall and the other is 12 ft tall, how far above the ground do the wires cross? [Note: The distance between the poles is not important; that is, x can be found regardless of the values of u and v.]

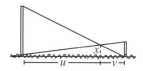

Exercise 44

45. A *pantograph* is an instrument used for reducing or enlarging a map or drawing. It consists of four bars hinged together at points A, B, C, and D so that $ABCD$ is a parallelogram and P, D, and E lie on the same line. Point P is attached to a drawing table and does not move. To produce a larger triangle 2 that is similar to triangle 1, a stylus that traces triangle 1 is inserted at D while a pen at E produces triangle 2. Although the angles of the parallelogram change, P, D, and E remain on the same line. Assume $PA = 6$ inches and $AB = 12$ inches. Answer the following:

a. Explain why $\triangle PAD \sim \triangle PBE$.

b. What is the value of $\dfrac{PA}{PB}$?

c. What is the value of $\dfrac{PD}{PE}$?

d. How do the sides of $\triangle 2$ compare to those of $\triangle 1$?

e. How would this pantograph be used to reduce a drawing?

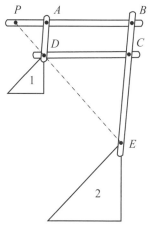

Exercise 45

5.3 Properties of Right Triangles

OBJECTIVES

1. Use theorems involving an altitude of a right triangle.
2. Prove theorem about median of right triangle.

Recall that a right triangle is a triangle that contains a right angle. The hypotenuse of a right triangle is the side opposite the right angle and the remaining two sides are called legs. In this section, we will prove similarity theorems that apply to right triangles. We begin by looking at the altitude from the right angle in a right triangle. It has an important property of similarity.

OBJECTIVE 1   Use theorems involving an altitude of a right triangle.

> **THEOREM 5.13**
>
> The altitude from the right angle to the hypotenuse in a right triangle forms two right triangles that are similar to each other and to the original triangle.

The plan for the proof of Theorem 5.13 is to redraw Figure 5.16 and from the three triangles, $\triangle ACD$, $\triangle CBD$, and $\triangle ABC$ pictured below. By showing these triangles have congruent angles, we can prove they are similar by the AA Theorem learned earlier in this chapter.

Given: Right triangle $\triangle ABC$ with altitude $\overline{CD}$ from right angle $\angle C$ (See Figure 5.16.)

Prove: $\triangle ACD \sim \triangle ABC$, $\triangle ABC \sim \triangle CBD$, and $\triangle ACD \sim \triangle CBD$

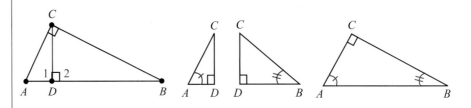

Figure 5.16

Proof

Statements	Reasons
1. $\overline{CD}$ is an altitude of right triangle $\triangle ABC$ drawn from right $\angle C$	1. Given
2. $\overline{CD} \perp \overline{AB}$	2. Def. of altitude
3. $\angle 1$ and $\angle 2$ are right angles	3. $\perp$ lines form rt.$\angle$'s
4. $\angle 1 \cong \angle ACB$	4. Rt. $\angle$'s are $\cong$
5. $\angle A \cong \angle A$	5. Reflexive law
6. $\triangle ACD \sim \triangle ABC$	6. AA
7. $\angle 2 \cong \angle ACB$	7. Rt. $\angle$'s are $\cong$
8. $\angle B \cong \angle B$	8. Reflexive law
9. $\triangle ABC \sim \triangle CBD$	9. AA
10. $\triangle ACD \sim \triangle CBD$	10. Using statements 6 and 9 and Transitive law for $\sim$

Theorem 5.13 has two important corollaries that often lead to working with quadratic equations.

COROLLARY 5.14

The altitude from the right angle to the hypotenuse in a right triangle is the geometric mean or mean proportional between the segments of the hypotenuse.

Given: Right triangle △*ABC* with altitude
$\overline{CD}$ from right angle ∠*C*
(See Figure 5.17.)

Prove: $\dfrac{AD}{CD} = \dfrac{CD}{BD}$

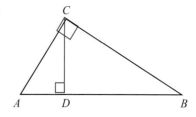

Figure 5.17

Proof _____

Statements	*Reasons*
1. △*ABC* is a right triangle with altitude $\overline{CD}$ from right angle ∠*C*	**1.** Given
2. △*ACD* ~ △*CBD*	**2.** Theorem 5.13
3. $\dfrac{AD}{CD} = \dfrac{CD}{BD}$	**3.** Corr. sides of ~ △'s are proportional

Note An alternative way to visually think of Corollary 5.14 is to examine the two small right triangles that are similar in Figure 5.17. Since the right triangles are similar, the ratio of the legs must be the same.

$$\text{In } \triangle ACD, \frac{\text{shorter leg} = AD}{\text{longer leg} = CD}. \quad \text{In } \triangle CBD, \frac{\text{shorter leg} = CD}{\text{longer leg} = BD}.$$

Thus, just remember that the ratios of the short leg to the long leg are equal.

EXAMPLE 1 Find the exact length of $\overline{CD}$ in Figure 5.18 if *DB* = 5 cm and *AD* = 16 cm. Then give the approximate length of $\overline{CD}$ to the nearest hundredth of a centimeter.

By Corollary 5.14, we have

$$\frac{AD}{CD} = \frac{CD}{DB}$$

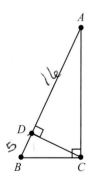

Figure 5.18

Let *x* = *CD*, substitute 16 for *AD* and 5 for *DB* and solve for *x*.

$$\frac{16}{x} = \frac{x}{5}$$

$x^2 = 16 \cdot 5$	Means-extremes property
$x = \pm\sqrt{16 \cdot 5}$	Take square root of both sides
$x = \pm 4\sqrt{5}$	Simplify

Because *CD* is a distance and cannot be negative, we discard $-4\sqrt{5}$. Thus, $CD = 4\sqrt{5}$ cm. Using a calculator we find $CD \approx 8.94$ cm.

> **COROLLARY 5.15**
>
> If the altitude is drawn from the right angle to the hypotenuse in a right triangle, then each leg is the geometric mean or mean proportional between the hypotenuse and the segment of the hypotenuse adjacent to the leg.

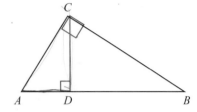

Given: Refer to Figure 5.17 (reprinted at right) showing right triangle $\triangle ABC$ with altitude $\overline{CD}$ from the right angle

Prove: $\dfrac{AD}{AC} = \dfrac{AC}{AB}$ and $\dfrac{BD}{CB} = \dfrac{CB}{AB}$

Figure 5.17

Proof

Statements	Reasons
1. $\triangle ABC$ is a right triangle with altitude $\overline{CD}$ from right angle $\angle C$	1. Given
2. $\triangle ACD \sim \triangle ABC$	2. Theorem 5.13
3. $\dfrac{AD}{AC} = \dfrac{AC}{AB}$	3. Corr. sides of $\sim$ $\triangle$'s are proportional
4. $\triangle CBD \sim \triangle ABC$	4. Theorem 5.13
5. $\dfrac{BD}{CB} = \dfrac{CB}{AB}$	5. Corr. sides of $\sim$ $\triangle$'s are proportional

Note There is an alternative way to think of Corollary 5.15 that is similar to the thinking for Corollary 5.14. Look at the small right triangle and the largest right triangle that are similar in Figure 5.17 ($\triangle ACD \sim \triangle ABC$). The ratios of the short leg to the hypotenuse are equal according to Corollary 5.15.

$$\text{In } \triangle ACD, \frac{\text{short leg} = AD}{\text{hypotenuse} = AC}. \quad \text{In } \triangle ABC, \frac{\text{short leg} = AC}{\text{hypotenuse} = AB}.$$

Thus, remember that the ratios of the short leg to the hypotenuse are equal.

EXAMPLE 2   Find the length of $\overline{PS}$ in Figure 5.19 if $QS = 12$ ft and $PR = 9$ ft.

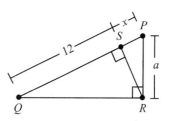

Figure 5.19

By Corollary 5.15 we have

$$\frac{PQ}{PR} = \frac{PR}{PS}.$$

Because $PQ = QS + PS$, $PQ = 12 + PS$. Let $x = PS$, and substitute 9 for PR and $12 + x$ for PQ.

$$\frac{12 + x}{9} = \frac{9}{x}$$

$$x(12 + x) = 81 \qquad \text{Means-extremes property}$$

$$12x + x^2 = 81 \qquad \text{Distributive law}$$

$$x^2 + 12x - 81 = 0$$

Using the quadratic formula, we determine $x = -6 \pm 3\sqrt{13}$. Because the length of a segment cannot be negative, we discard $-6 - 3\sqrt{13}$. Thus, $PS = (-6 + 3\sqrt{13})$ ft. In an applied problem, we would approximate this value. Using a calculator, we find $PS \approx 4.82$ ft, correct to the nearest hundredth of a foot.

An alternative way to solve this problem is to remember the ratio of the short leg to the hypotenuse is constant by Corollary 5.15, thus:

$$\frac{PS}{PR} = \frac{PR}{PQ}, \frac{x}{9} = \frac{9}{x + 12}$$

$$x(12 + x) = 81$$

thus, $x^2 + 12x - 81 = 0$

$$x = (-6 + 3\sqrt{13}) \text{ ft.}$$

OBJECTIVE 2   **Prove theorem about median of right triangle.**   The next theorem provides a useful property about the median from the right angle in a right triangle.

THEOREM 5.16

The median from the right angle in a right triangle is one-half the length of the hypotenuse.

Surveyors use an instrument called a *transit* to measure angles. Angle measurements are important in the construction of highways, bridges, tunnels, and shopping malls, where right triangles are used extensively.

PRACTICE EXERCISE **1**

Complete the proof of Theorem 5.16.

Given: $\triangle ABC$ is a right triangle where $\angle ACB$ is rt. $\angle$ and $\overline{CD}$ is a median.

Prove: $CD = \dfrac{1}{2}BA$

Auxiliary lines: Construct ℓ through D parallel to AC, thus $\overline{DE} \parallel \overline{AC}$

Proof

Statements	Reasons
1. $\triangle ABC$ is a right triangle where $\angle ACB$ is a rt. $\angle$ and $\overline{CD}$ is a median	1. _____
2. $\overline{DE} \parallel \overline{AC}$	2. By Construction
3. _____	3. Def. of median; def. $\cong$ seg.
4. $\dfrac{AD}{DB} = \dfrac{CE}{EB}$	4. _____
5. $CE = BE$, thus $\overline{CE} \cong \overline{BE}$	5. _____
6. $\overline{AC} \perp \overline{BC}$	6. If 2 lines form a right $\angle$, they are $\perp$.
7. $\overline{DE} \perp \overline{BC}$	7. _____
8. _____	8. $\perp$ lines form right $\angle$'s
9. $\triangle DBE$, $\triangle DCE$ are right $\triangle$'s	9. Def. rt. $\triangle$
10. $\overline{DE} \cong \overline{DE}$	10. _____
11. _____	11. LL
12. $\overline{BD} \cong \overline{CD}$, thus $BD = CD$	12. _____
13. $BD + DA = BA$	13. _____
14. $BD + BD = BA$	14. _____
15. $2BD = BA$	15. _____
16. _____	16. Substitution law
17. $CD = \dfrac{1}{2}BA$	17. _____

ANSWERS BELOW

Answers to Practice Exercises

1. 1. Given 3. $\overline{AD} \cong \overline{DB}$, thus $AD = DB$ 4. A line $\parallel$ to one side and intersecting two sides of a $\triangle$ divides the sides into proportional segments 5. From statements 3 and 4; def. $\cong$ seg. 7. A line $\perp$ to one of two $\parallel$ lines is $\perp$ to other 8. $\angle 1$ and $\angle 2$ are right angles 10. Reflexive law 11. $\triangle DEB \cong \triangle DEC$ 12. CPCTC; def. $\cong$ seg. 13. Seg. add. post. 14. Substitution law 15. Distributive law 16. $2CD = BA$ 17. Mult.-div. law

For Exercises 1–8, find the value of the variables in the figure.

1.

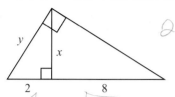

2   8

2.

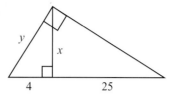

4   25

3.

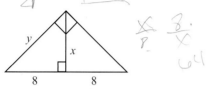

8   8

4.

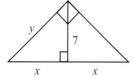

5.

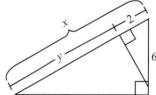

6.

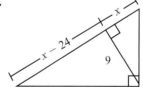

7.

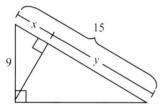

8.

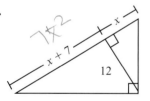

9. $\overline{AB}$ is a median

$\overline{AB}$ is a median

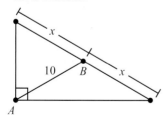

10. $\overline{AB}$ is a median

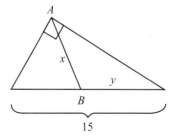

Exercises 11–26 refer to the figure below in which △*ABC*, △*ACD*, and △*BCD* are all right triangles, and *E* is the midpoint of $\overline{AB}$. When appropriate, give an approximate answer correct to the nearest hundredth of a unit.

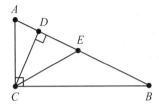

11. If *AB* = 20 cm, find *CE*.
13. If *AD* = 4 cm and *BD* = 9 cm, find *CD*.
15. If *CE* = 5 ft, find *AB*.
17. If *AD* = 3 cm and *CD* = 9 cm, find *BD*.
19. If *AB* = 32 yd and *AD* = 2 yd, find *AC*.
21. If *AB* = 10 ft and *AD* = 3 ft, find *AC*.
23. If *BD* = 8 cm and *AC* = 7 cm, find *AD*.
25. If *CE* = 5 yd and *BD* = 8 yd, find *BC*.

12. If *AB* = 68 yd, find *CE*.
14. If *AD* = 18 cm and *BD* = 50 cm, find *CD*.
16. If *CE* = 13 yd, find *AB*.
18. If *BD* = 18 ft and *CD* = 6 ft, find *AD*.
20. If *AB* = 50 cm and *BD* = 32 cm, find *BC*.
22. If *AB* = 14 yd and *BD* = 7 yd, find *BC*.
24. If *AD* = 2 ft and *BC* = 11 ft, find *BD*.
26. If *CE* = 7 cm and *AD* = 2 cm, find *AC*.

 27. In a paragraph, describe the differences and similarities of a median and an altitude of a triangle.

(5.4) The Pythagorean Theorem

OBJECTIVES

1. Prove the Pythagorean Theorem.
2. Apply the Pythagorean Theorem.
3. Prove the converse of Pythagorean Theorem.
4. Identify special cases of the Pythagorean Theorem.

OBJECTIVE 1   Prove the Pythagorean Theorem.   The Pythagorean Theorem is, perhaps, the most famous and most useful of all theorems in geometry. It has numerous applications in algebra, trigonometry, and calculus as well as many practical applications in everyday life. There is evidence that a special case of the theorem was known to the Egyptians as long ago as 2000 B.C. It is believed that Pythagoras, in about 525 B.C., gave the first deductive proof of the theorem, probably the one presented by Euclid in his book *Elements*. More than 250 different proofs have since been given, most of which involve finding areas of various figures. The proof we present now uses Corollary 5.15 and is one of the simplest of all the proofs of this theorem.

> **THEOREM 5.17   THE PYTHAGOREAN THEOREM**
>
> In a right triangle, the square of the length of the hypotenuse is equal to the sum of the squares of the lengths of the legs.

Given:	Right triangle $\triangle ABC$ with $c = AB$, $b = AC$, and $a = BC$ Let $x = DB$ (See Figure 5.20.)
Prove:	$a^2 + b^2 = c^2$
Auxiliary line:	Construct altitude $\overline{CD}$ from vertex C to hypotenuse $\overline{AB}$

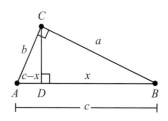

Figure 5.20

Proof

Statements	*Reasons*
1. $\triangle ABC$ is a right triangle	1. Given
2. $\overline{CD}$ is the altitude from vertex C to hypotenuse $\overline{AB}$	2. By construction
3. $\dfrac{c}{a} = \dfrac{a}{x}$ and $\dfrac{c}{b} = \dfrac{b}{c - x}$	3. Corollary 5.15
4. $a^2 = cx$ and $b^2 = c(c - x)$	4. Means-extremes prop.
5. $a^2 + b^2 = cx + c(c - x)$	5. Add.-subt. prop.
6. $a^2 + b^2 = c^2$	6. Distributive law and simplify result

Note Often we abbreviate the statement of the Pythagorean Theorem to say "In a right triangle, the sum of the squares of the legs equals the square of the hypotenuse." Also, for convenience, we usually designate the right angle in the right triangle C so we can use $a^2 + b^2 = c^2$ for the Pythagorean Theorem, with a the length of the side opposite $\angle A$, b the length of the side opposite $\angle B$, and c the length of the hypotenuse opposite $\angle C$. The hypotenuse of a right triangle is always opposite the right angle.

EXAMPLE 1 In right triangle $\triangle ABC$ (C is the right angle), $a = 12$ cm and $b = 7$ cm. Find the length of the hypotenuse, c.

Substituting into the Pythagorean Theorem we have

$$a^2 + b^2 = c^2$$
$$12^2 + 7^2 = c^2$$
$$144 + 49 = c^2$$
$$193 = c^2$$
$$\sqrt{193} = c$$

We use only the principal square root in this case because the length of a hypotenuse must be positive. We can also approximate $\sqrt{193}$ using a calculator to find $c \approx 13.9$ cm, correct to the nearest tenth of a centimeter.

PRACTICE EXERCISE ❶

In right triangle $\triangle ABC$, $c = 32$ ft and $a = 18$ ft, find b.

ANSWER ON PAGE 251

OBJECTIVE 2   **Apply the Pythagorean Theorem.**   Many applied problems can be solved using the Pythagorean Theorem.

EXAMPLE 2   A 100-foot tower is to be supported by four guy wires attached to the top of the tower and to points on the ground that are 35 ft from the base of the tower. Assume that each wire will require an extra 2 ft for attaching to the tower and to the points on the ground. How much wire will be needed for this project?

     We can make a sketch showing one of these wires and the tower as shown in Figure 5.21. The wire forms the hypotenuse of a right triangle with legs measuring 35 ft and 100 ft. Let x be the length of the wire as shown. Then by the Pythagorean Theorem,

$$x^2 = 35^2 + 100^2$$
$$= 1225 + 10{,}000$$
$$= 11{,}255.$$
$$x = \sqrt{11{,}225} \approx 105.95 \text{ ft}$$

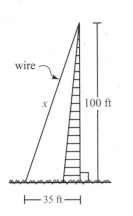

wire

x 100 ft

├── 35 ft ──┤

Figure 5.21

Add the 2 feet for attachment.

$$105.95 + 2 = 107.95 \text{ ft}$$

With four such wires, we would have

$$4(107.95) = 431.8 \text{ ft.}$$

Thus, about 432 ft of wire is required to secure the tower.

OBJECTIVE 3   **Prove the converse of the Pythagorean Theorem.**   The converse of the Pythagorean Theorem is also true and can be used to prove that a given triangle is a right triangle when its sides are given.

> **THEOREM 5.18**   **CONVERSE OF THE PYTHAGOREAN THEOREM**
>
> If the sides of a triangle have lengths a, b, and c, and $a^2 + b^2 = c^2$, then the triangle is a right triangle.

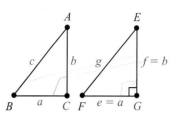

Given: $\triangle ABC$ with $a^2 + b^2 = c^2$
                     (See Figure 5.22.)

Prove: $\triangle ABC$ is a right triangle

Construction: Construct a right triangle
                     $\triangle EFG$ with $e = a$, $f = b$, and
                     $\angle G$ a right angle. (If we show
                     $\triangle ABC \cong \triangle EFG$, then $\angle C$ is
                     also a right angle making
                     $\triangle ABC$ a right triangle.)

Figure 5.22

Pythagoras (about 584–495 B.C.)

Pythagoras was one of the most remarkable mathematicians of all time, giving us many geometry proofs, including the theorem bearing his name. His society of mathematicians, the Pythagoreans, is credited with discovering the relationship between musical harmony and the length of the strings of a musical instrument and with developing the concept of irrational numbers.

Proof _____

Statements	**Reasons**
1. $\triangle ABC$ with $a^2 + b^2 = c^2$	1. Given
2. $\triangle EFG$ with $e = a$, $f = b$, and $\angle G$ a right angle	2. By construction
3. $e^2 + f^2 = g^2$	3. Pythagorean Theorem
4. $a^2 + b^2 = g^2$	4. Substitution law
5. $g^2 = c^2$	5. Trans. and sym. laws
6. $g = c$	6. The principal square roots of equal numbers are equal
7. $\triangle ABC \cong \triangle EFG$	7. SSS
8. $\angle C \cong \angle G$	8. CPCTC
9. $\angle C$ is a right angle	9. From statement 2 because $\angle G$ is a right angle
10. $\triangle ABC$ is a right triangle	10. Def. of rt. $\triangle$

EXAMPLE 3   A triangle has sides of 10 cm, 24 cm, and 26 cm. Determine if the triangle is a right triangle.

We know by the preceding theorem that a triangle is a right triangle if the sum of the squares of two sides is equal to the square of the other side. Because

$$10^2 = 100, 24^2 = 576, \text{ and } 26^2 = 676$$

and

$$10^2 + 24^2 = 100 + 576 = 676 = 26^2,$$

the triangle is a right triangle.

PRACTICE EXERCISE 2

Is a triangle with sides measuring 4 ft, 8 ft, and 9 ft a right triangle?

ANSWER ON PAGE 251

OBJECTIVE 4 | **Identify special cases of the Pythagorean Theorem.** Certain right triangles with acute angles of 45° and 45° (an isosceles right triangle), and of 30° and 60°, play an important role in the study of trigonometry.

These triangles are often referred to as a 45°-45°-90° triangle and a 30°-60°-90° triangle. The next two theorems present properties of the sides of these special triangles.

A baseball diamond is a square with sides 90 ft in length. Explain how you would use properties of a 45°-45°-90° triangle to find the distance from third base to first base.

THEOREM 5.19   45°-45°-90° THEOREM

In a 45°-45°-90° triangle, the hypotenuse is $\sqrt{2}$ times as long as each (congruent) leg.

Given: $\triangle ABC$ is a 45°-45°-90° triangle with legs of length a and hypotenuse of length c. (See Figure 5.23.)

Prove: $c = a\sqrt{2}$

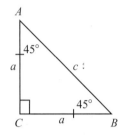

Figure 5.23

Proof _____

Statements	Reasons
1. $\triangle ABC$ is a 45°-45°-90° triangle with legs of length a and hypotenuse of length c.	1. Given
2. $a^2 + a^2 = c^2$	2. Pythagorean Theorem
3. $2a^2 = c^2$	3. Distributive law
4. $\pm\sqrt{2a^2} = c$	4. Take square root both sides
5. $\pm a\sqrt{2} = c$	5. $\sqrt{a^2} = a$
6. $c = a\sqrt{2}$	6. c cannot be negative

To summarize in figure form, if x is the length of a leg then

EXAMPLE 4   In Figure 5.24, find the lengths of the missing sides given △ABC is a 45°-45°-90° triangle.

The missing side is the hypotenuse, so its length is the length of the leg times $\sqrt{2}$, which is $6\sqrt{2}$.

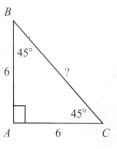

Figure 5.24

EXAMPLE 5   In Figure 5.25, find the length of the missing side.

The missing side is a leg so we write

$$5 = x\sqrt{2}$$

$$\frac{5}{\sqrt{2}} = x$$

$$x = \frac{5}{\sqrt{2}} \cdot \frac{\sqrt{2}}{\sqrt{2}}$$

$$x = \frac{5\sqrt{2}}{2}$$

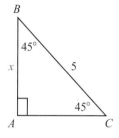

Figure 5.25

In the next theorem about the 30°-60°-90° triangle, some new terminology is used. The side opposite the 30° angle is the shortest side in the triangle so it is called the **short leg**. The side opposite the 60° angle is the longer of the two legs so it is called the **long leg**. Note the hypotenuse is always the longest side in any right triangle. The hypotenuse is not a leg of the triangle.

THEOREM 5.20   30°-60°-90° THEOREM

In a 30°-60°-90° triangle, the length of the hypotenuse is twice the length of the short leg, and the length of the long leg is $\sqrt{3}$ times as long as the length of the short leg.

An informal proof is presented. Copy △ABC (mirror image) across $\overline{BC}$ as shown in Figure 5.26 (△$ABC \cong$ △DBC). This creates △ABD, which is equiangular and, thus, equilateral by Corollary 2.8. Since △ABD is equilateral, $c = 2b$. The hypotenuse is twice as long as the short leg, and the first part of the theorem is proven. Since △ABC is a right triangle, $a^2 + b^2 = c^2$. By substituting $c = 2b$ into the equation, $a^2 + b^2 = (2b)^2$ thus $a^2 + b^2 = 4b^2$, so $a^2 = 3b^2$ and $a = b\sqrt{3}$. Thus, the long leg is $\sqrt{3}$ times the length of the short leg.

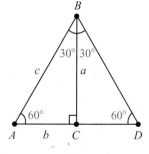

Figure 5.26

To summarize in figure form where x is the length of the short leg (opposite the 30° angle):

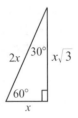

We can use the 30°-60°-90° triangle theorem to solve this applied problem.

EXAMPLE 6   A tightrope performer in a circus begins his act by walking up a wire to a platform that is 120 ft high. If the wire makes an angle of 30° with the horizontal, how far does he walk along the wire to reach the platform? Assume the pole with the platform is vertical.

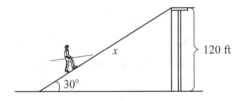

Figure 5.27

Figure 5.27 shows a sketch (not to scale) of the information given. We must find x. By Theorem 5.20, the hypotenuse is twice the short leg. Thus, we have

$$x = 2(120)$$
$$x = 240$$

Thus, the tightrope walker walks a distance of 240 ft to reach the platform.

EXAMPLE 7   Find the lengths of the missing sides of each right triangle.

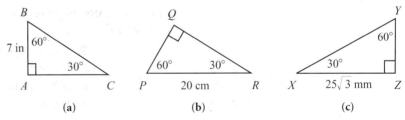

(a)   **(b)**   **(c)**

Answers

(a) $\overline{AB}$ is the short leg, so $BC = 2(7) = 14$ in.; $AC = 7\sqrt{3}$ in.

(b) $\overline{PR}$ is the hypotenuse, so $PQ = \frac{1}{2}(20) = 10$ cm; $QR = 10\sqrt{3}$ cm

(c) $\overline{XZ}$ is the long leg, so $YZ = 25$ mm; $XY = 50$ mm.

PRACTICE EXERCISE

Two airplanes leave the same airport at the same time, one flying due north and the other flying due east. If each is flying at a rate of 450 mph, use Theorem 5.19 to find the distance between the two after 2 hours, correct to the nearest tenth of a mile.

ANSWERS BELOW

Another way to investigate these special right triangles is with geometry software. Remember a 45°-45°-90° triangle is an isosceles right triangle.

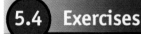 **Technology Connection**

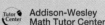

Geometric software is needed.

1. Draw a square and label it *PQRS* as in the diagram.
2. Construct diagonal $\overline{QS}$. You have now constructed an isosceles right triangle *PQS*. Explain why it's an isosceles right triangle.
3. Measure $\overline{PS}$, $\overline{PQ}$, and $\overline{QS}$. Find $\dfrac{QS}{PQ}$ and $\dfrac{QS}{PS}$. Is the ratio equal to approximately $\sqrt{2}$? Compare these results to Theorem 5.19.

Answers to Practice Exercises

1. $b = 10\sqrt{7}$ ft, which is approximately 26.5 ft   **2.** No, because $4^2 + 8^2 \neq 9^2$.
3. 1272.8 mi

5.4 Exercises

FOR EXTRA HELP: 📖 Student's Solutions Manual   Tutor Center   Addison-Wesley Math Tutor Center

In Exercises 1–8, use the Pythagorean Theorem to find the length of the missing side in right triangle $\triangle ABC$ with right angle *C*.

1. If $a = 3$ cm and $b = 4$ cm, find *c*.
2. If $a = 12$ yd and $b = 5$ yd, find *c*.
3. If $b = 25$ ft and $c = 65$ ft, find *a*.
4. If $a = 12$ cm and $c = 20$ cm, find *b*.
5. If $a = 6$ yd and $c = 11$ yd, find *b*.
6. If $b = 14$ ft and $c = 23$ ft, find *a*.
7. If $c = 2\sqrt{97}$ cm and $a = 8$ cm, find *b*.
8. If $c = 2\sqrt{130}$ cm and $b = 22$ cm, find *a*.

In Exercises 9–14, is the triangle with sides of the given lengths a right triangle?

9. 15 cm, 20 cm, 25 cm
10. 15 ft, 36 ft, 39 ft
11. 3 yd, 7 yd, $\sqrt{58}$ yd
12. $3\sqrt{3}$ cm, 6 cm, 3 cm
13. $\sqrt{7}$ ft, $\sqrt{2}$ ft, 9 ft
14. $\sqrt{11}$ yd, $\sqrt{5}$ yd, 16 yd

Exercises 15–26 refer to the 45°-45°-90° triangle shown below.

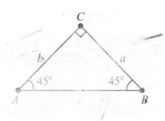

15. If $a = 10$ ft, find c.

16. If $b = 15$ cm, find c.

17. If $a = 3\sqrt{2}$ yd, find b.

18. If $b = 7\sqrt{2}$ ft, find a.

19. If $a = 3\sqrt{2}$ cm, find c.

20. If $b = 7\sqrt{2}$ yd, find c.

21. If $b = 3\sqrt{3}$ ft, find c.

22. If $a = 4\sqrt{5}$ cm, find c.

23. If $c = 6$ yd, find a.

24. If $c = 10$ ft, find b.

25. If $c = \dfrac{\sqrt{2}}{2}$ cm, find b.

26. If $c = \dfrac{\sqrt{3}}{3}$ yd, find a.

Exercises 27–38 refer to the 30°-60°-90° triangle shown below.

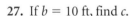

27. If $b = 10$ ft, find c.

28. If $b = 60$ cm, find c.

29. If $c = 16$ yd, find b.

30. If $c = 34$ ft, find b.

31. If $b = 7$ cm, find a.

32. If $b = 13$ yd, find a.

33. If $a = 2\sqrt{3}$ ft, find b.

34. If $a = 7\sqrt{3}$ cm, find b.

35. If $c = \sqrt{3}$ yd, find a.

36. If $c = 8\sqrt{3}$ ft, find a.

37. If $a = \sqrt{3}$ cm, find c.

38. If $a = 2\sqrt{6}$ yd, find c.

39. If the sides of a square are 4 inches long, what is the length of a diagonal?

40. If a rectangle has sides of 14 ft and 5 ft, what is the length of a diagonal?

41. A ladder 18 ft long is placed against the side of a building with the base of the ladder 6 ft from the building. To the nearest tenth of a foot, how far up the building will the ladder reach?

42. A telephone pole 35 ft tall has a guy wire attached to it 5 ft from the top and tied to a ring on the ground 15 ft from the base of the pole. Assume that an extra 2 feet of wire are needed to attach the wire to the ring and the pole. What length of wire is needed for the job? Give an answer to the nearest tenth of a foot.

43. A 400-foot tower has a guy wire attached to it that makes a 60°-angle with level ground. How far from the base of the tower is the wire anchored? Give an answer correct to the nearest tenth of a foot.

44. Two hikers leave their camp at the same time. When Dick is 6.5 mi due east of the camp, Vickie is due north of Dick and northeast of the camp. How far from the camp is Vickie? Give an answer correct to the nearest tenth of a mile.

45. Find the length of an altitude of an equilateral triangle with sides measuring 10 ft.

46. Prove that the area of an isosceles right triangle is one-fourth the square of the length of the hypotenuse.

47. Find the area of an equilateral triangle with sides measuring 10 ft.

48. Do some research on the Pythagorean Theorem. One idea is to find other applications of the theorem. This search may be done on the Internet or at the library. Write one page about your findings.

49. Find the length d of a diagonal of a cube with sides of length x. See the figure below.

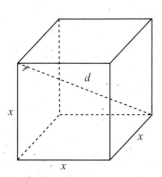

Exercise 49

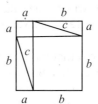

Exercise 50

50. Prove that the area of square $ABCD$ is half the area of square $ACEF$ in the figure above.

51. Draw a line segment about 1 inch in length. Construct an isosceles right triangle with legs equal in length to this segment. Then the hypotenuse is $\sqrt{2}$ times as long as the given segment. Using this hypotenuse as one leg, construct another right triangle with the second leg equal in length to the original segment. What is the length of the new hypotenuse? Can you continue this process to find a segment with length $\sqrt{4}$ times the length of the original segment? with $\sqrt{5}$ times the length of the original segment?

52. One proof of the Pythagorean Theorem involves expressing algebraically the areas of the two squares given below and equating the results. Notice that each contains copies of a right triangle with legs measuring a and b and hypotenuse measuring c. Note that both squares have sides of length $a + b$ making both areas $(a + b)^2$. Using this information, show that $a^2 + b^2 = c^2$.

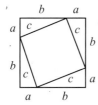

Exercise 52

53. In the left-hand figure for Exercise 52, you probably assumed that the "inside" quadrilateral with sides of length c was a square. Prove that this is indeed the case.

Student Activity

For this activity, a compass and ruler are needed.

54. **a.** Use the figure to the right to explain why the original statement of the Pythagorean Theorem was "In a right triangle,
the square *on* the hypotenuse is equal to the sum of the
squares *on* the legs."

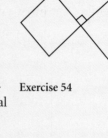

   b. Draw a right triangle and on each side construct a
semicircle with a diameter the length of the side. Compare
the sum of the areas of the semicircles on the legs with the
area of the semicircle on the hypotenuse. Remember the
area of a circle is $A = \pi r^2$.

   c. Draw a right triangle and on each side construct an equilat-   **Exercise 54**
eral triangle. Compare the sum of the areas of the equilateral
triangles on the legs with the area of the equilateral triangle
on the hypotenuse. [Hint: The area of an equilateral

   triangle is $A = \dfrac{x^2}{4}\sqrt{3}$ where x is the length of the side of

   the triangle.]

   d. State a conjecture about these areas from parts a–c and justify it.

55. Another proof of the Pythagorean Theorem uses the figure given below.
Refer to the figure to answer the following:

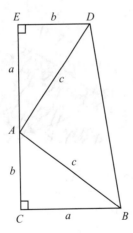

Exercise 55

a. Show that $\triangle ABD$ is a right triangle. [Hint: Show $\angle DAB$ is a right angle.]
b. Write three expressions for the areas of the three triangles.
c. Show that quadrilateral $BCED$ is a trapezoid.
d. Write an expression for the area of trapezoid $BCED$.
e. Equate an expression for the area of the trapezoid with an expression
formed by adding the areas of the three triangles and simplify the result
to obtain $a^2 + b^2 = c^2$.

5.5 Inequalities Involving Triangles (Optional)

OBJECTIVES

1. Review properties of inequalities.
2. Introduce properties of inequalities in triangles.
3. Use the SAS and SSS inequality theorems.

A water-diversion project in Arizona uses a canal to bring water from the Colorado River to central Arizona. One portion of the canal, located south of two towns, is in a straight line running west to east. An engineer wants to build a pumping station on the canal at a point from which water can be supplied to the two towns. To minimize the cost of building the two pipelines, he must find the point on the canal at which the sum of the distances from that point to the two towns is the least possible. Explain how the engineer can locate this point.

Up to now, our study of geometry has involved proving line segments and angles equal. In this section, we study relations between unequal line segments, unequal angles, and unequal arcs. We start by reviewing properties of inequalities from beginning algebra. Then we consider properties of inequalities related to triangles and circles.

The following applied problem depends on an understanding of topics introduced in this chapter, and its solution appears in Section 5.5, Example 3.

OBJECTIVE 1   **Review properties of inequalities.**   You probably studied properties of inequalities in beginning algebra. Recall that the symbol for "is less than" is $<$ and the symbol for "is greater than" is $>$. For example,

$$5 < 8, \quad 2 < 15, \quad 4 > 1, \quad \text{and} \quad 7 > 0$$

are all true statements involving inequalities.

For real numbers a and b, we could give a precise definition of the inequality $a < b$ and go on to prove several properties of inequalities. Instead, we'll assume a familiarity with inequalities and present these properties as postulates.

POSTULATE 5.2 *Trichotomy Law*

If a and b are real numbers, exactly one of the following is true:
$$a < b, a = b, \text{ or } a > b.$$

The trichotomy law states that given any two real numbers, either they are equal or one is less than the other.

POSTULATE 5.3 *Transitive Law*

If a, b, and c are real numbers with $a < b$ and $b < c$, then $a < c$.

For example, if we know that $x < 5$ and $5 < w + 1$, then $x < w + 1$.

POSTULATE 5.4 *Addition Properties of Inequalities*

If a, b, c, and d are real numbers with $a < b$ and $c < d$, then
$$a + c < b + c \quad \text{and} \quad a + c < b + d.$$

The first addition property states that if the same quantity is added to both sides of an inequality, the sums are unequal in the same order. The second addition property states that if unequal quantities are added to unequal quantities in the same order, the sums are also unequal in the same order.

POSTULATE 5.5 *Subtraction Properties of Inequalities*

If a, b, c, and d are real numbers with $a < b$ and $c = d$, then
$$a - c < b - c, a - c < b - d \quad \text{and} \quad c - a > d - b.$$

The first two subtraction properties are similar to the addition properties. The third property, however, states that if unequal quantities are subtracted from equal quantities, the differences are unequal but in the *reverse* order. For example, $5 < 12$ and $20 = 20$, so

$$20 - 5 > 20 - 12$$
$$15 > 8.$$

POSTULATE 5.6 *Multiplication Properties of Inequalities*

If a, b, and c are real numbers with $a < b$, then
$$ac < bc \text{ if } c > 0 \quad \text{and} \quad ac > bc \text{ if } c < 0.$$

The first multiplication property states that when both sides of an inequality are multiplied by a *positive* number, then the products are unequal in the *same* order. For example, $4 < 7$ and $3 \cdot 4 < 3 \cdot 7$ because $12 < 21$. The second property states that if both sides of an inequality are multiplied by a *negative* number, then the products are unequal in the *reverse* order. For example, $4 < 7$ and $(-3)(4) > (-3)(7)$ because $-12 > -21$.

In algebra class, Burford was told that a and b are two counting numbers with $a < b$. He then gave the following "proof" that $a > b$. What is wrong with Burford's "proof"?

$$a < b$$
$$a - b < 0$$
$$(a - b)^2 > 0$$
$$a^2 - 2ab + b^2 > 0$$
$$a^2 - 2ab > -b^2$$
$$a^2 + a^2 - 2ab > a^2 - b^2$$
$$2a^2 - 2ab > a^2 - b^2$$
$$2a(a - b) > (a + b)(a - b)$$
$$2a > a + b$$
$$2a - a > a - a + b$$
$$a > b$$

> **POSTULATE 5.7   Division Properties of Inequalities**
>
> If a, b, and c are real numbers with $a < b$, then
>
> $$\frac{a}{c} < \frac{b}{c} \text{ if } c > 0 \text{ and } \frac{a}{c} > \frac{b}{c} \text{ if } c < 0.$$

The division property is similar to the multiplication property. Notice that when both sides of an inequality are divided by the same *negative* number, the inequality symbol is *reversed*.

> **POSTULATE 5.8   The Whole Is Greater Than Its Parts**
>
> If a, b, and c are real numbers with $c = a + b$ and $b > 0$, then $c > a$.

Note In many algebra texts, Postulate 5.8 is presented as a definition of "less than" or "greater than" and the remaining postulates are proved as theorems. Note that although the postulates were stated using either $<$ or $>$, they are true for both inequality symbols.

The next example shows how the properties of inequalities are used to solve simple inequalities.

EXAMPLE 1   Solve the inequality.

$$2x - 3 < 3x + 5$$
$$2x - 3 + 3 < 3x + 5 + 3 \quad \text{Add 3 to both sides (Post. 5.4)}$$
$$2x < 3x + 8 \quad \text{Simplify}$$
$$2x - 3x < 3x - 3x + 8 \quad \text{Subtract } 3x \text{ from both sides (Post. 5.5)}$$
$$-x < 8$$
$$(-1)(-x) > (-1)(8) \quad \text{Multiply both sides by } -1 \text{ and reverse the inequality symbol (Post. 5.6)}$$
$$x > -8$$

Thus, the given inequality is true when x is any number greater than -8. We usually give the solution simply as $x > -8$.

> **PRACTICE EXERCISE 1**
>
> Solve $5(2x - 1) > 12x + 7$.
>
> **ANSWER ON PAGE 264**

OBJECTIVE 2 **Introduce properties of inequalities in triangles.** The next postulate seems simple, but it will be needed to prove the following theorems involving inequalities and triangles.

POSTULATE 5.9

The measure of each angle of a triangle is greater than 0°.

The next three theorems are about inequalities and triangles. Before looking at the theorems, try this experiment with a protractor and ruler or with geometry software.

 Technology Connection

Geometry software will be needed.

1. Draw an obtuse scalene triangle ABC, where angle A is the obtuse angle.
2. Extend $\overline{AB}$ to form the exterior angle CAR.
3. Measure $\angle BCA$, $\angle CBA$, $\angle CAB$, and $\angle CAR$. Which is larger $m\angle BCA$ or $m\angle CAR$? Which is larger $m\angle CBA$ or $m\angle CAR$?
4. Repeat steps 1 and 2 using a right scalene triangle and an acute scalene triangle as in the figure. Do not erase the obtuse scalene triangle.

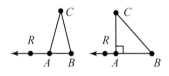

5. Is $m\angle CAR$ (the exterior angle) still greater than $m\angle BCA$ and $m\angle CBA$? Read Theorem 5.21. Do the drawings support the theorem?
6. Measure all the sides of the three triangles drawn.
7. In all three cases, was the smallest angle opposite the shortest side? Was the largest angle opposite the longest side? Read Theorems 5.22 and 5.23.
8. In all three cases, did the measure of the exterior angle equal the sum of the measures of the remote interior angles as previously studied?

THEOREM 5.21

The measure of an exterior angle of any triangle is greater than each remote interior angle.

Given: $\triangle ABC$ and exterior $\angle 1$
            (See Figure 5.28.)

Prove: $m\angle 1 > m\angle B$

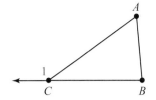

Figure 5.28

Proof

Statements	Reasons
1. $\triangle ABC$	1. Given
2. $m\angle 1 = m\angle B + m\angle A$	2. Ext. $\angle$ = sum of remote int. $\angle$'s
3. $m\angle A > 0°$	3. An $\angle$ of a $\triangle > 0°$
4. $m\angle 1 > m\angle B$	4. Whole is > each part

We could show $m\angle 1 > m\angle A$ in a similar manner.

We know that the angles opposite congruent sides in a triangle are congruent. The next theorem considers angles opposite unequal sides.

> **THEOREM 5.22**
>
> If the measures of two sides of a triangle are unequal, then the measures of the angles opposite those sides are unequal in the same order.

Given: $\triangle ABC$ with $AB > AC$
            (See Figure 5.29.)

Prove: $m\angle ACB > m\angle B$

Auxiliary line: Construct $\overline{AD}$ on $\overline{AB}$ such
                     that $AD = AC$ and draw $\overline{DC}$

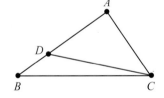

Figure 5.29

Proof

Statements	Reasons
1. $AB > AC$	1. Given
2. $AD = AC$	2. By construction
3. $m\angle ADC = m\angle ACD$	3. Angles opp. $\cong$ sides are $\cong$
4. $m\angle ACB = m\angle ACD + m\angle DCB$	4. Angle add. post.
5. $m\angle DCB > 0°$	5. An $\angle$ of a $\triangle > 0°$
6. $m\angle ACB > m\angle ACD$	6. Whole is > each part
7. $m\angle ACB > m\angle ADC$	7. Substitution law
8. $m\angle ADC > m\angle B$	8. Ext. $\angle$ of $\triangle$ is > a remote int. $\angle$
9. $m\angle ACB > m\angle B$	9. Transitive law

The converse of Theorem 5.22 is also true.

> **THEOREM 5.23**
>
> If the measures of two angles of a triangle are unequal, then the measures of the sides opposite those angles are unequal in the same order.

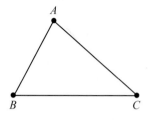

Figure 5.30

A paragraph-style indirect proof of this theorem is given making use of the trichotomy law and referring to $\triangle ABC$ in Figure 5.30.

PROOF: Suppose we are given that $m\angle B > m\angle C$. We must prove that $AC > AB$. By the trichotomy law, one of the following is true.

$$AC < AB \quad \text{or} \quad AC = AB \quad \text{or} \quad AC > AB.$$

If we show that the first two possibilities lead to contradictions, we know that $AC > AB$.

Assume that $AC < AB$. Then by Theorem 5.22, $m\angle B < m\angle C$, a contradiction, because we are given that $m\angle B > m\angle C$.

Assume that $AC = AB$. Then $m\angle B = m\angle C$ because angles opposite equal sides in a triangle are equal. This too is a contradiction because we are given that $m\angle B > m\angle C$.

Thus, $AC > AB$ because it is the only possibility that remains using the trichotomy law.

EXAMPLE 2   In $\triangle ABC$, $AB = 12$ cm, $BC = 10$ cm, and $AC = 14$ cm. Which angle of the triangle is the smallest? the largest?

By Theorem 5.22, because $AC > AB > BC$, and $\angle B$ is opposite $\overline{AC}$, $\angle C$ is opposite $\overline{AB}$, and $\angle A$ is opposite $\overline{BC}$, $m\angle B > m\angle C > m\angle A$. Thus, $\angle A$ is the smallest angle and $\angle B$ is the largest angle.

> **PRACTICE EXERCISE 2**
>
> In $\triangle PQR$, $m\angle P = 75°$ and $m\angle Q = 65°$. Which side is the shortest? Which side is the longest?
>
> ANSWERS ON PAGE 264

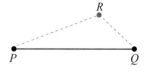

Figure 5.31 $PQ < PR + RQ$

In finding the shortest distance between two points such as P and Q in Figure 5.31, most of us would agree that it is a shorter distance to go from P to Q along the segment $\overline{PQ}$ than to go through a point off of $\overline{PQ}$. In other words, if R is a point *not* on PQ, $PQ < PR + RQ$ for any location of R. This example is a direct result of the next theorem.

> **THEOREM 5.24   The Triangle Inequality Theorem**
>
> The sum of the lengths of any two sides of a triangle is greater than the length of the third side.

Given:	$\triangle ABC$ (See Figure 5.32.)
Prove:	$AB + BC > AC$
Auxiliary lines:	Construct $\overline{BD}$ by extending $\overrightarrow{AB}$ and locating D on $\overrightarrow{BA}$ such that $BD = BC$; then construct $\overline{CD}$

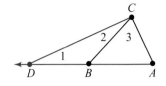

Figure 5.32

Proof _____

Statements	Reasons
1. $\triangle ABC$ is a triangle	1. Given
2. Point D is on $\overrightarrow{BA}$ with $BD = BC$	2. By construction
3. $\triangle DBC$ is a triangle	3. Def. of $\triangle$
4. $m\angle 1 = m\angle 2$	4. $\angle$'s opp. $\cong$ sides have $=$ measures
5. $m\angle ACD = m\angle 2 + m\angle 3$	5. Angle add. post.
6. $m\angle 3 > 0°$	6. An $\angle$ of a $\triangle > 0°$
7. $m\angle ACD > m\angle 2$	7. Whole is $>$ each part
8. $m\angle ACD > m\angle 1$	8. Substitution law
9. In $\triangle ACD$, $AD > AC$	9. Sides opp. $\neq \angle$'s are $\neq$ in same order
10. $AD = AB + BD$	10. Segment add. post
11. $AB + BD > AC$	11. Substitution law
12. $AB + BC > AC$	12. Substitution law

We could also show that $AC + CB > AB$ and $BA + AC > BC$ in a similar manner.

 **Student Activity**

You learned in Theorem 5.24, the sum of the lengths of two sides of a triangle is greater than the length of the third side. To explore this result a little further, use the manipulatives your instructor provides, which might be strips of paper, straws, or pipe cleaners. Make three lengths 2 inches, 5 inches, and 6 inches. Build a triangle from these pieces. Keep a record of the results. Now do the same thing with lengths 3 inches, 6 inches, and 8 inches and record the results. Repeat the experiment one more time with lengths 4 inches, 5 inches, and 9 inches. Record the results and answer the following questions.

(a) In your own words, explain the results of the experiment.
(b) Can any lengths be used to form a triangle? If not, which ones will work and explain why.
(c) Does there appear to be a minimum and maximum for the length of a third side of a triangle? If a triangle had sides of 10 inches and 14 inches, what would be the minimum and maximum length of the third side?

continued

(d) Describe, in general, how to find the minimum and maximum length of a third side of a triangle given the lengths of the other two sides.

(e) Using your rule from part (d) and the given lengths of two sides of a triangle, find the minimum and maximum length of the third side for each.
   1. 6 and 9 inches
   2. 10 and 17 inches
   3. 5 and 26 inches

We can now solve the applied problem given in the section introduction.

EXAMPLE 3

A water-diversion project in Arizona uses a canal to bring water from the Colorado River to central Arizona. One portion of the canal, located to the south of two towns, is in a straight line running west to east. An engineer wants to build a pumping station on the canal at a point from which water can be supplied to the two towns. To minimize the cost of building the two pipelines, he must find the point on the canal at which the sum of the distances from that point to the two towns is the least possible. Explain how the engineer can locate this point.

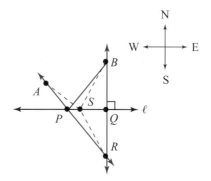

Figure 5.33

Figure 5.33 shows the information given in the problem with ℓ as the canal and A and B as the two towns. To locate the point P for the pumping station, the engineer constructs the line through B perpendicular to ℓ, locating point Q. He then determines point R on $\overleftrightarrow{BQ}$ such that $QR = BQ$. He draws line $\overrightarrow{AR}$ determining point P on ℓ, and constructs $\overline{PB}$. He then concludes that $\overline{AP}$ and $\overline{BP}$ determine the two pipelines.

To verify this, he chooses any other point on ℓ, for instance S, and shows that $AS + BS > AP + BP$. By the triangle inequality theorem, in $\triangle ASR$

$$AS + SR > AR = AP + PR.$$

But because ℓ is the perpendicular bisector of $\overline{BR}$, and S and P are on ℓ, $SR = BS$ and $BP = PR$. Substituting,

$$AS + BS > AP + BP.$$

Thus, P is the desired location for the pumping station.

OBJECTIVE 3   **Use the SAS and SSS inequality theorems.**   The next theorem considers inequalities relative to two triangles.

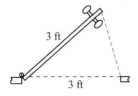

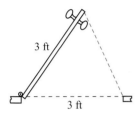

The SAS inequality theorem is sometimes called the Hinge Theorem. Consider the two top views of a door shown above. The door and the frame are each 3 ft wide. As the door opens, the angle at the hinge (the vertex of the angle) increases, and the opening between the edge of the door and the frame also increases.

> **THEOREM 5.25   SAS Inequality Theorem**
>
> If two sides of one triangle are equal in measure to two sides of another triangle, and the measure of the included angle of the first is greater than the measure of the included angle of the second, then the third side of the first triangle is greater than the third side of the second triangle.

This theorem is proved in three cases depending on whether the endpoint of a constructed segment is located on, inside, or outside the given triangle. You will be asked to verify each case in the exercises.

> **PRACTICE EXERCISE 3**
>
> In $\triangle ABC$, $AB = 8$ ft, $BC = 10$ ft, and $m\angle B = 36°$. In $\triangle DEF$, $DE = 8$ ft, $EF = 10$ ft, and $m\angle E = 42°$. What is the relationship between AC and DF?   **ANSWER ON PAGE 264**

> **THEOREM 5.26   SSS Inequality Theorem**
>
> If two sides of one triangle are equal in measure to two sides of another triangle, and the third side of the first is greater than the third side of the second, then the measure of the included angle of the first triangle is greater than the measure of the included angle of the second triangle.

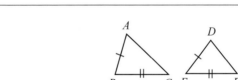

Figure 5.34

PROOF: The proof of Theorem 5.26 is indirect and uses the trichotomy law. Refer to Figure. 5.34.

Assume that $\triangle ABC$ and $\triangle DEF$ are given with $AB = DE$, $BC = EF$, and $AC > DF$ as shown in Figure 5.34. We must show that $m\angle B > m\angle E$. By the trichotomy law,

$$m\angle B < m\angle E,\ m\angle B = m\angle E, \quad \text{or} \quad m\angle B > m\angle E.$$

If we can show that the first two possibilities lead to contradictions, then we know that $m\angle B > m\angle E$, the desired conclusion.

Assume that $m\angle B < m\angle E$. By the SAS Inequality Theorem, we have $AC < DF$, a contradiction.

Assume that $m\angle B = m\angle E$, then $\triangle ABC \cong \triangle DEF$ by SAS, making $\overline{AC} \cong \overline{DF}$ by CPCTC, again a contradiction.

Because two of the possibilities cannot occur, we know that the third possibility, $m\angle B > m\angle E$, must be true by the trichotomy law.

EXAMPLE 4   In Figure 5.35, $AB = BC$ and $AD > DC$. What is the relationship between $m\angle ABD$ and $m\angle DBC$?

We can use the SSS inequality theorem on $\triangle ADB$ and $\triangle DBC$. Because $AB = BC$ and $BD = BD$, $AD > DC$ implies that $m\angle ABD > m\angle DBC$.

Answers to Practice Exercises

1. $x < -6$ **2.** The shortest side is $\overline{PQ}$, and the longest side is $\overline{QR}$.   **3.** $DF > AC$

Figure 5.35

 5.5   **Exercises**   *FOR EXTRA HELP:*   📖 Student's Solutions Manual   Tutor Center   Addison-Wesley Math Tutor Center

In Exercises 1–10, name the postulate that explains why each statement is true.

1. If $w < x$ and $x < 2$, then $w < 2$.

2. If $u < v$, then $u + 5 < v + 5$.

3. If $x < 12$, then $\dfrac{x}{2} < 6$.

4. If a is a real number, then $a < 10$, $a = 10$, or $a > 10$.

5. If $y = 5$, then $y - 2 > 5 - 4$.

6. If $w > 8$, then $5w > 40$.

7. If $x = z + 3$, then $x > z$.

8. If $\overline{AB}$ and $\overline{CD}$ are two segments, then $AB = CD$, $AB < CD$, or $AB > CD$.

9. If $m\angle A > m\angle B$ and $m\angle B > 90°$, then $m\angle A > 90°$.

10. If $\overline{AB}$, $\overline{BC}$, and $\overline{AC}$ are segments with $BC > 0$ and $AC = AB + BC$, then $AC > AB$.

In Exercises 11–18, solve each inequality and give a reason for each step in the solution.

11. $x + 3 < 7$

12. $-2y > 10$

13. $1 - 3x < 8$

14. $2y + 3 > 5y - 3$

15. $\dfrac{2x - 1}{3} < 5$

16. $\dfrac{y + 3}{-2} < 17$

17. $3(2x + 8) < 4(x - 3)$

18. $5(y + 3) + 1 > y - 4$

Exercises 19–30 refer to the figure to the right. Answer *true* or *false*. If the answer is false, explain why.

19. $m\angle A = m\angle 1$.

20. $m\angle A < m\angle 1$.

21. $m\angle 4 > m\angle 2$.

22. $m\angle 4 < m\angle A$.

23. If $AC = 7$ cm, $AB = 12$ cm, and $m\angle 2 = 110°$, then $m\angle 3 < 110°$.

24. If $AC = 6$ ft, $BC = 7$ ft, and $m\angle A = 33°$, then $m\angle 3 < 33°$.

25. If $m\angle A = 40°$, $m\angle 3 = 38°$, and $AC = 16$ yd, then $BC < 16$ yd.

26. If $m\angle 3 = 31°$, $m\angle 2 = 98°$, and $AB = 22$ cm, then $AC < 22$ cm.

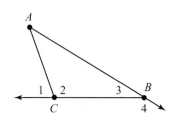

27. If $AB = 25$ ft, $AC = 15$ ft, and $BC = 20$ ft, then $m\angle 2 > m\angle A > m\angle 3$.

28. If $m\angle A = 34°$ and $m\angle 3 = 36°$, then $AB < AC < BC$.

29. If $AB = 10$ cm and $AC = 7$ cm, then $BC < 17$ cm.

30. If $AC = 7.2$ yd and $CB = 6.3$ yd, then $AB > 13.5$ yd.

Exercises 31–34 refer to the figure below.

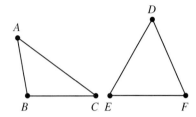

31. If $AB = EF$, $BC = DF$, and $m\angle B > m\angle F$, what is the relationship between AC and DE?

32. If $AC = DE$, $BC = DF$, and $m\angle C > m\angle D$, what is the relationship between AB and EF?

33. If $AC = 12$ cm, $BC = 8$ cm, $AB = 7$ cm, $DE = 12$ cm, $DF = 8$ cm, and $EF = 6$ cm, what is the relationship between $m\angle C$ and $m\angle D$?

34. If $AC = 10$ ft, $BC = 7$ ft, $AB = 6$ ft, $DE = 10$ ft, $DF = 8$ ft, and $EF = 6$ ft, what is the relationship between $m\angle A$ and $m\angle E$?

35. Explain why it is impossible to construct a triangle with sides 3 inches, 4 inches, and 8 inches.

36. The foreman of a ranch told his son Cal to measure the sides of a triangular pasture. Cal returned and told his father the sides are 10 mi, 12 mi, and 25 mi. Why was Cal sent to do the job again?

Exercises 37–39 present the proof of Theorem 5.25 by considering three cases.

Given: $\triangle ABC$, $\triangle DEF$, $AB = DE$, $BC = EF$,
and $m\angle ABC > m\angle E$

Auxiliary line: Construct $\overrightarrow{BP}$ such that $m\angle PBC = m\angle E$
and locate point Q on $\overrightarrow{BP}$ such that $BQ = ED$

37. Assume that Q is on $\overline{AC}$. Refer to the figure below.

Prove: $AC > DF$

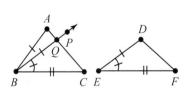

Exercise 37

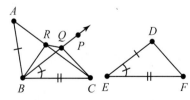

Exercise 38

38. Assume that Q is not on $\overline{AC}$ but outside $\triangle ABC$. Refer to the figure above. Construct the bisector of $\angle ABQ$ and call the point of intersection with $\overline{AC}$ point R. Construct $\overline{RQ}$ and $\overline{QC}$.

Prove: $AC > DF$

39. Assume that Q is not on $\overline{AC}$ but inside $\triangle ABC$. Refer to the figure to the right. Construct the bisector of $\angle ABQ$ and call the point of intersection with $\overline{AC}$ point R. Construct $\overline{RQ}$ and $\overline{QC}$.

Prove: $AC > DF$

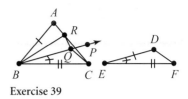

Exercise 39

40. *Given:* $\overline{CD}$ bisects $\angle ACB$

Prove: $m\angle 1 > m\angle 2$

41. *Given:* $m\angle A > m\angle B$ and $m\angle D > m\angle E$

Prove: $BE > AD$

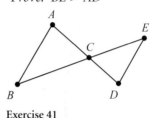

Exercise 40

Exercise 41

42. Prove that a diagonal of a rectangle is longer than any side.

43. Prove that the difference between the lengths of two sides of a triangle is less than the third side.

44. Prove that the perimeter of a quadrilateral is greater than the sum of its diagonals.

45. Prove that the sum of the lengths of the line segments drawn from any point inside a triangle to the vertices is greater than one-half the perimeter of the triangle.

46. Two cabins are located to the west of a stream that flows north to south as shown in the figure to the right. The owners of the cabins want to build a pumping station on the streambank so that the sum of the distances from the station to the cabins is the least possible. Explain how the owners should locate the point at which to build the pumping station.

Exercise 46

47. A light source at point P is reflected off the mirror M from P to Q. Prove that the path of the ray of light from P to B to Q is shorter than the path from P to C to Q, where C is any other point on line $\overline{AB}$ in the mirror. [*Hint:* $PA = AD$ because D is the perpendicular reflection of P in the mirror. Show $DB + BQ = DQ < DC + CQ$.]

48. Two sides of a triangular pasture are 3 mi and 5 mi. What is the possible range of values for the length of the third side? [*Hint:* Let x be the length of the third side, and solve the system of three inequalities obtained using the triangle-inequality theorem.]

49. The air distances from Phoenix to Denver, San Francisco, and Dallas are 589 mi, 651 mi, and 868 mi, respectively. Use the triangle-inequality theorem to find minimum and maximum air distances between **(a)** Denver and San Francisco, **(b)** San Francisco and Dallas, and **(c)** Denver and Dallas.

50. How are the SSS and SAS inequality theorems like the SSS and SAS congruence postulates? How are they different?

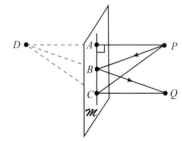

Exercise 47

Chapter (5) Review

Key Terms and Symbols

5.1 ratio
proportion
terms (of a proportion)
extremes
means

geometric mean
mean proportional
solving proportions
extended ratio
proportional segments

5.2 similar polygons (~)

5.3 approximately symbol is ≈

5.4 short leg
long leg

Chapter 5 Proof Techniques

To Prove:
Two Triangles Similar

1. Show two angles of one are equal to two angles of the other. (AA-Theorem 5.8)

2. Show the triangles are congruent. (Theorem 5.9)

A Triangle Is a Right Triangle

1. Show that it contains a right angle.

2. Show that it satisfies $a^2 + b^2 = c^2$. (Theorem 5.17)

Two Angles Unequal

1. Show one is an exterior angle of a triangle and the other is a remote interior angle. (Theorem 5.21)

2. Show they are angles opposite unequal sides in a triangle. (Theorem 5.22)

3. Show they are angles in two triangles included between equal corresponding sides but opposite unequal sides. (Theorem 5.26)

Two Segments Unequal

1. Show they are sides of a triangle opposite unequal angles of the triangle. (Theorem 5.23)

2. Show they are third sides of two triangles in which the remaining two sides of one are equal to the remaining two sides of the other, and the included angle of one is unequal to the included angle of the other. (Theorem 5.25)

Special Right Triangle Relationships

<table>
<tr><td align="center">45°-45°-90° Triangle</td><td align="center">30°-60°-90° Triangle</td></tr>
</table>

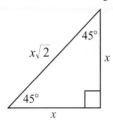

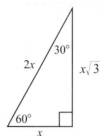

where x is the length of a leg.   where x is the length of the short leg
(opposite 30° angle)

Review Exercises

Section 5.1

In Exercises 1 and 2, write each ratio as a fraction and simplify.

1. 32 to 40

2. 300 mi in 5 hr

Solve each proportion in Exercises 3–6.

3. $\dfrac{a}{12} = \dfrac{1}{4}$ **4.** $\dfrac{4}{20} = \dfrac{8}{x}$ **5.** $\dfrac{y+5}{y} = \dfrac{21}{6}$ **6.** $\dfrac{a+3}{36} = \dfrac{a-3}{9}$

7. If 6 is to 5 as 24 is to x, find x.

8. Find the mean proportional or geometric mean between 16 and 25.

9. If $\dfrac{1}{4}$ inch on a map represents 20 mi, how many miles are represented by $2\dfrac{1}{4}$ inches?

Answer true or false in Exercises 10–12.

10. If $\dfrac{a}{b} = \dfrac{c}{d}$, then $ac = bd$.

11. If $\dfrac{a}{b} = \dfrac{c}{d}$, then $\dfrac{a+b}{b} = \dfrac{c+d}{d}$.

12. If $\dfrac{a}{b} = \dfrac{c}{d}$, then $\dfrac{a+c}{b+d} = \dfrac{a}{b}$.

Section 5.2

Exercises 13–16 refer to the quadrilaterals in the figure below.
Assume that ABCD ~ A′B′C′D′.

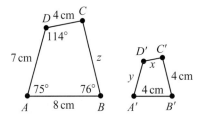

13. Find the value of *y*.   **14.** Find the value of *x*.   **15.** Find the value of *z*.   **16.** Find $m\angle C'$.

17. Are two isosceles right triangles always similar?

18. Is a right triangle ever similar to an equilateral triangle?

Exercises 19–22 refer to the figure below in which $\overline{DE} \parallel \overline{BC}$ and $\overline{AF}$ bisects $\angle A$.

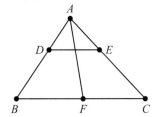

19. If $AD = 20$ ft, $DB = 15$ ft, and $AE = 28$ ft, find EC.

20. If $AC = 50$ ft, $AB = 48$ ft, $DB = 12$ ft, find AE and EC.

21. If $AB = 35$ cm, $AC = 50$ cm, and $BF = 14$ cm, find FC.

22. If $AB = 30$ ft, $AC = 80$ ft, and $BC = 77$ ft, find BF and FC.

23. *Given:* △ABC is isosceles with base $\overline{BC}$, and $\angle C$ is supplementary to $\angle 1$

Prove: $\dfrac{AD}{DB} = \dfrac{AE}{EC}$

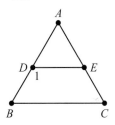

Exercise 23

24. *Given:* △ABC, $\overline{BD}$ bisects $\angle B$, and $\overline{ED} \parallel \overline{BC}$

Prove: $\dfrac{AE}{EB} = \dfrac{AB}{BC}$

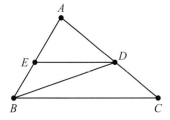

Exercise 24

25. Draw three segments $\overline{AB}$, $\overline{CD}$, and $\overline{EF}$ with AB about 3 cm, CD about 5 cm, and EF about 4 cm. Construct $\overline{UV}$ so that $\dfrac{AB}{CD} = \dfrac{EF}{UV}$.

Section 5.3

Exercises 26–30 refer to the figure below in which E is the midpoint of
$\overline{AB}$, $\overline{CD} \perp \overline{AB}$, *and* $\angle C$ *is a right angle in* $\triangle ABC$.

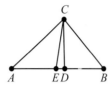

26. If $AB = 44$ cm, find CE.

27. If $AD = 25$ ft and $DB = 16$ ft, find CD.

28. If $AB = 28$ cm and $AD = 16$ cm, find AC.

29. If $BD = 5$ yd and $AC = 6$ yd, find AD.

30. If $AD = 36$ ft and $DB = 25$ ft, find the area of $\triangle ABC$.

Exercises 31 and 32 refer to the figure below in which ABCD is a rhombus with
diagonals $\overline{AC}$ *and* $\overline{BD}$.

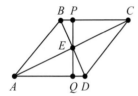

31. *Given:* $\overline{PQ} \perp \overline{BC}$
   Prove: $(PE)^2 = (BP)(PC)$

32. *Given:* $\angle EQA$ is a right angle
   Prove: $(AE)^2 = (AQ)(AD)$

Section 5.4

33. A pasture is in the shape of a right triangle with hypotenuse 100 yd and
one leg 80 yd. What is the area of the pasture?

In Exercises 34 and 35, use the Pythagorean Theorem to find the length of the
missing side in right triangle $\triangle ABC$ *with right angle C.*

34. $a = 11$ cm and $b = 8$ cm, find c.   **35.** $c = 2\sqrt{170}$ ft and $a = 14$ ft, find b.

36. Is the triangle with sides 6 yd, $\sqrt{11}$ yd, and 7 yd a right triangle?

Exercises 37–45 refer to the figure below.

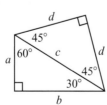

37. If $a = 30$ cm, find c.

38. If $a = 20$ yd, find b.

39. If $c = 8$ ft, find a.

40. If $c = 12$ cm, find b.

41. If $b = 3\sqrt{3}$ yd, find a.

42. If $b = 7$ ft, find c.

43. If $c = \dfrac{\sqrt{2}}{5}$ yd, find d.

44. If $d = 5\sqrt{2}$ cm, find c.

45. If $a = 3$ ft, find d.

46. Find the length of a diagonal of a rectangle with sides 18 ft and 7 ft. Give an answer correct to the nearest tenth of a foot.

47. A mountain road is inclined 30° with the horizontal. If a pickup truck drives 2 mi on this road, what change in altitude has been achieved?

48. Prove that the area of an equilateral triangle with side x is $\dfrac{\sqrt{3}}{4}x^2$.

Section 5.5

In Exercises 49–53, name the postulate that explains why each statement is true.

49. If $m\angle A < m\angle B$ and $m\angle B < m\angle C$, then $m\angle A < m\angle C$.

50. If $x < 8$, then $x - 3 < 8 - 3$.

51. If $-3x < 9$, then $x > -3$.

52. If $\overline{AB}$ and $\overline{PQ}$ are segments, then $AB = PQ$, $AB < PQ$, or $AB > PQ$.

53. If $m\angle A = m\angle B + m\angle C$ and $m\angle C = 20°$, then $m\angle A > m\angle B$.

In Exercises 54 and 55, solve each inequality and give a reason for each step in the solution.

54. $\dfrac{x+3}{-2} > 7$

55. $2(y-1) < 3y + 5$

Exercises 56–60 refer to the figure below. Answer **true** *or* **false**. *If the answer is false, explain why.*

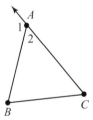

56. $m\angle B > m\angle 1$

57. $m\angle 1 > m\angle C$

58. If $AC = 10.2$ cm, $AB = 9.6$ cm, and $m\angle C = 42°$, then $m\angle B > 42°$.

59. If $m\angle C = 53°$ and $m\angle 2 = 64°$, then $BC < AC < AB$.

60. If $AC = 11.5$ yd and $BC = 10.8$ yd, then $AB < 22.3$ yd.

61. What is the relationship between AC and DE? In $\triangle ABC$ and $\triangle DEF$, $AB = EF$, $BC = DF$, and $m\angle B < m\angle F$.

62. Is it possible to have a triangle with sides measuring 10 ft, 12 ft, and 23 ft?

63. In $\triangle ABC$ median $\overline{CP}$ makes $m\angle APC > m\angle BPC$. Prove that $AC > BC$.

Chapter ⑤ Practice Test

1. Write the ratio 400 mi in 10 hr as a fraction and simplify.

2. Solve the proportion for a: $\dfrac{a}{a-2} = \dfrac{21}{15}$

3. Find the geometric mean between 4 and 36.

4. If 50 ft of wire weighs 65 lb, how much will 70 ft of the same wire weigh?

5. True or false: If $\dfrac{a}{b} = \dfrac{c}{d}$, then $\dfrac{a}{c} = \dfrac{b}{d}$.

6. If $\square ABCD \sim \square A'B'C'D'$, $AB = 18$ ft, $A'B' = 3$ ft, and $BC = 12$ ft, what is the value of $B'C'$?

7. Can two acute triangles be similar?

Exercises 8 and 9 refer to the figure below in which $\overline{DE} \parallel \overline{AB}$, $\overline{CF}$ bisects $\angle C$, $AF = 7$, $AD = 6$, $DC = 8$, $EC = 10$, $BE = x$, $BF = y$

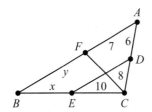

8. Find the value of x.

9. Find the value of y.

In Exercises 10 and 11, state whether the two polygons are always, sometimes, or never similar.

10. Two kites.

11. An obtuse triangle and a right triangle.

12. How tall is a tower if it casts a shadow 65 ft long at the same time that a building 160 ft tall casts a shadow 130 ft long?

13. In the figure below, $\overline{AB} \parallel \overline{ED}$. Prove $\triangle ABC \sim \triangle EDC$.

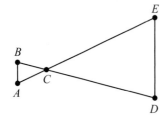

Exercise 13

14. Find the geometric mean between 15 and 20.

15. *Given:* Right $\triangle XYZ$ where $\overline{XW}$ is a median of the triangle and $YZ = 16$ inches.

    Find: WZ and XW

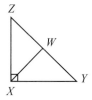

Exercise 15

Exercises 16–20 refer to the figure below in which $\overline{BD} \perp \overline{AC}$ and $\overline{AB} \perp \overline{BC}$.

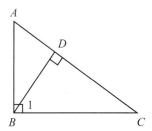

16. If $CD = 12$ cm and $AD = 5$ cm, find BD.

17. If $BC = 18$ yd and $AC = 24$ yd, find CD.

18. If $BD = 8$ ft and $AC = 20$ ft, find AD.

19. If $m\angle 1 = 45°$ and $BC = 6$ cm, find BD.

20. If $m\angle A = 60°$ and $AB = 14$ yd, find AD.

21. A garden is in the shape of a square 12 yd on a side. Mary wishes to place a picket fence across one diagonal. To the nearest tenth of a yard, how much fencing will she need?

22. In right triangle $\triangle ABC$, $\angle C$ is the right angle, $a = \sqrt{11}$ cm, and $c = 6$ cm, find b.

23. Find the area of an isosceles triangle with equal sides of length x and base angles 30°.

24. Name the postulate that explains why the following statement is true:
    If $30° < m\angle B < m\angle A$ then $m\angle B - 10° < m\angle A - 10°$.

25. Solve $4(x - 3) > 5(x - 2)$.

26. If $\angle E$ is an exterior angle of $\triangle ABC$ adjacent to $\angle A$, what is the relationship between $\angle E$ and $\angle B$?

27. In $\triangle ABC$ and $\triangle DEF$, $AC = DF$, $BC = EF$, and $AB < DE$. What is the relationship between $m\angle C$ and $m\angle F$?

28. In $\triangle ABC$, if $AB = 10$ ft and $BC = 8$ ft, then AC must be less than __?__ ft and more than __?__ ft.

6

Circles

In this chapter, we study properties of a circle and the arcs, lines, and angles associated with a circle. We'll also consider regular polygons as they are inscribed in and circumscribed around circles.

The circle is frequently used in architecture and design because of its symmetric properties. Other applications of circles are found in science and engineering, including the one given below, which is solved in Section 6.3, Example 1.

AN APPLICATION

Assume that a cross section of the Earth is a circle with radius 4000 mi. If a communications satellite is in orbit 110 mi above the Earth's surface, what is the approximate distance from the satellite to the horizon, the farthest point that can be seen on the surface of the Earth?

6.1 Circles and Arcs

OBJECTIVE 1 Define a circle and terms related to it.

One of the most familiar of all geometric figures is the *circle*. We'll begin by reviewing a few familiar terms and introducing some new ones.

> **DEFINITION:** *Circle*
>
> A **circle** is the set of all points in a plane that are located a fixed distance from a fixed point called its **center**. A line segment joining the center of a circle to one of its points is called the **radius** of the circle.

Note All radii (the plural of radius) of a circle are congruent. The symbol for a circle is $\odot$.

Figure 6.1 shows a circle with center O and radius $\overline{OP}$. Although the radius of a circle is a segment, it is common practice to call the radius the length of the segment and denote the radius by r. For example, if $OP = 5$ cm in Figure 6.1, we might say that the radius of the circle is $r = 5$ cm. The segment $\overline{QR}$ in Figure 6.1 passing through center O is called a **diameter** of the circle. We often use d to represent the diameter of a circle. In this case, $d = QR$, and it follows that $QR = QO + OR = r + r = 2r$, which proves the following theorem.

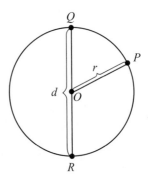

Figure 6.1

> **THEOREM 6.1**
>
> The diameter d of a circle is twice the radius r of the circle. That is, $d = 2r$.

Recall that two geometric figures are congruent if they can be made to coincide.

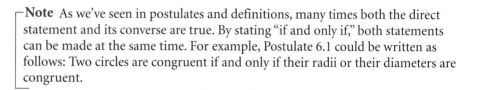

> **POSTULATE 6.1 Congruent Circles**
>
> If two circles are congruent, then their radii and diameters are congruent. Conversely, if the radii or diameters are congruent, then two circles are congruent.

Circles that lie in the same plane and have a common center are called *concentric circles*. The concentric circles of tree rings are used by foresters to study the climate and the ecology of the region in which the tree grew.

Note As we've seen in postulates and definitions, many times both the direct statement and its converse are true. By stating "if and only if," both statements can be made at the same time. For example, Postulate 6.1 could be written as follows: Two circles are congruent if and only if their radii or their diameters are congruent.

The following are more terms related to circles.

> **DEFINITION: Arcs and Semicircles**
>
> An **arc** of a circle forms a continuous part of the circle. An arc of a circle whose endpoints are the endpoints of a diameter of the circle is called a **semicircle**. An arc that is longer than a semicircle is called a **major arc** of the circle, and an arc that is shorter than a semicircle is called a **minor arc** of the circle.

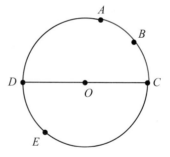

Figure 6.2 Arcs and Semicircles

Arcs of a circle can be named using three points on the arc. In Figure 6.2, minor arc ABC (shown in color), denoted by $\overparen{ABC}$ has endpoints A and C. Major arc $\overparen{ADC}$ (shown in black) has the same endpoints as minor arc $\overparen{ABC}$. If $\overline{DC}$ is a diameter of the circle, then $\overparen{DAC}$ and $\overparen{DEC}$ are both semicircles. If we use two letters to name an arc, such as $\overparen{AC}$, we always mean the minor arc with endpoints A and C. That is $\overparen{AC}$ names the same arc as $\overparen{ABC}$ in Figure 6.2.

OBJECTIVE 2   Define central and inscribed angles.

> **DEFINITION: Central Angle**
>
> An angle with sides that are radii of a circle and vertex the center of the circle is called a **central angle**.

Figure 6.3 shows central angle $\angle AOB$ that **intercepts** (cuts off) minor arc $\overset{\frown}{AB}$.

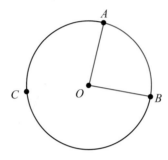

Figure 6.3   Central Angle $\angle AOB$

> **DEFINITION: Measure of an Arc**
>
> The **measure of an arc** is the number of degrees in the central angle that intercepts the arc. **Congruent arcs** are arcs with equal measure.

If $m\angle AOB = 84°$ in Figure 6.3, then the measure of $\overset{\frown}{AB}$ is 84°. We often abbreviate this statement and simply say that $m\overset{\frown}{AB} = 84°$. Because $360° - 84° = 276°$, $m\overset{\frown}{ACB} = 276°$. Notice that every minor arc has a measure less than 180°, every major arc has a measure greater than 180°, and every semicircle has a measure equal to 180°.

CAUTION The measure of an arc is not the same as the length of an arc. The length of an arc (meaning how long it is) will be discussed in Chapter 7. The measure of an arc is given using a central angle in a circle and is measured in degrees.

The next postulate is similar to Postulate 1.13 for segments. Refer to Figure 6.2.

> **POSTULATE 6.2   Arc Addition Postulate**
>
> Let A, B, and C be three points on the same circle with B between A and C. Then $m\overset{\frown}{AC} = m\overset{\frown}{AB} + m\overset{\frown}{BC}$, $m\overset{\frown}{BC} = m\overset{\frown}{AC} - m\overset{\frown}{AB}$, and $m\overset{\frown}{AB} = m\overset{\frown}{AC} - m\overset{\frown}{BC}$.

> **DEFINITION: Inscribed Angle**
>
> An angle whose vertex is on a circle and whose sides intersect the circle in two other points is called an **inscribed angle**.

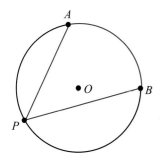

Figure 6.4 Inscribed Angle $\angle APB$

In Figure 6.4, $\angle APB$ is an inscribed angle in the circle with center O. We say that $\angle APB$ **intercepts** $\overset{\frown}{AB}$ and is **inscribed** in $\overset{\frown}{APB}$. The next theorem presents a property involving the measure of an inscribed angle.

OBJECTIVE 3 Determine the measure of an inscribed angle.

> **THEOREM 6.2**
> The measure of an inscribed angle is one-half the measure of its intercepted arc.

The complete proof of Theorem 6.2 can be accomplished by considering three cases.

Case 1: The center of the circle is on one of the sides of the inscribed angle.

Case 2: The center of the circle is in the interior of the inscribed angle.

Case 3: The center of the circle is in the exterior of the inscribed angle.

We will prove the theorem for Case 1. Proofs of Cases 2 and 3 can be accomplished by constructing a line segment through the center of the circle and following the proof of Case 1. Proof of Case 1:

Given:	Inscribed angle $\angle APB$ with O on $\overline{PB}$ (See Figure 6.5.)
Prove:	$m\angle APB = \frac{1}{2}m\overset{\frown}{AB}$
Auxiliary line:	Construct Segment $\overline{AO}$

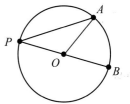

Figure 6.5

Proof _____

Statements	*Reasons*
1. O is on side $\overline{PB}$ of inscribed angle $\angle APB$	1. Given
2. $m\angle AOB = m\overset{\frown}{AB}$	2. Def. of measure of arc
3. $\overline{PO} \cong \overline{AO}$	3. Radii are congruent
4. $\angle APB \cong \angle A$ so $m\angle APB = m\angle A$	4. $\angle$'s opp. $\cong$ sides are $\cong$
5. $m\angle APB + m\angle A = m\angle AOB$	5. Ext. $\angle$ = sum of nonadj. int. $\angle$'s
6. $m\angle APB + m\angle APB = m\angle AOB$	6. Substitution law
7. $2m\angle APB = m\angle AOB$	7. Distributive law
8. $m\angle APB = \frac{1}{2}m\angle AOB$	8. Mult.-div. law
9. $m\angle APB = \frac{1}{2}m\overset{\frown}{AB}$	9. Substitution law

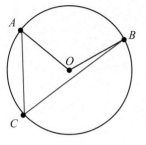

Figure 6.6

Student Activity

For this activity, the following equipment will be needed: two 6-inch squares of unlined, lightweight, white paper (like tracing paper), a compass and a straightedge.

1. With a compass, draw a circle on one piece of tracing paper and mark the center with a pencil. Draw an inscribed angle.
2. Draw a central angle intercepting the same arc as the inscribed angle.
3. Fold one side of the central angle on top of the other side and crease the paper. The fold will go through the vertex of the central angle.
4. On the second piece of paper, make a copy of the inscribed angle.
5. Slide the copy of the inscribed angle under the central angle on the original circle matching the vertices and one side of the angle. The fold will be the other side of the angle. What do you observe about the measure of the inscribed angle and the central angle? Do you see that the inscribed angle "fits" inside the central angle twice? Compare these results with Theorem 6.2.

EXAMPLE 1

In Figure 6.6, assume that $m\angle AOB = 116°$. What is the measure of $\angle ACB$?

Because the measure of central angle $\angle AOB = 116°$, $m\overarc{AB} = 116°$. With inscribed angle $\angle ACB$ intercepting $\overarc{AB}$, its measure is one-half the measure of $\overarc{AB}$. Thus,

$$m\angle ACB = \frac{1}{2}m\overarc{AB} = \frac{1}{2}(116°) = 58°.$$

PRACTICE EXERCISE 1

Assume that $m\angle ACB = 60°$ in Figure 6.6. What is the measure of $\overarc{AB}$? What is the measure of $\overarc{ACB}$?

ANSWERS ON PAGE 281

Theorem 6.2 has two useful corollaries.

COROLLARY 6.3

Inscribed angles that intercept the same or congruent arcs are congruent.

To investigate the next corollary, do the following activity.

To find the center of a given circle, a student places a sheet of paper with one corner on the circle at point P, locates points Q and R, and draws chord $\overline{QR}$.

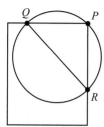

She then places the paper so that the corner is on another point S on the circle, locates points T and V, and draws chord $\overline{TV}$.

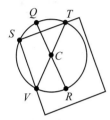

The point of intersection of $\overline{QR}$ and $\overline{TV}$, C, is the center of the circle. Can you explain why this procedure works?

Student Activity

For this activity, the following equipment will be needed: one 6-inch square of unlined, lightweight, white paper (like tracing paper), a compass, a protractor, and a straightedge.

1. Draw a circle on the paper.
2. Fold a diameter by folding the circle in half and crease the paper. Label the diameter $\overline{AB}$.
3. Make a fold that goes through point A and intersects the circle at another point. Label it C.
4. Fold $\overline{BC}$.
5. Measure $\angle ACB$. Draw $\overline{AC}$ and $\overline{BC}$ with a pencil, if needed.
6. Does your angle measure match others in the group? Will this always happen? See Corollary 6.4

COROLLARY 6.4

Every angle inscribed in a semicircle is a right angle.

EXAMPLE 2   Use Figure 6.7 to answer the following.

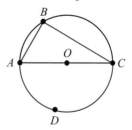

Figure 6.7

(a) What is the measure of $\overset{\frown}{ADC}$?
Because $\overset{\frown}{ADC}$ is a semicircle, $m\overset{\frown}{ADC} = 180°$

(b) What is the measure of $\angle ABC$?

Because $\angle ABC$ is inscribed in semicircle $\overset{\frown}{ABC}$, $\angle ABC$ is a right angle so $m\angle ABC = 90°$.

Answers to Practice Exercises

1. $m\overset{\frown}{AB} = 120°$; $m\overset{\frown}{ACB} = 240°$

6.1   Exercises

FOR EXTRA HELP: Student's Solutions Manual   Addison-Wesley Math Tutor Center

In Exercises 1–4, find the diameter of each circle with the given radius.

1. $r = 11$ in.   **2.** $r = 5.8$ cm   **3.** $r = \dfrac{3}{4}$ ft   **4.** $r = 13.25$ yd

In Exercises 5–8, find the radius of each circle with the given diameter.

5. $d = 16$ in.   **6.** $d = 4.8$ cm   **7.** $d = \dfrac{2}{3}$ ft   **8.** $d = 22.42$ yd

Match each part in the left column with a name in the right column, using the figure at the right. Each item in the right column is used only once.

9. $\overline{AO}$
10. $\overline{DB}$
11. $\overline{BC}$
12. $\angle AOB$
13. $\angle DBC$
14. $\angle C$
15. $\triangle AOD$
16. $\triangle BCD$
17. $\overset{\frown}{DAB}$
18. $\overset{\frown}{AB}$
19. $\overset{\frown}{ACD}$

(a) radius
(b) chord
(c) diameter
(d) inscribed angle
(e) major arc
(f) minor arc
(g) right angle
(h) isosceles triangle
(i) central angle
(j) semicircle
(k) right triangle

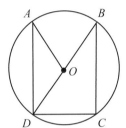

$\odot\, O$ with center at O

Exercises 9–19

20. *Given:* $\odot\, P$ with radius $PB = 5$ cm and $\overline{PB} \perp \overline{AC}$
   Find: AC and BC

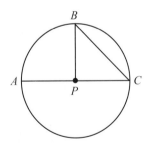

$\odot\, P$
Exercise 20

In Exercises 21–24, use the figure at the right.

Given: $\odot\, Q$ where $m\overset{\frown}{XY} : m\overset{\frown}{YZ} : m\overset{\frown}{ZX} = 5:6:7$

21. Find $m\overset{\frown}{XY}$, $m\overset{\frown}{YZ}$, $m\overset{\frown}{ZX}$
22. Find $m\angle 1$, $m\angle 2$, $m\angle 3$
23. Find $m\angle 4$
24. Find $m\angle 5$

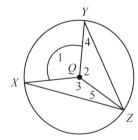

$\odot\, Q$
Exercises 21–24

25. Use the figure to the right to answer each question.
  a. What is ∠AOC called with respect to the circle?
  b. What is ∠ABC called with respect to the circle?
  c. What is the measure of $\overset{\frown}{AC}$?
  d. What is the measure of $\overset{\frown}{ABC}$?
  e. What is the measure of ∠ABC?
  f. What is the measure of the arc intercepted by ∠ABC?

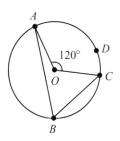

Exercise 25

26. Use the figure to the right to answer each question.
  a. What is ∠AOC called with respect to the circle?
  b. What is ∠ABC called with respect to the circle?
  c. What is the measure of ∠AOC?
  d. What is the measure of $\overset{\frown}{AC}$?
  e. What is the measure of $\overset{\frown}{ABC}$?
  f. What is the measure of the arc intercepted by ∠ABC?

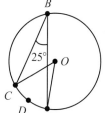

Exercise 26

27. Use the figure to the right to answer each question.
  a. What is the measure of $\overset{\frown}{ADC}$?
  b. What is the measure of $\overset{\frown}{ABC}$?
  c. What is the measure of ∠AOC?
  d. What is the measure of ∠AEC?

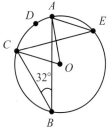

Exercise 27

28. Use the figure to the right to answer each question.
  a. What is the measure of ∠AOC?
  b. What is the measure of $\overset{\frown}{ABC}$?
  c. What is the measure of ∠ABC?
  d. What is the measure of ∠ADC?

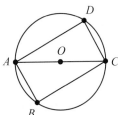

Exercise 28

29. *Given:* $\overline{AB} \perp \overline{BC}$
    Prove: $m\angle \overparen{ADC} = 180°$

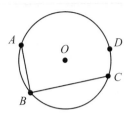

Exercise 29

30. *Given:* $\overline{AB} \parallel \overline{CD}$
    Prove: $m\overparen{AC} = m\overparen{BD}$

[Hint: Draw auxiliary segment $\overline{AD}$.]

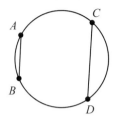

Exercise 30

31. *Given:* $\overline{AB}$ is a diameter
    $\overline{AC} \parallel \overline{OD}$
    Prove: $m\overparen{BD} = m\overparen{DC}$

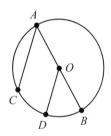

Exercise 31

32. *Given:* $\overline{AB}$ is a diameter
    $\overline{CD} \perp \overline{AB}$
    Prove: $(CD)^2 = (AD)(DB)$

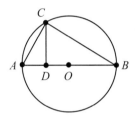

Exercise 32

 33. A classmate says the endpoints of any diameter of a circle are also endpoints of a semicircle. Do you agree? Explain your reasoning.

(6.2) Chords and Secants

OBJECTIVES

1. Define a chord of a circle.
2. Prove theorems about chords.
3. Define a secant of a circle.
4. Prove theorems about secants.

In Section 6.1, we defined the radius and diameter of a circle. Now we'll examine several properties of other segments and lines relative to a circle.

OBJECTIVE 1 Define a chord of a circle.

> **DEFINITION:** *Chord*
> A line segment joining two distinct points on a circle is called a **chord** of the circle.

In Figure 6.8, $\overline{AB}$ is a chord of the circle O. Notice that diameter $\overline{CD}$ is a special chord that passes through the center O. We say that minor arc $\overparen{AB}$ is the arc formed by chord $\overline{AB}$.

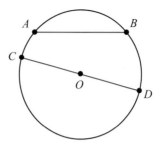

Figure 6.8 Chords of a Circle

OBJECTIVE 2   Prove theorems about chords.

> **THEOREM 6.5**
> When two chords of a circle intersect, the measure of each angle formed is one-half the sum of the measures of its intercepted arc and the arc intercepted by its vertical angle.

Given:	Chords $\overline{AB}$ and $\overline{CD}$ that intersect at point P (See Figure 6.9.)
Prove:	$m\angle 1 = \frac{1}{2}(m\overparen{BC} + m\overparen{AD})$
Auxiliary line:	Construct segment $\overline{AC}$

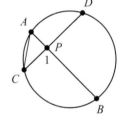

Figure 6.9

Proof _____

Statements	**Reasons**
1. Chords $\overline{AB}$ and $\overline{CD}$ intersect at point P	**1.** Given
2. $m\angle CAB = \frac{1}{2}m\overparen{BC}$ and $m\angle ACD = \frac{1}{2}m\overparen{AD}$	**2.** Measure of inscribed $\angle$ is $\frac{1}{2}$ measure of intercepted arc
3. $m\angle 1 = m\angle CAB + m\angle ACD$	**3.** Measure of ext. $\angle$ of $\triangle$ = sum of measures of nonadj. int. $\angle$'s
4. $m\angle 1 = \frac{1}{2}m\overparen{BC} + \frac{1}{2}m\overparen{AD}$	**4.** Substitution law
5. $m\angle 1 = \frac{1}{2}(m\overparen{BC} + m\overparen{AD})$	**5.** Distributive law

EXAMPLE 1

In Figure 6.10, assume that $m\widehat{AC} = 30°$ and $m\widehat{DB} = 52°$. Find the measure of $\angle 1$.

By Theorem 6.5,

$$m\angle 1 = \frac{1}{2}(m\widehat{AC} + m\widehat{BD})$$

$$= \frac{1}{2}(30° + 52°)$$

$$= \frac{1}{2}(82°) = 41°.$$

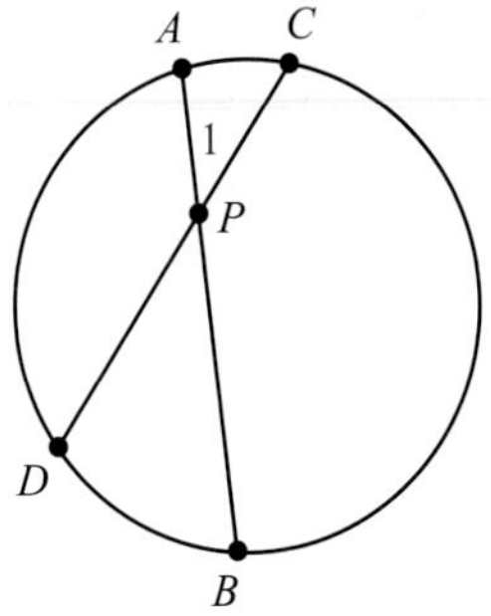

Figure 6.10

PRACTICE EXERCISE 1

In Figure 6.10, assume that $m\widehat{AD} = 112°$ and $m\widehat{BC} = 176°$.
Find the measure of $\angle 1$.

ANSWERS ON PAGE 295

THEOREM 6.6

In the same circle, the arcs formed by congruent chords are congruent.

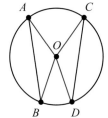

Figure 6.11

Given: $\overline{AB}$ and $\overline{CD}$ are chords with $\overline{AB} \cong \overline{CD}$ (See Figure 6.11.)

Prove: $\widehat{AB} \cong \widehat{CD}$

Auxiliary lines: Construct radii $\overline{AO}, \overline{BO}, \overline{CO},$ and $\overline{DO}$

Proof _____

Statements	Reasons
1. $\overline{AB}$ and $\overline{CD}$ are chords with $\overline{AB} \cong \overline{CD}$, so $AB = CD$	1. Given
2. $\overline{AO}, \overline{BO}, \overline{CO},$ and $\overline{DO}$ are radii	2. By construction
3. $\overline{AO} \cong \overline{BO} \cong \overline{CO} \cong \overline{DO}$	3. Radii are $\cong$
4. $\triangle AOB \cong \triangle COD$	4. SSS
5. $\angle AOB \cong \angle COD$	5. CPCTC
6. $\widehat{AB} \cong \widehat{CD}$	6. Def. of measure of an arc

The converse of Theorem 6.6 is also true, and its proof is requested in the exercises.

THEOREM 6.7

In the same circle, the chords formed by congruent arcs are congruent.

> **DEFINITION:** *Bisector of an Arc*
>
> A line that divides an arc into two arcs with the same measure is called a **bisector of the arc**.

> **THEOREM 6.8**
>
> A line drawn from the center of a circle perpendicular to a chord bisects the chord and the arc formed by the chord.

Given: Chord $\overline{AB}$ with $\overline{OD} \perp \overline{AB}$ (See Figure 6.12.)

Prove: $\overline{AD} \cong \overline{DB}$ and $\overset{\frown}{AC} \cong \overset{\frown}{CB}$

Auxiliary lines: Construct radii $\overline{OA}$ and $\overline{OB}$

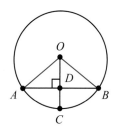

Figure 6.12

Proof

Statements	**Reasons**
1. $\overline{AB}$ is a chord with $\overline{OD} \perp \overline{AB}$	1. Given
2. $\overline{OA} \cong \overline{OB}$	2. Radii are $\cong$
3. $\angle ODA$ and $\angle ODB$ are rt. $\angle$'s	3. Def. $\perp$ lines
4. $\triangle ADO, \triangle BDO$ are rt. $\triangle$'s	4. Def. right $\triangle$'s
5. $\overline{OD} \cong \overline{OD}$	5. Reflexive law
6. $\triangle ADO \cong \triangle BDO$	6. HL
7. $\overline{AD} \cong \overline{DB}$	7. CPCTC
8. $\angle AOC \cong \angle BOC$	8. CPCTC
9. $\overset{\frown}{AC} \cong \overset{\frown}{CB}$	9. Def. of measure of an arc

Paper folding can be used to illustrate the converse of Theorem 6.8.

 Student Activity

For this activity, the following equipment will be needed: one 6-inch square of unlined, lightweight, white paper (like tracing paper), a compass, and a straightedge.

1. With a compass, draw a circle on the paper. Label the center O.
2. Draw a chord that is not a diameter of the circle. Label it $\overline{AB}$.
3. Fold the paper through point O while bringing points A and B together. The fold creates a diameter of the circle. Label it $\overline{CD}$. Label the intersection of $\overline{AB}$ and $\overline{CD}$, point R. What do you notice about point R?
4. What do you observe about $\overline{AB}$ and $\overline{CD}$? Is it true for the other members of your group?

Now read Theorem 6.9.

The converse of Theorem 6.8 is also true.

> **THEOREM 6.9**
>
> A line drawn from the center of a circle to the midpoint of a chord (not a diameter) or to the midpoint of the arc formed by the chord is perpendicular to the chord.

The proof of Theorem 6.9 is requested in the exercises.

EXAMPLE 2 In Figure 6.13, $\overline{CD} \perp \overline{OF}$, $AB = 14$ cm, $ED = 7$ cm, and $m\overset{\frown}{AB} = 84°$. Find CE and $m\overset{\frown}{FD}$.

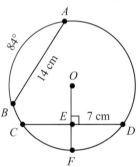

Figure 6.13

Because $\overline{OE} \perp \overline{CD}$, $\overline{OE}$ bisects chord $\overline{CD}$ and arc $\overset{\frown}{CD}$ by Theorem 6.8. Then $CE = ED$, so $CE = 7$ cm. Because chords $\overline{AB}$ and $\overline{CD}$ are both 14 cm in length, by Theorem 6.6, $m\overset{\frown}{AB} = m\overset{\frown}{CD} = 84°$. Then $m\overset{\frown}{FD}$ is one-half of $m\overset{\frown}{CD}$, so $m\overset{\frown}{FD} = 42°$.

> **THEOREM 6.10**
>
> In the same circle, congruent chords are equidistant from the center of the circle.

Given:	Chords $\overline{AB}$ and $\overline{CD}$ with $\overline{AB} \cong \overline{CD}$ (See Figure 6.14.)
Auxiliary lines:	Construct $\overline{OE} \perp \overline{AB}$, $\overline{OF} \perp \overline{CD}$, and radii $\overline{AO}$ and $\overline{CO}$
Prove:	$\overline{OE} \cong \overline{OF}$

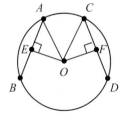

Figure 6.14

Sidebar (left margin):

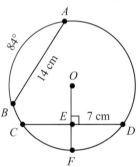

When an object is seen from two different positions, it seems to change location. The change is called *parallax.* Astronomers calculate the distance to a planet or star by using information resulting from the parallax. When a planet is viewed from two points on Earth, from *A* it appears in line with star *S*, but from *B* it is not. By measuring various angles in this figure, an astronomer can approximate the distance to the planet.

Proof

Statements	Reasons
1. $\overline{AB}$ and $\overline{CD}$ are chords with $\overline{AB} \cong \overline{CD}$	1. Given
2. $\overline{OE} \perp \overline{AB}, \overline{OF} \perp \overline{CD}$, and $\overline{AO}$ and $\overline{CO}$ are radii	2. By construction
3. $\angle AEO, \angle CFO$ are rt. $\angle$s and $\triangle AEO, \triangle CFO$ are rt. $\triangle$'s	3. Def. $\perp$ lines; def. rt. $\triangle$'s
4. $\overline{AO} \cong \overline{CO}$	4. Radii are congruent
5. $\overline{AE} \cong \overline{EB}$ and $\overline{CF} \cong \overline{FD}$ therefore $AE = EB$ and $CF = FD$	5. Line from center $\perp$ chord bisects the chord; def. $\cong$ segments
6. $AB = AE + EB$ and $CD = CF + FD$	6. Seg. add. post.
7. $AB = 2AE$ and $CD = 2CF$	7. Substitution and distributive laws
8. $2AE = 2CF$	8. Substitution
9. $AE = CF$ so $\overline{AE} \cong \overline{CF}$	9. Mult.-div. law and def. $\cong$ segments
10. $\triangle AEO \cong \triangle CFO$	10. HL
11. $\overline{OE} \cong \overline{OF}$	11. CPCTC

The converse of Theorem 6.10 is also true, and its proof is requested in the exercises.

> **THEOREM 6.11**
> In the same circle, chords equidistant from the center of the circle are congruent.

EXAMPLE 3 In Figure 6.15, $CF = 5$ ft, $OE = 7$ ft, and $AB = 10$ ft. Find OF.

Because $\overline{OF} \perp \overline{CD}$, F is the midpoint of $\overline{CD}$ by Theorem 6.8. Thus, $CF = FD = 5$ ft making $CD = 10$ ft. Because $AB = 10$ ft, $\overline{CD} \cong \overline{AB}$, and by Theorem 6.10, $\overline{OE} \cong \overline{OF}$. Thus, $OF = 7$ ft.

> **THEOREM 6.12**
> The perpendicular bisector of a chord passes through the center of the circle.

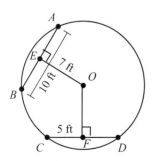

Figure 6.15

PRACTICE EXERCISE

Complete the proof of Theorem 6.12.

Given: $\overline{CE}$ is the perpendicular bisector of chord $\overline{AB}$

Prove: $\overline{CE}$ passes through O

Auxiliary line: Construct radii $\overline{AO}$ and $\overline{BO}$

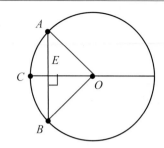

Proof _____

Statements	Reasons
1. $\overline{CE} \perp \overline{AB}$ and $AE = EB$	1. _____
2. _____	2. Radii are $\cong$
3. $\triangle AOB$ is isosceles	3. _____
4. _____	4. Def. of base of isos. $\triangle$
5. _____	5. $\perp$ bisector of base of isos. $\triangle$ passes through vertex O

ANSWERS ON PAGE 295

Technology Connection

Geometry software will be required.

1. Construct a circle and a chord that is not a diameter.
2. Label the chord $\overline{AB}$.
3. Construct the perpendicular bisector of the chord.
4. Drag point A. What do you notice?

Does this result support Theorem 6.12?

The next example shows how Theorem 6.12 can be used to determine the center of a given arc or circle.

EXAMPLE 4   A curve in a highway has the shape of an arc of a circle as shown in Figure 6.16. Explain how to find the center of this arc (the center of the circle that contains the arc).

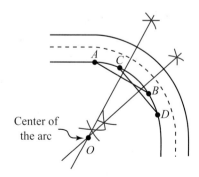

Figure 6.16

Select two points A and B on the arc and construct the perpendicular bisector of $\overline{AB}$. Now select two other points C and D on the arc and construct the perpendicular bisector of $\overline{CD}$. Because both of these bisectors pass through the center of the circle by Theorem 6.12, the point at which they intersect must be the center of the circle, hence the center of the arc.

Note All theorems considered thus far in this section are true for congruent circles as well as for the same circle.

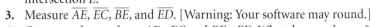

Technology Connection

Geometric software will be needed.

1. Draw a circle.
2. Choose any points A, B, C, and D on the circle and draw two intersecting chords using these four points as their endpoints. Label the point of intersection E.
3. Measure $\overline{AE}$, $\overline{EC}$, $\overline{BE}$, and $\overline{ED}$. [Warning: Your software may round.]
4. Compute the products $AE \cdot EC$ and $BE \cdot ED$. What do you observe? Compare this to Theorem 6.13.
5. Measure $\angle BEC$, $\overset{\frown}{BC}$, and $\overset{\frown}{AD}$. How do these compare? Refer to Theorem 6.5.

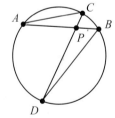

THEOREM 6.13

If two chords intersect inside a circle, the product of the lengths of the segments of one chord is equal to the product of the lengths of the segments of the other.

Given: Chords $\overline{AB}$ and $\overline{CD}$ that intersect at point P (See Figure 6.17.)

Prove: $(AP)(PB) = (CP)(PD)$

Auxiliary lines: Construct $\overline{AC}$ and $\overline{BD}$

Figure 6.17

Proof

Statements	Reasons
1. Chords $\overline{AB}$ and $\overline{CD}$ intersect at P	1. Given
2. $\overline{AC}$ and $\overline{BD}$ are chords	2. By construction
3. $\angle A \cong \angle D$ and $\angle C \cong \angle B$	3. Inscribed angles intercepting $\cong$ arcs are $\cong$
4. $\triangle APC \sim \triangle DPB$	4. AA
5. $\dfrac{AP}{PD} = \dfrac{CP}{PB}$	5. Corr. sides of $\sim \triangle$'s are proportional
6. $(AP)(PB) = (CP)(PD)$	6. Means-extremes prop.

PRACTICE EXERCISE 3

Refer to Figure 6.17 in which $AP = 3$ cm, $CP = 4$ cm, and
$PD = 6$ cm. Find PB.

ANSWER ON PAGE 295

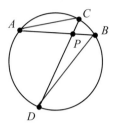

Figure 6.17 (repeated)

OBJECTIVE 3   Define a secant of a circle.   A line and a circle in the same plane can intersect in one or two points or none at all. Lines that intersect a circle in two points have several important properties.

> **DEFINITION: Secant**
>
> If a line intersects a circle in two points, the line is called a **secant**.

Line ℓ in Figure 6.18 is a secant that intersects the circle in points A and B. Notice that this secant determines chord $\overline{AB}$. All secants determine a chord.

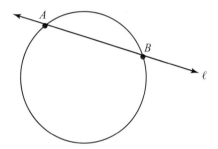

Figure 6.18 Secant

The following activity investigates the measure of an angle if the vertex is outside the circle.

 Technology Connection

Geometric software will be needed.

1. Construct a circle and draw two secants that intersect outside the circle. Label the points of intersection with the circle Q, R, S, and T. Label the point of intersection of the two secants, U. (See the diagram.)

2. Measure $\angle RUS$, $\overset{\frown}{RS}$, and $\overset{\frown}{QT}$.

3. Drag point U, but make sure the secants remain secants of the circle (not tangents or segments outside the circle). What is the $m\angle RUS$ now? Is $m\angle RUS = \frac{1}{2}(m\overset{\frown}{RS} - m\overset{\frown}{QT})$?

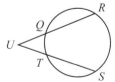

Make a conjecture about the measure of an angle formed by the intersection of two secants. Compare the conjecture to Theorem 6.14.

OBJECTIVE 4 Prove theorems about secants.

> **THEOREM 6.14**
> If two secants intersect forming an angle outside the circle, then the measure of this angle is one-half the difference of the measures of intercepted arcs.

Given: Secants $\overleftrightarrow{PA}$ and $\overleftrightarrow{PB}$ forming exterior angle $\angle APB$. (See Figure 6.19.)

Prove: $m\angle APB = \frac{1}{2}(m\widehat{AB} - m\widehat{CD})$

Auxiliary line: Construct $\overline{BC}$

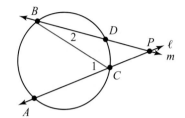

Figure 6.19

Proof

Statements	**Reasons**
1. Secants $\overleftrightarrow{PA}$ and $\overleftrightarrow{PB}$ forming $\angle APB$	1. Given
2. $m\angle 1 = m\angle 2 + m\angle APB$	2. Ext $\angle$ of $\triangle$ = sum of remote int. $\angle$'s
3. $m\angle APB = m\angle 1 - m\angle 2$	3. Add-subt. law
4. $m\angle 1 = \frac{1}{2}m\widehat{AB}$ and $m\angle 2 = \frac{1}{2}m\widehat{CD}$	4. Inscribed $\angle = \frac{1}{2}$ meas. of its intercepted arc
5. $m\angle APB = \frac{1}{2}m\widehat{AB} - \frac{1}{2}m\widehat{CD}$	5. Substitution law
6. $m\angle APB = \frac{1}{2}(m\widehat{AB} - m\widehat{CD})$	6. Distributive law

EXAMPLE 5 Refer to Figure 6.20. Assume that $m\widehat{DG} = 110°$ and $m\widehat{BF} = 40°$. Find $m\angle A$.

By Theorem 6.14,

$$m\angle A = \frac{1}{2}(m\widehat{DG} - m\widehat{BF})$$

$$= \frac{1}{2}(110° - 40°)$$

$$= \frac{1}{2}(70°) = 35°$$

Figure 6.20

Use technology to discover relationships between secants and arcs.

Technology Connection

Geometric software will be needed.

1. Draw a circle.
2. Choose any points A and B on the circle and a point D *outside* the circle.
3. Draw secants $\overline{BD}$ and $\overline{AD}$. Label the points of intersection with the circle as X and Y respectively.
4. Measure $\angle ADB$, $\overset{\frown}{AB}$, and $\overset{\frown}{XY}$. How do these compare? [Warning: Your software may round.] Refer to Theorem 6.14.
5. Measure $\overline{AY}$, $\overline{YD}$, $\overline{BX}$, and $\overline{XD}$.
6. Compute the products $AD \cdot YD$ and $BD \cdot XD$.

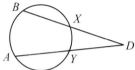

What do you observe. Compare this to Theorem 6.15.

THEOREM 6.15

If two secants are drawn to a circle from an external point, the product of the lengths of one secant segment and its external segment is equal to the product of the lengths of the other secant segment and its external segment.

Given:	Secants $\overleftrightarrow{PA}$ and $\overleftrightarrow{PB}$
	PA and PB are the lengths of the two secant segments (See Figure 6.21.)
Prove:	$(PA)(PC) = (PB)(PD)$
Auxiliary lines:	Construct segments $\overline{AD}$ and $\overline{BC}$

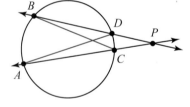

Figure 6.21

Proof _____

Statements	Reasons
1. PA and PB are the lengths of the secant segments formed by secants $\overleftrightarrow{PA}$ and $\overleftrightarrow{PB}$	1. Given
2. $\angle DBC \cong \angle DAC$	2. Both inscribed $\angle$'s intercept same arc
3. $\angle P \cong \angle P$	3. Reflexive law
4. $\triangle APD \sim \triangle BPC$	4. AA
5. $\dfrac{PA}{PB} = \dfrac{PD}{PC}$	5. Corr. sides of $\sim \triangle$'s are proportional
6. $(PA)(PC) = (PB)(PD)$	6. Means-extremes prop.

EXAMPLE 6   In Figure 6.21, assume that $BD = 8$ cm, $PD = 7$ cm, and $PC = 6$ cm. Find AC.

Let $x = AC$, then $PA = AC + PC = x + 6$. Also, $PB = BD + PD = 8 + 7 = 15$. Use Theorem 6.15 and substitute.

$$(PA)(PC) = (PB)(PD)$$
$$(x + 6)(6) = (15)(7)$$
$$6x + 36 = 105$$
$$6x = 69$$
$$x = 11.5$$

Thus, $AC = 11.5$ cm.

The table below summarizes the theorems from sections 1 and 2.

Some Properties of Circles from Sections 6.1 and 6.2

DIAGRAM	ANGLE RELATIONSHIPS	SEGMENT RELATIONSHIPS
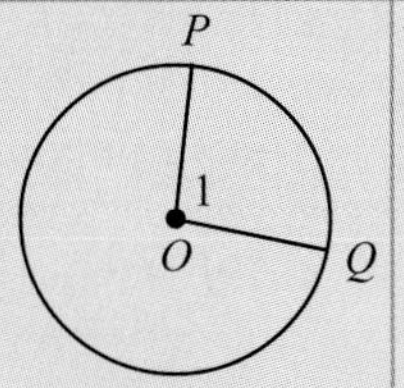 Central angle	$m\angle 1 = m\widehat{PQ}$ (O is center of circle)	$OP = OQ$ (both radii)
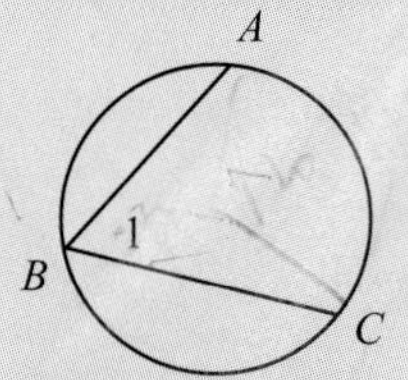 Inscribed angle	Theorem 6.2 $m\angle 1 = \frac{1}{2}m\widehat{AC}$ (vertex on circle)	(no segment relationship)
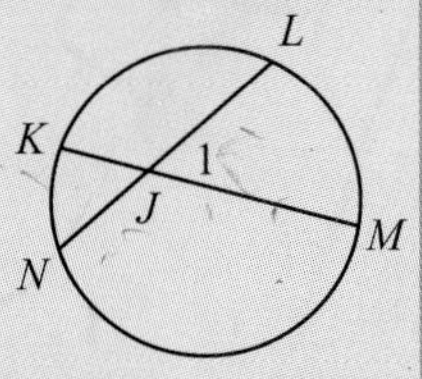	Theorem 6.5 $m\angle 1 = \frac{1}{2}(m\widehat{LM} + m\widehat{KN})$ (vertex inside circle)	Theorem 6.13 $(KJ)(JM) = (NJ)(JL)$
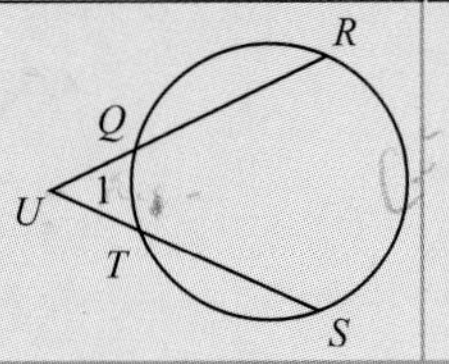	Theorem 6.14 $m\angle 1 = \frac{1}{2}(m\widehat{RS} - m\widehat{QT})$ (vertex outside circle)	Theorem 6.15 $(US)(UT) = (UR)(UQ)$

Answers to Practice Exercises

1. $36°$ **2.** 1. Given 2. $\overline{AO} \cong \overline{BO}$ 3. Def. of isos. $\triangle$ 4. $\overline{AB}$ is the base of $\triangle AOB$
5. $\overline{CE}$ passes through O **3.** 8 cm

Exercises 1–10 refer to the figure below.

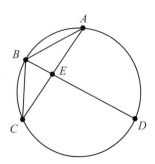

1. If $m\overset{\frown}{AB} = 58°$ and $m\overset{\frown}{CD} = 130°$, find $m\angle AEB$.
2. If $m\overset{\frown}{BC} = 64°$ and $m\overset{\frown}{AD} = 120°$, find $m\angle AED$.
3. If $\overline{AB} \cong \overline{BC}$ and $m\overset{\frown}{BC} = 60°$, find $m\overset{\frown}{AB}$.
4. If $\overline{AB} \cong \overline{BC}$ and $AB = 8$ cm, find BC.
5. If $\overline{BD}$ passes through the center of the circle and is perpendicular to $\overline{AC}$, and $AE = 12$ inches, find CE.
6. If $\overline{BD}$ passes through the center of the circle and is perpendicular to $\overline{AC}$, and $m\overset{\frown}{AB} = 55°$, find $m\overset{\frown}{BC}$.
7. If $\overline{BD}$ passes through the center of the circle, $AE = 5$ ft, and $CE = 5$ ft, find $m\angle AED$.
8. If $\overline{BD}$ passes through the center of the circle, $m\overset{\frown}{AB} = 63°$, and $m\overset{\frown}{BC} = 63°$, find $m\angle DEC$.
9. If $AE = 5$ cm, $EC = 6$ cm, and $BE = 3$ cm, find ED.
10. If $BE = 4$ ft, $ED = 6$ ft, and $AE = 3$ ft, find CE.

Exercises 11–16 refer to the figure to the right in which O is the center of the circle.

11. If $AB = 14$ ft, $CD = 14$ ft, and $OF = 11$ ft, find OE.
12. If $OE = 7$ cm, $OF = 7$ cm, and $AB = 9$ cm, find CD.
13. If $AE = 4$ cm, $OE = 5$ cm, and $OF = 5$ cm, find DF.
14. If $OE = 6$ ft, $OF = 6$ ft, and $m\overset{\frown}{AGB} = 60°$, find $m\overset{\frown}{CHD}$.
15. If $\overset{\frown}{AGB} \cong \overset{\frown}{DHC}$ and $OE = 15$ cm, find OF.
16. If $\overset{\frown}{AG} \cong \overset{\frown}{HC}$ and $OF = 9$ in., find OE.

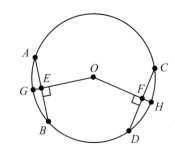

Exercises 11–16

Exercises 17–24 refer to the figure to the right.

17. If $m\overset{\frown}{AB} = 130°$ and $m\overset{\frown}{CD} = 50°$, find $m\angle P$.
18. If $m\overset{\frown}{AB} = 125°$ and $m\overset{\frown}{CD} = 47°$, find $m\angle P$.
19. If $m\overset{\frown}{AB} = 120°$ and $m\angle P = 45°$, find $m\overset{\frown}{CD}$.
20. If $m\overset{\frown}{CD} = 56°$ and $m\angle P = 42°$, find $m\overset{\frown}{AB}$.
21. If $AP = 12$ ft, $CP = 5$ ft, and $PB = 15$ ft, find PD.
22. If $BP = 25$ yd, $AP = 15$ yd, and $CP = 5$ yd, find PD.
23. If $AC = 5$ cm, $CP = 4$ cm, and $DP = 3$ cm, find BD.
24. If $BD = 3$ ft, $DP = 5$ ft, and $AC = 6$ ft, find CP.

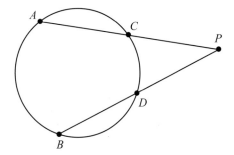

Exercises 17–24

Exercises 25 and 26 refer to the figure below.

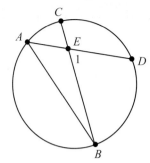

25. If $m\angle 1 = 40°$ and $m\angle DAB = 30°$, find $m\widehat{AC}$ and $m\angle ABC$.

26. If $m\angle 1 = 50°$ and $m\angle DAB = 35°$, find $m\widehat{AC}$ and $m\angle ABC$.

27. *Given:* $m\widehat{BC} + m\widehat{AD} = 180°$

   Prove: $\overline{AC} \perp \overline{BD}$

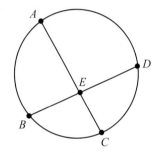

Exercise 27

28. If two equal chords of a circle intersect, prove that the segments of one are equal, respectively, to the segments of the other.

29. *Given:* $\overline{AB} \cong \overline{BC}$

   Prove: $m\angle P = \dfrac{1}{2}(m\widehat{AB} - m\widehat{AD})$

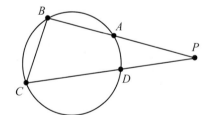

Exercise 29

30. If two equal chords $\overline{AB}$ and $\overline{CD}$ meet at point P when extended, prove that the secant segments $\overline{AP}$ and $\overline{CP}$ are equal.

31. Prove Theorem 6.7

32. Prove Theorem 6.9

33. Prove Theorem 6.11

34. How far is a 12 cm chord from the center of a circle with diameter 20 cm?

35. What is the radius of a circle in which a chord 10 inches long is 1 inch from the midpoint of the arc it forms?

36. Draw a circle with three points on it. (See the figure.) Draw $\overline{XY}$ and $\overline{YZ}$. With a compass, construct the perpendicular bisectors of the two chords. The perpendicular bisectors will intersect in a point. Where does that point appear to be? Justify your answer.

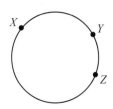

Exercise 36

6.3 **Tangents**

OBJECTIVES

1. Define the tangent of a circle.
2. Learn postulates about tangents.
3. Construct a tangent to a circle.
4. Find measures of angles formed by tangents.
5. Find lengths of segments involving tangents.
6. Investigate tangent circles.

In Section 6.2, we studied secants, which are lines that intersect a circle in two distinct points. We'll now consider lines that intersect a circle in exactly one point.

OBJECTIVE 1 Define the tangent of a circle.

> **DEFINITION: Tangent**
>
> If a line intersects a circle in one and only one point, the line is called a **tangent** to the circle. The point of intersection is called a **point of tangency**.

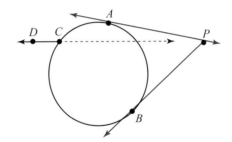

Figure 6.22 Tangent Lines

In Figure 6.22, $\overleftrightarrow{AP}$ is a tangent to the circle with A the point of tangency. We may say that ray $\overrightarrow{PB}$ and segment $\overline{PB}$ are tangent to the circle because they are contained in the tangent line $\overleftrightarrow{BP}$. $\overleftrightarrow{AP}$ and $\overleftrightarrow{BP}$ are said to be tangents to the circle from the same common external point P. Notice that $\overleftrightarrow{DC}$ is not a tangent because when $\overline{DC}$ is extended, it will intersect the circle in two points. Actually, $\overleftrightarrow{DC}$ is a secant.

OBJECTIVE 2 **Learn postulates about tangents.** The next two properties of a tangent can be proved using indirect proofs; however, here we'll consider them as postulates.

> **POSTULATE 6.3**
>
> If a line is perpendicular to a radius of a circle and passes through the point where the radius intersects the circle, then the line is a tangent.

> ### POSTULATE 6.4
>
> A radius drawn to the point of tangency of a tangent is perpendicular to the tangent.

In Figure 6.23, below, $\overline{OP}$ is a radius in circle O. If $\overleftrightarrow{AP}$ is tangent to circle O at point P, then $\overline{OP} \perp \overleftrightarrow{AP}$.

Postulate 6.3 provides a convenient way to construct a tangent to a circle at a point on the circle.

OBJECTIVE 3   Construct a tangent to a circle.

> ### CONSTRUCTION 6.1
>
> Construct a tangent to a circle at a given point on the circle.

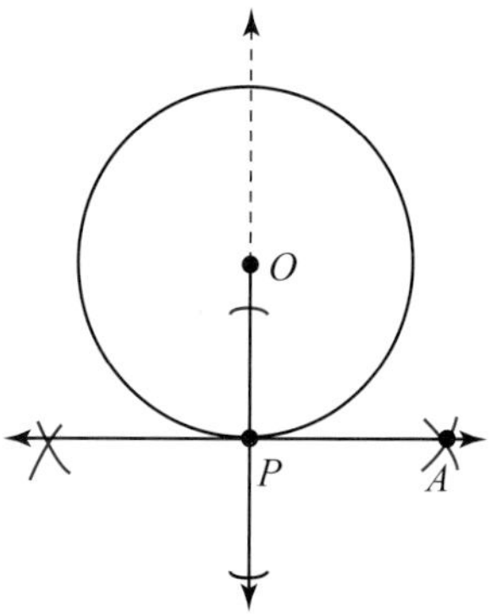

Given: P is a point on a circle with center O (See Figure 6.23.)

To Construct: $\overleftrightarrow{AP}$ tangent to the circle

Construction _____

1. Construct radius $\overline{OP}$ and extend it forming $\overleftrightarrow{OP}$.
2. Use Construction 1.4 to construct $\overleftrightarrow{AP}$ perpendicular to $\overleftrightarrow{OP}$ passing through P. By Postulate 6.3, $\overleftrightarrow{AP}$ is the desired tangent.

Postulate 6.3, together with Corollary 6.4, give us a way to construct a tangent to a circle from a point outside the circle.

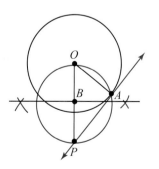

Figure 6.23

> ### CONSTRUCTION 6.2
>
> Construct a tangent to a circle from a point outside the circle.

Given: P is a point outside a circle with center O (See Figure 6.24.)

To Construct: $\overleftrightarrow{PA}$ tangent to the circle

Construction _____

1. Construct $\overline{OP}$ and use Construction 1.3 to locate the midpoint B of $\overline{OP}$.
2. Construct the circle with B as the center and PB as the radius.
3. Let A be a point of intersection of the two circles.
4. Construct $\overleftrightarrow{PA}$ and $\overline{OA}$. Then, by Corollary 6.4, $\angle OAP$ is a right angle because it is inscribed in a semicircle. Thus, $\overleftrightarrow{PA} \perp \overline{OA}$, and by Postulate 6.3, $\overleftrightarrow{PA}$ is a desired tangent.

Figure 6.24

The next example solves the applied problem given in the chapter introduction.

EXAMPLE 1

Assume that a cross section of the Earth is a circle with radius 4000 mi. If a communications satellite is in orbit 110 mi above the surface of the Earth, what is the approximate distance from the satellite to the horizon, the farthest point that can be seen on the surface of the Earth?

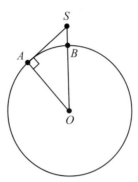

Figure 6.25

In Figure 6.25, S corresponds to the satellite, $\overline{OA}$ the radius of the Earth, and SB the height of the satellite above the surface of the Earth. The distance from the satellite to the horizon is SA, the length of segment $\overline{SA}$, which is included in tangent $\overleftrightarrow{SA}$ to the Earth's surface from S. Because $\triangle OAS$ is a right triangle, we can use the Pythagorean Theorem to obtain

$$(OS)^2 = (SA)^2 + (OA)^2.$$

Thus,

$$
\begin{aligned}
(SA)^2 &= (OS)^2 - (OA)^2 \\
&= [OB + SB]^2 - (OA)^2 \\
&= [4000 + 110]^2 - (4000)^2 \\
&= (4110)^2 - (4000)^2 \\
&= 892{,}100.
\end{aligned}
$$

Figuring the square root with a calculator, we have

$$SA \approx 945.$$

Thus, the distance to the horizon is about 945 miles.

PRACTICE EXERCISE 1

Repeat Example 1 for a space station located 95 mi above the surface of the Earth.

ANSWER ON PAGE 308

OBJECTIVE 4 **Find measures of angles formed by tangents.** We now consider two properties of angles, one formed by a tangent and a chord and the other formed by a tangent and a secant.

THEOREM 6.16

The angle formed by a tangent and a chord has a measure one-half the measure of its intercepted arc.

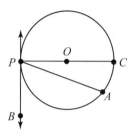

Given: ∠APB formed by tangent $\overleftrightarrow{PB}$ and chord $\overline{PA}$. (See Figure 6.26.)

Prove: $m\angle APB = \frac{1}{2}m\widehat{AP}$.

Auxiliary line: Construct diameter $\overline{CP}$

Figure 6.26

Proof

Statements	*Reasons*
1. ∠APB is formed by tangent $\overleftrightarrow{PB}$ and chord $\overline{PA}$	1. Given
2. $\overline{CP}$ is a diameter forming semicircle $\widehat{CAP}$	2. By construction
3. $m\angle APB = m\angle CPB - m\angle CPA$	3. Angle add. post.
4. $m\angle CPB = 90°$	4. Radius drawn to pt. of tangency is ⊥ to tan.
5. $m\angle CPA = \frac{1}{2}m\widehat{AC}$	5. Inscribed $\angle = \frac{1}{2}$ intercepted arc
6. $m\angle APB = 90° - \frac{1}{2}m\widehat{AC}$	6. Substitution law
7. $2m\angle APB = 180° - m\widehat{AC}$	7. Mult.-div. law
8. $m\widehat{CAP} = 180°$	8. Semicircle measures 180°
9. $2m\angle APB = m\widehat{CAP} - m\widehat{AC}$	9. Substitution law
10. $m\widehat{CAP} - m\widehat{AC} = m\widehat{AP}$	10. Arc add. post.
11. $2m\angle APB = m\widehat{AP}$	11. Transitive law
12. $m\angle APB = \frac{1}{2}m\widehat{AP}$	12. Mult.-div. law

EXAMPLE 2   In Figure 6.27, $\overleftrightarrow{AB}$ is a tangent and $m\angle CDB = 64°$. Find $m\angle CBA$.

Because ∠CDB is an inscribed angle intercepting arc $\widehat{BC}$, and $m\angle CDB = 64°$, $m\widehat{BC} = 128°$. Thus, by Theorem 6.16, $m\angle CBA = \frac{1}{2}(128°) = 64°$.

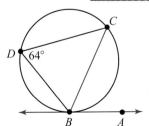

Figure 6.27

THEOREM 6.17

The angle formed by the intersection of a tangent and a secant has a measure one-half the difference of the measures of the intercepted arcs.

Given: Tangent $\overleftrightarrow{AP}$ and secant $\overleftrightarrow{PC}$ (See Figure 6.28.)

Prove: $m\angle P = \frac{1}{2}(m\widehat{AC} - m\widehat{AB})$

Auxiliary line: Construct chord $\overline{AC}$

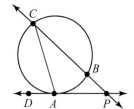

Figure 6.28

Proof _____

Statements	*Reasons*
1. $\overleftrightarrow{AP}$ is a tangent and $\overleftrightarrow{PC}$ is a secant forming $\angle P$	1. Given
2. $\overline{AC}$ is a chord	2. By construction
3. $m\angle CAD = m\angle ACB + m\angle P$	3. Measure ext. $\angle$ of $\triangle$ = sum remote int. $\angle$'s
4. $m\angle P = m\angle CAD - m\angle ACB$	4. Add.-subt. law
5. $m\angle CAD = \frac{1}{2}m\widehat{AC}$	5. Measure $\angle$ formed by tan. and chord $= \frac{1}{2}$ measure intercepted arc
6. $m\angle ACB = \frac{1}{2}m\widehat{AB}$	6. Measure of inscribed $\angle$
7. $m\angle P = \frac{1}{2}m\widehat{AC} - \frac{1}{2}m\widehat{AB}$	7. Substitution law
8. $m\angle P = \frac{1}{2}(m\widehat{AC} - m\widehat{AB})$	8. Distributive law

The proof of the next theorem is similar to that for Theorem 6.17 and is left for you to do as an exercise.

THEOREM 6.18

The angle formed by the intersection of two tangents has a measure one-half the difference of the measures of the intercepted arcs.

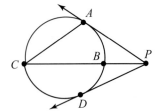

Figure 6.29

In Figure 6.29, Theorem 6.18 says $m\angle APD = \frac{1}{2}(m\widehat{ACD} - m\widehat{ABD})$.

EXAMPLE 3   Refer to Figure 6.29. If $m\angle ACB = 30°$ and $m\angle APB = 34°$, find $\overset{\frown}{AC}$.
By Theorem 6.17,

$$m\angle APB = \frac{1}{2}(m\overset{\frown}{AC} - m\overset{\frown}{AB}).$$

Because $m\angle ACB = 30°$, $m\overset{\frown}{AB} = 60°$. Substituting, we have

$$34° = \frac{1}{2}(m\overset{\frown}{AC} - 60°)$$

$$68° = m\overset{\frown}{AC} - 60° \qquad \text{Multiply both sides by 2}$$

$$128° = m\overset{\frown}{AC}. \qquad \text{Add } 60° \text{ to both sides}$$

Thus, $m\overset{\frown}{AC} = 128°$.

PRACTICE EXERCISE 2

Refer to Figure 6.29. If $m\overset{\frown}{ACD} = 236°$, find $m\angle APD$.

ANSWERS ON PAGE 308

OBJECTIVE 5   Find lengths of segments involving tangents.

 **Student Activity**

For this activity, the following equipment will be needed: one 6-inch square of unlined, lightweight, white paper (like tracing paper), a compass, and a straightedge.
1. Using the compass, draw a circle on the paper.
2. Draw a point outside the circle and label it X.
3. Fold the paper so that a tangent is formed from X to one side of the circle. Crease the paper. Label the point of tangency Y.
4. Open the paper and draw $\overline{XY}$ using a straightedge.
5. Fold the paper for another tangent from X to the other side of the circle. The two tangent lines intersect at X. Label the new point of tangency, Z.
6. Open the paper and draw $\overline{XZ}$ using a straightedge.
7. Bring points Y and Z together. Make a new fold that goes through X while the two tangents cover each other.

Compare XY and XZ. Read Theorem 6.19.

Two tangent segments to a circle from the same point have equal lengths.

DEFINITION:

Tangent Segment

A *tangent segment* is a segment that is part of a tangent line with an endpoint at the point of tangency. In Figure 6.30, $\overline{PA}$ and $\overline{PB}$ are referred to as tangent segments.

Given:	$\overleftrightarrow{PA}$ and $\overleftrightarrow{PB}$ are tangents to a circle with center O (See Figure 6.30.)
Prove:	$PA = PB$
Auxiliary lines:	Construct radii $\overline{OA}$ and $\overline{OB}$ and segment $\overline{OP}$

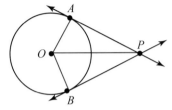

Figure 6.30

Proof —————————————————————

Statements	**Reasons**
1. $\overleftrightarrow{PA}$ and $\overleftrightarrow{PB}$ are tangents to a circle with center O	1. Given
2. $\overline{OA}$ and $\overline{OB}$ are radii	2. By construction
3. $\overline{OA} \perp \overline{AP}$ and $\overline{OB} \perp \overline{PB}$	3. Radii are $\perp$ to tangents
4. $\triangle OAP$ and $\triangle OBP$ are right triangles	4. Def. rt. $\triangle$
5. $\overline{OA} \cong \overline{OB}$	5. Radii are $\cong$
6. $\overline{OP} \cong \overline{OP}$	6. Reflexive law
7. $\triangle AOP \cong \triangle BOP$	7. HL
8. $\overline{PA} \cong \overline{PB}$	8. CPCTC
9. $PA = PB$	9. Def. $\cong$ segments

If a secant and a tangent are drawn to a circle from an external point, the length of the tangent segment is the geometric mean between the length of the secant segment and its external segment.

DEFINITION:

Secant Segment

A *secant segment* is a segment that contains a chord of a circle. An external secant segment is the part of a secant segment that lies outside the circle. In Figure 6.31, $\overline{PC}$ is a secant segment and $\overline{PB}$ is its external segment.

Given:	$\overleftrightarrow{AP}$ is a tangent and $\overline{PC}$ is a secant to circle with center O (See Figure 6.31.)
Prove:	$\dfrac{PC}{PA} = \dfrac{PA}{PB}$
Auxiliary lines:	Construct segments $\overline{AB}$ and $\overline{AC}$

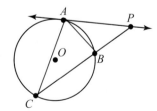

Figure 6.31

Proof _____

Statements	Reasons
1. $\overleftrightarrow{AP}$ is a tangent and $\overleftrightarrow{PC}$ is a secant	1. Given
2. $m\angle C = \frac{1}{2}m\widehat{AB}$	2. Measure of inscribed $\angle$
3. $m\angle PAB = \frac{1}{2}m\widehat{AB}$	3. The measure of $\angle$ formed by tan. and chord $= \frac{1}{2}$ measure intercepted arc
4. $m\angle C = m\angle PAB$	4. Sym. and trans. laws
5. $\angle C \cong \angle PAB$	5. Def. $\cong$ $\angle$'s.
6. $\angle P \cong \angle P$	6. Reflexive law
7. $\triangle APB \sim \triangle CPA$	7. AA
8. $\dfrac{PC}{PA} = \dfrac{PA}{PB}$	8. Corr. sides of $\sim$ $\triangle$'s are proportional

EXAMPLE 4

The gears in a watch illustrate many properties of tangent circles.

Refer to Figure 6.31. Assume that $PA = 10$ cm and $PB = 6$ cm. Find BC.

Let $x = BC$. Then $PC = BC + PB = x + 6$.
By Theorem 6.20,
$$\frac{PC}{PA} = \frac{PA}{PB}.$$

Substituting, we obtain

$$\frac{x + 6}{10} = \frac{10}{6}.$$

$6(x + 6) = (10)(10)$	Means-extremes property
$6x + 36 = 100$	Distributive law
$6x = 64$	Subtract 36 from both sides
$x = 10.666\ldots$	Divide by 6

When a decimal has a repeating block of digits, such as the digit 6 in $10.666\ldots$, we usually write the answer as $10.\overline{6}$, placing a bar over the repeating digit or digits. Thus, $BC = 10.\overline{6}$ cm.

OBJECTIVE 6 **Investigate tangent circles.** We have seen that a line and a circle can intersect in one or two points or none at all. A similar situation exists for two circles. If two circles intersect in exactly one point, the circles are said to be **tangent** to each other. Two possibilities exist. In Figure 6.32(a), the circles are **tangent internally** with point of tangency P, and in Figure 6.32(b), the circles are **tangent externally** with point of tangency Q.

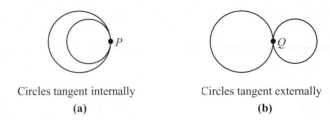

Circles tangent internally
(a)

Circles tangent externally
(b)

Figure 6.32

If two circles do not intersect, the circles can have a **common tangent**. If the circles are located on the same side of a common tangent, as in Figure 6.33(a), the tangent is called a **common external tangent**. If the circles are located on opposite sides of a common tangent, as in Figure 6.33(b), the tangent is called a **common internal tangent**.

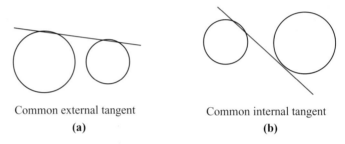

Common external tangent
(a)

Common internal tangent
(b)

Figure 6.33

DEFINITION: Line of Centers

The line passing through the centers of two circles is called their **line of centers**.

The proof of the following theorem follows from Postulates 6.6 and 1.17 and is left for you to do as an exercise.

THEOREM 6.21

If two circles are tangent internally or externally, the point of tangency is on their line of centers.

The next theorem also involves the line of centers of two circles.

THEOREM 6.22

If two circles intersect in two points, then their line of centers is the perpendicular bisector of their common chord.

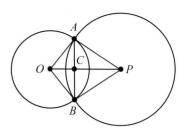

Given: Two circles with centers O and P that intersect at points A and B. (See Figure 6.34.)

Prove: $\overline{OP} \perp \overline{AB}$ and $AC = BC$

Auxiliary lines: Construct radii $\overline{OA}, \overline{OB}, \overline{PA},$ and $\overline{PB}$

Figure 6.34

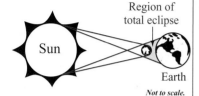

Region of total eclipse

Sun

Earth

Not to scale.

A solar eclipse occurs when the Moon passes between Earth and the sun and blocks the sun's rays. Some areas of the Earth experience no eclipse at all while others experience a partial eclipse, and a small area will be in total eclipse. The lines drawn tangent to the sun and moon in the figure show the areas affected by the eclipse. A total eclipse occurs in the area between the two external tangents.

Proof _____

Statements	Reasons
1. Two circles centered at O and P intersect at points A and B.	1. Given
2. $\overline{OA} \cong \overline{OB}$ and $\overline{PA} \cong \overline{PB}$	2. Radii are $\cong$
3. $\overline{OP} \cong \overline{OP}$	3. Reflexive law
4. $\triangle AOP \cong \triangle BOP$	4. SSS
5. $\angle AOP \cong \angle BOP$	5. CPCTC
6. $\overline{OC} \cong \overline{OC}$	6. Reflexive law
7. $\triangle AOC \cong \triangle BOC$	7. SAS
8. $\overline{AC} \cong \overline{BC}$ so $AC = BC$	8. CPCTC; def. $\cong$ seg.
9. $\angle ACO \cong \angle BCO$	9. CPCTC
10. $\angle ACO$ and $\angle BCO$ are adjacent angles	10. Def. adj. $\angle$'s
11. $\overline{OP} \perp \overline{AB}$	11. Def. of $\perp$ lines

We'll conclude this section with two constructions involving common external and internal tangents to a circle.

CONSTRUCTION 6.3

Construct a common external tangent to two given circles that are not congruent.

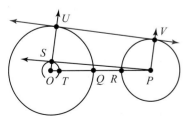

Figure 6.35

Given: Two noncongruent circles with centers O and P (See Figure 6.35.)

To Construct: $\overleftrightarrow{UV}$, an external tangent to both circles

Construction _____

1. Construct $\overline{OP}$ obtaining points Q and R, the intersections of the circles and the segment.
2. Construct $\overline{QT}$ on $\overline{QO}$ such that $QT = RP$.
3. Construct a circle centered at O with radius OT.

4. Use Construction 6.2 to construct a tangent to the circle in step 3 from point P, $\overleftrightarrow{PS}$.

5. Construct $\overrightarrow{OS}$ intersecting the original circle at point U.

6. Construct $\overrightarrow{PV}$ through P parallel to $\overrightarrow{OU}$ using Construction 3.1.

7. Construct $\overleftrightarrow{UV}$. Because $\square USPV$ is a rectangle, $\overleftrightarrow{UV}$ is the desired tangent.

CONSTRUCTION 6.4

Construct a common internal tangent to two given circles.

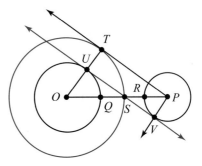

Figure 6.36

Given:	Two circles with centers O and P (See Figure 6.36.)
To Construct:	$\overleftrightarrow{UV}$, an internal tangent to both circles

Construction

1. Construct $\overline{OP}$ obtaining points Q and R, the intersections of the circles and the segment.

2. Construct $\overline{QS}$ on $\overline{QP}$ such that $QS = RP$.

3. Construct a circle centered at O with radius OS.

4. Use Construction 6.2 to construct a tangent to this circle from P, $\overleftrightarrow{PT}$.

5. Construct $\overline{OT}$ with point U the intersection of $\overline{OT}$ with the original circle.

6. Use Construction 3.1 to construct line $\overleftrightarrow{PV}$ through P parallel to $\overline{OT}$ intersecting the original circle at V.

7. Construct $\overleftrightarrow{UV}$. Because $\square UTPV$ is a rectangle, $\overleftrightarrow{UV}$ is the desired tangent.

Answers to Practice Exercises

1. approximately 877 mi   2. 56°

The table below summarizes theorems from this section.

Some Properties of Circles from Section 6.3

Diagram	Angle Relationships	Segment Relationships
	Theorem 6.16 $m\angle 1 = \frac{1}{2}m\widehat{BC}$ (vertex on circle)	(no segment relationship)
	Theorem 6.17 $m\angle 1 = \frac{1}{2}(m\widehat{JM} - m\widehat{JL})$ (vertex outside circle)	**Theorem 6.20** $\dfrac{KM}{JK} = \dfrac{JK}{KL}$
	Theorem 6.18 $m\angle 1 = \frac{1}{2}(m\widehat{YVZ} - m\widehat{YWZ})$ (vertex outside circle)	**Theorem 6.19** $XY = XZ$

6.3 Exercises

FOR EXTRA HELP: Student's Solutions Manual   Tutor Center Addison-Wesley Math Tutor Center

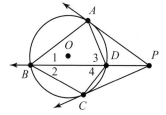

Exercises 1–22 refer to the figure to the right.

1. If $m\widehat{AD} = 70°$, find $m\angle PAD$.

2. If $m\widehat{CD} = 60°$, find $m\angle PCD$.

3. If $m\angle 1 = 36°$, find $m\angle PAD$.

4. If $m\angle 2 = 29°$, find $m\angle PCD$.

5. If $m\angle PAD = 40°$, find $m\angle 1$.

6. If $m\angle PCD = 35°$, find $m\angle 2$.

7. If $m\widehat{AB} = 140°$ and $m\widehat{AD} = 70°$, find $m\angle APD$.

8. If $m\widehat{BC} = 85°$ and $m\widehat{CD} = 61°$, find $m\angle CPD$.

9. If $m\angle 1 = 36°$ and $m\angle 3 = 70°$, find $m\angle APD$.

10. If $m\angle 2 = 30°$ and $m\angle 4 = 50°$, find $m\angle CPD$.

11. If $m\angle APD = 40°$ and $m\widehat{AB} = 138°$, find $m\angle 1$.

12. If $m\angle CPD = 30°$ and $m\widehat{BC} = 80°$, find $m\angle 2$.

13. If $m\widehat{ADC} = 130°$, find $m\angle APC$.

14. If $m\widehat{ABC} = 240°$, find $m\angle APC$.

15. If $m\angle 1 = 35°$ and $m\angle 2 = 30°$, find $m\angle APC$.

16. If $m\angle 3 = 70°$ and $m\angle 4 = 50°$, find $m\angle APC$.

17. If $AP = 17$ cm, find CP.

18. If $PC = 24$ yd, find PA.

19. If $AP = 12$ cm and $BP = 18$ cm, find DP.

20. If $PC = 8$ ft and $PB = 12$ ft, find PD.

21. If $AP = 15$ yd and $DP = 10$ yd, find BD.

22. If $PC = 24$ cm and $PD = 18$ cm, find BD.

23. If two circles do not intersect and neither is inside the other, how many common internal tangents do the circles have? Explain your reasoning.

 24. If two circles do not intersect and neither is inside the other, how many common external tangents do the circles have? Explain your reasoning.

25. *Given:* $\overrightarrow{PA}$ and $\overrightarrow{PB}$ are tangents to the circle

Prove: $m\angle P + m\widehat{ACB} = 180°$

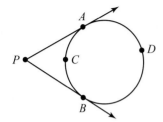

Exercise 25

26. *Given:* $\overrightarrow{PA}$ is tangent to the circle with center O

Prove: $\triangle APB \sim \triangle CPA$

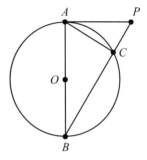

Exercise 26

27. *Given:* $\overleftrightarrow{AB}$ and $\overleftrightarrow{CD}$ are common external tangents to the circles that are not congruent. [Hint: Extend $\overleftrightarrow{AB}$ and $\overleftrightarrow{CD}$ to meet at point P.]

Prove: $AB = CD$

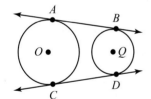

Exercise 27

28. *Given:* $\overleftrightarrow{AB}$ and $\overleftrightarrow{CD}$ are common internal tangents to the circles

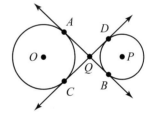

Exercise 28

29. Prove Theorem 6.18.

30. Prove Theorem 6.21.

31. Construct a tangent to a circle at a point on the circle.

32. Construct a tangent to a circle from a point outside the circle.

33. Construct a common external tangent to two circles that do not intersect and have unequal radii.

34. Construct a common external tangent to two circles that do not intersect and have equal radii.

35. Construct a common internal tangent to two circles that do not intersect and have unequal radii.

36. Construct a common internal tangent to two circles that do not intersect and have equal radii.

37. From a balloon 1 mile high, how far away, to the nearest tenth of a mile, is the horizon, the farthest point that can be seen on the surface of the Earth? [Hint: See Example 1.]

38. Draw three circles with one common tangent.

39. Draw two externally tangent circles. How many common tangents can be drawn?

Student Activity

40. For each figure below, explain why the given fact is true. Assume O is the center of each circle and if a segment looks like a radius, tangent, or chord, it is. Do not assume anything else. Discuss your results with two other students.

(a)

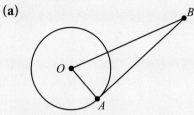

△*AOB* is a right triangle

(b)

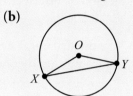

△*XOY* is an isosceles triangle

(c)

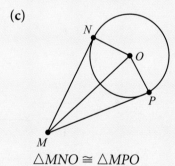

△*MNO* ≅ △*MPO*

(d)

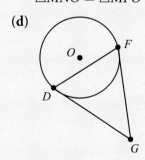

△*DFG* is an isosceles triangle

6.4 Circles and Regular Polygons

OBJECTIVES

1. Define inscribed and circumscribed figures.
2. Learn relationships involving inscribed and circumscribed figures.
3. Construct inscribed and circumscribed figures.
4. Identify regular polygons and their parts.

In this section, we will study relationships between circles and polygons.

OBJECTIVE 1 Define inscribed and circumscribed figures.

> **DEFINITION:** *Inscribed and Circumscribed Circles and Polygons*
> If a polygon has its vertices on a circle, the polygon is **inscribed in the circle** and the circle is **circumscribed around the polygon**. If each side of a polygon is tangent to a circle, the polygon is **circumscribed around the circle** and the circle is **inscribed in the polygon**.

In Figure 6.37(a), quadrilateral $ABCD$ is inscribed in the circle O and circle O is circumscribed around the quadrilateral. In Figure 6.37(b), $\triangle QRS$ is circumscribed around circle P and circle P is inscribed in the triangle.

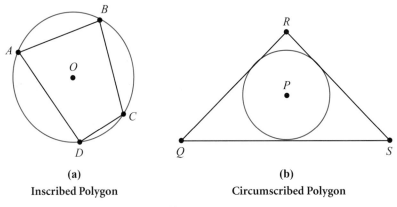

(a)	(b)
Inscribed Polygon	Circumscribed Polygon

Figure 6.37

OBJECTIVE 2 Learn relationships involving inscribed and circumscribed figures.

> **THEOREM 6.23**
> If a quadrilateral is inscribed in a circle, the opposite angles are supplementary.

**Karl Fredrick Gauss
(1777–1855)**

Karl Fredrick Gauss made many major contributions to arithmetic, number theory, algebra, astronomy, biology, physics, and, of course, geometry. At the age of 19, Gauss proved by considering points on a circle that a regular polygon with seventeen sides can be constructed using a straightedge and a compass. He considered this to be one of his greatest achievements.

Given: ABCD is a quadrilateral inscribed in a circle (See Figure 6.38.)

Prove: $\angle A$ and $\angle C$ are supplementary and $\angle B$ and $\angle D$ are supplementary

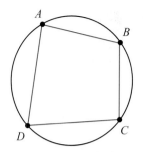

Figure 6.38

Proof _____

Statements	*Reasons*
1. Quadrilateral ABCD is inscribed in a circle	1. Given
2. $m\angle A = \frac{1}{2}m\widehat{BCD}$ and $m\angle C = \frac{1}{2}m\widehat{DAB}$	2. Measure of inscribed $\angle$
3. $m\widehat{BCD} + m\widehat{DAB} = 360°$	3. Arc add. post.
4. $m\angle A + m\angle C = \frac{1}{2}m\widehat{BCD} + \frac{1}{2}m\widehat{DAB}$	4. Add.-subt. law
5. $m\angle A + m\angle C = \frac{1}{2}(m\widehat{BCD} + m\widehat{DAB})$	5. Distributive law
6. $m\angle A + m\angle C = \frac{1}{2}(360°)$	6. Substitution law
7. $m\angle A + m\angle C = 180°$	7. Simplify
8. $\angle A$ and $\angle C$ are supplementary	8. Def. of supp. $\angle$'s

We could show that $\angle B$ and $\angle D$ are supplementary in a similar manner.

  Technology Connection

Geometry software will be needed.

1. Draw a circle.
2. Locate four points on the circle and label them A, B, C, D.
3. Construct the quadrilateral ABCD. [Note: The circle circumscribes the quadrilateral.]
4. Measure $\angle A$, $\angle B$, $\angle C$, and $\angle D$.
5. Drag A. What do you notice about the relationship between $\angle A$ and $\angle C$, and between $\angle B$ and $\angle D$?
6. Do you find the same result if you drag B, C, or D? Is your result consistent with Thereom 6.23?

The proof of the next theorem follows from Theorem 6.23 and the definition of a rectangle. The proof is requested in the exercises.

THEOREM 6.24
If a parallelogram is inscribed in a circle, then it is a rectangle.

Recall a regular polygon has all sides congruent and all angles congruent. Circles with regular polygons have many special properties. The following theorems introduce some of them.

> **THEOREM 6.25**
>
> If a circle is divided into n equal arcs, $n > 2$, then the chords formed by these arcs form a regular n-gon.

The proof of Theorem 6.25 follows by noting that equal arcs have equal chords, and that each angle of the n-gon formed is one-half the sum of four equal arcs.

> **THEOREM 6.26**
>
> If a circle is divided into n equal arcs, $n > 2$, and tangents are constructed to the circle at the endpoints of each arc, then the figure formed by these tangents is a regular n-gon.

The following paragraph-style proof to this theorem shows 6 equal arcs, but note that the same proof can be applied to any number of arcs n, $n > 2$. Refer to Figure 6.39.

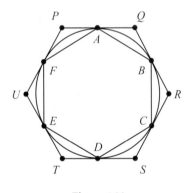

Figure 6.39

PROOF: Because $m\widehat{AB} = m\widehat{BC} = m\widehat{CD} = m\widehat{DE} = m\widehat{EF} = m\widehat{FA}$, the corresponding chords are also congruent, making $\overline{AB} \cong \overline{BC} \cong \overline{CD} \cong \overline{DE} \cong \overline{EF} \cong \overline{FA}$. Because the tangent segments to a circle from a point outside the circle are equal, $\triangle AQB$, $\triangle BRC$, $\triangle CSD$, $\triangle DTE$, $\triangle EUF$, and $\triangle FPA$ are all isosceles triangles. Because angles formed by a chord and a tangent each measure one-half the intercepted arc, and because the arcs are all congruent, the base angles of the six triangles are all congruent. Thus, all six triangles are congruent by ASA. Hence, $\angle P \cong \angle Q \cong \angle R \cong \angle S \cong \angle T \cong \angle U$ because they are corresponding parts of congruent triangles, making all angles of the hexagon congruent. Also, because $\overline{PQ}, \overline{QR}, \overline{RS}, \overline{ST}, \overline{TU},$ and $\overline{UP}$ are all formed by adding two equal segments, $\overline{PQ} \cong \overline{QR} \cong \overline{RS} \cong \overline{ST} \cong \overline{TU} \cong \overline{UP}$. Thus, because all angles and all sides are congruent, $PQRSTU$ is a regular hexagon.

OBJECTIVE 3   **Construct inscribed and circumscribed figures.**   Given a circle and a way to determine n equal arcs on the circle, Theorem 6.25 can be used to inscribe a regular n-gon in the circle. If a regular n-gon was to circumscribed around the circle, then use Theorem 6.26.

EXAMPLE 1   Inscribe a regular octagon in a circle.

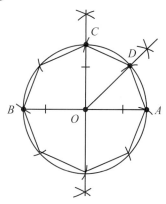

Figure 6.40

Consider the circle in Figure 6.40. Choose a point A on the circle and construct radius $\overline{OA}$, extended to form diameter $\overline{BA}$. Use Construction 1.4 to form the perpendicular bisector of $\overline{BA}$ containing $\overline{OC}$, and use Construction 1.6 to bisect $\angle AOC$ obtaining $\overline{OD}$. Using length AD, mark off eight equal arcs around the circle starting at point A. The chords formed by these arcs give the desired regular octagon by Theorem 6.25.

PRACTICE EXERCISE 1

Circumscribe a regular octagon around the circle given in Example 1. Use Theorem 6.26.

ANSWER ON PAGE 318

We now consider the problem of circumscribing a circle around a given regular polygon.

CONSTRUCTION 6.5

Construct a circle that is circumscribed around a given regular polygon.

The construction uses a regular pentagon, but note that you can use any regular polygon.

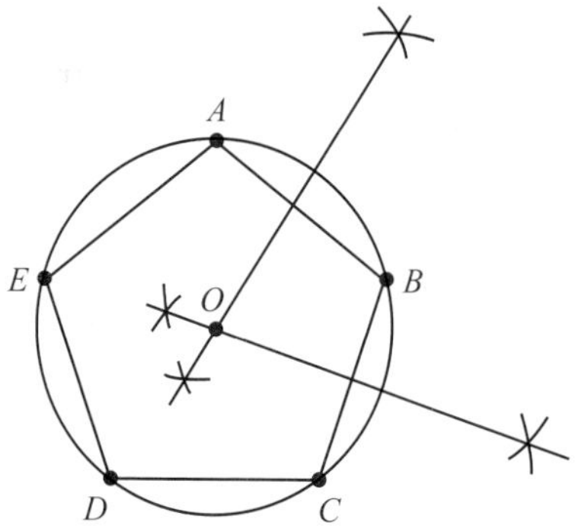

Given: Regular pentagon $ABCDE$ (See Figure 6.41.)

To Construct: A circle that is circumscribed around $ABCDE$

Construction _____

1. Use Construction 1.4 to form the perpendicular bisectors of two adjacent sides, for instance $\overline{AB}$ and $\overline{BC}$. The bisectors intersect in point O.
2. Use O as the center and AO as the radius and construct a circle. This is the desired circle circumscribed around $ABCDE$.

Figure 6.41

Next, inscribe a circle inside a given regular polygon.

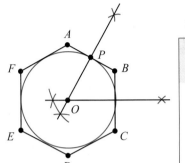

Figure 6.42

> **CONSTRUCTION 6.6**
>
> Construct a circle that is inscribed in a given regular polygon.

The construction uses a regular hexagon, but note that you can use any regular polygon.

Given: Regular hexagon *ABCDEF* (See Figure 6.42.)

To Construct: A circle that is inscribed in *ABCDEF*

Construction

1. Use Construction 1.4 to form the perpendicular bisector of two adjacent sides, for instance $\overline{AB}$ and $\overline{BC}$. The bisectors intersect in point *O*.
2. Let *P* be the midpoint of $\overline{AB}$. Use *O* as the center and *OP* as the radius and construct a circle. This is the desired circle inscribed in *ABCDEF*.

Note In Construction 6.5, the radius of the circumscribed circle is the segment from a *vertex* of the regular polygon. In Construction 6.6, the radius of the inscribed circle is the segment from the *midpoint* of a side of the regular polygon to the intersection of the perpendicular bisectors.

OBJECTIVE 4   Identify regular polygons and their parts.

> **DEFINITION:**   *Center of a Regular Polygon*
>
> The **center of a regular polygon** is the center of the circle circumscribed around the polygon.

In view of Constructions 6.5 and 6.6, the center of a regular polygon is also the center of the circle that is inscribed in the polygon because the center points coincide. In Figure 6.43, *O* is the center of the regular polygon.

> **DEFINITION:**   *Radius of a Regular Polygon*
>
> A **radius of a regular polygon** is the segment joining the center of the polygon to one of its vertices.

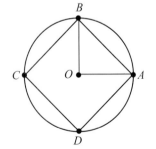

Figure 6.43

In Figure 6.43, $\overline{OA}$ and $\overline{OB}$ are radii of the regular polygon.

> **THEOREM 6.27**
>
> All radii of a regular polygon are equal in length.

The proof of Theorem 6.27 is requested in the exercises.

> **DEFINITION: *Central Angle of a Regular Polygon***
> A **central angle of a regular polygon** is an angle formed by two radii to two adjacent vertices.

In Figure 6.43, $\angle AOB$ is a central angle of the regular polygon.

> **THEOREM 6.28**
> All central angles of a regular polygon have the same measure.

The proof of Theorem 6.28 is requested in the exercises.

Because there are n equal central angles in a regular n-gon and these angles sum to $360°$, the proof of the formula given in the next theorem should be obvious.

> **THEOREM 6.29**
> The measure a of each central angle in a regular n-gon is determined with the formula
> $$a = \frac{360°}{n}.$$

EXAMPLE 2   Find the measure of each central angle in a regular octagon.

Because an octagon has 8 sides, by Theorem 6.29 we substitute 8 for n in the formula

$$a = \frac{360°}{n}.$$

$$a = \frac{360°}{8} = 45°$$

Thus, in Figure 6.44, central angle $\angle AOB$ has measure $45°$.

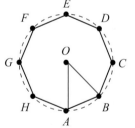

Figure 6.44

> **PRACTICE EXERCISE** 2
>
> Find the number of sides in a regular polygon if each central angle measures $24°$.
>
> **ANSWER ON PAGE 318**

In groups of three students, complete the following table. Remember, the central angle of a regular polygon is formed by two adjacent radii. To find the central angle measure use Theorem 6.29. An interior angle of the polygon is formed by two adjacent chords of the circumscribed circle. To find the measure of an interior angle of a regular polygon use Corollary 3.16.

Central angle   **Interior angle**

REGULAR POLYGON	CENTRAL ANGLE MEASURE	INTERIOR ANGLE MEASURE
Equilateral triangle		
Square		
Regular pentagon		
Regular hexagon		
Regular octagon		
Regular nonagon (9 sides)		
Regular n-gon		

Answers to Practice Exercises

1. Begin by dividing the circle into 8 equal arcs using the method shown in Example 1. Use Construction 6.1 to construct the tangent to the circle at each of these eight points. The points of intersection of these tangents form the vertices of the circumscribed octagon.   **2.** 15 sides

6.4 **Exercises**

Exercises 1–8 refer to the figure below in which quadrilateral *ABCD* is inscribed in the circle centered at *O*.

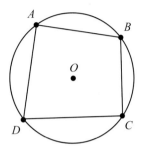

1. If $m\angle A = 86°$, what is the measure of $\angle C$?
2. If $m\angle B = 97°$, what is the measure of $\angle D$?
3. Assume that $m\angle A + m\angle B + m\angle C = 276°$. Find $m\angle B$.
4. Assume that $m\angle B + m\angle C + m\angle D = 269°$. Find $m\angle C$.
5. If $\overline{AB} \parallel \overline{DC}$ and $\overline{AB} \cong \overline{DC}$, find $m\angle A$.
6. If $\overline{AB} \cong \overline{DC}$ and $\overline{AD} \cong \overline{BC}$, find $m\angle B$.
7. If $\angle A \cong \angle C$, find $m\angle C$.
8. If $\overline{AB} \cong \overline{BC} \cong \overline{CD} \cong \overline{DA}$, find $m\angle D$.
9. Draw a circle and divide it into four equal arcs by constructing the perpendicular bisector of a diameter. Inscribe a square in the circle.
10. Circumscribe a square around a given circle. [Hint: Refer to Exercise 9.]
11. Follow these directions to draw a regular hexagon.
    a. Construct a circle with a compass. Mark the center with a pencil.
    b. Label any point *on* the circle, *P*.
    c. Using the *same* radius setting on the compass, starting at point *P*, draw an arc *on* the circle. Label this point *Q*.
    d. Place the point of the compass on *Q* and draw an arc that intersects the circle. Label it point *R*.
    e. Repeat the previous step to find points *S, T,* and *U*.
    f. Draw $\overline{PQ}, \overline{QR}, \overline{RS}, \overline{ST}, \overline{TU},$ and $\overline{UP}$.
    g. Is polygon *PQRSTU* equilateral? Explain.
    h. Is polygon *PQRSTU* equiangular? Explain. [Hint: What type of triangles are formed by the 6 radii of the polygon?]
12. Circumscribe a regular hexagon around a given circle.
13. Draw a circle and divide it into three equal arcs. Inscribe an equilateral triangle in the circle. [Hint: You may want to divide the circle into six equal arcs first. The length of a side of a regular inscribed hexagon is equal to the radius.]
14. Circumscribe an equilateral triangle around a given circle. [Hint: Refer to Exercise 13.]
15. Construct a square and inscribe a circle in it.
16. Construct a square and circumscribe a circle around it.
17. Construct an equilateral triangle and circumscribe a circle around it.
18. Construct an equilateral triangle and inscribe a circle in it.

In Exercises 19–24, find the measure of each central angle of the given regular polygon.

19. equilateral triangle   **20.** square   **21.** regular pentagon
22. regular hexagon   **23.** regular nonagon   **24.** regular decagon

In Exercises 25–28, find the number of sides in a regular polygon if each central angle has the given measure.

25. $45°$ **26.** $30°$ **27.** $20°$ **28.** $15°$

29. Prove Theorem 6.24.
30. Prove that Construction 6.5 provides the required circumscribed circle.
31. Prove that Construction 6.6 provides the required inscribed circle.
32. Prove Theorem 6.27.
33. Prove Theorem 6.28.
34. Page 313 contains a brief biographical sketch of the great mathematician Karl Gauss. Using the Internet or the library, do a brief biographical sketch of another individual who contributed to the study of geometry. Be sure to include the name of your subject, the dates he or she lived, where he or she lived, describe his or her professional life including contributions to math, science, and geometry in particular. Include your sources.
35. Find the length of a side of an equilateral triangle that is inscribed in a circle with radius 4 cm.
36. To inscribe a regular pentagon $ABCDE$ in a circle centered at O (see the figure below), select a point A on the circle, construct radius $\overline{OA}$, and construct radius $\overline{OP} \perp \overline{OA}$. Construct the midpoint Q of $\overline{OP}$, and draw segment $\overline{QA}$. Bisect $\angle OQA$ to obtain R on $\overline{OA}$. Construct $\overline{BR} \perp \overline{OA}$. Then $\overline{AB}$ is one side of the desired pentagon. Use AB to mark off five equal arcs on the circle to complete the pentagon. Draw a circle with radius about 2 inches in length and inscribe a regular pentagon in the circle. Then circumscribe a regular pentagon around the circle by constructing lines perpendicular to radii drawn to each vertex.

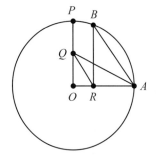

Exercise 36

6.5 Inequalities Involving Circles (Optional)

OBJECTIVES

1. Prove theorems involving inequalities and arcs of a circle.
2. Prove theorems involving inequalities and chords of a circle.

In this section, we will study inequalities that involve circles.

OBJECTIVE 1 Prove theorems involving inequalities and arcs of a circle.

> **THEOREM 6.30**
> In the same circle or in congruent circles, the greater of two central angles intercepts the greater of two arcs.

Given: Circle with center O, central angles $\angle 1$ and $\angle 2$, with $m\angle 1 > m\angle 2$ (See Figure 6.45.)

Prove: $m\widehat{AB} > m\widehat{CD}$

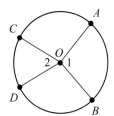

Figure 6.45

Proof

Statements	**Reasons**
1. $\angle 1$ and $\angle 2$ are central angles of the circle with center O and $m\angle 1 > m\angle 2$	1. Given
2. $m\widehat{AB} = m\angle 1$ and $m\widehat{CD} = m\angle 2$	2. Def. of measure of an arc
3. $m\widehat{AB} > m\widehat{CD}$	3. Substitution law

The proof of Theorem 6.30 for the case involving congruent circles is left for you to do as an exercise. The converse of this theorem is also true.

> **THEOREM 6.31**
> In the same circle or in congruent circles, the greater of two arcs is intercepted by the greater of two central angles.

The proof of Theorem 6.31 is left for you to do as an exercise.

In the design of automobile headlights, engineers use properties of arcs intercepted by central angles to determine the size of the illuminated area.

OBJECTIVE 2   Prove theorems involving inequalities and chords of a circle.

Technology Connection

Geometric software will be needed.

1. Construct a circle and draw two chords. For clarity, make them nonintersecting chords. Label them $\overline{AB}$ and $\overline{CD}$.
2. Measure $\overline{AB}$ and $\overline{CD}$. Which chord is greater?
3. Measure $\overarc{AB}$ and $\overarc{CD}$. Which arc is greater?

Does the larger chord create the larger arc? Is the reverse true? Which of the chords is visibly closer to the center of the circle? Compare these results with Theorems 6.32, 6.33, 6.34, and 6.35.

THEOREM 6.32

In the same circle or in congruent circles, the greater of two chords forms the greater arc.

Given: Circle centered at O with chords $\overline{AB}$ and $\overline{CD}$ such that $AB > CD$ (See Figure 6.46.)

Prove: $m\overarc{AB} > m\overarc{CD}$

Auxiliary lines: Construct radii $\overline{OA}, \overline{OB}, \overline{OC},$ and $\overline{OD}$

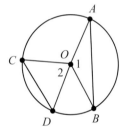

Figure 6.46

Proof _____

Statements	**Reasons**
1. $\overline{AB}$ and $\overline{CD}$ are chords of the circle with center O such that $AB > CD$	1. Given
2. $\overline{OA} \cong \overline{OB} \cong \overline{OC} \cong \overline{OD}$	2. Radii are all $\cong$
3. $m\angle 1 > m\angle 2$	3. SSS ineq. thm.
4. $m\overarc{AB} > m\overarc{CD}$	4. The $>$ of 2 central $\angle$'s intercepts the $>$ arc

The proof of Theorem 6.32 for the case involving congruent circles is left for you to do as an exercise. The converse of this theorem is also true.

> **THEOREM 6.33**
>
> In the same circle or in congruent circles, the arc with the greater measure has the greater chord.

PRACTICE EXERCISE ❶

Complete the proof of Theorem 6.33.

Given: Arcs $\overset{\frown}{AB}$ and $\overset{\frown}{CD}$ in the circle with center O such that $m\overset{\frown}{AB} > m\overset{\frown}{CD}$

Prove: $AB > CD$

Auxiliary lines: Construct radii $\overline{OA}$, $\overline{OB}$, $\overline{OC}$, and $\overline{OD}$ and chords $\overline{AB}$ and $\overline{CD}$

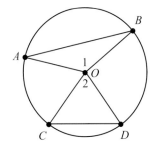

Proof _____

Statements	Reasons
1. $\overset{\frown}{AB}$ and $\overset{\frown}{CD}$ are arcs in the circle with $m\overset{\frown}{AB} > m\overset{\frown}{CD}$	1. _____
2. _____	2. Def. of arc measure and the substitution law
3. _____	3. Radii are $\cong$
4. $AB > CD$	4. _____

ANSWERS ON PAGE 325

The proof of the remaining part of Theorem 6.33 for congruent circles is left for you to do as an exercise.

> **THEOREM 6.34**
>
> In the same circle or in congruent circles, the greater of two unequal chords is nearer the center of the circle.

For Theorem 6.34, we will give the proof for congruent circles and leave the proof for the same circle for you to do as an exercise.

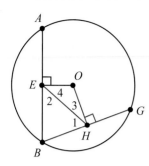

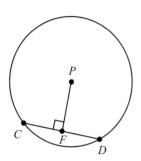

Figure 6.47

Given: Congruent circles with centers O and P, chords $\overline{AB}$ and $\overline{CD}$ with $AB > CD$, $\overline{OE} \perp \overline{AB}$ and $\overline{PF} \perp \overline{CD}$ (See Figure 6.47.)

Prove: $OE < PF$

Auxiliary lines: Construct chord $\overline{BG}$ with $BG = CD$, construct $\overline{OH}$ such that $\overline{OH} \perp \overline{BG}$, and construct $\overline{EH}$

Proof _____

Statements	Reasons
1. $\overline{AB}$ and $\overline{CD}$ are chords in congruent circles with $AB > CD$, $\overline{OE} \perp \overline{AB}$, and $\overline{PF} \perp \overline{CD}$	1. Given
2. $\overline{BG}$ is a chord with $BG = CD$, $\overline{OH} \perp \overline{BG}$, and $\overline{EH}$ is a segment	2. By construction
3. $AB > BG$	3. Substitution law
4. $\frac{1}{2}AB > \frac{1}{2}BG$	4. Mult. prop. of $>$
5. $EB = \frac{1}{2}AB$ and $BH = \frac{1}{2}BG$	5. Line $\perp$ to chord through the center bisects the chord
6. $EB > BH$	6. Substitution law
7. $m\angle 1 > m\angle 2$	7. Using statement 6 and $\angle$'s opp. $\neq$ sides are $\neq$ in same order
8. $\angle OHB$ and $\angle OEB$ are rt. $\angle$'s	8. Def. $\perp$ lines
9. $m\angle OHB = m\angle OEB$	9. Measures rt. $\angle$'s are $=$
10. $m\angle 1 + m\angle 3 = m\angle 2 + m\angle 4$	10. $\angle$ add. post. and substitution law
11. $m\angle 3 < m\angle 4$	11. Using statements 7 and 10 and the subt. prop. of $>$
12. $OE < OH$	12. Sides opp. $\neq$ $\angle$'s are $\neq$ in same order
13. $OH = PF$	13. Equal chords are equidistant from center of $\odot$.
14. $OE < PF$	14. Substitution law

The converse of Theorem 6.34 is also true and can be proved by reversing the order of the statements in the proof above. The proof is left for you to do as an exercise.

> **THEOREM 6.35**
> In the same circle or in congruent circles, if two chords have unequal distances from the center of the circle, the chord nearer the center is greater.

Note Theorem 6.34 can also be stated using "the *smaller* of two unequal chords is *farther from* the center"; and Theorem 6.35 using "the chord *farther from* the center is *smaller*."

The next corollary follows directly from Theorem 6.35 using the fact that a diameter passes through the center of a circle making its distance from the center zero.

> **COROLLARY 6.36**
> Every diameter of a circle is greater than any other chord that is not a diameter.

Answers to Practice Exercise

1. 1. Given 2. $m\angle 1 > m\angle 2$ 3. $\overline{OA} \cong \overline{OB} \cong \overline{OC} \cong \overline{OD}$ 4. SAS ineq. thm.

6.5 Exercises

FOR EXTRA HELP:  Student's Solutions Manual   Addison-Wesley Math Tutor Center

In Exercises 1–8, answer *true* or *false*. If you answer false, explain why.

1. If $\angle A$ and $\angle B$ are two central angles in a circle with $m\angle A > m\angle B$, then the arc intercepted by $\angle A$ is greater than the arc intercepted by $\angle B$.

2. If $\overline{XY}$ and $\overline{WZ}$ are two chords in a circle with $WZ > XY$, then $m\overset{\frown}{XY} > m\overset{\frown}{WZ}$.

3. If $\overline{AB}$ and $\overline{CD}$ are two chords in a circle with $AB > CD$, then $\overline{AB}$ is nearer the center than $\overline{CD}$.

4. If $\overline{PQ}$ and $\overline{RS}$ are two chords in a circle with $\overline{PQ}$ nearer the center, then $PQ > RS$.

5. If $\overset{\frown}{AB}$ and $\overset{\frown}{CD}$ are two arcs in a circle with $m\overset{\frown}{AB} > m\overset{\frown}{CD}$, then chord $\overline{CD}$ is nearer the center of the circle.

6. The length of any chord in a circle is less than or equal to the length of a diameter.

7. If a central angle in a circle measures $45°$, the chord determined by the angle is shorter than a radius. [Hint: What angle determines a chord equal to a radius?]

8. If a central angle in a circle measures $70°$, the chord determined by the angle is shorter than a radius.

Exercises 9–16 refer to the figure at the right.

9. If $m\angle 1 = 130°$ and $m\angle 2 = 110°$, what is the relationship between $\overset{\frown}{AB}$ and $\overset{\frown}{CD}$?

10. If $m\overset{\frown}{AC} = 60°$ and $m\overset{\frown}{BD} = 70°$, what is the relationship between $\angle BOD$ and $\angle COA$?

11. If $AB = 13.5$ cm and $CD = 12.5$ cm, what is the relationship between $m\overset{\frown}{AB}$ and $m\overset{\frown}{CD}$?

12. If $BD = 6.5$ ft and $AC = 6.0$ ft, what is the relationship between $m\overset{\frown}{AC}$ and $m\overset{\frown}{BD}$?

13. If $m\overset{\frown}{AB} = 125°$ and $m\overset{\frown}{CD} = 120°$, what is the relationship between $\overline{AB}$ and $\overline{CD}$?

14. If $AC = 5.5$ cm and $BD = 6.2$ cm, which of $\overline{AC}$ and $\overline{BD}$ is nearer center O?

15. If $AB = 10.2$ yd and $CD = 9.8$ yd, which of $\overline{AB}$ and $\overline{CD}$ is farther from center O?

16. If $\overline{BC}$ is a diameter of the circle and $m\angle 1 < 180°$, which of $\overline{AB}$ and $\overline{BC}$ is nearer the center O?

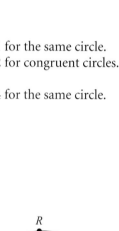

Exercises 9–16

17. Prove Theorem 6.30 for congruent circles.

18. Prove Theorem 6.31 for the same circle.

19. Prove Theorem 6.31 for congruent circles. (Hint: Use Exercise 18.)

20. Prove Theorem 6.32 for congruent circles.

21. Prove Theorem 6.33 for congruent circles.

22. Prove Theorem 6.34 for the same circle.

23. Prove Theorem 6.35.

24. Prove that in a circle a chord determined by an arc of 180° is twice as long as a chord determined by an arc of 60°.

25. If $\triangle ABC$ is inscribed in a circle and $m\angle A > m\angle B$, prove that $m\overset{\frown}{BC} > m\overset{\frown}{AC}$.

26. If $\triangle ABC$ is an isosceles triangle with base $\overline{AB}$ inscribed in a circle and $m\angle A > 60°$, prove that $\overline{BC}$ is nearer the center than $\overline{AB}$.

27. *Given:* $\overline{AB}$ and $\overline{CD}$ are chords in a circle centered at O, $AB > CD$, and $\overline{AB}$ and $\overline{CD}$ intersect at P. $\overline{EF}$ is a diameter containing P

    Prove: $m\angle 1 < m\angle 2$

    Auxiliary lines: Construct $\overline{RS}$ through P with $m\angle RPO = m\angle 2$.

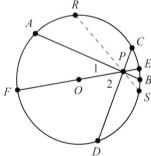

Exercise 27

6.6 Locus and Basic Theorems (Optional)

OBJECTIVES

1. Define locus of points.
2. Define distance from a point to a circle.
3. Prove theorems involving locus of points.

Earlier in this chapter, we defined a circle as the set of all points in a plane a given distance r, the radius, from a point O, the center of the circle. Another way to describe a circle is to say "a circle is the *locus of points* in a plane a given distance r from a fixed point O."

OBJECTIVE 1   Define locus of points.

> **DEFINITION: Locus of Points**
> A **locus** consists of all those points and only those points that satisfy one or more conditions.

Unless we say otherwise, we will agree to restrict loci (plural of locus) to a plane. In each example, we will make a sketch and find a few points on the locus before determining all the points on the locus.

EXAMPLE 1   What is the locus of all points (in a plane) equidistant from two parallel lines?

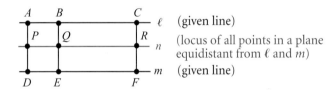

Figure 6.48

Referring to Figure 6.48, we can reason that points such as P, Q, and R that are each equidistant from the two parallel lines ℓ and m will lie on a line between the given lines ℓ and m. Also, because $PA = PD$, $QB = QE$, and $RC = RF$, the desired locus is the line n parallel to both ℓ and m and midway between ℓ and m.

Notice that Figure 6.48 can also be used in a discussion of the locus of all points equidistant from a given line. For this locus, let n be the given line. Then, ℓ and m would be the required locus because the points equidistant from the given line n form two lines parallel to n at the same distance on either side of n.

We can also give an example of a locus of points involving circles, but first consider the following definition.

OBJECTIVE 2   Define distance from a point to a circle.

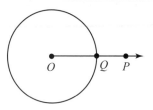

Figure 6.49

> **DEFINITION: *Distance from a Point to a Circle***
>
> Consider a circle with center O, a point P not on the circle, and point Q that is the intersection of $\overrightarrow{OP}$ and the circle. The **distance from P to the circle** is PQ. See Figure 6.49.

EXAMPLE 2   Determine the locus of all points equidistant from two **concentric circles** with the same center O, and radii r and R.

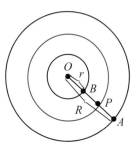

Figure 6.50

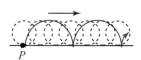

An example of an interesting locus is a *cycloid*, the path of a fixed point on a circle as the circle rolls along a straight line. As the circle rolls from left to right in the figure above, point P traces out a cycloid.

The figure above shows one arch of a cycloid. If an object is released from point A, it will travel to point C along the curve in less time than along any other path from A to C. This is why a cycloid is referred to as the "curve of fastest descent."

In Figure 6.50, the circles with radii $OA = R$ and $OB = r$ are the given circles. If $AP = BP$, then P is a point on the desired locus. All such points form a circle with center at O and radius OP. OP can be determined in terms of r and R.

$$OP = OB + BP$$
$$= r + \frac{1}{2}(R - r)$$
$$= r + \frac{1}{2}R - \frac{1}{2}r$$
$$= \frac{1}{2}(r + R)$$

Thus, the locus of all points equidistant from two concentric circles centered at O with radii r and R is a circle with center O and radius $\frac{1}{2}(r + R)$.

PRACTICE EXERCISE 1

Determine the locus of all points a given distance g from a circle with center O and radius r where $r > g$.

ANSWERS ON PAGE 332

Notice that in the definition, a locus consists of *all those points and only those points* satisfying given conditions. This means that to prove that a given figure is a locus we must show that any point satisfying the conditions is on the figure *and* that any point on the figure satisfies the conditions. Thus, each of the following proofs is in two parts.

OBJECTIVE 3   Prove theorems involving locus of points.

> **THEOREM 6.37**
> The locus of all points equidistant from two given points *A* and *B* is the perpendicular bisector of $\overline{AB}$.

Part I:

Given: *C* is any point equidistant from *A* and *B* and does not lie on $\overleftrightarrow{AB}$; *D* is the midpoint of $\overline{AB}$ (See Figure 6.51.)

Prove: $\overline{CD}$ is the perpendicular bisector of $\overline{AB}$

Construction: Draw $\overline{AC}$ and $\overline{BC}$

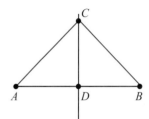

Figure 6.51

Proof

Statements	Reasons
1. *C* is equidistant from *A* and *B*	1. Given
2. $\overline{AC} \cong \overline{BC}$	2. By construction and the definition of "equidistant from 2 pts"
3. $\overline{CD} \cong \overline{CD}$	3. Reflexive law
4. *D* is the midpoint of $\overline{AB}$	4. Given
5. $\overline{AD} \cong \overline{BD}$	5. Def. of midpt. of seg.
6. *ACD* and *BCD* are triangles	6. Def. of $\triangle$
7. $\triangle ACD \cong \triangle BCD$	7. SSS
8. $\angle ADC \cong \angle BDC$	8. CPCTC
9. $\angle ADC$ and $\angle BDC$ are adjacent angles	9. Def. of adj. $\angle$'s
10. $\overline{CD} \perp \overline{AB}$	10. From statements 8 and 9 and def. of $\perp$ lines
11. $\overline{CD}$ is the $\perp$ bisector of $\overline{AB}$	11. From statements 4 and 10 and def. of $\perp$ bisector

Thus, we have shown that any point equidistant from *A* and *B* is on the perpendicular bisector of $\overline{AB}$. Now we must show that any point on the perpendicular bisector of $\overline{AB}$ is equidistant from *A* and *B*.

Part II:

Given: C is any point on the perpendicular bisector $\overline{CD}$ of $\overline{AB}$

Prove: C is equidistant from A and B

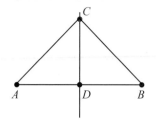

Figure 6.51 (repeated)

Proof

Statements	Reasons
1. $\overline{CD}$ is ⊥ bisector of $\overline{AB}$	1. Given
2. $\overline{AD} \cong \overline{BD}$	2. Def. of bisector
3. $\angle ADC$ and $\angle BDC$ are adjacent angles	3. Def. of adj. ∠'s
4. $\angle ADC \cong \angle BDC$	4. Def. of ⊥ lines
5. $\overline{CD} \cong \overline{CD}$	5. Reflexive law
6. $\triangle ACD \cong \triangle BCD$	6. SAS
7. $\overline{AC} \cong \overline{BC}$	7. CPCTC
8. C is equidistant from A and B	8. Def. of "equidistant from 2 pts."

> **THEOREM 6.38**
>
> The locus of all points equidistant from the sides of an angle is the angle bisector.

Part I:

Given: D is any point equidistant from $\overrightarrow{BA}$ and $\overrightarrow{BC}$, the sides of $\angle ABC$ (See Figure 6.52.)

Prove: $\overrightarrow{BD}$ is the bisector of $\angle ABC$

Construction: Construct $\overline{AD} \perp \overline{AB}$ and $\overline{CD} \perp \overline{BC}$

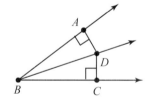

Figure 6.52

Proof

Statements	Reasons
1. $\overline{AD} \perp \overline{AB}$ and $\overline{CD} \perp \overline{BC}$	1. By construction
2. $\overline{AD} \cong \overline{CD}$	2. Def. of "equidistant from 2 lines"
3. $\angle BAD$ and $\angle BCD$ are right angles	3. ⊥ lines form rt. ∠'s
4. $\triangle ADB$ and $\triangle CDB$ are right triangles	4. Def. of rt. △
5. $\overline{BD} \cong \overline{BD}$	5. Reflexive law
6. $\triangle ADB \cong \triangle CDB$	6. HL
7. $\angle ABD \cong \angle CBD$	7. CPCTC
8. $\overrightarrow{BD}$ is the bisector of $\angle ABC$	8. Def. of ∠ bisector

Thus, we have shown that any point equidistant from the sides of $\angle ABC$ is on the bisector of $\angle ABC$. Now we must show that any point on the bisector of $\angle ABC$ is equidistant from the sides of $\angle ABC$.

PRACTICE EXERCISE 2

Complete the proof of Theorem 6.38.

Part II:

Given: D is any point on $\overrightarrow{BD}$, the bisector of $\angle ABC$ (see Figure 6.52)

Prove: D is equidistant from $\overrightarrow{BA}$ and $\overrightarrow{BC}$

Construction: Construct $\overline{AD} \perp \overline{AB}$ and $\overline{CD} \perp \overline{BC}$

Proof _____

Statements	Reasons
1. _____	1. Given
2. $\angle ABD \cong \angle CBD$	2. _____
3. $\overline{AD} \perp \overline{AB}$ and $\overline{CD} \perp \overline{BC}$	3. By construction
4. $\angle DAB$ and $\angle DCB$ are right angles	4. _____
5. $\triangle ABD$ and $\triangle CBD$ are right triangles	5. _____
6. _____	6. Reflexive law
7. $\triangle ABD \cong \triangle CBD$	7. HA
8. _____	8. CPCTC
9. _____	9. Def. of "equidistant from 2 lines" with statements 3 and 8 ANSWERS ON PAGE 332

A telephone jack is located in the center of a wall that is 20 ft long. What is the locus of the positions to place a telephone if the connecting cable is 7 ft long?

The definition of a locus includes the possibility of more than one condition being given to determine the locus. The next example illustrates this.

EXAMPLE 3 Describe the locus of points that is on both a circle of radius r and a circle with a larger radius R.

Figure 6.53 illustrates the possibilities. Figure 6.53 (a) shows exactly two points in the locus; Figure 6.53 (b) and (c) shows one point in the locus; and Figure 6.53 (d) and (e) show no point in the locus. Thus, depending on the relationship of the circles, the locus will contain exactly two points, exactly one point, or no points.

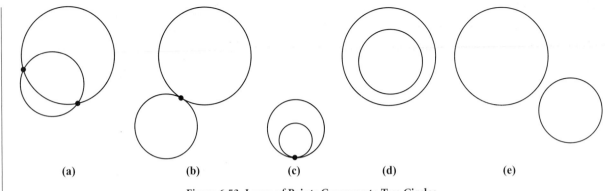

(a) (b) (c) (d) (e)

Figure 6.53 Locus of Points Common to Two Circles

Answers to Practice Exercises

1. The points on the circles centered at O, with radii $r - g$ and $r + g$. **2.** 1. $\overrightarrow{BD}$ is the bisector of $\angle ABC$ 2. Def. of $\angle$ bisector 4. $\perp$ lines form rt. $\angle$'s 5. Def. of rt. $\triangle$ 6. $\overline{BD} \cong \overline{BD}$ 8. $\overline{AD} \cong \overline{CD}$ 9. D is equidistant from $\overrightarrow{BA}$ and $\overrightarrow{BC}$

6.6 Exercises   FOR EXTRA HELP: Student's Solutions Manual   Tutor Center   Addison-Wesley Math Tutor Center

Describe and draw a sketch of each locus in Exercises 1–12.

1. All points 5 units from a given point P.
2. All points 5 units from a given line m.
3. All points 3 units from a circle with radius 4 units.
4. All points 4 units from a circle with radius 3 units.
5. All points equidistant from parallel lines that are 3 units apart.
6. All points equidistant from two concentric circles with radii 6 and 8 units.
7. All points that lie on a given line m and that are also on a given circle.
8. All points that lie on a given triangle and that are also on a given line m.
9. In a circle, all points that are midpoints of all chords that are parallel to a given chord.
10. In a circle, all points that are midpoints of all chords of a given length.
11. All points that are centers of circles tangent to both of two parallel lines and that are also on a line intersecting the parallel lines.
12. All points that are centers of circles tangent to both sides of $\angle ABC$ and that are also on the circle that has radius 5 units.
13. Prove that the locus of all points that are centers of circles tangent to both of two parallel lines is the line equidistant from the two given lines.
14. Prove that the locus of all points that are centers of circles tangent to both sides of $\angle ABC$ is the bisector of the angle.
15. The city council wishes to place a statue in a rectangular park. The statue is to be equidistant from the corners of the park. What is the locus of all points satisfying this condition?
16. A TV station has a broadcast range of 70 mi. What is the locus of all points that can receive the signal from the station?
17. What is the locus of points in your classroom that is equidistant from the side walls and equidistant from the back and front walls? Explain.

Chapter 6 Review

Key Terms and Symbols

6.1 circle
center
radius
diameter
semicircle
major arc
minor arc
central angle
intercepts
measure of an arc

congruent arcs
inscribed angle

6.2 chord
bisector of an arc
secant

6.3 tangent
point of tangency
tangent internally
tangent externally
common tangent

common external
tangent
common internal
tangent
line of centers

6.4 inscribed polygon
circumscribed
polygon
inscribed circle
circumscribed circle

center of a regular
polygon
radius of a regular
polygon
central angle of a
regular polygon

6.6 locus
distance from P to
the circle
concentric circles

Properties of Circles

Type of Angle	Diagram	Angle Relationships	Segment Relationships
central angle		$m\angle 1 = m\widehat{AB}$ O is the center of the circle (Definition of measure of arc)	$OA = OB$
inscribed angle		$m\angle 1 = \frac{1}{2} m\widehat{AC}$ vertex on circle (Theorem 6.2)	no segment relationship
angles formed by intersecting chords		$m\angle 1 = \frac{1}{2}(m\widehat{AB} + m\widehat{CD})$ vertex inside circle (Theorem 6.5)	$(AE)(EC) = (BE)(ED)$ (Theorem 6.13)
angle formed by two secants		$m\angle 1 = \frac{1}{2}(m\widehat{AE} - m\widehat{BD})$ vertex outside circle (Theorem 6.14)	$(CA)(CB) = (CE)(CD)$ (Theorem 6.15)
angle formed by tangent and a chord		$m\angle 1 = \frac{1}{2}(m\widehat{BC})$ vertex on circle (Theorem 6.16)	no segment relationship
angle formed by secant and tangent		$m\angle 1 = \frac{1}{2}(m\widehat{AD} - m\widehat{AC})$ vertex outside circle (Theorem 6.17)	$\dfrac{BD}{AB} = \dfrac{AB}{BC}$ (Theorem 6.20)
angle formed by two tangents		$m\angle 1 = \frac{1}{2}(m\widehat{BCD} - m\widehat{BED})$ vertex outside circle (Theorem 6.18)	$AB = AD$ (Theorem 6.19)

Review Exercises

Section 6.1

1. Find the diameter of the circle with radius 2.6 cm.

2. Find the radius of the circle with diameter $\frac{4}{5}$ yd.

3. Find $m\angle P$ if $m\overset{\frown}{PQ} = 70°$.

4. Find $m\angle PQR$.

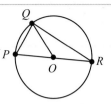

$\odot O$, diameter $\overline{PR}$
Exercises 3 and 4

5. Refer to the figure to the right to answer each question.
   (a) What is $\angle AOB$ called with respect to the circle?
   (b) What is $\angle ACB$ called with respect to the circle?
   (c) What is the measure of $\overset{\frown}{AB}$?
   (d) What is the measure of $\overset{\frown}{ACB}$?
   (e) What is the measure of $\angle ACB$?

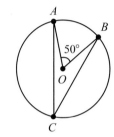

Exercise 5

6. *Given:* $m\angle AOB = 55°$
   $\triangle OBC$ is equilateral;
   $\overline{OA}$, $\overline{OB}$, $\overline{OC}$, and $\overline{OD}$ are radii of circle O.

   Find:
   (a) $m\angle BCD$
   (b) $m\overset{\frown}{AB}$
   (c) $m\angle ADB$
   (d) $m\overset{\frown}{BC}$
   (e) $m\overset{\frown}{DC}$
   (f) $m\angle DOC$
   (g) $m\angle ODC$
   (h) $m\overset{\frown}{AD}$

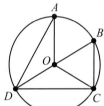

Exercise 6

Section 6.2

Exercises 7–10 refer to the figure to the right.

7. If $m\overset{\frown}{BC} = 56°$ and $m\overset{\frown}{AD} = 132°$, find $m\angle BPC$.

8. If $CD = DA$ and $m\overset{\frown}{CD} = 135°$, find $m\overset{\frown}{DA}$.

9. If $\overline{BD} \perp \overline{CA}$ and $\overline{BD}$ passes through the center of the circle, find CP if $AP = 17$ cm.

10. If $AP = 5$ ft, $CP = 6$ ft, and $BP = 3$ ft, find PD.

11. If two chords in a circle are equal, what can be said about their distance from the center of the circle?

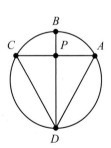

Exercises 7–10

Exercises 12–14 refer to the figure to the right.

12. If $m\overset{\frown}{BD} = 68°$ and $m\overset{\frown}{AC} = 28°$, find $m\angle P$.

13. If $m\overset{\frown}{AC} = 30°$ and $m\angle P = 24°$, find $m\overset{\frown}{BD}$.

14. If $PB = 12$ ft, $PA = 3$ ft, and $PC = 4$ ft, find PD.

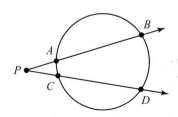

Exercises 12–14

15. *Given:* $ABCD$ is a rectangle
Prove: $\overset{\frown}{AB} \cong \overset{\frown}{CD}$

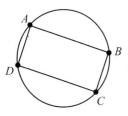

Exercise 15

16. *Given:* $\overline{AD} \perp \overline{BC}$ and $\overline{AD}$ contains the center O
Prove: $\overset{\frown}{AB} \cong \overset{\frown}{AC}$

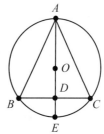

Exercise 16

Section 6.3

Exercises 17–22 refer to the figure to the right.

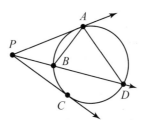

17. If $m\overset{\frown}{AB} = 66°$, find $m\angle PAB$.

18. If $m\angle ADB = 35°$, find $m\angle PAB$.

19. If $m\overset{\frown}{AD} = 130°$ and $m\overset{\frown}{AB} = 64°$, find $m\angle APD$.

20. If $m\overset{\frown}{ABC} = 130°$, find $m\angle APC$.

21. If $AP = 38$ cm, find CP.

22. If $AP = 20$ ft and $DP = 30$ ft, find BP.

23. *Given:* $m\overset{\frown}{BD} = 2m\overset{\frown}{AC}$
  Prove: $\overline{PC} \cong \overline{BC}$

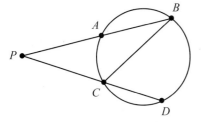

Exercise 23

24. *Given:* $\overline{AD}$ and $\overline{DC}$ are tangent to the circle and $ABCD$ is a parallelogram
  Prove: $ABCD$ is a rhombus

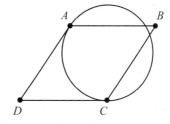

Exercise 24

25. *Given:* $\overline{AB}$ is a common internal tangent to the circles centered at O and P, and $\overleftrightarrow{OP}$ is the line of centers
  Prove: $\angle O \cong \angle P$

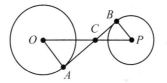

Exercise 25

26. Prove that if two circles are tangent externally, tangents to the circles from a point on their common internal tangent are equal in length.

27. Construct a tangent to a circle at a point on the circle.

28. *Given:* $\odot O$ where $\overline{GD}$ and $\overline{GA}$ are tangents and $m\angle BGC = 40°$;
$m\widehat{DC} = 75°$; $m\widehat{ED} = 55°$; $m\widehat{AH} = 20°$; $m\widehat{HF} = 30°$;
$m\widehat{FE} = 20°$; $HK = 1.4$ cm; $FK = 1.5$ cm; $KB = 7$ cm;
$GE = 2.7$ cm; $EC = 7.4$ cm

Find:
(a) $m\widehat{BC}$
(b) $m\widehat{AB}$
(c) $m\angle EGD$
(d) $m\angle AGF$
(e) $m\angle BKC$
(f) KC
(g) GD (rounded to nearest tenth)
(h) AG (rounded to nearest tenth)

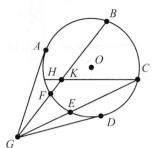

Exercise 28

Section 6.4

Exercises 29 and 30 refer to the figure to the right.

29. If $m\angle D = 88°$, find $m\angle B$.

30. If $\overline{AD} \parallel \overline{BC}$ and $AD = BC$, find $m\angle D$.

31. Inscribe a square in a circle. Bisect a side of the square and use the result to inscribe a regular octagon in the circle.

32. Construct a regular hexagon and inscribe a circle in it.

33. What is the measure of each central angle in a regular 18-gon?

34. How many sides does a regular polygon have if each central angle has measure 10°?

35. Find the length of a side of a regular hexagon that is inscribed in a circle with radius 14 cm.

36. Prove that a radius of a regular polygon bisects an interior angle of the polygon.

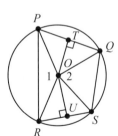

Exercises 29 and 30

Section 6.5 (Optional)

Exercises 37–41 refer to the figure to the right. Answer true *or* false. *If the answer is false, explain why.*

37. If $m\angle 1 > m\angle 2$, then $m\widehat{PR} > m\widehat{QS}$.
38. If $m\angle 1 > m\angle 2$, then $PR < QS$.

39. If $PQ > RS$, then $TO > UO$.
40. If $m\widehat{PQ} > m\widehat{RS}$, then $PQ < RS$.

41. If $TO > UO$, then $m\widehat{PQ} < m\widehat{RS}$.

Section 6.6 (Optional)

Determine the locus of points described in Exercises 42–45.

42. All points 2 units from a circle with radius 10 units.

43. All points 6 units from a circle with radius 4 units.

44. All points equidistant from the sides of $\angle ABC$ that are also on a circle with center at B.

45. All points that are on a square and also on a line m.

1. Find the radius of a circle with diameter 1.3 feet.

Exercises 2–5 refer to the figure below in which O is the center of the circle,
$m\angle BOD = 130°$, $m\angle ADC = 30°$, $BE = 6$ cm, $EC = 2$ cm, $AE = 4$ cm.

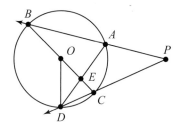

2. Find $m\angle BAD$.

3. Find $m\angle P$.

4. Find DE.

5. Find $m\angle AEC$.

Exercises 6 and 7 refer to the figure below in which O is the center of the circle.

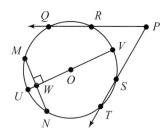

6. If $\overline{MN} \perp \overline{UV}$ and $MW = 4$ cm, find NW.

7. If $PQ = 12$ cm, $PR = 6$ cm, and $PT = 9$ cm, find PS.

8. If $\overline{PA}$ and $\overline{PB}$ are two tangents to a circle at points A and B from common point P outside the circle, and $PA = 15$ ft, find PB.

9. Construct a tangent to a circle from a point outside the circle.

10. Quadrilateral $ABCD$ is inscribed in a circle with $\angle A$ opposite $\angle C$. If $m\angle B = 100°$, find $m\angle D$.

11. Construct an equilateral triangle and inscribe a circle in it.

12. What is the measure of each central angle in a regular octagon?

Exercises 13–16 refer to the figure below. (Optional Section 6.5)

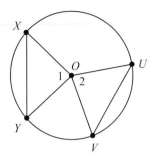

13. If $m\widehat{XY} > m\widehat{UV}$, what is the relationship between $m\angle 1$ and $m\angle 2$?
14. If $m\widehat{XY} < m\widehat{UV}$, which of $\overline{XY}$ or $\overline{UV}$ is nearer center O?
15. If $m\angle 1 > m\angle 2$, what is the relationship between $\overline{XY}$ and $\overline{UV}$?
16. Prove that the diagonals of a rhombus that is not a square are unequal.

Determine the locus of points in Exercises 17–19. (Optional Section 6.6.)
17. All points equidistant from parallel lines 6 units apart.
18. All points that are on a circle of radius 4 units and also on a circle of radius 6 units.
19. All points that are midpoints of all chords of a circle that are parallel to one diameter of the circle.
20. What is the locus of points that a dog can reach when it is tied at the center of a 30-foot fence with an 8-foot leash?

7

Areas of Polygons and Circles

*I*n Section 4.3, we discovered that every polygon has associated with it a measure called its perimeter. Now we consider another measure that is associated with a polygon called its *area*. To be precise, a polygon really has no area, rather, it encloses an area. However, for simplicity, and because it is common practice, we shall use the terminology *area of a polygon*. The area of a geometric figure is often used in architecture, design, construction, and engineering and has many everyday applications. One of these applications, presented below, is solved later in Section 1.

In this chapter, we assume the use of a scientific calculator.

AN APPLICATION

The canvas mat of a boxing ring is a square with an area of 400 ft^2 (square feet). What is the length of each side of the ring? If the rope is set 1 foot in from the four outer edges of the canvas, how much rope is needed to go around the ring once?

7.1 Areas of Quadrilaterals

OBJECTIVE 1   Apply the area of a rectangle formula.   To measure the length of a segment, we determine how many times it contains a particular unit (such as inch, centimeter, foot, and so forth). The same is true for angles in which the common unit of measure is the degree. In a similar manner, we can measure the area of a polygon by determining the number of particular units it contains. For example, suppose we consider the area of a rectangular tabletop that has sides measuring 3 ft and 5 ft as shown in Figure 7.1.

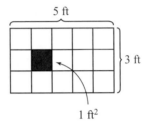

Figure 7.1

The tabletop can be divided into squares measuring 1 ft on each side. If we say that each of these squares has an area of 1 square foot, we can use this unit of measure to describe the area of the rectangle. In fact, since the rectangle contains 15 of these squares in its interior, we say that the rectangle has area 15 square feet. We often abbreviate *square feet* with the symbol ft^2. Thus, the area of the tabletop is 15 ft^2.

EXAMPLE 1

Figure 7.2 shows the sail on a sailboat. Estimate the number of square yards of material in the sail.

By counting the number of complete squares in the triangle and approximating the parts of those not totally contained in the triangle, you should come up with about 32 squares. Thus, the area of the sail is about 32 yd^2.

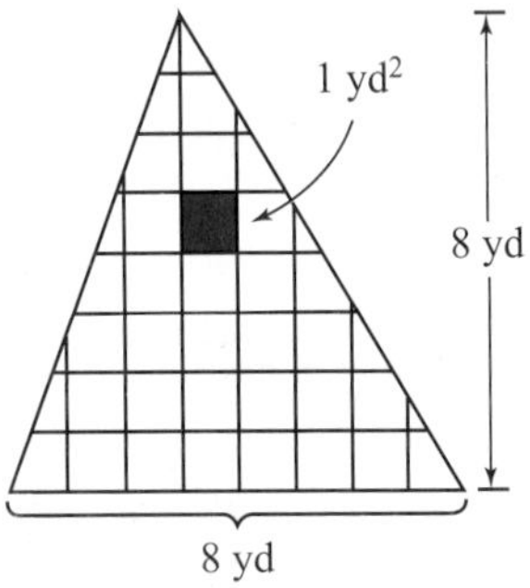

Clearly, it is difficult to count the number of square units (square inches, square centimeters, square feet) in polygons whose sides "cut off" parts of squares. As a result, it is appropriate for us to search for other ways to find areas. We begin with a rectangle. Notice in Figure 7.1 that the area of the rectangle, 15 ft^2, can be found by multiplying the lengths of the sides, 3 ft and 5 ft, because $3 \cdot 5 = 15$. This observation leads to our next postulate. For convenience, we'll give names to two parts of a rectangle. Consider $\square ABCD$ in Figure 7.3. The length of $\overline{AB}$, AB, is called the **base** of the rectangle and is denoted by b. The length of $\overline{BC}$, BC, is called the **height** of the rectangle and is denoted by h. Notice that $b = AB = DC$ and $h = BC = AD$. It is common practice to call the

Figure 7.2

8 yd

1 yd^2

8 yd

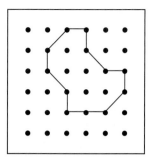

A geoboard is a square piece of wood with nails arranged in rows and columns. If there are 6 rows and 6 columns for a total of 36 nails, the geoboard has dimension 6 × 6. By placing a rubber band around the nails, various polygonal regions can be formed.

We can calculate the area enclosed by the rubber band using *Pick's Theorem*: If x is the number of nails on the border of the polygon, y is the number of nails on the interior of the polygon, the area A, of the polygon is given by
$$A = \frac{x}{2} + y - 1.$$
What is the area of the polygon shown here?

larger of the two numbers, the *base* and the smaller of the two numbers, the *height* of the rectangle.

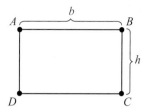

Figure 7.3 Base and Height of a Rectangle

⌐**Note** It is also common to describe the dimensions of a rectangle as its length
⌊(l) and width (w). An alternate formula for the area of a rectangle is $A = lw$.

POSTULATE 7.1

The **area of a rectangle** with base b and height h is determined with the formula $A = bh$.

EXAMPLE 2 Find the area of $\square ABCD$ if $AB = 7$ ft and $BC = 3$ ft.
We are given that $b = 7$ ft and $h = 3$ ft, so

$$A = bh \qquad \text{Postulate 7.1}$$
$$= 7 \cdot 3 \qquad \text{Substitute 7 for } b \text{ and 3 for } h$$
$$= 21 \qquad \text{Multiply}$$

Thus, the area of the rectangle is 21 ft².

Because a square is a rectangle, the proof of the following corollary comes directly from the definition of a square and Postulate 7.1.

COROLLARY 7.1

The **area of a square** with sides of length s is determined with the formula $A = s^2$.

EXAMPLE 3 Find the area of a square with sides of length 5 cm. Because $s = 5$ cm, we have

$$A = s^2$$
$$= 5^2 \qquad 5^2 = 5 \cdot 5 = 25$$
$$= 25 \text{ cm}^2.$$

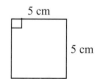

Figure 7.4

PRACTICE EXERCISE 1

A canvas mat of a boxing ring is a square with an area of 400 ft². What is the length of each side of the ring? If the rope is set 1 foot in from the four outer edges of the canvas, how much rope is needed to go around the ring once?

$s^2 = 400$ so $s = $ **1.** _____ and the length of each side of the ring is **2.** _____ feet.

Now that you know the length of each side of the ring, make a drawing showing the square ring, with the rope set in 1 foot from each side. The next question is a perimeter problem (not area). Recall from Section 2.1 perimeter is the sum of the lengths of the sides of the figure. The length of rope on one side of the ring is **3.** _____. The length of rope needed to go around the ring once is **4.** _____.

ANSWERS ON PAGE 351

👥 Student Activity

For this activity, use a copy of the figure below from your instructor.
1. Complete as many one-by-one squares *inside the figure* as possible.
2. Count the number of squares and triangles *in the figure*.
3. The area of one of the squares will be called *1 square unit*. The area of one triangle will be $\frac{1}{2}$ a square unit. Count the number of square units in the figure.
4. Use Pick's Theorem as described to the left of Example 2, to find the area of this figure.
5. Compare the answers of steps 3 and 4.

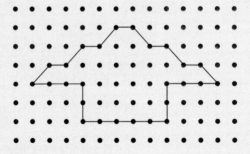

When a given figure can be divided into nonoverlapping parts, the total area of the figure is the sum of the areas of the parts. The next postulate affirms this.

POSTULATE 7.2 Additive Property of Areas

If lines divide a given area into several smaller nonoverlapping areas, the given area is the sum of the smaller areas.

For example, the area in Figure 7.5 is equal to the sum of the areas of □*ABFG* and □*CDEF*.

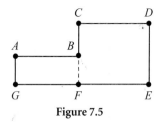

Figure 7.5

Two polygons can have the same area and not be congruent. For example, a square with sides of length 6 inches has area 36 in.², the same area as a rectangle with length 9 inches and width 4 inches. Clearly, the square and the rectangle cannot be made to coincide. However, two polygons that are congruent have the same area.

> **POSTULATE 7.3**
>
> Two congruent polygons have the same area.

Thus far, the area of two quadrilaterals, a rectangle and a square, have been investigated. The area of another quadrilateral, a parallelogram, will be analyzed using the idea of altitude of a parallelogram.

OBJECTIVE 2   **Prove the formula for area of parallelogram.**   The following definition will help us find the area of a parallelogram.

> **DEFINITION: Altitude and Base of a Parallelogram**
>
> An **altitude** of a parallelogram is a segment from a vertex of the parallelogram perpendicular to a nonadjacent side (possibly extended). The length of an altitude is called the **height** of the parallelogram and the side to which it is drawn is called the **base** of the parallelogram.

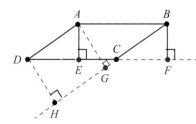

Figure 7.6

In Figure 7.6, $\overline{AE}$ and $\overline{BF}$ are altitudes of □*ABCD* with base $\overline{DC}$. The height of □*ABCD* is *AE* (which is equal to *BF*). Alternatively, if $\overline{BC}$ is extended, we see that $\overline{AG}$ and $\overline{DH}$ are also altitudes of □*ABCD* with base $\overline{BC}$. In this case, the height of □*ABCD* is *AG* (which is equal to *DH*).

Student Activity

For this activity, the following equipment will be needed: scissors, a straightedge, and a parallelogram drawn on a sheet of paper from your instructor.

1. Cut out the parallelogram.
2. Cut along one altitude of the parallelogram. What is the shape of the figure you cut off?
3. Matching the hypotenuses of both right triangles, place the cut triangle on the opposite side from which you cut it. What shape is formed? What do you observe about the opposite sides of this shape?

The original parallelogram was transformed into a rectangle, thus, the area of the parallelogram must be the area of the rectangle, $A = bh$.

THEOREM 7.2

The **area of a parallelogram** with length of base b and height h is determined with the formula $A = bh$.

The idea for the proof of the next theorem is to move the triangle that is formed by drawing the height of the parallelogram to the base ($\triangle BEC$ in Figure 7.7). It is moved to the other side of the parallelogram. A rectangle is formed ($\square ABEF$). Now apply Postulate 7.1 that says the area of a rectangle is base times height.

Given: $\square ABCD$ with altitude $\overline{BE}$, height $h = BE$, and length of base $b = DC$
(See Figure 7.7.)

Prove: The area of $\square ABCD$ is determined using the formula $A = bh$

Auxiliary lines: Extend $\overline{DE}$ and construct $\overline{AF} \perp \overline{DE}$

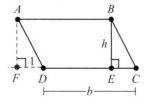

Figure 7.7

Proof _____

Statements	Reasons
1. $ABCD$ is a parallelogram	1. Given
2. $\overline{AD} \cong \overline{BC}$	2. Opp. sides of $\square$ are $\cong$
3. $\overline{AD} \parallel \overline{BC}$	3. Opp. sides of $\square$ are $\parallel$
4. $\angle C \cong \angle 1$	4. Corr. $\angle$'s are $\cong$
5. $\overline{AF} \perp \overline{FD}$	5. By construction
6. $\overline{BE} \perp \overline{EC}$	6. Def. of altitude
7. $\angle AFD$ and $\angle BEC$ are right angles	7. $\perp$ lines form rt. $\angle$'s
8. $\angle AFD \cong \angle BEC$	8. Rt. $\angle$'s are $\cong$
9. $\triangle AFD \cong \triangle BEC$	9. AAS
10. Area of $\triangle AFD$ = Area of $\triangle BEC$	10. $\cong$ polygons have = area
11. $AB = b$	11. Opp. sides of $\square$ are $\cong$, thus = in measure
12. Area of $\square ABEF = bh$	12. Area of a $\square$ (Post. 7.1)
13. Area of $\square ABEF$ = Area of trapezoid $ABED$ + Area of $\triangle AFD$	13. Add. prop. of areas (Post. 7.2)
14. Area of $\square ABCD$ = Area of trapezoid $ABED$ + Area of $\triangle BEC$	14. Add. prop. of areas (Post. 7.2)
15. Area of $\square ABCD$ = Area of trapezoid $ABED$ + Area of $\triangle AFD$	15. Substitution law (statements 10 and 14)
16. Area of $\square ABCD$ = Area of $\square ABEF$	16. Sym. and trans. laws (statements 13 and 15)
17. Area of $\square ABCD = A = bh$	17. Substitution law

EXAMPLE 4 Find the area of $\square ABCD$ given in Figure 7.8.

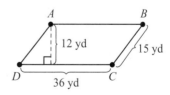

Figure 7.8

The height of the parallelogram is 12 yd and the base is 36 yd. Thus, the area is given by

$$A = bh$$
$$= 36 \cdot 12 \quad \text{Substitute 36 for } b \text{ and 12 for } h$$
$$= 432 \text{ yd}^2.$$

Notice that the length of side $\overline{BC}$ is 15 yd, but this fact is not used to find the area. It is used to find the perimeter of the parallelogram, 102 yd.

We leave quadrilaterals for a short time and consider triangles.

OBJECTIVE 3   **Prove the formula for area of a triangle.**   The next theorem provides a way to determine the area of a triangle. We use the term **height** to represent the length of an altitude drawn to a **base** much like that which was done for a parallelogram.

> **THEOREM 7.3**
>
> The **area of a triangle** with length of base b and height h is determined with the formula $A = \frac{1}{2}bh$.

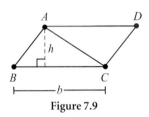

Figure 7.9

Given: $\triangle ABC$ (See Figure 7.9.)
Prove: $A = \frac{1}{2}bh$
Auxiliary lines: Construct the line through $A \parallel$ to $\overline{BC}$ and the line through $C \parallel$ to $\overline{AB}$

Proof

Statements	**Reasons**
1. $\overline{AD} \parallel \overline{BC}$ and $\overline{DC} \parallel \overline{AB}$	1. By construction
2. $ABCD$ is a parallelogram	2. Def. of $\square$
3. $\triangle ABC \cong \triangle CDA$	3. Diag. of $\square$ form $\cong \triangle$'s
4. Area of $\triangle ABC =$ Area of $\triangle CDA$	4. $\cong$ polygons have $=$ areas (Post. 7.3)
5. Area of $\square ABCD =$ Area of $\triangle ABC +$ Area of $\triangle CDA$	5. Add. prop. of areas (Post. 7.2)
6. Area of $\square ABCD =$ Area of $\triangle ABC +$ Area of $\triangle ABC$	6. Substitution law
7. Area of $\square ABCD = 2$ (Area of $\triangle ABC$)	7. Distributive law
8. Area of $\square ABCD = bh$	8. Formula for area of $\square$ (Thm 7.2)
9. 2(Area of $\triangle ABC$) $= bh$	9. Substitution law
10. Area of $\triangle ABC = A = \frac{1}{2}bh$	10. Mult.-div. law

EXAMPLE 5 Find the area of a triangle whose base is 4.2 cm and altitude is 6.5 cm.
Substitute 4.2 for b and 6.5 for h in the following formula:

$$A = \frac{1}{2}bh.$$

$$= \frac{1}{2}(4.2)(6.5)$$

$$= \frac{1}{2}(27.3) = 13.65$$

Thus, the area is 13.65 cm².

OBJECTIVE 4 **Use Heron's formula.** If the lengths of the sides of a triangle are known but the height is unknown, the area cannot be found with the previous formula. A mathematician named Heron of Alexandria (in the first century A.D.) derived a formula to compute the area of a triangle given the lengths of the three sides. To use the formula the semiperimeter must first be calculated.

The semiperimeter of a triangle is half of its perimeter. If the three sides of the triangle are represented by a, b, and c, then the semiperimeter, s, is found by adding the sides and dividing by 2.

$$s = \frac{a + b + c}{2} \quad \text{or} \quad \frac{1}{2}(a + b + c)$$

Because the proof of this theorem requires some complicated algebra, **Heron's formula** will be stated without proof.

> **THEOREM 7.4** *Heron's Formula*
> If the three sides of a triangle have lengths a, b, and c, the area is
> $$A = \sqrt{s(s - a)(s - b)(s - c)},$$
> where $s = \dfrac{a + b + c}{2}$

EXAMPLE 6 Find the area of a triangle with sides of lengths 16, 52, and 60 ft. Refer to Figure 7.10. Let $a = 16$, $b = 52$, and $c = 60$.

$$s = \frac{16 + 52 + 60}{2} = 64$$

$$A = \sqrt{64(64 - 16)(64 - 52)(64 - 60)} = \sqrt{147,456} = 384 \text{ ft}^2$$

Heron's formula is used to derive a very useful formula for the area of an equilateral triangle. The formula $A = \frac{1}{2}bh$ will certainly also work to find the area, but for this formula we do not need to know the height.

52 ft   16 ft

60 ft

Figure 7.10

COROLLARY 7.5

The **area of an equilateral triangle** with sides length a is

$$A = \frac{a^2\sqrt{3}}{4}$$

Given: An equilateral triangle with sides of length a.
(See Figure 7.11.)

Prove: $A = \dfrac{a^2\sqrt{3}}{4}$

Figure 7.11

Proof ——————————————————————————————

The semiperimeter is $s = \dfrac{a + a + a}{2} = \dfrac{3a}{2}$

Using Heron's formula, $A = \sqrt{s(s - a)(s - b)(s - c)}$

$$A = \sqrt{\frac{3a}{2}\left(\frac{3a}{2} - a\right)\left(\frac{3a}{2} - a\right)\left(\frac{3a}{2} - a\right)}$$

$$A = \sqrt{\frac{3a}{2}\left(\frac{a}{2}\right)\left(\frac{a}{2}\right)\left(\frac{a}{2}\right)}$$

$$A = \sqrt{\frac{3a^4}{16}} = \frac{a^2\sqrt{3}}{4}$$

■

EXAMPLE 7 Find the exact area of an equilateral triangle with sides that measure 8 inches. Do not approximate the answer.

$$A = \frac{8^2\sqrt{3}}{4} = \frac{64\sqrt{3}}{4} = 16\sqrt{3} \text{ in.}^3$$

OBJECTIVE 5 **Prove the formula for area of a trapezoid.** In Chapter 4, a trapezoid was defined as a quadrilateral with exactly one pair of parallel sides (called bases).

An **altitude** of a trapezoid is a segment from a vertex of the trapezoid perpendicular to the nonadjacent base. The length of an altitude is the **height** of the trapezoid.

THEOREM 7.6

The **area of a trapezoid** with length of bases b and b' and height h is determined with the formula $A = \dfrac{1}{2}(b + b')h$.

[Note: b' (read b prime) means the second base where b is the first base.]

PRACTICE EXERCISE 2

Complete the proof of Theorem 7.6.

Given: Trapezoid $ABCD$

Prove: $A = \dfrac{1}{2}(b + b')h$

Auxiliary line: Construct diagonal $\overline{BD}$

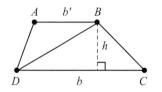

Proof _____

Statements	**Reasons**
1. _____	1. Given
2. Area of $ABCD$ = Area of $\triangle BCD$ + Area of $\triangle ABD$	2. _____
3. Area of $\triangle ABD = \dfrac{1}{2}b'h$	3. _____
4. _____	4. Formula for area of $\triangle$
5. Area of $ABCD = \dfrac{1}{2}bh + \dfrac{1}{2}b'h$	5. _____
6. _____	6. Distributive law

ANSWERS ON PAGE 351

EXAMPLE 8

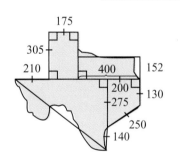

The area of an irregularly shaped region can be approximated by dividing it up into rectangular, trapezoidal, and triangular subregions. Using the information given in miles, approximate the area of the state of Texas.

The figure shows the state of Texas divided into two rectangles, one triangle, and one trapezoid. Add the four areas for an approximation for the area of Texas. The formula for the area of the two rectangles is $A = bh$.

$$A = (305)(175) = 53{,}375 \text{ square miles}$$
$$A = (400)(152) = 60{,}800 \text{ square miles}$$

The formula for the area of the triangle is $A = \dfrac{1}{2}bh$.

$$A = \dfrac{1}{2}(210 + 175 + 200)(275 + 140) = \dfrac{1}{2}(585)(415)$$
$$= 121{,}387.5 \text{ square miles}$$

The formula for the area of the triangle is $A = \dfrac{1}{2}(b + b')h$.

$$A = \dfrac{1}{2}(275 + 130)(200) = 40{,}500 \text{ square miles}$$

The sum of the areas of the subregions is $53{,}375 + 60{,}800 + 121{,}387.5 + 40{,}500 = 276{,}062.5$.

Thus, an approximation for the area of Texas is $276{,}062.5 \text{ mi}^2$. According to the 2002 edition of *World Book Encyclopedia*, the actual area of Texas is 266,874 square miles.

Technology can show another way to arrive at the formula for the area of a trapezoid.

 **Technology Connection**

Geometric software will be needed.

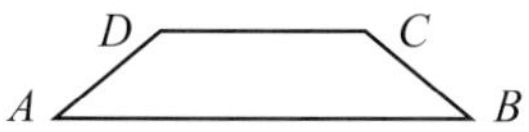

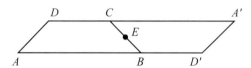

1. Construct a trapezoid and label it *ABCD* as in the figure.
2. Measure the area of the trapezoid *ABCD*.
3. Construct the midpoint of $\overline{BC}$ and label it *E*. This will be the center of rotation.
4. Rotate trapezoid *ABCD* about *E* by 180°. Label the new corresponding vertices *D'* and *A'*.

5. What shape is formed by the two trapezoids together, *ABCD* and *A'CBD'*?
6. Measure the area of the new figure *ADA'D'*. Recall the area of a parallelogram is *A = bh*. In the above problem, what is the base of the parallelogram?
    Can you make a general formula for the area of one trapezoid based on what you noticed in steps 2 and 6 when *AB = b*, *CD = b'*, and *h* is the height of the trapezoid?

Compare your results to Theorem 7.6.

EXAMPLE 9   An archway in a building is shaped like a rectangle topped by an isosceles trapezoid as shown in Figure 7.12. What is the area of the opening?

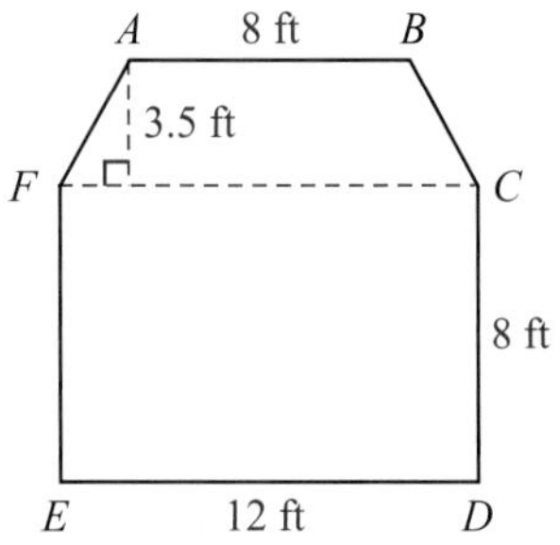

To find the total area, we find the area of the rectangle ($\square FCDE$), the area of the trapezoid (*ABCF*), and add the results using the additive property of areas.

$$\text{Area of } FCDE = bh = (12)(8) = 96 \text{ ft}^2 \quad b = 12 \text{ and } h = 8$$

$$\text{Area of } ABCF = \frac{1}{2}(b + b')h$$

$$= \frac{1}{2}(12 + 8)(3.5) \quad\quad b = 12, b' = 8, \text{ and } h = 3.5$$

$$= \frac{1}{2}(20)(3.5)$$

$$= (10)(3.5) = 35 \text{ ft}^2$$

Thus, the area of the archway is 96 ft² + 35 ft² = 131 ft².

Figure 7.12

OBJECTIVE 6   **Use the formula for area of a rhombus.**   Because a rhombus is a parallelogram, the area of a rhombus can be determined by using the formula *A = bh*. However, the next theorem provides an alternative way to find the area of a rhombus by using its diagonals. The proof of this theorem is left for you to do as an exercise.

> **THEOREM 7.7**
> The **area of a rhombus** with diagonals of length d and d' is determined with the formula $A = \frac{1}{2}dd'$.

Answers to Practice Exercises

1. 1. 20 2. 20 3. 18 ft 4. 72 ft   **2.** 1. *ABCD* is a trapezoid 2. Add. prop. of areas 3. Formula for area of $\triangle$ 4. Area of $\triangle BCD = \frac{1}{2}bh$ 5. Substitution law 6. $A = \frac{1}{2}(b + b')h$

SUMMARY

FIGURE	AREA
Rectangle	$A = bh$
Square	$A = s^2$
Parallelogram	$A = bh$
Triangle	$A = \frac{1}{2}bh$
Triangle (Heron)	$A = \sqrt{s(s-a)(s-b)(s-c)}$, where $s = \dfrac{a+b+c}{2}$
Equilateral triangle	$A = \dfrac{a^2\sqrt{3}}{4}$
Trapezoid	$A = \frac{1}{2}(b + b')h$
Rhombus	$A = \frac{1}{2}dd'$

7.1 Exercises

FOR EXTRA HELP: 📖 Student's Solutions Manual   Tutor Center   Addison-Wesley Math Tutor Center

In Exercises 1–18, find the area and perimeter of each figure.

1.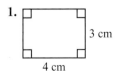
3 cm
4 cm

2.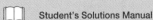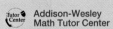
4 ft
16 ft

3.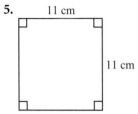

10 yd

5 yd

6 yd

4.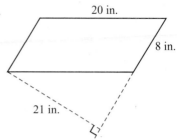

20 in.

8 in.

21 in.

5.

11 cm

11 cm

6.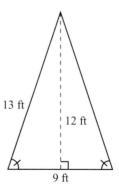

4.5 ft

4.5 ft

7.

7 yd

6 yd

11 yd

12 yd

8.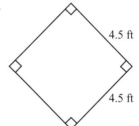

22.5 in.

9 in. 6 in.

15 in.

9.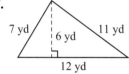

9 cm

4 cm

16 cm

10.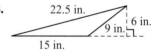

13 ft

12 ft

9 ft

11.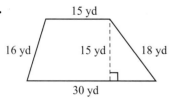

15 yd

16 yd

15 yd

18 yd

30 yd

12.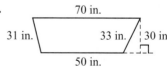

70 in.

31 in.

33 in.

30 in.

50 in.

13.

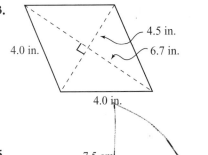

4.5 in.

6.7 in.

4.0 in.

4.0 in.

14.

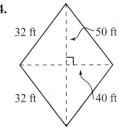

32 ft

50 ft

32 ft

40 ft

15.

7.5 cm

5 cm

3 cm

12 cm

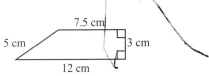

16.

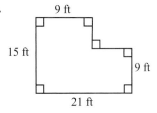

9 ft

15 ft

9 ft

21 ft

17.

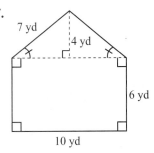

7 yd

4 yd

6 yd

10 yd

18.

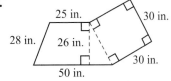

25 in.

30 in.

28 in.

26 in.

30 in.

50 in.

In Exercises 19 and 20, find the exact area of each figure. Do not round the answer.

19.

5 ft

5 ft

3 ft

3 ft

5 ft

20. *ABCD* is a kite

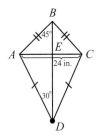

B

45°

E

A

C

24 in.

30°

D

21. Complete the proof of Theorem 7.7.

Given: $ABCD$ is a rhombus with diagonals of length $AC = d$ and $BD = d'$

Prove: $A = \frac{1}{2}dd'$

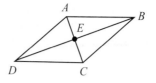

Exercise 21

Proof _____

Statements	*Reasons*
1. $ABCD$ is a rhombus with $AC = d$ and $BD = d'$	**1.** _____
2. _____	**2.** Diag. of rhombus are $\perp$
3. Area of $ABCD$ = Area of $\triangle ADC$ + Area of $\triangle ABC$	**3.** _____
4. Area of $\triangle ADC = \frac{1}{2}(d)(DE)$	**4.** _____
5. _____	**5.** Form. for area of $\triangle$
6. Area of $ABCD = \frac{1}{2}(d)(DE) + \frac{1}{2}(d)(BE)$	**6.** _____
7. Area of $ABCD = \frac{1}{2}d(DE + BE)$	**7.** _____
8. _____	**8.** Seg. add. post.
9. _____	**9.** Substitution law

For Exercises 22–24, use Heron's formula or Corollary 7.5 and a calculator to find the area of the following triangles. If the answer is an irrational number, express the answer in decimal form rounded to the nearest hundredth.

22. 9, 10, and 17 cm

23. 12, 13, and 14 in.

24. 16, 16, and 16 ft

25. Find the area of a parallelogram with base 18 ft and height 12 ft.

26. Find the area of a rectangle with base 25 cm and height 13 cm.

27. Find the area of a right triangle with legs 3.6 ft and 5.2 ft.

28. Find the area of a triangle with base 4.3 yd and height 6.4 yd.

29. Find the area of a trapezoid with height 7.2 inches and bases 6.3 inches and 11.5 inches.

30. Find the area of a rhombus with diagonals measuring 8.2 ft and 7.1 ft.

31. The base of a rectangle is 24 cm and its area is 168 cm². What is the height of the rectangle?

32. The base of a triangle is 16 ft and its area is 248 ft². What is the height of the triangle?

33. A wall is 41 ft long and 9 ft high. There are two windows in the wall measuring 2 ft by 3 ft. If a single gallon of paint will cover 400 ft², and Randy wants to give the wall two coats of paint, how many gallons of paint will she need?

34. The floor of a family room is rectangular in shape measuring 9 yd by 7 yd. In the center of the room is a firepit measuring 1.5 yd by 1.5 yd. If carpeting costs $19.95 per square yard, neglecting waste, how much will it cost to carpet the room?

Exercises 35–38, refer to the following diagram of the Xang's lot and home.

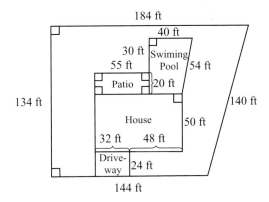

Exercises 35–38

35. The Xang's plan to put outdoor carpeting on their patio. If carpeting costs $15.95 per square yard, neglecting waste, how much will the project cost? (Hint: 9 ft² = 1 yd²)

36. How much will it cost for a pool cover made of insulated vinyl costing $12.50 per square yard? Ignore any waste.

37. How much will it cost to seal the Xang's driveway if one gallon of sealer costs $15.95 and covers 300 ft²?

38. The Xang's plan to put sod on all areas of their lot not already covered. If the sod costs $0.30 a square foot, how much will the project cost?

39. Use the Internet or the library to write one page about the life of Heron of Alexandria. Include an interesting fact about his personal life or mathematical work (other than he was a mathematician who lived in the first century A.D.).

40. The cross section of an I-beam is shown at the right. If the width of the upper and lower rectangles is 1.5 inches, find the area of the cross section.

Exercise 40

The areas of certain figures can be found by subtracting areas from larger areas. Use this hint in Exercises 41–42 and find the area of each shaded region.

41.

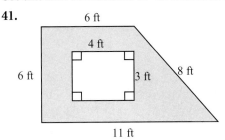

42.

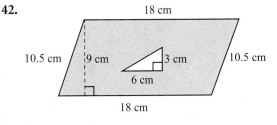

Exercises 43 and 44 require the ability to solve a quadratic equation using techniques learned in beginning algebra.

43. The area of a rectangle is 120 ft². If the base of the rectangle is 2 ft more than the height, find the base and height.

44. The area of a triangle is 44 cm². If the base of the triangle is 3 cm less than the height, find the base and height.

Exercises 45–47 refer to the figure in which $\overline{CD}$ is an altitude of right triangle $\triangle ABC$.

45. If $AD = 4$ cm and $BD = 16$ cm, find the area of $\triangle ABC$.
46. If $AD = 9$ ft and $BD = 25$ ft, find the area of $\triangle ABC$.
47. If $AD = 4$ cm and $BD = 16$ cm, find the area of $\triangle ADC$.
48. A right triangle has hypotenuse 13 cm and one leg 12 cm. Find the area of the triangle.

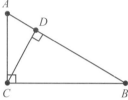

Exercises 45–47

7.2 Circumference and Area of Circles

OBJECTIVES

1. Define the circumference of a circle.
2. Define the area of a circle.

As we learned in Chapter 6, every circle has a radius and diameter. The diameter of a circle is twice its radius, that is $d = 2r$.

OBJECTIVE 1   Define the circumference of a circle.   The **circumference** of a circle is the distance around the circle (similar to the perimeter of a polygon). The ratio of the circumference C of any circle to its diameter d is the constant irrational number **pi**, denoted by π. Thus,

$$\pi = \frac{C}{d}.$$

We often approximate π using 3.14, but the actual value is the unending non-repeating decimal 3.14159265358. . . . This is an irrational number.

 Technology Connection

Geometric software will be needed.

1. Draw a circle.
2. Draw a diameter of the circle.
3. Measure the diameter and the circumference of the circle.
4. Using the software, find the ratio of the circumference to the diameter.
5. Make the circle larger. What happened to the ratio of circumference to diameter?
6. Make the circle smaller and observe the ratio again.

These results indicate the ratio of circumference to diameter will always be the same (approximately π). Read Postulate 7.4.

> **POSTULATE 7.4 Circumference of a Circle**
>
> The circumference C of any circle with radius r and diameter d is determined with the formula $C = 2\pi r = \pi d$.

EXAMPLE 1 Find the circumference of a circle with radius 3.50 cm. Give an approximate value of the circumference, correct to the nearest hundredth of a centimeter, using the π key on your calculator.

Because $C = 2\pi r$, substituting 3.50 for r, we have $C = 2\pi(3.50) = 7\pi$ cm as the **exact** value of the circumference. Using the calculator, $7\pi \approx 21.991148\ldots$ Rounding to the nearest hundredth $C \approx 21.99$ cm.

PRACTICE EXERCISE 1

What is the radius of a circle, correct to the nearest tenth of a foot, if the circumference is 23.0 ft?

ANSWER ON PAGE 358

OBJECTIVE 2 **Define the area of a circle.** The next postulate gives the formula for the area of a circle. Figure 7.13 helps to show why the formula is true.

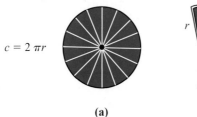
$c = 2\pi r$
(a)

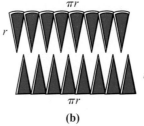
(b)

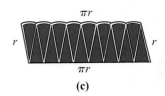
(c)

Figure 7.13

Suppose we cut the circle in Figure 7.13(a) along all the radii shown. When we place the top half of the circle above the bottom half as in Figure 7.13(b) and then slide them together as in Figure 7.13(c), the figure formed is approximately a parallelogram with height r. The length of the base of the "parallelogram" is πr because $2\pi r$ is the circumference, and half the circle is on the top and half is on the bottom. Because the area of a parallelogram is determined with the formula $A = bh$, the area of the circle is

$$A = bh = (\pi r)(r) = \pi r^2.$$

> **POSTULATE 7.5 Area of a Circle**
>
> The **area of a circle** with radius r is determined with the formula $A = \pi r^2$.

EXAMPLE 2 Find the area of a circle with radius 4.2 cm. Give the exact answer by leaving it in terms of π.

Since $A = \pi r^2$, substitute 4.2 for r, we have

$$A = \pi(4.2)^2 = \pi(17.64) = 17.64\pi \text{ cm}^2$$

Note We have multiplied centimeters by centimeters so the unit of measure for the area is square centimeters. It is abbreviated cm^2.

PRACTICE EXERCISE 2

The diameter of a circle is 6.4 inches. Find the area of the circle leaving the answer in terms of π.

ANSWER BELOW

EXAMPLE 3 In the machine part shown in Figure 7.14, each circular hole has radius 3 cm. Using the $\boxed{\pi}$ key on a calculator, find the area of the remaining metal. Round the answer to the nearest tenth of a centimeter.

$$A = \frac{1}{2}(b + b')h$$

$$= \frac{1}{2}(16 + 24)(14)$$

$$= \frac{1}{2}(40)(14)$$

$$= (20)(14) = 280 \text{ cm}^2$$

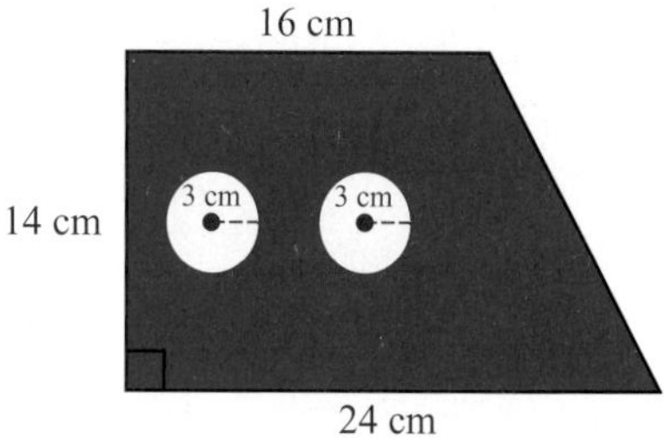

Now find the area of each circular hole.

$$A = \pi r^2 = \pi(3)^2 = \pi(9) \approx 28.3 \text{ cm}^2$$

Thus, the two circles have a combined area of approximately

$$2(28.3) = 56.6 \text{ cm}^2.$$

The area of the remaining metal is the area of the trapezoid minus the area of the circles, $280 - 56.6 = 223.4$. Thus, the remaining area is approximately 223.4 cm^2.

16 cm

14 cm

3 cm   3 cm

24 cm

Figure 7.14

PRACTICE EXERCISE 3

A triangle with base 10 inches and height 5 inches has three holes drilled through it, each with diameter 2 inches. What is the remaining area of the triangle after the holes are drilled? Round the answer to the nearest tenth of a square inch.

ANSWER BELOW

Answers to Practice Exercises

1. ≈ 3.7 ft   **2.** 10.24π in.^{2} **3.** ≈ 15.6 in.^{2}

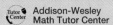

Classify the statements as *true* or *false*.

1. $C = 2\pi r$ and $C = \pi r^2$ are both correct formulas for the circumference of a circle.

2. The ratio $\dfrac{C}{d}$ is the same for all circles.

3. π is exactly equivalent to 3.14.

4. In the given figure where the square is inscribed in the circle, the radius of the circle is 10 inches.

5. The diameter of a circle is twice the radius.

Exercise 4

In Exercises 6–9, find the circumference and area of each circle with the given radius or diameter. Leave the answer in terms of π.

6. $r = 7$ yd

7. $r = 6.2$ mi

8. $d = \dfrac{4}{3}$ cm

9. $d = 12.78$ ft

In Exercises 10–13, find the approximate circumference and area of each circle with the given radius or diameter using the calculator to approximate the answer to the nearest hundredth.

10. $r = 4.10$ ft

11. $r = \dfrac{3}{4}$ cm

12. $d = 12.00$ mi

13. $d = 18.36$ yd

14. How much more cross-sectional area is there for water to pass through in a $\dfrac{3}{4}$-inch-diameter water hose than there is in a $\dfrac{1}{2}$-inch-diameter water hose? Round to the nearest hundredth.

15. A circular garden has radius 9 m. If a 1-meter-wide circular walk surrounds it, what is the area of the walk? [Hint: Find the area of a circle with 10-meter radius and subtract the area of the garden.] Round to the nearest hundredth.

16. A 12-inch-diameter pizza costs $10.00. A 16-inch-diameter pizza costs $12.00. Which pizza costs less per square inch? Use the $\boxed{\pi}$ key on the calculator if necessary.

17. A patio is in the shape of a trapezoid with bases 8.1 yd and 6.7 yd and height 5.8 yd. A circular dining area in the center of the patio has diameter 3.2 yd and is covered with Mexican tile. Assuming no waste, how much will it cost to the nearest dollar, to cover the remainder of the patio with outdoor carpeting that costs $18.50 per square yard?

In Exercises 18–21, find the area of metal remaining on each machine part with circular holes drilled in it. Round answers to the nearest hundredth of a unit.

18.

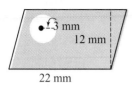

19.

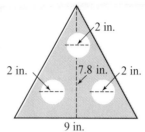

20.

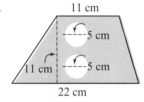

21.

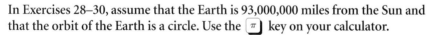

22. Find the length of the radius of a circle whose area is 246.49π cm^2.

23. Find the length of the diameter of a circle whose circumference is $\frac{5}{8}\pi$ in.

24. The running track at the high school is shaped as in the figure. If each straight part is 100 yd and the total length of the track is 440 yd, what is the radius of each semicircle shown? Approximate the answer to the nearest whole number.

25. A washer is a piece of hardware formed by a flat piece of metal as shown in the figure. Using the measurements in the figure, find the area of the washer rounded to the nearest tenth.

26. Find the exact area of the shaded region in the figure. The figure inside the circle is a square with each side measuring 6 cm.

27. An oak tree has a circumference of 52 inches. If we assume it is perfectly round, what is the tree's approximate radius? Explain your reasoning.

In Exercises 28–30, assume that the Earth is 93,000,000 miles from the Sun and that the orbit of the Earth is a circle. Use the $\boxed{\pi}$ key on your calculator.

28. What distance does the Earth travel during one year? Round the answer to the nearest hundred thousand miles.

29. What distance does the Earth travel during one day? [Note: Use 365 days in a year.] Round the answer to the nearest hundred thousand miles.

30. What distance does the earth travel during one minute? Round the answer to the nearest hundred miles.

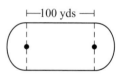

Exercise 24

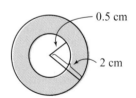

Exercise 25

Exercise 26

<table>
<tr><td>

7.3

</td><td>

Area and Arc Length of a Sector

</td></tr>
</table>

OBJECTIVES

1. Define a sector of a circle.
2. Find the area of a sector.
3. Define arc length.

OBJECTIVE 1   Define a sector of a circle.   If we cut a piece from a pizza as shown in Figure 7.15, the slice is an example of a *sector* of a circle.

Figure 7.15 Sector of a Circle

One way to form a mental image of statistical data is to display the data using a *circle graph*. The size of a sector of the circle can visually convey the information presented. The circle graph above shows the distribution of Mr. Whitney's monthly income. You have probably seen graphs like this in newspapers and magazines.

Take-home pay
52%

Credit union
savings
14%

Social Security tax
7%

State income tax
3%

Federal income
tax
24%

> **DEFINITION: Sector**
>
> A **sector** of a circle is a region bounded by two radii of the circle and the arc of the circle determined by the radii.

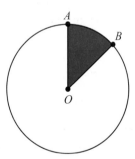

Figure 7.16 Sector

OBJECTIVE 2   Find the area of a sector.   The sector of the circle shown in color in Figure 7.16, denoted by sector *AOB*, is formed by radii $\overline{OA}$ and $\overline{OB}$ and arc $\overset{\frown}{AB}$. Assume that the measure of $\overset{\frown}{AB}$ is $45°$, that is $m\angle AOB = 45°$. Then because $45°$ is one-eighth of $360°$, the area of the sector is one-eighth the area of the circle. In view of this, it is reasonable to accept the next postulate.

> **POSTULATE 7.6   Area of a Sector**
>
> The area of a sector of a circle with radius *r* whose arc has measure $m°$ is determined with the formula
> $$A = \frac{m}{360}\pi r^2.$$

EXAMPLE 1 What is the area of a slice of pizza cut from a pizza with diameter 20 inches if the arc of the slice measures 30°?

Substitute 30 for m and 10 for r in the formula

$$A = \frac{m}{360}\pi r^2.$$

$$= \frac{30}{360}\pi(10)^2$$

$$= \frac{1}{12}\pi(100)$$

$$= \frac{25}{3}\pi$$

The actual area is $\frac{25}{3}\pi$ in.^{2}, which can be approximated by 26.2 in.^{2}, using a calculator and rounding to the nearest tenth of a square inch.

OBJECTIVE 3 **Define arc length.** A sector is a piece of a circle. Now consider finding the length of the arc that is a boundary of the sector.

CAUTION The *length* of an arc is different from the *measure* of the arc. The *measure* of an arc is the number of degrees in the central angle that intercepts the arc (refer to Section 6.1). The *length* of an arc refers to a part of the circumference of the entire circle. Remember, the circumference is the distance around a circle.

Suppose an angle measures 60°. Its corresponding arc also *measures* 60°. Since there are 360° in a circle, the ratio used to determine the part of the circle is $\frac{60}{360} = \frac{1}{6}$, thus, the *length* of the arc is $\frac{1}{6}$ the circumference of the circle.

POSTULATE 7.7 Arc Length

The **length of an arc** measuring $m°$ in a circle with radius r is determined with the formula

$$L = \frac{m}{360}2\pi r = \frac{m}{180}\pi r.$$

EXAMPLE 2 What is the length of the blue arc in Figure 7.17 if the central angle measures 30° and the radius of the circle is 15 cm? Also find the perimeter of the sector bounded by the two radii and the blue arc. Find the actual measurements (leave the answer in terms of π) and approximated measurements to the nearest tenth of a unit.

30°
15 cm

Figure 7.17

Substitute 30 for m and 15 for r in the formula

$$L = \frac{m}{180}\pi r.$$

$$= \frac{30}{180}\pi(15)$$

$$= \frac{5}{2}\pi$$

The actual length is $\frac{5}{2}\pi$ cm, which can be approximated by 7.9 cm, using a calculator and rounding to the nearest tenth of a centimeter.

To find the perimeter of the sector add arc length and the two radii.

$$P = \frac{5}{2}\pi + 15 + 15$$

$$P = \frac{5}{2}\pi + 30$$

The actual perimeter is $\left(\frac{5}{2}\pi + 30\right)$ cm, which can be approximated by 37.9 cm using a calculator and rounding to the nearest tenth of a centimeter.

Another important construction in a circle is a *segment*.

DEFINITION: Segment of a Circle

A **segment** of a circle is a region bounded by a chord of the circle and the arc formed by the chord.

The color region in Figure 7.18 is a segment of the circle. To find the area of this segment, find the area of sector AOB and subtract the area of $\triangle AOB$.

EXAMPLE 3   Find the area of the segment of the circle in Figure 7.18 if $m\angle AOB = 60°$ and $AO = 20$ cm to the nearest tenth of a square centimeter.

First, find the area of the triangle. Since $\overline{OB}$ and $\overline{OA}$ are radii of the circle, $\triangle AOB$ is isosceles and $\angle A \cong \angle B$. Thus, all angles in $\triangle AOB$ are 60° and the triangle is equiangular and, therefore, equilateral. This makes $AB = 20$ cm. Construct altitude $\overline{OD}$. This creates $\triangle BOD$, which is a 30°-60°-90° triangle since the altitude $\overline{OD}$ is perpendicular to the base. Using Theorem 5.20, $DB = 10$ cm and $OD = 10\sqrt{3}$ cm because the hypotenuse $\overline{OB}$ is 20 cm.

$$\text{Area } \triangle AOB = \frac{1}{2}(AB)(OD).$$

$$= \frac{1}{2}(20)(10\sqrt{3})$$

$$= 100\sqrt{3} \text{ cm}^2$$

Alternatively, Corollary 7.5 could also be used to find the area of $\triangle AOB$

The area of sector AOB is $\frac{60}{360}\pi(20)^2 = \frac{200}{3}\pi$ cm^2.

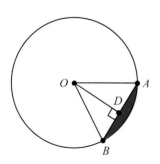

Figure 7.18 Segment of a Circle

Thus, the area of the segment is

$$\left(\frac{200}{3}\pi - 100\sqrt{3}\right) cm^2$$

which can be approximated by 36.2 cm².

7.3 Exercises

FOR EXTRA HELP: Student's Solutions Manual   Tutor Center   Addison-Wesley Math Tutor Center

In Exercises 1–4, find the *actual* area and the *actual* perimeter of each shaded region. This means leave the answer in terms of π; do not round it.

1.

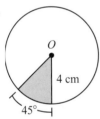

2.

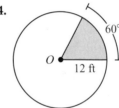

3.

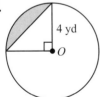

4.

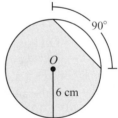

In Exercises 5–10, find the area of each shaded region of the circle. Use the π key on your calculator and give the answer correct to the nearest tenth.

5.

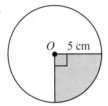

6.

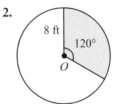

7.

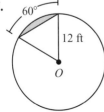

8.

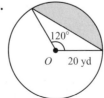

9.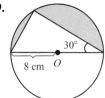

10.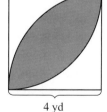

11. Find the area of a sector of a circle if the diameter of the circle is 8.2 cm and the arc of the sector is 30°. Give answer correct to the nearest tenth of a square centimeter.

12. Find the area of a sector of a circle if the diameter of the circle is 21.5 ft and the arc of the sector is 60°. Give answer correct to the nearest tenth of a square foot.

13. The area of a sector of a circle is 24π yd². If the arc of the sector is 60°, find the diameter of the circle.

14. The area of a sector of a circle is 50π cm². If the arc of the sector is 45°, find the diameter of the circle.

15. Find the area of a sector of a circle if the radius is 12.6 inches and the arc of the sector is 25°. Give the answer correct to the nearest tenth of a square inch.

16. Find the area of a sector of a circle if the radius is 25.7 cm and the arc of the sector is 42°. Give the answer correct to the nearest tenth of a square centimeter.

17. If the area of a circle is 720 cm², what is the area of the sector if its central angle measures 12°?

18. If the area of a circle is 540 in.², what is the area of the sector if its central angle measures 50°?

19. Find the perimeter and area of the shaded region in the figure at the right.

In Exercises 20 and 21, find the area of the shaded region. Give answer correct to the nearest tenth.

Exercise 19

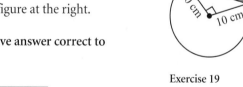

Exercise 20

4 yd

20. The arcs have their centers at the vertices of the equilateral triangle.
21. The arcs have their centers at opposite vertices of the square.

22. In the figure below, the externally tangent circles with centers O and P have radii 6 ft and 2 ft, respectively. What is the area of the shaded region, correct to the nearest tenth of a square foot? [Hint: Look at the trapezoid.]

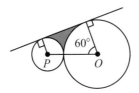

Exercise 22

23. Suppose you could split open and flatten an ice cream cone. Before doing that, you measure the circumference around the top of the cone and find it is about 9.4 inches. What is the radius of the opening of the cone approximated to the nearest tenth of an inch? Explain how you found the answer. Draw a picture of the flattened cone. What does the circumference of the cone correspond to on your drawing?

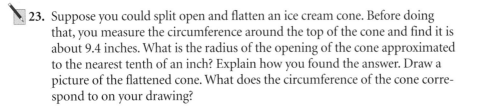

7.4   Area of Regular Polygons

OBJECTIVES

1. Define the apothem of a regular polygon.
2. Prove formula for area of a regular polygon.
3. Apply the area formula.

Previously, we considered areas of triangles and certain quadrilaterals such as parallelograms, rectangles, and trapezoids. We'll now show how to find the area of any regular polygon using the notion of an *apothem* of a regular polygon.

OBJECTIVE 1   Define the apothem of a regular polygon.

> **DEFINITION:**   *Apothem of a Regular Polygon*
>
> An **apothem of a regular polygon** is a line segment from the center of the polygon perpendicular to one of its sides.

Recall from Section 6.4, the center of a regular polygon is the center of the circle circumscribed around the polygon.

In Figure 7.19, $\overline{OP}$ and $\overline{OQ}$ are apothems of hexagon $ABCDEF$ inscribed in the circle with center O.

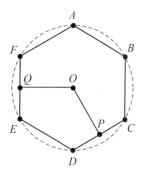

Figure 7.19 Apothem

> ### THEOREM 7.8
> Every apothem of a regular polygon has the same length.

The proof of Theorem 7.8 is a direct result of Theorem 6.10. Use Figure 7.19 as a model and construct any regular polygon. By Construction 6.5, a circle can be circumscribed around the regular polygon. Notice each side of the polygon is a chord of the circle and they are all congruent to each other (for example, $\overline{EF} \cong \overline{DC}$). Since they are congruent chords, Theorem 6.10 says they are equidistant from the center of the circle (for example, $\overline{OQ} \cong \overline{OP}$). Thus, every apothem of a regular polygon has the same length.

Since every apothem of a regular polygon has the same length, we often say that any apothem of a regular polygon is *the* apothem of the regular polygon.

> ### THEOREM 7.9
> The apothem of a regular polygon bisects its respective side.

Theorem 7.9 is a direct result of Theorem 6.8.

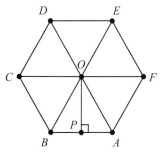

Figure 7.20

Given: Regular polygon $ABCDEF$ and $\overline{OP}$ an apothem of the polygon. (See Figure 7.20.)

Prove: $\overline{OP}$ bisects $\overline{AB}$

Proof _____

Statements	Reasons
1. Regular polygon $ABCDEF$ and apothem $\overline{OP}$	**1.** Given
2. $\overline{OP}$ bisects $\overline{AB}$	**2.** A line drawn from the center of a circle perpendicular to a chord bisects the chord (Thm 6.8)

The next theorem is stated without proof.

> **THEOREM 7.10**
> Every radius of a regular polygon bisects the angle at the vertex to which it is drawn.

EXAMPLE 1 Find $m\angle ABO$ and $m\angle AOB$ in the given regular octagon $ABCDEFGH$ where $\overline{AO}$ and $\overline{BO}$ are radii of the polygon. Refer to Figure 7.21.

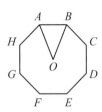

Figure 7.21

Answers

By Theorem 3.16, $m\angle ABC = \dfrac{(8-2)180°}{8} = 135°$.

By Theorem 7.10, $m\angle ABO = \dfrac{135°}{2} = 67.5°$.

By Theorem 6.29, $m\angle AOB = \dfrac{360°}{8} = 45°$.

OBJECTIVE 2 **Prove formula for area of a regular polygon.** Consider the regular hexagon $ABCDEF$ in Figure 7.20 with apothem $\overline{OP}$. The area of the hexagon is the sum of the areas of triangles $\triangle ABO$, $\triangle BCO$, $\triangle CDO$, $\triangle DEO$, $\triangle EFO$, and $\triangle FAO$. These triangles are all congruent by SSS. The area of $\triangle ABO$ is

$$\frac{1}{2}(AB)(OP).$$

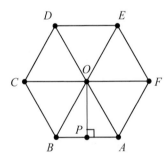

Figure 7.20

Similarly, we can find the area of each of the five remaining triangles. If we add these areas together and simplify, we obtain

$$\frac{1}{2}(OP)(AB + BC + CD + DE + EF + FA).$$

Let p be the perimeter of $ABCDEF$, then

$$p = AB + BC + CD + DE + EF + FA.$$

Let a be the length of the apothem. Then the sum of the areas of the triangles, which is equal to the area of the regular hexagon, simplifies to

$$A = \frac{1}{2}ap.$$

A similar argument can be given for any regular polygon, which would provide the proof of the next theorem.

> **THEOREM 7.11** *Area of a Regular Polygon*
> The **area of a regular polygon** with apothem of length a and perimeter p is determined with the formula
>
> $$A = \frac{1}{2}ap.$$

EXAMPLE 2   Find the length of a radius of $\square ABCD$ if the apothem measures 7 cm. (See Figure 7.22.)

Since the radius of a regular polygon bisects the angle to the vertex, $m\angle OCE = 45°$. $\triangle OEC$ is a 45°-45°-90° triangle, therefore the radius is $7\sqrt{2}$ cm.

Figure 7.22

OBJECTIVE 3   Apply the area formula.

EXAMPLE 3   Find the area of a regular hexagon with sides measuring 12 cm.

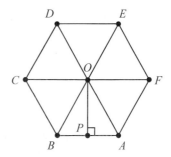

Figure 7.23

Refer to Figure 7.23. The measure of central angle $\angle AOB$ is $\dfrac{360°}{6} = 60°$ by Theorem 6.29. Then $m\angle AOP = 30°$ so $\triangle AOP$ is a 30°-60°-90° triangle. $AB = 12$ cm and $AP = 6$ cm, hypotenuse $AO = 12$ cm because it is twice as long as the short leg. Thus, the side opposite the 60° angle, the apothem, has length $\dfrac{\sqrt{3}}{2}(12) = 6\sqrt{3}$ cm. The perimeter of the hexagon is $6(12) = 72$ cm. Substitute $6\sqrt{3}$ for a and 72 for p in the formula

$$A = \frac{1}{2}ap.$$

$$A = \frac{1}{2}(6\sqrt{3})(72)$$
$$= 216\sqrt{3}$$

Thus, the exact area is $216\sqrt{3}$ cm², which can be approximated by 374.1 cm², correct to the nearest tenth of a square centimeter.

PRACTICE EXERCISE 1

Find the area of a regular pentagon with an apothem that measures 4.5 inches and each side is 6.5 inches.

ANSWER BELOW

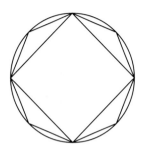

Figure 7.24

In Postulate 7.5, we said that the area of a circle is determined with the formula $A = \pi r^2$. The following example shows how the formula for the area of a regular polygon can lead us to the formula for the area of a circle. Suppose we are given a circle. If we inscribe a square in the circle, the area of the square would give an approximation for the area of the circle. It would not be a very good approximation because the area of the four segments of the circle would not be included. If we bisect the four equal arcs and construct the inscribed octagon as shown in Figure 7.24, the area of the octagon would clearly be a better approximation for the area of the circle.

If we repeat this process forming a regular 16-gon, the area of the 16-gon would be an even better approximation for the area of the circle. Continuing in this manner, we would obtain a better and better approximation of the area of the circle. The area of each regular polygon is

$$A = \frac{1}{2}ap,$$

and as the process continues, a is approaching the radius of the circle r, and the perimeter p is approaching the circumference of the circle, $2\pi r$. Thus, the area is approaching

$$A = \frac{1}{2}r(2\pi r) = \pi r^2.$$

Note Because it is not always easy to find the length of the apothem of a regular polygon with many sides, we often approximate the area of such a polygon with the area of its circumscribed circle. If the apothem is known but the perimeter is not, approximate the perimeter using $r = a$ for the inscribed circle.

Answer to Practice Exercise

1. 73.125 in.^{2}

7.4   Exercises

FOR EXTRA HELP: Student's Solutions Manual   Addison-Wesley Math Tutor Center

---

1. A regular polygon has perimeter 80 yd and apothem 10 yd. Find its area.
2. A regular polygon has perimeter 144 cm and apothem 18 cm. Find its area.
3. Find the lengths of the apothem and the side of a regular hexagon whose radius measures 8 in.

4. Find the lengths of the radius and the side of an equilateral triangle whose apothem measures 8 cm.

5. Find the lengths of the radius and the apothem of a square whose side measures 6 cm.

6. Find the lengths of the radius and the apothem of a square whose side measures 10 in.

7. Find the area of a regular hexagon with sides 16 ft.

8. Find the area of a regular hexagon with sides 12 ft.

9. The area of a regular hexagon is $1350\sqrt{3}$ cm². Find the length of a side.

10. The area of a regular hexagon is $864\sqrt{3}$ yd². Find the length of a side.

11. Find the area of a square whose apothem measures 5 cm.

12. Find the area of an equilateral triangle whose apothem measures 8 in.

13. In a regular octagon, the measure of each apothem is 5.76 centimeters and each side is 6 cm. Find the area.

14. In a regular octagon, the measure of each radius is 29 in. and each apothem is 21 in. Find the area.

15. Estimate to the nearest tenth of a unit, the area of a regular 16-gon with apothem 12 ft. (Hint: Find the area of the inscribed circle using $A = \pi r^2$.)

16. Estimate to the nearest tenth of a unit, the area of a regular 20-gon with an apothem 8.5 centimeters.

17. Find the perimeter of a regular polygon if the apothem is 9 m and the area is 388.6 m². Round the answer to the nearest tenth of a meter.

18. Use the formula $A = \frac{1}{2}ap$ to find the area of a square. Use s as the length of a side of the square. Compare the result to the formula given in Section 7.1.

19. The outside edge of a puzzle is in the shape of a regular octagon. If one side measures 4.25 inches and the area of the puzzle is 85 square inches, find the length of the apothem.

20. The outside edge of a puzzle is in the shape of a regular hexagon. If one side measures 4 inches and the area of the puzzle is 42 square inches, find the length of the apothem.

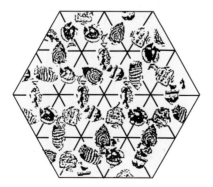

Exercise 20

21. Find the area of the shaded region given the regular hexagon in the figure where the apothem measures 8.7 mm and one side of the hexagon measures 10 mm.

Exercise 21

22. Corollary 7.5 shows the formula for the area of an equilateral triangle is $A = \dfrac{s^2\sqrt{3}}{4}$, where s is the measure of the side of the triangle. Use the formula $A = \dfrac{1}{2}ap$ to show the same formula is true for any equilateral triangle with side s.

23. Brahmagupta was an East Indian mathematician who lived during the seventh century. He discovered a formula for the area of a quadrilateral inscribed within a circle. As we learned in Chapter 6, this means the vertices of the quadrilateral lie on the circle.

$A = \sqrt{(s-a)(s-b)(s-c)(s-d)}$, where a, b, c, and d are the lengths of the sides of the quadrilateral and $s = \dfrac{a+b+c+d}{2}$

This formula looks similar to Heron's formula for the area of a triangle. Discuss the similarities and differences. Show a numerical example of how the formula works. Research more about the life of Brahmagupta.

Exercise 23

Chapter 7 Review

Key Terms and Symbols

7.1
base (of a rectangle)
height (of a rectangle)
area of a rectangle
area of a square
altitude (of a parallelogram)
height (of a parallelogram)
base (of a parallelogram)
area of a parallelogram
height (of a triangle)

base (of a triangle)
area of a triangle
Heron's formula
area of an equilateral triangle
altitude (of a trapezoid)
height (of a trapezoid)
area of a trapezoid
area of a rhombus

7.2
circumference
pi (π)
area (of a circle)

7.3
sector
area (of a sector)
length of an arc
segment (of a circle)

7.4
apothem
area of regular polygon

Formulas to Remember

7.1
area of a rectangle $A = bh$

area of a square $A = s^2$

area of a parallelogram $A = bh$

area of a triangle $A = \dfrac{1}{2}bh$

Heron's formula $A = \sqrt{s(s-a)(s-b)(s-c)}$, where $s = \dfrac{a+b+c}{2}$

area of an equilateral triangle $A = \dfrac{a^2\sqrt{3}}{4}$

area of a trapezoid $A = \dfrac{1}{2}(b + b')h$

area of a rhombus $A = \dfrac{1}{2}dd'$

7.2
circumference of a circle $C = 2\pi r = \pi d$

area of a circle $A = \pi r^2$

7.3
area of a sector $A = \dfrac{m}{360}\pi r^2$

length of an arc $L = \dfrac{m}{360}2\pi r = \dfrac{m}{180}\pi r$

7.4
area of a regular polygon $A = \dfrac{1}{2}ap$

Review Exercises

Section 7.1

1. Find the area and perimeter of the given figure to the nearest hundredth.

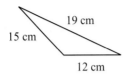

19 cm

15 cm

12 cm

Exercise 1

2. Find the area and perimeter of the given figure.

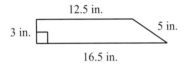

12.5 in.

3 in.

5 in.

16.5 in.

Exercise 2

3. Find the area of an equilateral triangle with a perimeter of 36 inches.

4. If the side of a square is doubled, how does the area change?

5. A room has four rectangular walls each measuring 14 ft by 8 ft. Two of the walls have windows measuring 4 ft by 3 ft. If the walls are to be given two coats of paint and 1 gallon of paint covers 300 ft², how many gallons will be needed for the job? Round up to the next gallon.

6. Find the area of a rhombus with diagonals measuring 14 cm and 18 cm.

7. Find the area and perimeter of the figure below.

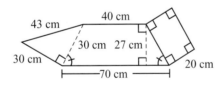

43 cm   40 cm

30 cm   27 cm

30 cm

70 cm

20 cm

Exercise 7

Section 7.2

8. Find the diameter of a circle with radius 2.6 cm.

9. Find the radius of a circle with diameter $\frac{4}{5}$ yd.

10. Find the actual area and circumference of a circle with diameter $\frac{5}{8}$ m. Leave the answer in terms of π.

11. Find the approximate circumference and area of the circle with diameter 9.2 ft. Round the approximation to the nearest hundredth.

12. A machine part is in the shape of an equilateral triangle 10 inches on a side. A hole with diameter 3 inches is drilled in the center of the part. To the nearest tenth, what is the area of the remaining metal?

13. A compact disc (CD) has a diameter of 12 cm. The hole in the center has a diameter of $\frac{3}{2}$ cm. Find the area of the surface of the CD rounded to the nearest tenth of a square centimeter.

Exercise 13

Section 7.3

14. Find the area of a sector of a circle, correct to the nearest tenth of a square inch if the radius of the circle is 11.4 in. and the arc of the sector is 18°.

15. Find the area of a segment of a circle formed by two radii measuring 10 cm that form a central 60° angle. Give the answer correct to the nearest tenth of a square centimeter.

16. To the nearest tenth, what is the length of an arc in a circle of radius 5.2 ft formed by a central angle measuring 40°?

17. Find the actual area and perimeter of the shaded sector. Do not round the answer.

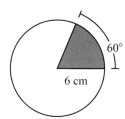

18. The windshield wiper on the back hatch of an SUV rotates through a 120° angle from beginning to end of its sweep. The wiper blade touches the windshield 8 in. from the point of rotation. The actual blade is 12 in. long. (See the figure.) Find the area cleaned by the wiper blade.

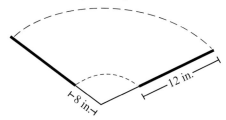

Exercise 18

Section 7.4

19. Find the area of a square whose apothem measures 6 cm.

20. A regular polygon has a perimeter of 120 ft and an apothem of 7 ft. Find its area.

21. Assume a stop sign is a regular octagon. A side measures 12.5 in. and a radius is 16.25 in. What is the actual area of the surface of the stop sign?

22. Find the area of a regular hexagon with sides 30 yd. Give the area correct to the nearest tenth of a square yard.

23. Estimate the area of a regular 30-gon with apothem 15.5 ft to the nearest tenth of a square foot.

Exercise 21

Chapter **7** **Practice Test**

1. Find the area of a parallelogram with base 20 ft and height 17 ft.

2. The base of a triangle is 15 yd and the area is 60 yd². What is the height?

3. The diagram (right) shows the family room in a house. The square in the center of the room corresponds to a fireplace. If carpeting costs $31.95 per square yard installed, assuming no waste, what will carpet cost for the entire room?

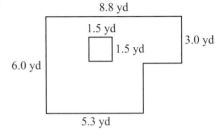

Exercise 3

4. Find the area, rounded to the nearest tenth of a square millimeter, of a triangle with sides measuring 5, 6, and 7 mm.

5. Find the area of a rhombus whose perimeter is 40 inches and one diagonal is 12 inches.

6. Find the measurement of the height of a trapezoid if the bases are 13 inches and 7 inches and the area is 40 square inches.

7. Find the area of each polygon below. Compare the areas. Do the results mean the two figures are congruent? Explain your answer.

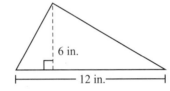

Exercise 7

8. Suppose a yield sign is a regular polygon measuring 33 inches on a side. Find the area of the sign. Round the answer to the nearest tenth of a square inch.

9. Find the actual circumference and area of a circle with diameter of 12.6 cm. Also find an approximate circumference and area rounded to the nearest hundredth.

10. Find the area of the shaded region rounded to the nearest tenth of an inch.

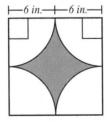

Exercise 10

Exercise 8

11. Find the circumference of a circle if its area is 25π cm².

12. Find the area of the shaded region to the nearest tenth of a unit.

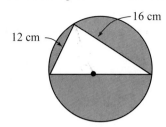

Exercise 12

13. Find the exact area and perimeter of the shaded sector if the radius is 12 m.

Exercise 13

14. If a folding fan is open 120° and has a radius of 18 cm, find the length of the arc at the outer edge of the fan.

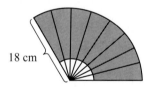

Exercise 14

15. Find the area of a regular polygon with perimeter 56 yd and apothem 7 yd.

16. Find the area of a segment of a circle formed by the chord joining the endpoints of two radii each measuring 8 centimeters and forming a central angle of 60°. Round the answer to the nearest tenth of a unit.

17. Find the exact area of a regular hexagon with sides 16 feet.

18. Estimate to one decimal place the area of a regular 20-gon with an apothem of 12.2 centimeters.

Chapters (4–7) Cumulative Review

Chapter 4

For Exercises 1–6, answer **true** *or* **false.**

1. Every rhombus is a square.

2. The opposite sides of a parallelogram are congruent.

3. In a kite, both diagonals bisect their respective angles.

4. A square is a rectangle.

5. The diagonals of a parallelogram bisect each other. The diagonals of a rhombus, rectangle, and square also bisect each other.

6. In a trapezoid, exactly one pair of opposite sides are parallel.

7. Given parallelogram *ABCD*, where $m\angle A = x + 10$ and $m\angle B = 2x - 4$, find *x* and find the measures of the four angles.

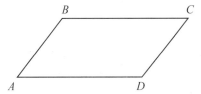

8. Given trapezoid *ABCD*, where $\overline{AD} \parallel \overline{BC}$ and points *X* and *Y* are the midpoints of their respective legs. If $BC = 3x + 1$, $XY = 6x + 7$, and $AD = 12x - 2$, find *x*.

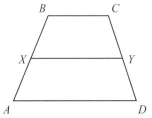

Chapter 5

9. *Given:* $\triangle ABC$ and $\overline{AB} \parallel \overline{DE}$

   Prove: $\dfrac{EC}{BC} = \dfrac{DC}{AC}$

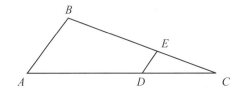

For Exercises 10–12, use the given right triangle ABC as in the figure to the right where ∠ACB is a right angle, E is the midpoint of AB, and $\overline{CD} \perp \overline{AB}$.

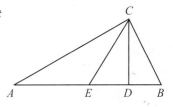

Exercises 10–12

10. If $CE = 16$ cm, find AE.

11. If $AD = 16$ ft and $DB = 4$ ft, find CD.

12. If $AC = 16$ in. and $AB = 34$ in., find CB.

13. In the given 30°-60°-90° triangle, one leg is 10 mm, find the measure of the other leg and the hypotenuse.

14. In a 45°-45°-90° triangle where the hypotenuse measures $8\sqrt{2}$ m, find the area of the triangle.

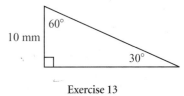

Exercise 13

15. Answer the following question *true* or *false*. If it is false, explain why. In △ABC below, if $m\angle C > m\angle A > m\angle B$ then $AB > AC > BC$.

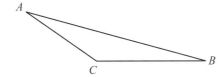

Chapter 6

In Exercises 16–27, use the given figure where $\overline{OA}$, $\overline{OB}$, $\overline{OC}$, and $\overline{OD}$ are radii of the circle, $m\angle BOC = 40°$, $m\widehat{AD} = 120°$, and points A and D are point of tangency. Find the following:

16. $m\widehat{BC}$

17. $m\angle BAC$

18. $m\widehat{AB}$

19. $m\angle ABC$

20. $m\angle ODA$

21. $m\widehat{CD}$

22. $m\angle AED$

23. $m\angle FAB$

24. $m\angle AGE$

25. Explain why $AB > AD$.

26. Explain why $AC > AB$.

27. If $AG = 20$ and $CG = 5$, find GD.

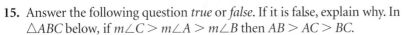

28. A stop sign is a regular octagon. Find the measure of each central angle in the sign.

29. Answer the following question *true* or *false*. If it is false, explain why. If $\overline{AB}$ and $\overline{CD}$ are two chords in a circle with $CD > AB$, then $m\overparen{CD} > m\overparen{AB}$.

30. Describe and draw a sketch of the locus of points that lie on a given pentagon and that are also a given line *m*.

Chapter 7

31. The base of a triangle is 15 yd and the area is 60 yd². What is the height of the triangle?

32. Find the exact circumference of a circle with a radius of 25 in.

33. Find the area of a segment of a circle formed by two radii measuring 20 mm that form a central angle of 60°. Round the answer to the nearest tenth of a square millimeter.

34. Find the actual area of a regular hexagon with sides 20 yd.

35. Find the area of a sector of a circle, correct to the nearest tenth of a square inch if the radius of the circle is 12.8 in. and the arc of the sector is 36°.

36. A square and an equilateral triangle have the same perimeter. The area of the square is 36 square inches. Find the area of the triangle.

37. In the given figure, a large square with one side measuring 4 inches is placed next to a small square measuring 1 inch on a side. Find the area of the shaded region. Note that $\overline{ACE}$ is **not** a straight line.

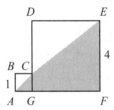

8

Solid Geometry

*T*o this point, we have discussed only figures that lie completely in a single plane. In this chapter, we consider planes, lines, and figures that can be formed in space.

One of our main considerations will be determining the surface area and volume of solid figures. Many applications, such as the one that follows, are based on surface area and volume. The solution to the problem below appears in Example 3 of Section 8.2.

The use of a scientific calculator is assumed throughout the entire chapter.

AN APPLICATION

A bird feeder has six sides with exactly the same shape. They are perpendicular to the base. The base of the feeder is a regular six-sided figure, a hexagon. If the part of the feeder that holds birdseed measures 18 inches in height and each side is 2 inches in length, how much birdseed can the feeder hold when full? See Example 3 in Section 8.2 for the solution.

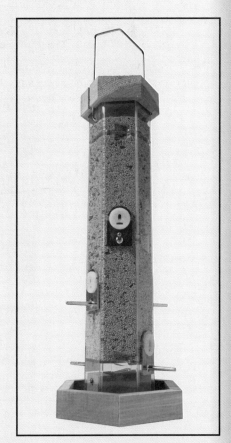

8.1 Planes and the Polyhedron

OBJECTIVES

1. Determine the behavior of lines and planes in space.
2. Define a polyhedron.
3. Identify the platonic solids.

OBJECTIVE 1   Determine the behavior of lines and planes in space.   In a plane, two distinct lines are either parallel or they intersect. When expanding to three-dimensions, a similar property exists for a line and a plane. Figure 8.1 illustrates a line parallel to a plane. Figure 8.2 shows one possible way a line can intersect a plane.

> **DEFINITION: Line Parallel to a Plane**
> A line is **parallel** to a plane if it does not intersect the plane.

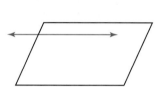
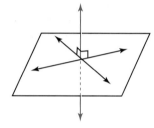

Figure 8.1 Line Parallel to Plane   Figure 8.2 Line Perpendicular to Plane

> **DEFINITION: Perpendicular to a Plane**
> A line is **perpendicular** to a plane if each line in the plane that passes through the point of intersection is perpendicular to the line.

In Figure 8.2, the blue line is perpendicular to the plane. Continuing the study of geometry in three dimensions; consider two planes. They will either be parallel or intersect.

> **DEFINITION: Parallel Planes**
> Two planes are **parallel** if they do not intersect.

Refer to Figure 8.3 for parallel planes. The next postulate is about intersecting planes.

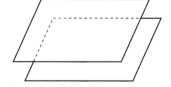

Figure 8.3 Parallel Planes

> **POSTULATE 8.1**
> The intersection of two distinct planes is a line.

Refer to Figure 8.4(a) for the following definition.

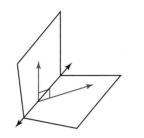

Figure 8.4(a) Perpendicular Planes

> ### DEFINITION: *Perpendicular Planes*
> Two planes are **perpendicular** if either plane contains a line that is
> perpendicular to the other plane.

There are, of course, planes that intersect but that are not perpendicular.
See Figure 8.4(b).

> ### DEFINITION: *Oblique Planes*
> If two planes or a line and a plane intersect but are not perpendicular,
> they are called **oblique**.

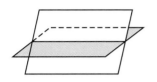

Figure 8.4(b) Oblique Planes

OBJECTIVE 2   Define a polyhedron.   With this background, we can now
consider solid figures formed by the intersection of planes.

> ### DEFINITION: *Polyhedron*
> A solid figure formed by the intersection of planes is called a **polyhedron**.

Some of the simplest solid figures are polyhedrons. For example, the box
shown in Figure 8.5 fits the definition of a polyhedron.

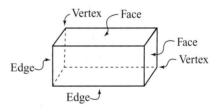

Figure 8.5 Polyhedron

The plane surfaces of a polyhedron are called **faces**, the lines of intersection of the planes are called **edges**, and the intersections of edges are called
vertices (plural of **vertex**).

> ### DEFINITION: *Regular Polyhedron*
> A **regular polyhedron** is a solid figure in which all faces are congruent
> *regular* polygons.

OBJECTIVE 3 **Identify the platonic solids.** By considering the nature of the intersecting faces in polyhedrons, it can be shown that there are only five possible regular polyhedrons, which are also called the *platonic solids*. They are shown and labeled in Figure 8.6.

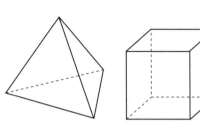 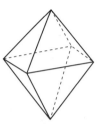 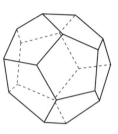 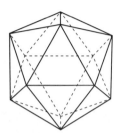

Name	Tetrahedron	Hexahedron (Cube)	Octahedron	Dodecahedron	Icosahedron
Each Face	Equilateral triangle	Square	Equilateral triangle	Regular pentagon	Equilateral triangle

Figure 8.6 Regular Polyhedrons (Platonic Solids)

Student Activity

Using Figure 8.6, complete the table below.

Platonic Solids	Number of Faces (f)	Number of Vertices (v)	Number of Edges (e)
Tetrahedron			
Hexahedron (cube)			
Octahedron			
Dodecahedron			
Icosahedron			

1. Looking at the data in the table, find a relationship between f, v, and e. It can be found by adding or subtracting f, v, and e (or some combination of these operations). This equation is called Euler's equation.
2. The relationship found in question 1 will apply to any polyhedron, not just regular ones. For the given figure to the left show the relationship works.

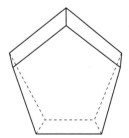

EXAMPLE 1 Find the number of vertices of a polyhedron that has 12 faces and 30 edges.
Substitute $f = 12$ and $e = 30$ in Euler's formula found in the preceding Student Activity and solve for v.

$$f + v - e = 2 \quad \text{Euler's formula}$$
$$12 + v - 30 = 2$$
$$v - 18 = 2$$
$$v = 20$$

Thus, the polyhedron has 20 vertices.

8.1 Exercises   *FOR EXTRA HELP:*   Student's Solutions Manual   Tutor Center   Addison-Wesley Math Tutor Center

In the figure below, line ℓ is the intersection of planes $\mathcal{P}$, $\mathcal{Q}$, and $\mathcal{R}$. Lines k and m are in $\mathcal{P}$, lines n and s are in $\mathcal{R}$, $m \perp \ell$, $m \perp n$, $\ell \parallel k$, and $\ell \parallel s$. Use this information to answer *true* or *false* in Exercises 1–12.

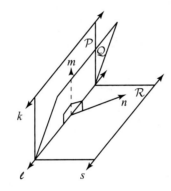

1. $k \parallel \mathcal{R}$	**2.** $k \parallel s$	**3.** $\mathcal{P} \parallel \mathcal{R}$
4. $\mathcal{P} \parallel \mathcal{Q}$	**5.** $m \perp \mathcal{R}$	**6.** $m \perp \mathcal{Q}$
7. $\mathcal{P} \perp \mathcal{Q}$	**8.** $\mathcal{P} \perp \mathcal{R}$	**9.** m and $\mathcal{Q}$ are oblique
10. $\mathcal{R}$ and $\mathcal{Q}$ are oblique	**11.** $m \perp k$	**12.** $n \parallel k$

In Exercises 13–18, determine the number of faces f, vertices v, and edges e in each polyhedron. Check to see that $f + v - e = 2$.

13.

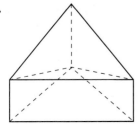

14.

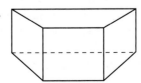

15.

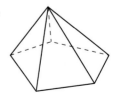

16.   **17.** 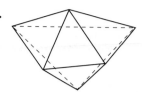 **18.**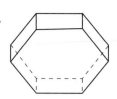

19. A polyhedron has 12 vertices and 30 edges. How many faces does it have?
20. A polyhedron has 14 faces and 9 vertices. How many edges does it have?

Drawing a cube

Drawing three-dimensional figures can be fun and challenging. This skill is very important in computer graphics. It may be helpful to do the drawing on graph paper.

Drawing a Cube

(a) To draw a cube from a front view, first draw the square at the top of the cube, oriented as in the picture at the right.

(b) Then draw three line segments to represent the three edges that are visible from our front-view perspective.

(c) Third, draw the bottom face that is congruent to the top face. In mathematics, we use dashed lines to show edges that are not visible. This produces the illusion of three dimensions.

(d) Connect the vertices that are hidden using dashed lines.

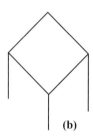

21. Using the above instructions, draw a rectangular prism (a six-sided figure with rectangle on the top and bottom) where the bases *appear* to be parallelograms.

22. Draw a triangular prism that is a five-sided figure in which the top and bottom are triangles and the sides are rectangles.

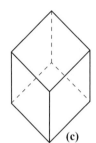

Drawing a Cylinder

(a) Draw a circle for the top.

(b) Draw two line segments down for the sides of the cylinder.

(c) Draw a circle for the base that is congruent to the top. Use a dashed line for the "back" of the circle because it is hidden.

23. Draw a tube lying on its side following the instruction for the cylinder.

24. Draw a cone (like an upside down ice cream cone) where a dashed line is used to draw the part of the bottom hidden from view.

Drawing a cylinder

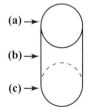

(a) →
(b) →
(c) →

Drawing a Pyramid

(a) Draw a base (the figure it "sits" on) using dashed lines to represent the edges that are hidden. In the example, we have chosen a triangular base but it may be any shape.

(b) Choose a point somewhere above the base.

(c) Draw solid lines from that point to the endpoints of the solid line on the base to represent the edges that are visible.

(d) Draw a dashed line to represent the last hidden edges.

25. Draw a pyramid with a square base.

Drawing a pyramid

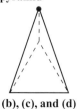

(a)   (b), (c), and (d)

Solve each applied problem in Exercises 26 and 27.

26. A regular dodecahedron is to be constructed using metal tubing for the edges. If each edge is to be 3 m long, how much tubing will be required for the project?

27. A storage tank in the shape of a cube is to be insulated using material that costs $2.50 per square foot. If the edge of the cube is 12 ft, how much will the insulation cost?

A **diagonal** of a polyhedron is a line segment joining two vertices that are not on the same face. Use this definition in Exercises 28 and 29.

28. How many diagonals does a cube have?

29. Use the Pythagorean theorem to derive a formula for the length of the diagonal of a cube with edge of length e.

30. For whom were the platonic solids named? Write a paragraph about the person's name, background, and any other mathematical achievements.

31. Investigate Leonhard Euler. How is Euler's equation related to the platonic solids?

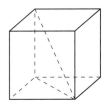

Hexahedron (Cube)

Exercises 28 and 29

(8.2) Prisms

OBJECTIVES

1. Define a prism.
2. Define lateral area of a prism.
3. Define and calculate the surface area of a prism.
4. Calculate the volume of a prism.

The cube drawn in the exercises of the last section is an example of a polyhedron. More specifically, it is a prism because the top and bottom are congruent figures in parallel planes. See Figure 8.7 below.

Cube

Figure 8.7

In this section, we will determine formulas for calculating the surface area and volume of prisms.

OBJECTIVE 1   Define a prism.

> **DEFINITION: Prism**
>
> The solid figure formed by joining two congruent polygonal regions in parallel planes is called a **prism**. The polygonal regions are called **bases** and the other surfaces are **lateral faces**.

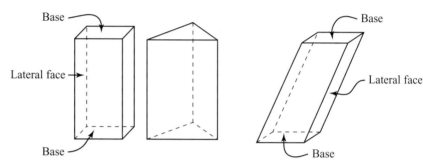

Figure 8.8 Right Prisms   **Figure 8.9 Oblique Prism**

Notice that the lateral faces are parallelograms. If the lateral faces are rectangles as in Figure 8.8, the prism is a **right prism**; otherwise it is called an **oblique prism** as in Figure 8.9.

OBJECTIVE 2   **Define lateral area of a prism.**   **Lateral area** of a solid is the sum of the area of the lateral faces. To develop a formula for the lateral area, we restrict ourselves to *right* prisms. Consider the right prism shown in Figure 8.10, where *h* is the **height** of the *prism*, the length of an **altitude**, and *a*, *b*, *c*, *d*, and *e* are lengths of the sides of a base. Because the sides of a right prism are rectangles, the lateral area is given by

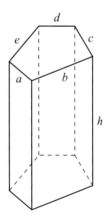

Figure 8.10

$$LA = ah + bh + ch + dh + eh$$
$$= (a + b + c + d + e)h \qquad \text{Distributive property}$$
$$= ph,$$

where *p* is the perimeter of a base. This idea is applicable no matter how many sides the base has. This is an informal proof of the following theorem.

> **THEOREM 8.1   Lateral Area of a Right Prism**
>
> The lateral area (*LA*) of a right prism is determined with the formula
> $$LA = ph,$$
> where *p* is the perimeter of a base and *h* is the height of the prism.

OBJECTIVE 3   **Define and calculate the surface area of a prism.**   Surface area is the sum of the areas of the surfaces of a solid figure. The **surface area** (*SA*) of the prism is the lateral area plus the area of the two bases. Thus, the total area of a prism is its surface area.

$$SA = LA + 2B,$$

where *B* represents the area of one base of the prism.

EXAMPLE 1   How many square inches of glass are needed to make the fish tank in Figure 8.11 if the top is left open and the sides and base are glass?

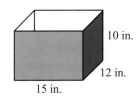

Figure 8.11

We must find the lateral area and the area of one of the bases. The base area is

$$(15 \text{ in.})(12 \text{ in.}) = 180 \text{ in.}^2.$$

To find the lateral area, first determine the perimeter.

$$p = 15 \text{ in.} + 12 \text{ in.} + 15 \text{ in.} + 12 \text{ in.}$$
$$= 54 \text{ in.}$$
$$LA = ph$$
$$= (54 \text{ in.})(10 \text{ in.})$$
$$= 540 \text{ in.}^2$$

Thus, the surface area of glass required is

$$SA = 180 \text{ in.}^2 + 540 \text{ in.}^2$$
$$= 720 \text{ in.}^2.$$

PRACTICE EXERCISE 1

The outside walls of a garage are to be covered with siding that costs $1.50 per square foot. How much will it cost if the building is 16 ft wide, 18 ft long, and 8 ft high?

ANSWER ON PAGE 393

Some of the prisms in this section will be regular prisms. A **regular prism** is a prism whose base is a regular polygon. Thus, in a regular right prism the bases are congruent regular polygons and the lateral faces are rectangles.

EXAMPLE 2

Figure 8.12 is a right regular hexagonal prism. Notice the two bases are regular hexagons and the lateral faces are rectangles. Find the surface area of the prism given the dimensions in Figure 8.12.

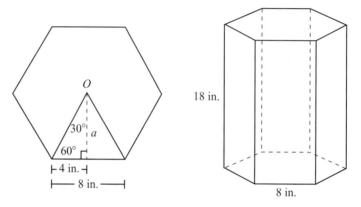

Figure 8.12

The area of a regular hexagon is $A = \frac{1}{2}ap$ (from Section 7.4), where a is the apothem and p is the perimeter of the base.

$$A = \frac{1}{2}(4\sqrt{3})(48) = 96\sqrt{3} \text{ in.}^2 \quad \text{area of the base } (B)$$

$$LA = ph = (48)(18) = 864 \text{ in.}^2$$

$$SA = LA + 2B = 864 + 2(96\sqrt{3})$$

$$SA = (864 + 192\sqrt{3}) \text{ in.}^2 \text{ or approximately } 1196.6 \text{ in.}^2$$

OBJECTIVE 4 **Calculate the volume of a prism.** In our study of plane geometry, we used units such as 1 cm and 1 inch to measure length and 1 square centimeter (cm²) and 1 square inch (in.²) to measure area. For measuring the volume of a solid, we'll need cubic units such as 1 cubic centimeter (cm³) and 1 cubic inch (in.³) (Volume is the measurement of the amount of space within a solid figure.) Figure 8.13 illustrates these units using one of the most basic prisms, the cube.

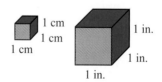

Figure 8.13 Unit Volume

To see how we find the volume of other prisms in cubic units, consider Figure 8.14. The prism in Figure 8.14(a) has volume 20 cm³ because there are 10 cm³ on the bottom layer and 10 cm³ on the top layer. In Figure 8.14(b), there is one more layer (10 cm³) than in Figure 8.14(a), and the volume is 30 cm³.

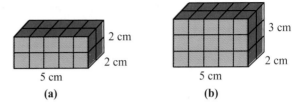

Figure 8.14 Volume of a Right Prism

The volumes in Figure 8.14(a) and 8.14(b) can be found by multiplying the area of the base of each prism by its height.

$$20 \text{ cm}^3 = \underbrace{5 \text{ cm} \cdot 2 \text{ cm}}_{\text{area of base}} \cdot \underbrace{2 \text{ cm}}_{\text{height}}$$

$$30 \text{ cm}^3 = \underbrace{5 \text{ cm} \cdot 2 \text{ cm}}_{\text{area of base}} \cdot \underbrace{3 \text{ cm}}_{\text{height}}$$

Theorem 8.2 follows from this information.

THEOREM 8.2   Volume of a Right Prism

The volume of a right prism is determined with the formula.

$$V = Bh,$$

where B is the area of a base and h is the height.

EXAMPLE 3

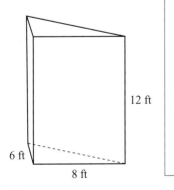

Figure 8.15

A chemical is to be stored in the tank (a right prism) shown in Figure 8.15. The base is a right triangle with legs 6 ft and 8 ft, and the tank is 12 ft deep. How many cubic feet of chemical will the tank hold?

Because the base is a right triangle,

$$B = \frac{1}{2}(6 \text{ ft})(8 \text{ ft}) = 24 \text{ ft}^2.$$

Thus, the volume of the right prism is

$$V = Bh$$
$$= (24 \text{ ft}^2)(12 \text{ ft}) = 288 \text{ ft}^3.$$

The tank will hold 288 ft³ of chemical.

PRACTICE EXERCISE 2

A highway construction truck has a bed 5.5 yd long, 2.5 yd wide, and 2.0 yd deep. How many cubic yards of gravel will it hold?

ANSWER ON PAGE 393

EXAMPLE 4

The bird feeder presented at the opening of this chapter is a right hexagonal prism. The base of the bird feeder is a regular 6-sided figure, a hexagon. If the part of the feeder that holds birdseed measures 18 inches in height and each side is 2 inches in length, how much birdseed can the feeder hold when full? Refer to Figure 8.16.

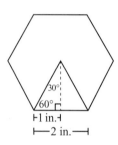

Figure 8.16

Finding the amount of feed for the bird feeder is a volume problem where the area of the base is needed.

$$V = Bh$$

$$A = \frac{1}{2}ap \qquad \text{area of base } (B)$$

$$= \frac{1}{2}(\sqrt{3})(12)$$

$$= 6\sqrt{3} \text{ in.}^2 \qquad \text{area of base } (B)$$

$$V = (6\sqrt{3})(18)$$

$$= 108\sqrt{3} \text{ in.}^3 \text{ or approximately } 187.1 \text{ in.}^3$$

 **Technology Connection**

Use a spreadsheet or a graphing calculator to investigate these cubes by using the fact that a cube is a right prism with all sides congruent.

1. Given an edge of a cube, find a formula for the surface area.
2. Given an edge of a cube, find a formula for the volume.
3. Complete the following table.
4. Describe the rate at which the surface area and volume increase as the edge of the cube gets larger. On the same screen of a graphing calculator, graph y_1 = "your formula for the surface area" along with y_2 = "your formula for the volume." Compare the two graphs in an appropriate window corresponding to the table values.

EDGE (cm)	SURFACE AREA (cm²)	VOLUME (cm³)
2		
4		
6		
8		
10		
12		

Answers to Practice Exercises

1. $816   2. 27.5 yd³

8.2 Exercises

FOR EXTRA HELP: Student's Solutions Manual   Addison-Wesley Math Tutor Center

In Exercises 1–3, determine whether the statement is *true* or *false* using the figure.

1. For a prism with rectangular base of dimensions l and w and with height h, the surface area is $SA = 2lw + 2lh + 2wh$.
2. For a prism with rectangular base of dimensions l and w and with height h, the volume is $V = lwh$.
3. The lateral surfaces of a right prism are rectangles.

Exercises 1–3

Determine the surface area (*SA*) of each right prism described in Exercises 4–11.

4. Cube with edge 5 ft
5. Cube with edge 1.5 m
6. Rectangular base 2 ft by 6 ft; height 5 ft
7. Rectangular base 6.4 cm by 15.5 cm; height 20.2 cm
8. Equilateral triangular base with side 2.8 yd; height 1.5 yd
9. Right isosceles triangular base with legs 6 cm; height 11 cm
10. Regular hexagon base with side 26 inches; height 14 inches
11. Right triangular base with one leg 9 mm and hypotenuse 14 mm; height 22 mm

Determine the volume of each right prism described in Exercises 12–19.

12. Cube with edge 8 m
13. Cube with edge 10.2 cm
14. Rectangular base 2.3 cm by 1.2 cm; height 4.5 cm
15. Rectangular base 3.4 inches by 9.5 inches; height 8.75 inches
16. Right isosceles triangular base with legs 10.5 ft; height 20.4 ft
17. Equilateral triangular base with side 8 cm; height 24 cm
18. Right triangular base with hypotenuse 32 m and one leg 20 m; height 9 m
19. Regular hexagon base with side 3.6 yd; height 4.8 yd
20. A pentagonal right prism has height 10 inches. Its base edges are 4, 6, 7, 8, and 10 inches. Find the prism's lateral area.
21. The surface area of a cube is 294 cm². Find the length of one edge and the volume of the cube.
22. If the figure was cut out and folded along the dashed lines, it would be a cube. Find the surface area and volume of the cube.

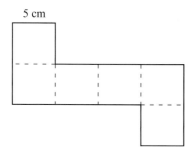

Exercise 22

23. Find the volume of the given rectangular prism that has a cut out that measures 2 feet by 3 feet.

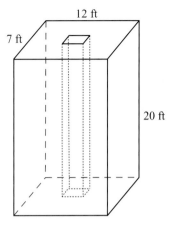

Exercise 23

Solve the applied problems in Exercises 24–28.

24. A highway construction truck has a bed 6.5 yd long, 3.0 yd wide, and 2.4 yd deep. How many cubic yards of gravel will it hold?

25. A tank is in the shape of a cube. How many grams of water will it hold if the edge of the cube is 22 cm and water weighs 1 g per cubic centimeter?

26. How many square centimeters of glass will it take to form the bottom and sides of the tank in Exercise 25?

27. Elizabeth Wright owns a rectangular building 30 ft long, 12 ft wide, and 8 ft high. How much will it cost to put siding on the building if the siding costs $0.75 per square foot?

28. Sol is building a planting bed in the shape of a trapezoid. See the figure at right. The bed will be 9 inches deep. The bases of the trapezoid are 2 ft and 8 ft and the sides are both 5 ft long. How many cubic feet of topsoil is needed to fill the bed?

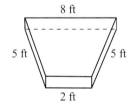

Exercise 28

29. The rectangular base of a right prism has dimensions l and w and height of the prism is h. Derive a formula for the length d of a diagonal. (A diagonal joins vertices that are not on the same face.) See the figure at right.

30. Use the formula derived in Exercise 29 to find the length of the diagonal of a rectangular box that is 17.0 inches by 11.0 inches by 14.0 inches.

31. Prisms are seen throughout nature. Honeycombs, quartz, and salt crystals are a few examples. Select an example of a prism in nature. Research how the geometric shape is beneficial or why the example is shaped as it is. Make a sketch or model that shows your example's geometric characteristics.

32. Write a paragraph describing the different examples of prisms you find around your home and school.

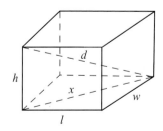

Exercise 29

8.3 Pyramids

OBJECTIVES

1. Define a pyramid.
2. Determine the formula for lateral area of a pyramid.
3. Calculate the surface area of a pyramid.
4. Calculate the volume of a pyramid.

OBJECTIVE 1 Define a pyramid. We now consider another polyhedron, the *pyramid*.

> **DEFINITION: Pyramid**
> The solid figure formed by connecting a polygon with a point not in the plane of the polygon is called a **pyramid**. The polygonal region is called the base and the point is the **vertex**. The line segment from the vertex perpendicular to the plane of the base is the **altitude**.

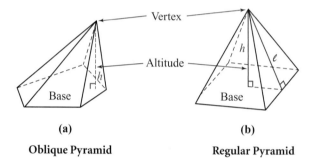

(a)   **(b)**

Oblique Pyramid   **Regular Pyramid**

Figure 8.17 Pyramids

Figure 8.17(a) shows an **oblique pyramid** and Figure 8.17(b) shows a *regular pyramid* like those built by the Egyptians. A **regular pyramid** has a regular polygon for a base, congruent isosceles triangles for the lateral surfaces, and an altitude passing through the center of the base. In this section, we will study only regular pyramids.

OBJECTIVE 2   Determine the formula for lateral area of a pyramid.   In Figure 8.17(b), the distance ℓ is called the **slant height** of the lateral surfaces of a regular pyramid. The distance ℓ is also the height of the triangular **face** of the pyramid. Thus, we can determine the area of each face using one-half the base length multiplied by the slant height.

Consider "flattening" a pyramid so that the sides are in the same plane as the base as in Figure 8.18. Since the lateral faces are all triangles, the area of a

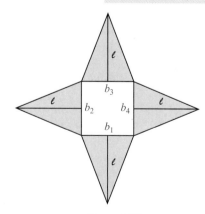

Figure 8.18

face is calculated using the formula $A = \frac{1}{2}bh$. Let the length of the bases of the triangles be $b_1, b_2, b_3,$ and b_4. The area of all the faces in Figure 8.18 is

$$A = \frac{1}{2}b_1\ell + \frac{1}{2}b_2\ell + \frac{1}{2}b_3\ell + \frac{1}{2}b_4\ell$$

$$= \frac{1}{2}\ell(b_1 + b_2 + b_3 + b_4) \qquad \text{Distributive property}$$

$$= \frac{1}{2}\ell p$$

$$= \frac{1}{2}p\ell,$$

where p is perimeter of the base of the pyramid. This leads to Theorem 8.3.

THEOREM 8.3   Lateral Area of a Regular Pyramid

The lateral area of a regular pyramid is determined with the formula

$$LA = \frac{1}{2}p\ell,$$

where p is the perimeter of the base and ℓ is the slant height.

⌐**Note** Because the lateral surfaces of a regular pyramid are congruent isosceles triangles, another method to find the lateral area of the pyramid is to find the area of one lateral surface and multiply by the number of lateral faces.

EXAMPLE 1   Find the lateral area of a regular pyramid if its base is a regular hexagon. The slant height of the pyramid is 13 inches, the altitude of the pyramid is 12 inches, and one edge of the base is 5.8 in. Refer to Figure 8.19.

Each face is an isosceles triangle with an altitude of 13 inches because the slant height of the pyramid is the same as the altitude of the face. The 12 inches is a piece of information not needed for this problem.

$$p = 6(5.8) = 34.8 \text{ in.}$$

$$LA = \frac{1}{2}(34.8)(13) = 226.2 \text{ in.}^2$$

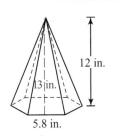

Figure 8.19

OBJECTIVE 3   **Calculate the surface area of a pyramid.**   Similar to the prism, the surface area of a pyramid is the lateral area plus the area of the base.

$$SA = \frac{1}{2}p\ell + B$$

EXAMPLE 2

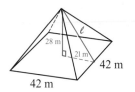

Figure 8.20

Find the surface area of a regular pyramid if each side of its square base is 42 m and the altitude of the pyramid has length 28 m.

The area of the square base is $(42)^2$ square meters, but before we can find the lateral surface area, we must use the height and base dimensions to find the slant height. In Figure 8.20, the right triangle with ℓ as its hypotenuse will have one leg 28 m (the height) while the other leg is one-half the length of the side of the base, 21 m. By the Pythagorean theorem,

$$\ell^2 = (28)^2 + (21)^2$$
$$\ell = \sqrt{(28)^2 + (21)^2}$$
$$= \sqrt{784 + 441}$$
$$= \sqrt{1225} = 35 \text{ m}.$$

The perimeter of the base is

$$p = 4(42) = 168 \text{ m}.$$

Finally, the surface area is

$$SA = \frac{1}{2}p\ell + B$$
$$= \frac{1}{2}(168)(35) + (42)^2$$
$$= 2940 + 1764 = 4704 \text{ m}^2.$$

PRACTICE EXERCISE 1

Find the surface area of a pyramid that has four equilateral triangles as faces, a base edge $3\sqrt{3}$ m long, and the slant height is 4.5 m. Round the answer to the nearest tenth of a square meter.

ANSWER ON PAGE 400

OBJECTIVE 4

Calculate the volume of a pyramid. The formula for the volume of a pyramid can be derived by comparing the prism and the pyramid. Because of the complexity of the proof, we will state the following theorem without proof.

> **THEOREM 8.4   Volume of a Regular Pyramid**
>
> The volume of a regular pyramid is determined with the formula
>
> $$V = \frac{1}{3}Bh,$$
>
> where B is the area of the base and h is the height.

The volume formula is true for all pyramids, but we will continue to emphasize regular pyramids.

The Transamerica Tower in San Francisco, California, was designed using the form of a pyramid.

EXAMPLE 3   If rock was used to build the pyramid in Example 2 and it weighs 1200 kg per cubic meter, what is the total weight of the pyramid?

We must first calculate the volume of the pyramid and then multiply by 1200. Recall from Example 2 that the square base had a side 42 m and the height of the pyramid is 28 m. Thus,

$$V = \frac{1}{3}Bh$$

$$= \frac{1}{3}(42)^2(28)$$

$$= 16{,}464 \text{ m}^3.$$

To find the total weight, multiply by 1200.

$$\text{Total Weight} = (16{,}464)(1200)$$

$$= 19{,}756{,}800 \text{ kilograms}$$

EXAMPLE 4   Find the volume of the regular pyramid in Example 1. Its base is a regular hexagon and the slant height of the pyramid is 13 inches. The altitude of the pyramid is 12 inches and one edge of the base is 5.8 inches. Round the final answer to the nearest tenth of a square inch. Refer to Figure 8.21.

The formula needed is $V = \frac{1}{3}Bh$. To find the area of a regular hexagon, use Theorem 7.10 $\left(A = \frac{1}{2}ap \right)$. Refer to Example 1 in Section 7.4 for an example. In the 30°-60°-90° triangle, the long leg is $\sqrt{3}$ times the short leg, so the apothem is $2.9\sqrt{3}$.

$$p = 6(5.8)$$

$$= 34.8 \text{ in.}$$

$$B = \frac{1}{2}ap$$

$$= \frac{1}{2}(2.9\sqrt{3})(34.8)$$

$$V = \frac{1}{3}Bh$$

$$= \frac{1}{3}\left[\frac{1}{2}(2.9\sqrt{3})(34.8) \right](12)$$

$$\approx 349.6 \text{ in.}^3$$

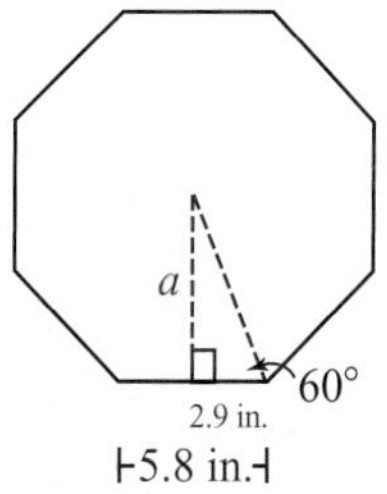

Figure 8.21

Note When the final answer is to be approximated, be sure to round only the final answer not the calculations needed to find the final answer.

PRACTICE EXERCISE 2

Find the volume of a pyramid that has four equilateral triangles as faces and a square base with each edge measuring $3\sqrt{3}$ meters long. The pyramid has a height of 3.7 m. Round the answer to the nearest tenth of a cubic meter.

ANSWER BELOW

SUMMARY

	LATERAL AREA (LA)	SURFACE AREA (SA)	VOLUME (V)
Right prism	ph	$LA + 2B$	Bh
Regular pyramid	$\frac{1}{2}p\ell$	$LA + B$	$\frac{1}{3}Bh$

CAUTION h means the height of the solid, *not* the height of a face. The height of a face is its slant height, ℓ.

Note LA means lateral area
B means area of the base
ℓ means slant height
p means perimeter of the base

Answers to Practice Exercises

1. 73.8 m² **2.** 33.3 m³

8.3 Exercises

FOR EXTRA HELP: Student's Solutions Manual Addison-Wesley Math Tutor Center

In Exercises 1–4, find the lateral surface area of each regular pyramid described below.

1. Equilateral triangle base with side 22 m; slant height 28 m
2. Square base with side 6.8 ft; slant height 4.5 ft
3. Square base with side 124 cm; height of pyramid is 86 cm
4. Regular hexagon base with side 17.8 inches; height of pyramid is 32.2 inches

In Exercises 5–18, round the answers to the nearest tenth of a unit.

5. Find the surface area of the square pyramid in the figure.

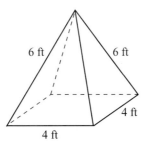

6 ft   6 ft

4 ft

4 ft

6. Find the surface area of the regular pentagonal pyramid in the figure. The apothem of the pentagon is 5.5 cm and $AB = 10$ cm.

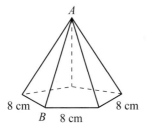

A

8 cm   8 cm

B   8 cm

7. Find the surface area of the regular hexagonal pyramid in the figure. The slant height is 24.8 mm and length of a side of the base is 7 mm.

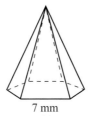

7 mm

8. A regular tetrahedron is a regular triangular pyramid. All four sides of this pyramid are congruent equilateral triangles. Recall Corollary 7.5 gives the formula for area of an equilateral triangle. If one edge of the tetrahedron measures 5 inches, find the surface area of the figure.

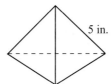

5 in.

In Exercises 9–12, find the volume of each regular pyramid described.

9. Square base with side 22 yd; height 15 yd
10. Equilateral triangle base with side 8.2 m; height 4.6 m
11. Regular hexagon base with side 16 cm; slant height 24 cm
12. Square base with side 30.6 ft; slant height 46.8 ft
13. Using the information in Exercise 5, find the volume of the pyramid if the height of the pyramid is $2\sqrt{7}$ ft.
14. Using the information in Exercise 6, find the volume of the pyramid if the height of the pyramid is 7.33 centimeters.
15. Using the information in Exercise 7, find the volume of the pyramid if the height of the pyramid is 24.0 millimeters.
16. Two friends owned different tents. They are planning a backpacking trip and deciding which tent to bring. Kate owns a tee-pee style tent with a height of 6 ft. The bottom of the tent looks like a regular hexagon where the radius is 4 ft. Terri has a standard pup tent with dimensions shown in the figure.
    a. Which tent has the greatest volume?
    b. Which tent has the greatest floor space?
    c. Which of the tents would be best for backpacking?
    Explain your reasoning and show the calculations needed.

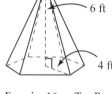

Exercise 16a—Tee-Pee

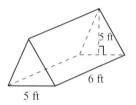

Exercise 16b—Pup Tent

17. A sculpture in a park has the shape of a regular pyramid with a square base 100 ft on a side. The height of the structure is 25 ft. If one gallon of paint will cover 220 ft^2, how many gallons will be required to paint the lateral exterior surface of the structure?
18. A perfume bottle is a regular pyramid with square base. If the base has side 4 cm and slant height 3 cm, how many cubic centimeters of perfume will the bottle hold?
19. A candle manufacturer wants to make a candle in the shape of a regular pyramid with a square base. One side of the base will measure 10 centimeters and the height of the candle will be 15 centimeters. How much wax will be needed to manufacture the candle?

Exercise 19

20. A manufacturer is designing special packaging for a product in the shape of a truncated pyramid. A truncated pyramid is a pyramid with the top cut off. The model he is building is shown in the figure with the measurements. The bases are not regular figures. The lateral sides are trapezoids. The height of each trapezoidal face is 12 cm. Determine the amount of material needed to build the model (assuming no overlap on the sides).

Exercise 20

Student Activity

21. Make a tetrahedron using a small envelope. You will need an envelope that measures approximately $3\frac{5}{8}$ inches by $6\frac{1}{2}$ inches, a straightedge, and a pair of scissors.

    (a) Seal the envelope.
    (b) Fold the envelope in half the short way. Make a very sharp crease. Open the envelope flat.

    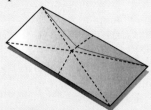

    (c) Draw the two diagonals of the envelope.
    (d) Fold the envelope along the diagonals, one at a time, creasing the folds sharply. Open the envelope flat.
    (e) With a scissors, cut as shown in the diagram. Remove this piece.

    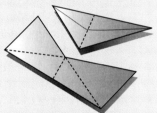

    (f) Open the envelope and fold it along the first fold made as in the diagram.

    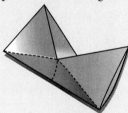

    (g) Tuck one side of the envelope into the other side as shown in the diagram.

    You now have a tetrahedron. Point out the vertices to a partner. How many faces are there? Do all the faces appear congruent?

Cylinders and Cones

OBJECTIVES

1. Define a right circular cylinder.
2. Define and calculate the lateral area and surface area of a cylinder.
3. Calculate the volume of cylinder.
4. Define a right circular cone.
5. Calculate the lateral area and the surface area of cone.
6. Calculate the volume of a cone.

A cylinder has the same general shape as a prism but has circles for its bases instead of polygons. In this section, a scientific calculator will continue to be used. Most of the calculations will involve π because of the circular bases. The $\boxed{\pi}$ key on the calculator will be used rather than using the approximation of 3.14.

OBJECTIVE 1 Define a right circular cylinder.

> **DEFINITION: Cylinder**
> The solid figure formed by joining two congruent circles in parallel planes is called a **cylinder**.

The cylinder shown in Figure 8.22 is a **right circular cylinder** because the line joining the centers of the circles is perpendicular to the circles. This line is called the **height (altitude)** of the cylinder and the circles are called **bases**. There are other cylinders besides circular ones but we will restrict our work to right circular cylinders and from now on will use the term "cylinder" to mean "right circular cylinder."

OBJECTIVE 2 Define and calculate the lateral area and surface area of a cylinder. To find the lateral area of a cylinder, consider a can without top or bottom as shown in Figure 8.23a. If it is cut along the side and pressed flat, a parallelogram is formed with a height of h (height of the can) and the length of $2\pi r$ (circumference of the base circle). Figure 8.23b shows a *perpendicular* cut forming a *rectangle*. The length of the rectangle is the circumference of the base ($2\pi r$) and width of the rectangle is the height of the can, h.

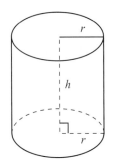

Figure 8.22 Right Circular Cylinder

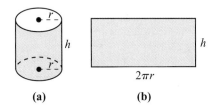

(a)   (b)

Figure 8.23 Surface Area of a Cylinder

Thus, the lateral area of the can is the area of the rectangle, which is $2\pi rh$.

Felix Klein (1849–1925)

Felix Klein, a German mathematician, was instrumental in formulating a theory that unified the concepts of Euclidean and non-Euclidean geometries. Three famous problems in geometry served as the basis for a series of lectures Klein gave to acquaint students with new developments in geometry. The three problems are the duplication of a cube (constructing a cube with volume twice that of a given cube); the quadrature of a circle (constructing a square with the same area as a given circle); and the trisection of an angle. All three problems proved impossible. Klein is perhaps most famous for his work in topology, sometimes called rubber-sheet geometry.

> **THEOREM 8.5   Lateral Area of a Right Circular Cylinder**
>
> The **lateral area** of a right circular cylinder is determined with the formula
>
> $$LA = 2\pi rh,$$
>
> where r is the radius of a base and h is the height, the length of the altitude.

The surface area of a right circular cylinder is the sum of the lateral area and the area of the two bases.

$$SA = LA + 2B \text{ or } 2\pi rh + 2\pi r^2$$

Postulate 7.5 stated the area of the circle is πr^2. Since both bases are circles, their area is $2\pi r^2$.

Notice the formulas for right lateral area and surface area of a right circular cylinder are analogous to the right prism.

	LATERAL AREA (*LA*)	SURFACE AREA (*SA*)
Right Circular Cylinder	$(2\pi r)h$ Ch	$2\pi rh + 2(\pi r^2)$ $LA + 2B$
Right Prism	ph	$LA + 2B$

where p is the circumference (C) of the circular base and B is the area of the base

EXAMPLE 1   Find to the nearest square foot, the surface area of a cylinder with radius 7 ft and height 9 ft.

$$\begin{aligned}
SA &= 2\pi rh + 2\pi r^2 \\
&= 2\pi(7)(9) + 2\pi(7)^2 \\
&= 126\pi + 98\pi \\
&= 224\pi \\
&\approx 704 \text{ ft}^2 \qquad \text{to the nearest square foot}
\end{aligned}$$

PRACTICE EXERCISE 1

Find the surface area of a cylinder with height 22.3 cm and radius 12.8 cm. Give your answer to the nearest tenth of a square centimeter.

ANSWER ON PAGE 410

EXAMPLE 2

The outside and top of a cylindrical water tank are to be painted. The tank has radius 3 m and height 8 m. Approximately how many liters of paint will be required if a liter covers 8 m²?

To find the lateral area, find $2\pi rh$ or ph

$$2\pi rh = 2(\pi)(3)(8) = 48\pi$$

To find the area of the top, find πr^2.

$$\pi r^2 = (\pi)(3)^2 = 9\pi$$

Thus, the approximate area to be painted is

$$48\pi + 9\pi = 57\pi$$

To find the number of liters of paint, divide by 8 m² per liter.

$$\frac{57\pi}{8} \approx 22.4 \text{ liters} \qquad \text{to the nearest tenth}$$

The job will require 23 liters of paint.

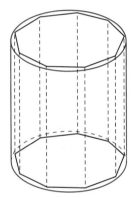

Figure 8.24

OBJECTIVE 3 **Calculate the volume of a cylinder.** The volume of a cylinder is related to the volume of a prism. Theorem 8.2 says the volume of a right prism is $V = Bh$. Suppose a regular prism is inscribed in a cylinder as shown in Figure 8.24. Imagine the number of sides in the inscribed prism increases. The volume of the prism will get very close to the volume of the cylinder.

THEOREM 8.6 **Volume of a Right Circular Cylinder**

The **volume** of a right circular cylinder is determined with the formula

$$V = \pi r^2 h,$$

where r is the radius of a base and h is the height.

EXAMPLE 3

A can of blueberry pie filling has diameter 3.0 inches and height 4.5 inches. How many cans of filling are needed to fill a 9-inch diameter pan 1 inch deep?

First find the volume of pie filling in one can.

$$V = \pi r^2 h$$
$$= (\pi)(1.5)^2(4.5) = 10.125\pi \qquad r = 1.5 \text{ inches because the diameter is 3 inches}$$

Now find the volume of the 9-inch diameter (4.5-inch radius) pan.

$$V = \pi r^2 h$$
$$= (\pi)(4.5)^2(1) \qquad \text{Filling is 1 inch deep}$$
$$= 20.25\pi \qquad \text{Volume of pie pan}$$

Next, divide the volume of the pan by the volume of the can of pie filling.

$$\frac{20.25\pi}{10.125\pi} = 2 \text{ cans}$$

It will take 2 cans of filling to fill the pan.

All warm-blooded animals lose heat during sleep in the same amount per square unit of surface area of skin. As a result, the necessary amount of food that must be consumed is proportional to the animal's surface area and not its weight (or volume). This is why a small animal must consume large amounts of food, sometimes more than half its weight, on a daily basis.

PRACTICE EXERCISE 2

Water is stored in a cylindrical tank with base diameter 12 m and height 10 m. How many water trucks having cylindrical tanks with diameter 1.6 m and length 5 m will be needed to remove all the water from the storage tank?

ANSWER ON PAGE 410

OBJECTIVE 4 **Define a right circular cone.** A solid figure that is closely related to the pyramid is the *cone*.

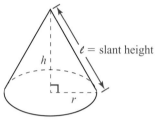

Figure 8.25 Right Cone

> **DEFINITION: Cone**
>
> The solid figure formed by connecting a circle with a point (vertex) not in the plane of the circle is called a **cone**.

Our work with cones will emphasize **right circular cones** like the one shown in Figure 8.25. Notice the **height (altitude)** starts at the **vertex** of the cone, passes through the center of the **base** (circle) and is perpendicular to the base. The **slant height** is the distance ℓ indicated in the figure. A right triangle is formed by r, h, and ℓ.

OBJECTIVE 5 **Calculate the lateral area and the surface area of a cone.** To find the lateral area and volume of a right circular cone, we make an argument similar to that for the volume of a cylinder. A cone is related to a pyramid. Imagine the number of sides of the pyramid increasing. The lateral area of the pyramid will get very close to the lateral area of a cone. The lateral area of a pyramid is $LA = \frac{1}{2}p\ell$, where p is the perimeter of the base and ℓ is the slant height. The perimeter of the cone's base is its circumference, $2\pi r$.

$$LA = \frac{1}{2}p\ell$$
$$= \frac{1}{2}(2\pi r)\ell$$
$$= \pi r\ell$$

> **THEOREM 8.7   Lateral Area of a Right Circular Cone**
>
> The lateral area of a right cone is determined with the formula
>
> $$LA = \pi r\ell,$$
>
> where r is the radius of the base and ℓ is the slant height.

The formula for the surface area of a pyramid is $SA = LA + B$, thus the surface area of a cone will be analogous.

$$SA = LA + B$$

or

$$SA = \pi rl + \pi r^2$$

EXAMPLE 4

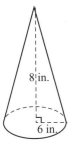

8 in.

6 in.

Figure 8.26

For the right circular cone in Figure 8.26 with a radius of 6 inches and a height of 8 inches, find the approximate surface area. Round the answer to the nearest tenth of a square inch.

The formula to be used is $SA = LA + B$. To find the $LA \left(\frac{1}{2}p\ell \right)$, the slant height is needed.

$$6^2 + 8^2 = \ell^2$$
$$36 + 64 = \ell^2$$
$$100 = \ell^2$$
$$10 = \ell$$
$$p = 2\pi(6)$$
$$= 12\pi$$
$$LA = \frac{1}{2}p\ell$$
$$LA = \frac{1}{2}(12\pi)(10)$$
$$= 60\pi.$$

To find the surface area of the cone, the area of the base is also needed.

$$B = \pi r^2$$
$$= \pi(6)^2$$
$$= 36\pi.$$
$$SA = LA + B$$
$$= 60\pi + 36\pi$$
$$= 96\pi \approx 301.6 \text{ in.}^2 \qquad \text{Rounded to nearest tenth}$$

OBJECTIVE 6   **Calculate the volume of a cone.**   The formula for the volume of a right pyramid is $V = \frac{1}{3}Bh$. The formula for the volume of a right cone will be analogous but the base is a circle.

$$V = \frac{1}{3}Bh \quad \text{or} \quad V = \frac{1}{3}\pi r^2 h$$

> **THEOREM 8.8   Volume of a Right Circular Cone**
> The volume of a right cone is determined with the formula
> $$V = \frac{1}{3}Bh \text{ or } \frac{1}{3}\pi r^2 h,$$
> where r is the radius of the base, B is the area of the base, and h is the height of the cone.

EXAMPLE 5   A pyramid with square base and height 12 cm is inscribed in a cone. If the square base has side 10 cm, find the volume and the surface area of the cone. Refer to Figure 8.27. Round the final answers to the nearest tenth of a centimeter.

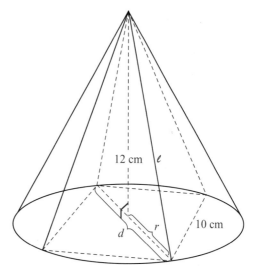

Figure 8.27

The height of the cone is the same as the height of the pyramid, 12 cm. The radius of the base of the cone will be one-half the length of a diagonal d of the square. By the Pythagorean Theorem,

$$d^2 = 10^2 + 10^2$$
$$d = \sqrt{10^2 + 10^2}$$
$$= \sqrt{2 \cdot 10^2}$$
$$= 10\sqrt{2}$$
$$\frac{d}{2} = r = 5\sqrt{2} \text{ cm}$$

With h and r known, we can find the volume of the cone.

$$V = \frac{1}{3}\pi r^2 h$$

$$= \frac{1}{3}(\pi)(5\sqrt{2})^2(12)$$

$$= \frac{1}{3}(\pi)(50)(12) = 200\pi$$

$$\approx 628.3 \text{ cm}^3 \qquad \text{Rounded to nearest tenth}$$

The formula for the surface area of the cone is $LA + B$. We can determine the area of the base of the cone.

$$B = \pi r^2$$

$$= (\pi)(5\sqrt{2})^2$$

$$= \pi\,(50)$$

Before we can find the lateral area, we must determine the slant height. Use the Pythagorean theorem.

$$\ell^2 = r^2 + 12^2$$

$$\ell = \sqrt{r^2 + 12^2}$$

$$= \sqrt{(5\sqrt{2})^2 + (12)^2}$$

$$= \sqrt{50 + 144}$$

$$= \sqrt{194}$$

Now find the lateral area.

$$LA = \pi r \ell$$

$$= \pi(5\sqrt{2})(\sqrt{194})$$

The surface area is

$$SA = LA + B$$

$$= \pi(5\sqrt{2})(\sqrt{194}) + \pi(50)$$

$$\approx 466.5 \text{ cm}^2 \qquad \text{Rounded to nearest tenth}$$

PRACTICE EXERCISE 3

A conical structure has height 16.2 ft and a base radius of 7.4 ft. How many square feet of plastic will be required to cover its lateral area? Round to the nearest tenth of a square foot.

ANSWER BELOW

Answers to Practice Exercises

1. 2822.9 cm^2 **2.** 113 trucks   **3.** 414.0 ft^2

SUMMARY

	LATERAL AREA (LA)	SURFACE AREA (SA)	VOLUME (V)
Right Cylinder	ph or $2\pi rh$	$LA + 2B$ or $2\pi rh + 2\pi r^2$	Bh or $\pi r^2 h$
Right Cone	$\frac{1}{2}p\ell$ or $\pi r\ell$	$LA + B$ or $\pi r\ell + \pi r^2$	$\frac{1}{3}Bh$ or $\frac{1}{3}\pi r^2 h$

where B means area of the base
$\quad\quad h$ means height of the solid
$\quad\quad \ell$ means slant height
$\quad\quad p$ means perimeter
$\quad\quad r$ means radius

8.4 Exercises

FOR EXTRA HELP: Student's Solutions Manual   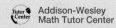 Addison-Wesley Math Tutor Center

In Exercises 1 and 2, determine whether the statement is *true* or *false*.

1. If a cylinder and a cone have the same radius and height, the volume of the cylinder is three times the volume of the cone.
2. If the height of a cylinder is doubled and the radius kept constant, the total surface area of the cylinder is doubled.

In Exercises 3–8, find the surface area of each cylinder. Round answer to tenth of a square unit.

3. $r = 6$ cm, $h = 5$ cm
4. $r = 9$ cm, $h = 15$ cm
5. $r = 12.2$ in., $h = 30.0$ in.
6. $r = 3.4$ m, $h = 10.5$ m
7. $d = 11.4$ m, $h = 4.4$ m
8. $d = 20.4$ in., $h = 8.5$ in.

In Exercises 9–14, find the volume of each cylinder. Round to nearest tenth.

9. $r = 7$ cm, $h = 8$ cm
10. $r = 14$ cm, $h = 19$ cm
11. $r = 14.5$ in., $h = 35.5$ in.
12. $r = 4.9$ m, $h = 17.8$ m
13. $d = 12.6$ m, $h = 16.2$ m
14. $d = 28.6$ in., $h = 10.7$ in.

In Exercises 15–20, find the surface area of each cone. Round to nearest tenth.

15. $r = 3$ yd, $\ell = 5$ yd
16. $r = 7$ cm, $\ell = 17$ cm
17. $r = 28.4$ in., $\ell = 42.6$ in.
18. $r = 17.8$ m, $\ell = 58.5$ m
19. $r = 16$ ft, $h = 23$ ft
20. $r = 8.8$ cm, $h = 17.2$ cm

In Exercises 21–26, find the volume of each cone. Round to the nearest tenth.

21. $r = 3$ in., $h = 4$ in.   **22.** $r = 19$ m, $h = 29$ m   **23.** $r = 6.9$ cm, $h = 14.2$ cm

24. $r = 18.5$ yd, $h = 27.2$ yd   **25.** $r = 20$ m, $\ell = 40$ m   **26.** $r = 8.9$ in, $\ell = 21.6$ in.

For Exercises 27–40, round to the nearest tenth if necessary.

27. If a cylinder has volume 26.8 cm³, what is the volume of a cylinder with the same radius but twice the height? with the same height but twice the radius?

28. If a cylinder has volume 26.8 cm³, what is the volume of a cylinder with the same height but one-half the radius? with the same radius but one-half the height?

29. Find the surface area and volume of a cylinder that has a prism with square base inscribed in it. The side of the base of the prism is 24 inches and the height is 16 inches.

30. Find the surface area and volume of a cone that has a pyramid with square base inscribed in it. The side of the base of the pyramid is 6.8 m and the height is 9.8 m.

Solve each applied problem in Exercises 31–36.

31. A can of cherry pie filling has diameter 8 cm and height 12 cm. How many cans are needed to fill a 20-cm diameter pan 3 cm deep?

32. A lab stores mercury in a cylinder of radius 16 cm and height 20 cm. How many cylindrical tubes with radius 0.5 cm and height 100 cm can be filled from a full supply?

33. A cylindrical storage tank has radius 3.8 ft and height 9.8 ft. How many gallons of paint are needed to paint the tank (including top and bottom) if one gallon covers 150 ft²?

34. A cylindrical tank is made from sheet metal. If the tank is to have radius 2.2 ft and height 12.0 ft, what is the total cost if the price of the metal is $16.50 per square foot?

35. An inverted cone used as a funnel for grain has radius 4.6 ft and height 22.2 ft. How many gallons of paint are required to paint the outside lateral surface if one gallon of paint covers 180 ft²?

36. A solid aluminum cube measuring 15 inches on a side is melted and rolled into an aluminum wire that is 0.2 inches in diameter. What will be the length of this wire?

37. Which container holds more cereal?
   a. a cylindrical container (like an oatmeal container) that is 9.5 in high and has a diameter of 5 inches
   b. a rectangular prism–shaped box that is 11 inches high, 2.25 inches wide, and 6.5 inches long

38. A homeowner has purchased a lawn roller as shown to the side. The roller's steel drum measures 18 inches in diameter with a width of 20 inches. The homeowner is going to fill the roller's drum with water so it becomes heavy enough to smooth out some uneven ground in the yard. How much water will be needed to fill the roller?

39. Jamal is making bird feeders to sell like the one in the figure. It is 18 inches high and has a diameter of 2 inches. The sides are made of fine wire fencing while the top and bottom are plastic disks. How much wire fencing will Jamal need to make the feeder? Do not consider overlapping of the fencing.

40. The highway department is repainting a large number of traffic cones. The base is a weighted square (so it does not blow over in the wind) measuring 15 inches on a side. The hole in the base of the square for the cone to fit into is 7 inches in diameter. The slant height of the cone is 26.25 inches. To determine the amount of paint needed for the entire job, the supervisor must know the number of square inches to be painted on the outside of one cone. Find this amount.

Exercise 38

Exercise 39

Exercise 40

Student Activity

41. The following equipment will be needed for this activity: a ruler, scissors, an empty cardboard tube from a roll of gift wrap, paper towels, or toilet paper. Measure the tube's diameter and length and calculate the lateral area. With a pair of scissors, cut along the spiral "glue line" of the tube and flatten it. What is the shape? Find the area of this shape. How does it compare to the lateral area of the original tube?

(8.5) Spheres and Composite Figures

OBJECTIVES

1. Define a sphere.
2. Calculate the surface area and volume of a sphere.
3. Present applications with composite figures.

In this section, we will consider the *sphere* and work with applications that involve composition of the figures we have studied. A scientific calculator will continue to be used for all calculations.

OBJECTIVE 1 Define a sphere.

> **DEFINITION: Sphere**
>
> A **sphere** is the set of all points in space a given distance, called the **radius**, from a given point, called the **center**.

Figure 8.28 shows a sphere. If we consider all the points in the interior of the sphere, we have a solid figure that is also called a sphere. This is the figure to which we refer when we discuss the volume of a sphere. Spheres you are familiar with are the planets and many varieties of balls.

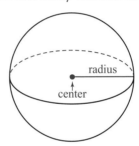

radius

center

Figure 8.28 Sphere

OBJECTIVE 2 Calculate the surface area and volume of a sphere.

The formulas for determining the surface area and volume of a sphere are postulated.

Saturn is spherical in shape, but its most distinguishing features are its rings. Scientists have discovered seven major rings, but many smaller ones probably also exist. It is believed that the rings are composed of ice and dust particles.

> **POSTULATE 8.2 Surface Area and Volume of a Sphere**
>
> For a sphere with radius r, the surface area is
>
> $$SA = 4\pi r^2$$
>
> and the volume is
>
> $$V = \frac{4}{3}\pi r^3.$$

If a plane is passed through the center of a sphere, two hemispheres are formed. A **hemisphere** is one-half a sphere.

In Example 1, we'll work with the **diameter** of a sphere, which is twice its radius.

EXAMPLE 1   Find the surface area and volume of a sphere with diameter 18 cm. Refer to Figure 8.29. Round the answers to the nearest tenth of a unit.

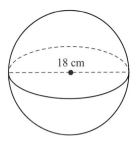

18 cm

Figure 8.29

Because the diameter is 18 cm, the radius is $r = \dfrac{18}{2}$ cm $= 9$ cm. We will need this number to determine both the surface area and volume.

$$SA = 4\pi r^2$$
$$= 4(\pi)(9)^2 = 324\pi$$
$$\approx 1017.9 \text{ cm}^2 \qquad \text{Rounded to the nearest tenth}$$

The surface area is approximately 1017.9 cm^2.

$$V = \frac{4}{3}\pi r^3$$
$$= \frac{4}{3}(\pi)(9)^3 = 972\pi$$
$$\approx 3053.6 \text{ cm}^3 \qquad \text{Rounded to the nearest tenth}$$

Thus, the volume is approximately 3053.6 cm^3.

EXAMPLE 2   The leather outer covering of a soccer ball consists of a repeating pattern of two hexagons with a common edge and a pentagon at the endpoints of the common edge. A size-5 ball has a circumference of 27 inches. The surface area and volume of the ball can be approximated using the properties of a sphere. Find the number of square inches of leather required to make the covering (disregard the seam allowances). How much air is needed to inflate the ball? See Figure 8.30. Round all answers to a tenth of a unit.

First, the radius of the ball is needed. We know $2\pi r = C$ and $C = 27$.

$$2\pi r = 27$$
$$r = \frac{27}{2\pi}$$

Figure 8.30

Now we can calculate surface area.

$$SA = 4\pi r^2$$
$$= 4\pi\left(\frac{27}{2\pi}\right)^2 \approx 232.0 \qquad \text{Rounded to the nearest tenth}$$

Thus, it will take approximately 232.0 in.^{2} of leather to make the covering.
To calculate the volume, use the formula:

$$V = \frac{4}{3}\pi r^3$$

$$= \frac{4}{3}\pi\left(\frac{27}{2\pi}\right)^3$$

$$\approx 332.4 \qquad \text{Rounded to nearest tenth}$$

Thus, approximately 332.4 in.^{3} of air is needed to inflate the soccer ball.

PRACTICE EXERCISE❶

A spherical tank with radius 3.2 m is to be insulated. If the material costs $2.50 per square meter, how much will it cost to cover the surface of the sphere?

ANSWER ON PAGE 417

OBJECTIVE 3 **Present applications with composite figures.** Many applications involve figures fashioned from two or more solid figures. These are called **composite solid figures**.

EXAMPLE 3

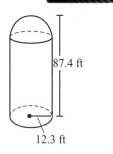

87.4 ft

12.3 ft

Figure 8.31

A grain storage silo is a cylindrical structure topped with a hemisphere. The radius of both the cylinder and the hemisphere is 12.3 ft, and the total height of the structure is 87.4 ft. How many cubic feet of grain will the silo hold? Round to the tenth of a cubic foot. Refer to Figure 8.31.

We must find the volume of the cylinder and add the volume of the hemisphere. The height of the cylinder must be found by subtracting the radius of the hemisphere (12.3 ft) from the total height of the structure.

$$h = 87.4 - 12.3 = 75.1 \text{ ft}$$

Now, we know that for the cylinder $r = 12.3$ ft and $h = 75.1$ ft.

$$V_{cyl} = \pi r^2 h$$

$$= (\pi)(12.3)^2(75.1)$$

$$= 11{,}361.879\pi$$

Also, the volume of the hemisphere is one-half the volume of a sphere with radius 12.3 ft.

$$V_{hsph} = \left(\frac{1}{2}\right)\left(\frac{4}{3}\right)\pi r^3$$

$$= \left(\frac{1}{2}\right)\left(\frac{4}{3}\right)(\pi)(12.3)^3$$

$$= 1240.578\pi$$

The total number of cubic feet of grain the silo will hold is

$$\text{Total volume } (V) = V_{cyl} + V_{hsph}$$

$$V = 11{,}361.879\pi + 1240.578\pi = 12{,}602.457\pi$$

$$\approx 39{,}591.8 \text{ ft}^3 \quad \text{Rounded to nearest tenth}$$

EXAMPLE 4

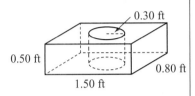

0.30 ft

0.50 ft

0.80 ft

1.50 ft

Figure 8.32

Find the volume and surface area of a block of wood with a hole bored in it as shown in Figure 8.32. Round to the nearest hundredth of a unit.

To find the volume, we must determine the volume of the prism and subtract the volume of the cylinder.

$$V_{prism} = Bh$$

$$= (1.50)(0.80)(0.50)$$

$$= 0.60 \text{ ft}^3$$

For the cylinder $r = 0.30$ ft and $h = 0.50$ ft, the height of the prism.

$$V_{cyl} = \pi r^2 h$$

$$= (\pi)(0.30)^2(0.50) = 0.045\pi$$

The approximate volume of the block with the hole is $V = V_{prism} - V_{cyl}$.

$$V = 0.60 - 0.045\pi$$

$$V \approx 0.46 \text{ ft}^3 \quad \text{Rounded to nearest hundredth}$$

Now determine surface area of the block by first finding the surface area of the prism, subtracting the area of both bases of the cylinder, and adding the lateral area of the cylinder.

$$SA_{prism} = 2(1.50)(0.80) + [2(1.50) + 2(0.80)](0.50)$$

$$= 4.70 \text{ ft}^2$$

$$2B_{cyl} = 2\pi(0.30)^2 = 0.18\pi$$

$$LA_{cyl} = 2(\pi)(0.30)(0.50) = 0.30\pi$$

$$SA_{block} = SA_{prism} - 2B_{cyl} + LA_{cyl}$$

$$SA = 4.70 - 0.18\pi + 0.30\pi$$

$$\approx 5.08 \text{ ft}^2 \quad \text{Rounded to nearest hundredth}$$

Answer to Practice Exercise

1. $321.70

For all exercises approximate answers to the tenth of a unit where appropriate.
In Exercises 1–6, find the surface area of each sphere.

1. $r = 12$ m
2. $r = 30$ in.
3. $r = 0.75$ ft
4. $r = 0.40$ cm
5. $r = 5.6$ yd
6. $r = 9.8$ yd

In Exercises 7–12, find the volume of each sphere.

7. $r = 7$ ft
8. $r = 11$ in.
9. $r = 0.90$ m
10. $r = 0.64$ cm
11. $r = 13.8$ yd
12. $r = 21.6$ cm

13. A cube and a sphere each have a surface area of 150 in.² Find the volume of each and determine which has the larger volume.

14. A cube and a sphere each have a volume of 216 m³. Find the surface area of each and determine which has the larger surface area.

15. If the surface area and volume of a sphere have the same numerical value, find the radius of the sphere.

Solve each applied exercise.

16. Jahanara is shopping at a produce market and finds oranges from Texas and Florida. It appears the Texas oranges have about twice the diameter as the Florida ones but cost 7 times as much. Which kind gives her more fruit for her money?

17. An apple pie recipe requires 5 apples that are 4 inches in diameter. The only apples at the market are 3 inches in diameter. How many 3-inch apples will be needed?

18. The mooring buoy in the figure at the right has an outer shell made of high density polyethylene. It is filled with foam so it will float. The buoy measures 12 inches in diameter. What is the surface area of the buoy?

19. A hanging pot is designed to hang from a tree and have flowers planted in it. The pot is a hemisphere and is made from plastic. The diameter of the pot is 14 inches. What is the surface area of the outside of the pot?

20. A plastic ball has a circumference of 70 centimeters. How much air is in the ball?

Exercise 18

Exercise 19

Exercise 20

21. A spherical tank with radius 32 ft is to be filled with gas. How many cubic feet of gas will it hold?

22. If one liter is 1000 cm³, how many liters of a liquid can be stored in a sphere with radius 65 cm?

23. A glue bottle has the shape of a cylinder with a cone on the top. The bottom of the bottle has a radius of 3 centimeters. See the figure to the right for other measurements. How much glue is in the bottle if it is filled to the top?

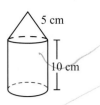

5 cm

10 cm

Exercise 23

24. A chemical-storage tank is a cylinder with a hemisphere cap on each end. If the height of the cylindrical portion is 16.2 ft and the radius of the cylinder and hemispheres is 2.8 ft, how many cubic feet of a chemical will the tank hold?

25. The chemical tank in Exercise 24 is to be insulated with material costing $1.25 per square foot. What will be the total cost of the insulating material?

26. A silo is a cylinder with a hemisphere of the same radius on top. The total height of the silo is 23.5 m and the radius is 3.8 m. Find the number of cubic meters of grain the silo will hold.

27. An ice cream cone is filled and topped with a hemisphere of ice cream. The radius of the cone and the hemisphere are both 1.2 inches and the overall height of the cone and hemisphere is 6.4 inches. Find the volume of ice cream served.

28. A machine part is a solid right prism with base 6.4 cm by 5.8 cm and height 2.3 cm. There is a cylindrical hole with radius 1.8 cm drilled vertically through the center of the prism. If the metal weighs 1.5 g per cubic centimeter, what is the weight of the machine part?

29. The finishing operation on the machine part in Exercise 28 costs $0.26 per square centimeter of surface. What is the total cost of the finishing operation?

30. Many new drugs that benefit our pets are currently being researched. To determine safe dosages, animal testing is conducted before the drug is released for use by veterinarians. Many of these same drugs are also prescribed for human use. The animal dosage is calculated based on the surface area of the animal. The human dosage would be proportional to the surface area of the person. How would you gather data to estimate the surface area of people with a given height and given weight?

Chapter 8   Review

Key Terms and Symbols

8.1   parallel line and plane
perpendicular line and
    plane
parallel planes
perpendicular planes
oblique planes
polyhedron
faces
edges
vertex (vertices)
regular polyhedron

8.2   prism
base of prism
lateral face of prism
right prism
oblique prism

lateral area (LA)
height of a prism
surface area (SA)
base area (B)
regular prism

8.3   pyramid
base of a pyramid
vertex
altitude of a pyramid
oblique pyramid
regular pyramid
slant height of a pyramid

8.4   cylinder
right circular cylinder
height of a cylinder

altitude of a cylinder
bases
lateral area of a cylinder
cone
right circular cone
height (altitude)
vertex
slant height of a cone

8.5   sphere
radius
center
diameter
hemisphere
composite solid figure

SUMMARY

	LATERAL AREA (LA)	SURFACE AREA (SA)	VOLUME (V)
Right Prism	ph	$LA + 2B$	Bh
Right Cylinder	ph or $2\pi rh$	$LA + 2B$ or $2\pi rh + 2\pi r^2$	Bh or $\pi r^2 h$
Right Pyramid	$\frac{1}{2}p\ell$	$LA + B$	$\frac{1}{3}Bh$
Right Cone	$\frac{1}{2}p\ell$ or $\pi r\ell$	$LA + B$ or $\pi r\ell + \pi r^2$	$\frac{1}{3}Bh$ or $\frac{1}{3}\pi r^2 h$
Sphere	n/a	$4\pi r^2$	$\frac{4}{3}\pi r^3$

where B means area of the base
    h means height of the solid
    ℓ means slant height
    p means perimeter
    r means radius

Review Exercises

Section 8.1

In the figure below, line s is the intersection of planes $\mathcal{P}$ *and* $\mathcal{Q}$*, and line v is the intersection of* $\mathcal{Q}$ *and* $\mathcal{R}$*. Line t is in* $\mathcal{P}$*, u is in* $\mathcal{Q}$*, w is in* $\mathcal{R}$*, t* $\perp$ *u, u* $\perp$ *s, u* $\perp$ *v, and u* $\perp$ *w. Use this information to answer true or false in Exercises 1–6.*

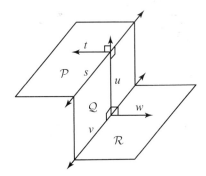

1. $s \parallel \mathcal{R}$ **2.** $w \parallel \mathcal{P}$ **3.** $\mathcal{P} \parallel \mathcal{R}$

4. $\mathcal{Q} \perp \mathcal{P}$ **5.** $\mathcal{Q}$ and $\mathcal{R}$ are oblique.   **6.** u and $\mathcal{P}$ are oblique.

Determine the number of faces f, vertices v, and edges e in each polyhedron in Exercises 7 and 8. Check to see that $f + v - e = 2$.

7.

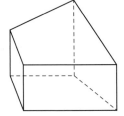

8.

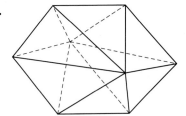

9. A polyhedron has 14 vertices and 36 edges. How many faces does it have?

10. The edges of a regular octahedron are to be constructed with tubing that costs $1.65 per foot. If each edge is 2.5 ft in length, how much will all the tubing required for the project cost?

For the remaining review problems, round answers to the tenth of a unit where appropriate.

Section 8.2

Find the surface area of each right prism described in Exercises 11 and12.

11. Equilateral triangular base with side 12.6 cm; height 21.3 cm

12. Rectangular base with sides 9.6 ft and 5.8 ft; height 2.6 ft

In Exercises 13 and 14, determine the volume of each right prism described.

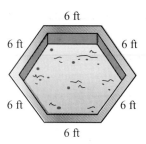

6 ft

6 ft   6 ft

6 ft   6 ft

6 ft

13. Cube with edge 18.6 in.

14. Regular hexagon base with side 5.2 m; height 7.4 m

15. Samuel is building a flower bed in the shape of a regular hexagon. Each side of the bed is 6 feet long and it is 2 inches deep. How much dirt will be needed to fill the flower bed to the top of the border?

Section 8.3

In Exercises 16 and 17, find the lateral area of each regular pyramid described.

16. Equilateral triangle base with side 251 cm; slant height 314 cm

17. Regular hexagon base with side 9.8 ft; height of the pyramid is 7.2 ft

In Exercises 18–20, find the volume of each regular pyramid described.

18. Equilateral triangular base with side 16.8 in.; height 25.5 in.

19. Square base with side 42.6 m; slant height 61.4 m

20. A decorative bottle has the shape of a regular pyramid with an equilateral triangular base. If the base has side 2.2 inches and the altitude of the pyramid is 4.1 inches, how many cubic inches of liquid will the bottle hold?

Section 8.4

In Exercises 21 and 22, find the surface area of each cylinder.

21. $r = 2.3$ ft, $h = 6.5$ ft

22. $r = 21.2$ cm, $h = 19.5$ cm

In Exercises 23 and 24, find the volume of each cylinder.

23. $r = 165$ in, $h = 214$ in

24. $r = 42.6$ m, $h = 15.7$ m

In Exercises 25 and 26, find the surface area of each cone.

25. $r = 6.2$ cm, $\ell = 9.5$ cm

26. $r = 18.9$ ft, $h = 26.8$ ft

In Exercises 27 and 28, find the volume of each cone.

27. $r = 12.8$ m, $h = 19.2$ m

28. $r = 25.8$ in, $\ell = 39.2$ in

29. A mechanic is changing oil in a car. Using a large, cone-shaped funnel the mechanic pours the used motor oil back into a container for recycling. The large funnel has a diameter of 1 foot and is 10 inches tall excluding the spout. Calculate the amount of oil the funnel can hold (excluding the spout).

30. A hollow metal tube is 2 m long. The outer diameter is 30 cm and the inner diameter is 20 cm. How many cubic centimeters of metal were used to make the tube? [Hint: Before calculating, change to a common unit of measure. 100 cm = 1 m]

Exercise 30

Section 8.5

In Exercises 31 and 32, find the surface area of each sphere.

31. $r = 3.9$ ft

32. $r = 26.8$ m

In Exercises 33 and 34, find the volume of each sphere.

33. $r = 0.92$ cm

34. $r = 12.9$ in.

35. A hollow plastic ball has a circumference of 30 inches. How much plastic will it take to manufacture the ball?

36. A sphere has a surface area of 100 m^2. Find the volume to the nearest unit.

37. A spherical tank with radius 15.8 cm is to be filled with a liquid weighing 0.9 grams per cubic centimeter. How many grams of the liquid will the tank hold?

38. A glue bottle has the shape of a cylinder with a cone on the top. The bottom of the bottle has a diameter of 4 centimeters. The height of the cylindrical part is 7 cm and the height of the cone-shaped top is 3 cm. How much glue does the bottle hold?

39. A water tank is a cylinder with a hemisphere cap on each end. The overall height of the tank is 9.9 m. If the radius of the cylinder and hemispheres is 1.8 m, how many kg of water will the tank hold if water weighs 1000 kg per cubic meter?

40. A machine part is in the shape of a cone with a hemisphere covering its base. The radius of the cone and hemisphere is 2.3 cm and the slant height of the cone is 12.2 cm. How much will it cost to finish the part if the finishing operation costs $2.65 per square centimeter?

Chapter (8) Practice Test

In the figure to the right, line u is the intersection of P and R, line v is the inter-section of Q and R, m and n are in Q, k is in R, ℓ is in P, k ⊥ u, k ⊥ ℓ, k ⊥ v, k ⊥ n, and n ⊥ v. Use this information to answer **true** *or* **false** *in Exercises 1–4.*

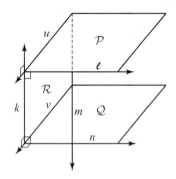

1. $P \parallel Q$
2. $R \perp Q$
3. $m \parallel P$
4. m and R are oblique.

5. A spherical tank has radius 8.6 m. If a rustproofing material costs $1.50 per square meter, what is the cost to rustproof the tank? Round the answer to the nearest cent.

Exercises 1–4

For Exercises 6–18, where appropriate, round answers to the nearest tenth of a unit.

6. Find the surface area of a right prism with height 8.2 centimeters and an equilateral triangular base with one side measuring 12.6 centimeters.

7. Find the volume of a right prism with a rectangular base 4.1 meters by 9.2 meters and the height of the prism is 5.5 meters.

8. Find the volume of a cone with radius 28.6 inches and slant height 42.5 inches.

9. Find the lateral area of a regular pyramid with height 18.3 meters and a square base with one side measuring 32.2 meters.

10. Find the volume of a regular pyramid with height 18.3 meters and a square base with one side measuring 32.2 meters.

11. Find the surface area of a cylinder with radius 1.2 feet and height 6.7 feet.

12. Find the volume of a cylinder with radius 1.2 feet and height 6.7 feet.

13. Find the surface area of a sphere with diameter 14 centimeters.

14. The figure to the right is a regular hexagonal prism and has a height of 12 centimeters. One edge of the base measures 8 centimeters. Find the volume of the prism.

Exercise 14

15. A plastic ice cream cone used as a store display is 24 inches high and has a diameter of 14 inches. How many square inches of plastic is needed to manufacture the ice cream cone?

16. A silo is a cylinder with a hemisphere on top. The radius of the cylinder and hemisphere is 9.8 meters and the overall height of the silo is 57.2 meters. How many cubic meters of grain will the silo hold when it's filled to capacity?

17. Find the surface area of a cone with radius 3 inches and height 4 inches.

18. A size-5 soccer ball has a circumference of 27 inches. If the ball is fully inflated, how much air is in the ball?

9 Introduction to Analytic Geometry

*I*n this chapter, we combine algebra with our study of geometry. Algebra and geometry remained separate areas of study until René Descartes, a seventeenth-century French mathematician, joined them into what is now called analytic geometry.

The following example presents an applied problem that involves both algebra and geometry. Its solution appears in Section 9.2, Example 7.

9.1 THE CARTESIAN COORDINATE SYSTEM, DISTANCE, AND MIDPOINT FORMULAS

9.2 SLOPE, EQUATION OF A LINE

9.3 PROOFS INVOLVING POLYGONS

AN APPLICATION

During a spring day in eastern Hong Kong, the pollution index increased from approximately 20 parts per million (ppm) at 8 A.M. to around 29 ppm at 2 P.M. Assuming the increase in pollution (*y*) per hour since 8 A.M. (*x*) can be approximated by a linear equation, find this equation where *x* is in terms of hours since 8 A.M. Interpret the meaning of the slope in the context of this problem. Interpret the meaning of the *y*-intercept in terms of this problem. Can you use this model to predict the pollution index at noon? If so, what is the index? If not, why not? Can you predict the pollution index for 6 P.M.? If so, what is the index? If not, why not?

$$\boxed{9.1}$$ **The Cartesian Coordinate System, Distance, and Midpoint Formulas**

OBJECTIVES

1. Learn to plot points.
2. Graph lines using a table.
3. Graph lines using *x*- and *y*-intercepts.
4. Calculate the length of a line segment.
5. Calculate the midpoint of a line segment.

We introduced and worked with number lines in Chapter 1. When two number lines are placed together, as in Figure 9.1, so that the origins coincide and the lines are perpendicular, the result is a **Cartesian** or **rectangular coordinate system** or a **coordinate plane.**

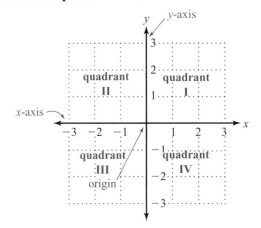

Figure 9.1 Cartesian Coordinate System

The horizontal number line is called the ***x*-axis**, and the vertical number line is called the ***y*-axis**. The point of intersection of the axes is called the **origin** $(0, 0)$.

The axes in a coordinate system separate the plane into four sections called **quadrants**. The first, second, third, and fourth quadrants are identified by the Roman numerals I, II, III, and IV, respectively, in the coordinate plane, as shown in Figure 9.1.

Recall that there is one and only one point on a number line associated with each real number. A similar situation exists for points in a plane and **ordered pairs** of numbers.

The positions taken by the members of a marching band during a half-time performance can be shown using a coordinate system superimposed on a football field. The yard lines can serve to give one coordinate.

POSTULATE 9.1

Associated with each point in the plane there is one and only one ordered pair of numbers.

OBJECTIVE 1   **Learn to plot points.**   The ordered pair $(2, 3)$ can be identified with a point in the coordinate plane in the following way. The first number 2, called the ***x*-coordinate** of the point, is associated with a value on the horizontal axis, or *x*-axis. The second number 3, called the ***y*-coordinate** of the point, is associated with a value on the vertical axis, or *y*-axis. The pair $(2, 3)$ is associated with the point at which the vertical line through 2 on the *x*-axis intersects the horizontal line through 3 on the *y*-axis, as shown in

Figure 9.2. The ordered pairs $(3, 2)$, $(-1, 3)$, $(-3, -2)$, and $(2, -2)$ are plotted in Figure 9.3.

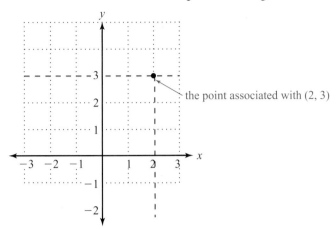

the point associated with (2, 3)

Figure 9.2

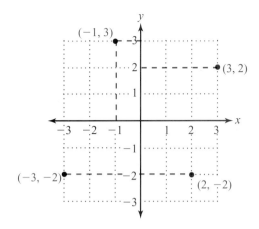

Figure 9.3

We often use (x, y) to refer to a general ordered pair of numbers. Thus, the point in the plane associated with the pair (x, y) has x as its x-coordinate, the first number in the ordered pair, and y as its y-coordinate, the second number in the ordered pair. When we identify the point in a plane associated with the given pair, we say that we **plot** the point. When referring to "the point P corresponding to the ordered pair (x, y)," we sometimes simply say "the point (x, y)," or write $P(x, y)$.

In analytic geometry, we plot points that satisfy an equation in two variables in order to construct the geometric figure that corresponds to the equation. We start our discussion with equations of the form

$$ax + by + c = 0,$$

which are associated with a line and are called **linear** or **first-degree equations**. This form is called the **general form**, where a, b, c are integers and a and b do not both equal 0. An ordered pair (x, y) that satisfies an equation is called a **solution**. If we could plot all the solutions to a linear equation, we would construct the **graph** of the equation in a coordinate system.

Suppose we graph the equation

$$3x - y - 2 = 0.$$

The ordered pair $(2, 4)$ is one solution because if x is replaced by 2 and y is replaced by 4, the resulting equation is true.

$$3x - y - 2 = 0$$
$$3(2) - (4) - 2 = 0 \qquad x = 2 \text{ and } y = 4$$
$$6 - 4 - 2 = 0$$
$$0 = 0 \qquad \text{True}$$

You can verify that $(0, -2)$ and $(-1, -5)$ are also solutions, whereas $(-2, 4)$ is not.

**René Descartes
(1596–1650)**

The contributions of René Descartes to philosophy and science were well received in his day. The Cartesian coordinate system was the basis of his discovery of analytic geometry. By combining algebra with geometry, he was able to broaden the scope of this ancient discipline.

OBJECTIVE 2   **Graph lines using a table.**   A **table of values** is a collection of solutions to a linear equation. One way to make such a table is to solve the equation for y, choose several values for x, substitute these values into the equation, and compute the corresponding values of y. For example, if we solve $3x - y - 2 = 0$ for y, we get

$$y = 3x - 2.$$

We can then construct the table of values.

SUBSTITUTION	RESULT IN $y = 3x - 2$
$x = 0$	$y = 3(0) - 2 = -2$
$x = 1$	$y = 3(1) - 2 = 1$
$x = -1$	$y = 3(-1) - 2 = -5$
$x = 2$	$y = 3(2) - 2 = 4$
$x = -2$	$y = 3(-2) - 2 = -8$
$x = 3$	$y = 3(3) - 2 = 7$
$x = -3$	$y = 3(-3) - 2 = -11$

TABLE OF VALUES	
x	y
0	-2
1	1
-1	-5
2	4
-2	-8
3	7
-3	-11

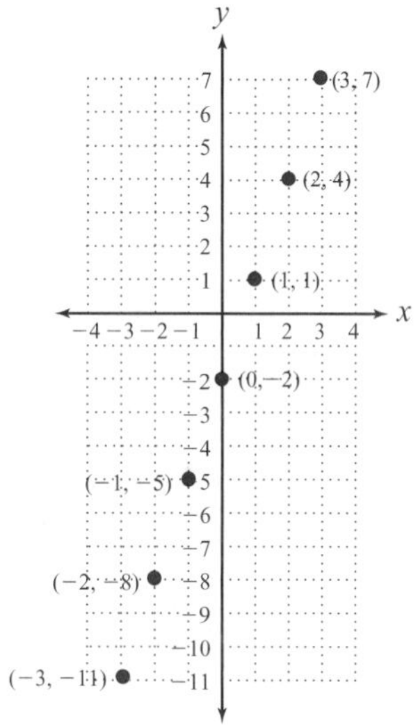

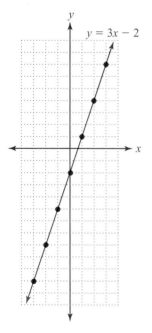

Figure 9.4

The table above lists seven of the many solutions to the equation $y = 3x - 2$.

$$(0, -2), \quad (1, 1), \quad (-1, -5), \quad (2, 4), \quad (-2, -8), \quad (3, 7), \quad (-3, -11)$$

Now we can plot the points that correspond to these ordered-pair solutions in a Cartesian coordinate system, as shown in Figure 9.4. It appears that all seven points do lie on a line. Thus, the graph of this equation is the line passing through these seven points in Figure 9.5.

OBJECTIVE 3   **Graph lines using x- and y-intercepts.**   Knowing that the graph of a linear equation is a line (by Postulate 1.1, which says two points determine a line), we need only two solutions to graph it. In most cases, the two pairs that are easiest to determine are the **intercepts**. The points at which a line crosses the x-axis and the y-axis are called the **x-intercept** and the **y-intercept**, respectively. Because the y-intercept is a point on the y-axis, it has x-coordinate 0. Similarly, the x-intercept is a point on the x-axis and has y-coordinate 0.

Thus, to find the x-intercept and y-intercept for any equation use the following.

FINDING INTERCEPTS

To find the x-intercept, set $y = 0$ and solve the equation. The ordered pair for the x-intercept is ($\_\_\_\_$, 0)

To find the y-intercept, set $x = 0$ and solve the equation. The ordered pair for the y-intercept is (0, $\_\_\_\_$).

Figure 9.5

$y = 3x - 2$

EXAMPLE 1 Graph $3x - 5y - 15 = 0$.

Find the x- and y-intercepts by completing the following table.

x	y
0	
	0

When $x = 0$, $3(0) - 5y - 15 = 0$, $-5y = 15$, so that $y = -3$. When $y = 0$, $3x - 5(0) - 15 = 0$, $3x = 15$, so that $x = 5$.

The completed table is

x	y	
0	-3	y-intercept
5	0	x-intercept

If we plot these two points and draw the line through them, we have the graph of $3x - 5y - 15 = 0$ as shown in Figure 9.6.

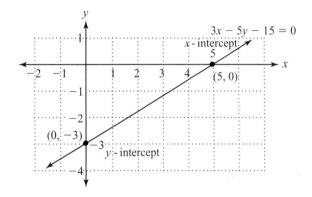

Figure 9.6

PRACTICE EXERCISE **1**

Graph $2x - 3y = 6$.

ANSWER ON PAGE 433

There are special cases in which using only the intercepts will not give the two points necessary to graph the equation. For example, $3y = 4x$ has only one intercept $(0, 0)$. In an equation like this, we must determine another point. The graph of $3y = 4x$ is shown in Figure 9.7 where an additional point $(3, 4)$ is plotted.

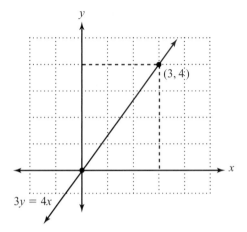

Figure 9.7

Other special cases include **horizontal lines** whose equations do not contain an x value. For example, $3y + 5 = 0$ is the same as $0x + 3y + 5 = 0$ (in the general form $ax + by + c = 0$).

Solving for y: $y = -\dfrac{5}{3}$. In general, a horizontal line is of the form $y = b$, where b is a constant.

Also, **vertical lines** have equations with no y value. For example, $x - 4 = 0$ is the same as $x + 0y - 4 = 0$.

Solving for x: $x = 4$. In general, a vertical line is of the form $x = a$, where a is a constant. The graph of horizontal line $y = -\dfrac{5}{3}$ is shown in Figure 9.8 and vertical line $x = 4$ is shown in Figure 9.9.

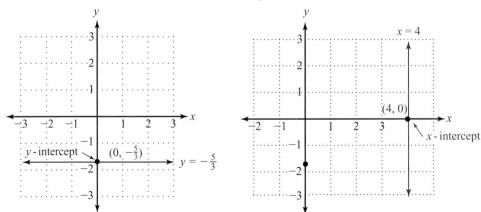

Figure 9.8   **Figure 9.9**

OBJECTIVE 4   **Calculate the length of a line segment.**   Since we now know how to graph lines, we will turn our attention to finding the length of part of a line (a line segment). For example, suppose the length of the segment formed by the points $(0, 0)$ and $(3, 4)$ is needed as shown in the figure below. The Pythagorean Theorem, discussed in Chapter 5, will be used to find its length. Look at the right triangle formed by the x-axis, the vertical line $x = 3$, and the line from the given equation $3y = 4x$. The vertices of the right triangle are the points $(0, 0)$, $(3, 0)$, and $(3, 4)$. The length of the horizontal segment from $(0, 0)$ to $(3, 0)$ is 3 units and the length of the vertical segment from $(3, 0)$ to $(3, 4)$ is 4 units, Next, use the Pythagorean Theorem to find the length of the segment from $(0, 0)$ to $(3, 4)$, which will be called c.

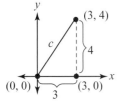

$$3^2 + 4^2 = c^2$$
$$9 + 16 = c^2$$
$$25 = c^2$$
$$\pm 5 = c$$

Thus, the length of the desired segment is 5 units. The solution of -5 is rejected because a segment must have a positive length.

To find the length of the line segment in the previous problem, the Pythagorean Theorem was used. The results will now be generalized into a formula that can be used to find the length of *any* line segment. Before reading Theorem 9.1, the symbols (x_1, y_1) and (x_2, y_2) need to be explained. The numbers $_1$ and $_2$ are called *subscripts*. They are read "x sub one" and "x sub two." These subscripts are used to distinguish the first x-coordinate from the second x-coordinate in the two points. The same is true for subscripts shown with the y-coordinates.

> **THEOREM 9.1   *Distance Formula***
>
> The **distance** between two points with coordinates (x_1, y_1) and (x_2, y_2) is determined with the formula.
>
> $$d = \sqrt{(x_2 - x_1)^2 + (y_2 - y_1)^2}.$$

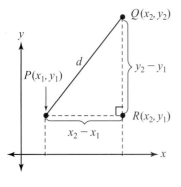

Figure 9.10 Distance Formula

Calculating the distance between two given points is a direct application of the Pythagorean Theorem. Suppose the two points, $P(x_1, y_1)$ and $Q(x_2, y_2)$, are in quadrant I, as in Figure 9.10. Then the numbers $x_2 - x_1$ and $y_2 - y_1$ are the lengths of the legs of the triangle with vertices P, Q, and R. Applying the Pythagorean Theorem, we obtain

$$d^2 = (x_2 - x_1)^2 + (y_2 - y_1)^2,$$

and because d is positive,

$$d = \sqrt{(x_2 - x_1)^2 + (y_2 - y_1)^2}$$

gives the length of the hypotenuse of the triangle—that is, the distance between the points (x_1, y_1) and (x_2, y_2). Because $(x_2 - x_1)$ and $(y_2 - y_1)$ are both squared, $(x_1 - x_2)^2 = (x_2 - x_1)^2$ and $(y_1 - y_2)^2 = (y_2 - y_1)^2$, and it is not important which point we label (x_1, y_1) and which point we label (x_2, y_2). Although our particular points are in the first quadrant, the same results hold regardless of the location of the two points.

We have just given a paragraph-style proof for Theorem 9.1.

EXAMPLE 2 (a) Find the distance between point $(6, 8)$ and point $(2, 5)$.

Let $x_1 = 6, y_1 = 8, x_2 = 2$, and $y_2 = 5$ and substitute into the distance formula.

$$d = \sqrt{(x_2 - x_1)^2 + (y_2 - y_1)^2}$$
$$= \sqrt{(2 - 6)^2 + (5 - 8)^2}$$
$$= \sqrt{(-4)^2 + (-3)^2} = \sqrt{16 + 9} = \sqrt{25} = 5$$

The distance between the given points is 5 units.

> **Note** If we let $x_1 = 2, y_1 = 5$, and $x_2 = 6, y_2 = 8$, the distance is still 5.

(b) Find the distance between point $(2, 2)$ and point $(5, 5)$.

Substituting into the distance formula, we have

$$d = \sqrt{(5 - 2)^2 + (5 - 2)^2}$$
$$= \sqrt{3^2 + 3^2} = \sqrt{2 \cdot 3^2} = 3\sqrt{2}$$

or approximately 4.24 rounded to the nearest hundredth of a unit.

OBJECTIVE 5 **Calculate the midpoint of a line segment.** Another formula, the **midpoint formula**, can be established using similar triangles. It can be shown that the coordinates of a point midway between two given points are found by averaging the corresponding x-coordinates and y-coordinates. The midpoint formula is stated in the next theorem without proof.

THEOREM 9.2 *Midpoint Formula*

The midpoint of a line segment joining (x_1, y_1) and (x_2, y_2) is

$$\left(\frac{x_1 + x_2}{2}, \frac{y_1 + y_2}{2} \right).$$

That is, the x-coordinate of the midpoint is the average of the x-coordinates of the points, and the y-coordinate of the midpoint is the average of the y-coordinates of the points.

┌─ **Note** In the distance formula, we subtract x-coordinates and y-coordinates, whereas, in the midpoint formula we add x-coordinates and y-coordinates. Do not confuse these operations when working with the two formulas.
└─

EXAMPLE 3   Find the midpoint between the points $(-2, 1)$ and $(6, -3)$.

Substitute into the midpoint formula.

$$\left(\frac{x_1 + x_2}{2}, \frac{y_1 + y_2}{2}\right) = \left(\frac{-2 + 6}{2}, \frac{1 + (-3)}{2}\right)$$

$$= \left(\frac{4}{2}, \frac{-2}{2}\right) = (2, -1)$$

The midpoint is $(2, -1)$.

PRACTICE EXERCISE

Find the length of the segment and the midpoint of the segment from $(2, 3)$ to $(-3, 5)$.

ANSWER BELOW

▌**Answers to Practice Exercises**

1. The graph is a straight line with intercepts $(3, 0)$ and $(0, -2)$.
2. $d = \sqrt{29}$, midpoint is $\left(-\frac{1}{2}, 4\right)$.

9.1 **Exercises**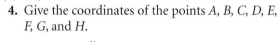

FOR EXTRA HELP: 📖   Student's Solutions Manual   Addison-Wesley Math Tutor Center

---

1. Plot the points corresponding to the ordered pairs $A(1, 4)$, $B(4, -2)$, $C(-3, 2)$, $D(-3, 0)$, $E(3, 0)$, $F(0, 0)$, $G(-3, -3)$, and $H(0, -2)$.
2. Plot the points corresponding to the ordered pairs $A(2, 5)$, $B(3, -1)$, $C(-4, 1)$, $D(-2, 0)$, $E(4, 0)$, $F(-2, -2)$, $G(0, 0)$, and $H(0, -5)$.
3. Give the coordinates of the points A, B, C, D, E, F, G, and H.
4. Give the coordinates of the points A, B, C, D, E, F, G, and H.

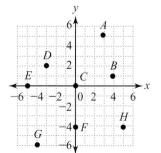

Exercise 3

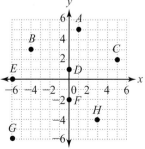

Exercise 4

5. Plot the points associated with the ordered pairs

$$J\left(\frac{1}{2}, 2\right), K\left(-\frac{5}{2}, 1\right), L\left(-2, -\frac{7}{4}\right), \text{ and } M\left(3, -\frac{3}{4}\right).$$

6. Plot the points associated with the ordered pairs

$$P\left(-\frac{1}{2}, 3\right), Q\left(\frac{7}{2}, 4\right), R\left(-3, -\frac{3}{4}\right), \text{ and } S\left(2, -\frac{7}{3}\right).$$

7. In what quadrants are the points J, K, L, and M of Exercise 5 located?
8. In what quadrants are the points P, Q, R, and S of Exercise 6 located?

In Exercises 9–14, find the missing coordinate so that each ordered pair is a solution to the equation.

9. $x + y + 2 = 0$;   (a) $(0, ?)$; (b) $(?, 0)$; (c) $(1, ?)$;   (d) $(?, -2)$
10. $2x - y + 1 = 0$;   (a) $(0, ?)$; (b) $(?, 0)$; (c) $(-1, ?)$; (d) $(?, 3)$
11. $x + 3 = 0$;   (a) $(0, ?)$; (b) $(?, 0)$; (c) $(2, ?)$;   (d) $(?, -4)$
12. $2x = 5$;   (a) $(0, ?)$; (b) $(?, 0)$; (c) $(-2, ?)$; (d) $(?, 1)$
13. $3y + 1 = 0$;   (a) $(0, ?)$; (b) $(?, 0)$; (c) $(3, ?)$;   (d) $(?, -1)$
14. $y = -4$;   (a) $(0, ?)$; (b) $(?, 0)$; (c) $(-5, ?)$; (d) $(?, 2)$

In Exercises 15–30, determine the intercepts and graph each linear equation.

15. $x + y + 2 = 0$ 16. $x + y - 2 = 0$ 17. $3x + y = 6$ 18. $x + 2y = 4$
19. $x - y = 2$ 20. $y - x = 1$ 21. $x - y = 0$ 22. $y - x = 0$
23. $3x - 7 = 0$ 24. $2x + 3 = 0$ 25. $y = -1$ 26. $y = 2$
27. $3x - 2y = 0$ 28. $2y - 3x = 0$ 29. $3x + 2y = 6$ 30. $3x - 2y = 6$

31. What are the intercepts of the line $x = 0$? What is its graph?
32. What are the intercepts of the line $y = 0$? What is its graph?

33. Graph $y = 2x + 1$, $y = \frac{1}{2}x + 1$, $y = 0 \cdot x + 1$, $y = -\frac{1}{2}x + 1$, and $y = -2x + 1$

in the same coordinate system. What common characteristic do all of the lines possess?

34. Graph $y = 2x + 3$, $y = 2x + 1$, $y = 2x + 0$, $y = 2x - 1$, and $y = 2x - 3$ in the same coordinate system. What common characteristic do all the lines possess?

In Exercises 35–40, find the distance between the points given.

35. $(3, 2)$ and $(-2, -10)$ 36. $(3, 6)$ and $(-3, -2)$ 37. $(0, 2)$ and $(-3, 0)$
38. $(a, 0)$ and $(0, b)$ 39. $(5, 2)$ and $(5, -3)$ 40. $(7, -1)$ and $(-2, -1)$
41. The distance between $(4, 3)$ and $(4, f)$ is 10. Find all the possible values of f.
42. The distance between $(-3, 2)$ and $(d, 2)$ is 6. Find all the possible values of d.

In Exercises 43–48, find the midpoint of the line segment between the points given.

43. $(-2, -1)$ and $(4, 1)$ **44.** $(3, -4)$ and $(1, 6)$ **45.** $(6, 11)$ and $(5, -3)$

46. $(5, -3)$ and $(-2, -1)$ **47.** $(2k, 2n)$ and $(2p, 2q)$ **48.** $(a, 0)$ and $(0, b)$

49. Suppose the origin is the midpoint of a segment and one endpoint of the segment is $(2, -4)$. Find the coordinates of the other endpoint.

50. Suppose the origin is the midpoint of a segment and one endpoint of the segment is (x, y). Find the coordinates of the other endpoint.

51. Given right triangle ABC, where M is the midpoint of $\overline{AC}$.

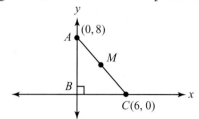

    a. Find the coordinates of M.

    b. Find and compare the lengths of $\overline{AM}$, $\overline{MC}$, and $\overline{MB}$. Compare the results to Theorem 5.16.

52. Given right triangle DEF, where M is the midpoint of $\overline{DF}$.

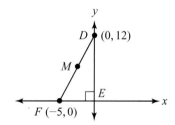

    a. Find the coordinates of M.

    b. Find and compare the lengths of $\overline{MD}$, $\overline{MF}$, and $\overline{ME}$. Compare the results to Theorem 5.16.

53. Plot the points $A(5, 2)$, $B(1, 2)$, and $C(5, 6)$. Connect the points to form a triangle.

    a. Calculate the length of each side of the triangle.

    b. What kind of triangle is it? Justify your answer.

54. Plot the points $A(3, -2)$, $B(6, -5)$, and $C(2, -6)$. Connect the points to form a triangle.

    a. Calculate the length of each side of the triangle.

    b. What kind of triangle is it? Justify your answer.

55. Research the connection between René Descartes and the Cartesian coordinate system.

9.2 Slope, Equation of a Line

OBJECTIVES

1. Define and calculate the slope of a line.
2. Use point-slope form of a line.
3. Use slope-intercept form of a line.
4. Determine slopes of parallel and perpendicular lines.

OBJECTIVE 1   Define and calculate the slope of a line.   We can expand our discussion of linear equations by defining the *slope* of a line. **Slope** means the pitch or slant of a line. The lowercase m is used to denote slope. The slope of a line is the ratio of change in the vertical direction to the change in the horizontal direction. The **rise** is the vertical change $[y_2 - y_1]$ in the line. The **run** is the horizontal change in the line $[x_2 - x_1]$. We use these terms to define a line's slope. Suppose that the points $P(x_1, y_1)$ and $Q(x_2, y_2)$ are two points on a line, shown in Figure 9.11.

DEFINITION: The Slope of a Line

Let $P(x_1, y_1)$ and $Q(x_2, y_2)$ be two points on a *nonvertical* line. The **slope** of the line is

$$m = \frac{y_2 - y_1}{x_2 - x_1} = \frac{\text{rise}}{\text{run}} = \frac{\text{change in } y\text{-coordinates}}{\text{change in } x\text{-coordinates}}, \text{ for } x_2 \neq x_1$$

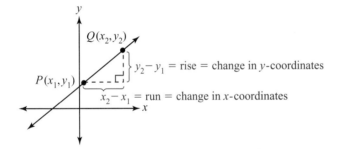

Figure 9.11

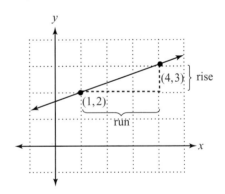

Figure 9.12

EXAMPLE 1   Find the slope of the line passing through the two points $(4, 3)$ and $(1, 2)$.
    Refer to the graph in Figure 9.12. Suppose we identify point $P(x_1, y_1)$ with $(1, 2)$ and point $Q(x_2, y_2)$ with $(4, 3)$. The slope will then be determined with

$$m = \frac{y_2 - y_1}{x_2 - x_1} = \frac{3 - 2}{4 - 1} = \frac{1}{3}.$$

What happens if we identify $P(x_1, y_1)$ with $(4, 3)$ and $Q(x_2, y_2)$ with $(1, 2)$? In this case, we have

$$m = \frac{y_2 - y_1}{x_2 - x_1} = \frac{2 - 3}{1 - 4} = \frac{-1}{-3} = \frac{1}{3}.$$

Thus, we see that the slope is the same regardless of how the two points are identified.

EXAMPLE 2 Find the slope of the line passing through the two points $(-3, 2)$ and $(1, -1)$.

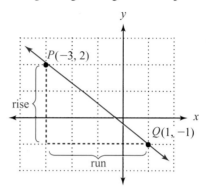

Figure 9.13

Finding the slope of a line can be useful in many practical situations. In the design of a drawbridge, the amount of slope of the bridge when in an open position determines the clearance for ships that pass beneath it.

Let us identify $P(x_1, y_1)$ with $(-3, 2)$ and $Q(x_2, y_2)$ with $(1, -1)$ as in Figure 9.13. The slope is determined with

$$m = \frac{y_2 - y_1}{x_2 - x_1} = \frac{(-1) - (2)}{(1) - (-3)} \qquad \text{Watch the signs}$$

$$= \frac{-3}{1 + 3} = -\frac{3}{4}.$$

PRACTICE EXERCISE 1

Find the slope of the line passing through the points $(1, -3)$ and $(-2, 4)$.

ANSWER ON PAGE 443

Figure 9.14

The line passing through the points $(3, 2)$ and $(-1, 2)$ is horizontal and its slope is 0. Refer to Figure 9.14.

$$m = \frac{y_2 - y_1}{x_2 - x_1} = \frac{2 - 2}{-1 - 3} = \frac{0}{-4} = 0.$$

Figure 9.15

If we try to calculate the slope of the vertical line passing through $(-1, 4)$ and $(-1, -2)$, we find that it is undefined. Refer to Figure 9.15.

$$m = \frac{y_2 - y_1}{x_2 - x_1} = \frac{-2 - 4}{-1 - (-1)} = \frac{-6}{-1 + 1} = \frac{-6}{0} \qquad \text{Undefined because we cannot divide by zero}$$

OBJECTIVE 2 **Use point-slope form of a line.** An understanding of slope will help with the graphing of a line. A line may be graphed given its slope and any point on the line.

EXAMPLE 3

Graph the line through $(-3, 1)$ with a slope $\frac{1}{2}$. Refer to Figure 9.16.

First graph the point $(-3, 1)$.

$$\text{slope} = \frac{1}{2} = \frac{rise}{run} \quad \begin{array}{l} \text{up} \\ \text{right} \end{array}$$

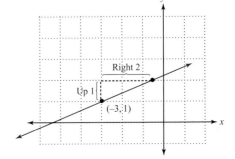

Figure 9.16

To locate another point on the line, count one unit 1 and then 2 units to the right. Draw the line through the given point $(-3, 1)$ and this new point.

In Section 9.1, the general form of a linear equation was defined as $ax + by + c = 0$. Slope helped us graph lines, now slope will help us find the equation of lines. Consider the *point-slope form* and the *slope-intercept form*.

DEFINITION: *Point-Slope Form of the Equation of a Line*

The equation of a line with slope m and passing through the point (x_1, y_1) is determined using the **point-slope form**

$$y - y_1 = m(x - x_1).$$

The length of a ski run can be computed using the concept of the slope of a line. It is interesting that we often use the term "ski slope" to describe a ski facility.

EXAMPLE 4 Find the point-slope form of the equation of the line from Example 3 with slope $\frac{1}{2}$ passing through the point $(-3, 1)$. Also, write the general form of the equation of this line.

We have $(x_1, y_1) = (-3, 1)$ and $m = \frac{1}{2}$, so by substituting we obtain the point-slope form.

$$y - y_1 = m(x - x_1)$$

$$y - 1 = \frac{1}{2}(x - (-3)) \qquad \text{Watch the sign}$$

To determine the general form of the equation, eliminate fractions, remove parentheses, and then collect all terms on the left side of the equation.

$$y - 1 = \frac{1}{2}(x + 3) \qquad \text{Point-slope form}$$

$$2(y - 1) = 2 \cdot \frac{1}{2}(x + 3) \qquad \text{Multiply both sides by 2}$$

$$2y - 2 = x + 3 \qquad \text{Remove parentheses}$$

$$2y - 2 - x - 3 = x + 3 - x - 3 \qquad \text{Subtract } x \text{ and } 3$$

$$-x + 2y - 5 = 0 \qquad \text{Multiply by } -1$$

$$x - 2y + 5 = 0 \qquad \text{General form}$$

PRACTICE EXERCISE 2

Find the general form of the equation of a line passing through $(2, -3)$ with slope -3 and graph the line.

ANSWER ON PAGE 443

EXAMPLE 5 Graph the line through $(0, 7)$ with a slope -2. Refer to Figure 9.17.

First graph the point $(0, 7)$. Notice it is on the y-axis so it is the y-intercept of the line.

Write the slope as $m = \dfrac{rise}{run} = -2 = \dfrac{-2}{1}$

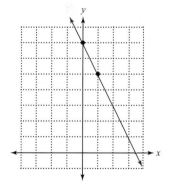

Figure 9.17

To locate another point on the line, count down 2 and right 1. (We could have written the slope as $\dfrac{2}{-1}$ instead. In this case, move up 2 and left 1 from $(0, 7)$.)

OBJECTIVE 3   **Use slope-intercept form of a line.**   To find the *equation* of the line in Example 5, the point-slope form can be used, but there is another alternative since the given point is the *y*-intercept.

> **DEFINITION: *Slope-Intercept Form of the Equation of a Line***
>
> The equation of a line with slope *m* and *y*-intercept $(0, b)$ is determined using the **slope-intercept form**
> $$y = mx + b.$$

EXAMPLE 6   Find the slope-intercept form of the equation of the line with slope -2 and *y*-intercept $(0, 7)$.

Substitute -2 for *m* and 7 for *b* in the slope-intercept form

$$y = mx + b$$
$$y = -2x + 7.$$

The importance of the slope-intercept form is that once this form is obtained, the slope (the coefficient of *x*) can be read directly, as can the *y*-coordinate of the *y*-intercept (the constant term). Given any form of the equation of a line, if the equation is solved for *y*, the coefficient of *x* is the slope and the constant term is the *y*-coordinate of the *y*-intercept.

 Technology Connection

Geometry software will be needed.

1. Draw a line that is neither horizontal nor vertical on the *xy*-grid.
2. Label two points on the line as points *A* and *B*.
3. Label the *y*-intercept of $\overline{AB}$, point *C*.
4. Measure the slope of $\overline{AB}$ and find the coordinates of *C*.
5. From the slope-intercept form, predict the equation of the line and then check this equation by having the software measure/calculate the equation of the line.
6. Drag point *A* to another quadrant and repeat steps 4 and 5. Do this several times. Did the slope-intercept form help you predict the equation correctly each time?

The next example solves the applied problem given in the chapter introduction by using a linear equation.

EXAMPLE 7

During a spring day in eastern Hong Kong, the pollution index increased from approximately 20 parts per million (ppm) at 8 A.M. to around 29 ppm at 2 P.M. Assuming the increase in pollution (y) per hour since 8 A.M. (x) can be approximated by a linear equation, find the equation where x is in terms of hours since 8 A.M. Interpret the meaning of the slope in the context of this problem. Interpret the meaning of the y-intercept in terms of this problem. Can the pollution index model be used to predict the pollution index at noon? At 6 P.M.? If so, what is the index? If not, why not?

Suppose 8 A.M. is identified as $x = 0$, 9 A.M. as $x = 1$ and so on. Then, 2 P.M. corresponds to $x = 6$ when the pollution index (y) is 29. At 8 A.M. the pollution index is 20, thus, there are two points: $(6, 29)$ and $(0, 20)$.

$$m = \frac{29 - 20}{6 - 0} = \frac{9}{6} = \frac{3}{2}$$

The slope-intercept form is now used, where $m = \frac{3}{2}$ and the y-intercept is $(0, 20)$. [Note: The point-slope form could also be used.]

$$y = \frac{3}{2}x + 20$$

The slope of $\frac{3}{2}$ in the problem means every two hours the pollution index rises 3 ppm between the hours of 8 A.M. and 2 P.M. The y-intercept equals the pollution index at the beginning time (8 A.M.). To find the pollution index at noon substitute $x = 4$ (four hours since 8 A.M.) and calculate y.

$$y = \frac{3}{2}(4) + 20$$

$$y = \frac{12}{2} + 20$$

$$y = 26 \text{ ppm}$$

Thus, the pollution index at noon is 26 ppm. This model only works for the hours between 8 A.M. and 2 P.M. so the pollution index at 6 P.M. cannot be calculated with this model.

OBJECTIVE 4   **Determine slopes of parallel and perpendicular lines.**   The slopes of parallel lines have a relationship as do the slopes of perpendicular lines. The next activity investigates these relationships.

Geometry software will be needed.

Remember: The software will round slopes and angle measures.

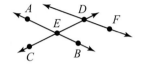

1. Construct two intersecting lines. Label them $\overleftrightarrow{AB}$ and $\overleftrightarrow{CD}$. Label the intersection point E.
2. Measure the slope of $\overleftrightarrow{AB}$ and $\overleftrightarrow{CD}$.
3. Measure $\angle DEB$.
4. Drag point D until the $m\angle DEB = 90°$.
5. What do you observe about the product of the slopes of the two lines?
6. Draw another line through point D. Label another point on this line F.
7. Measure the slope of $\overleftrightarrow{DF}$.
8. Drag F until the slope of $\overleftrightarrow{DF}$ is equal to the slope of $\overleftrightarrow{AB}$.
9. Measure $\angle FDE$ and $\angle BEC$. What do you observe?

The following theorems can be proven using similar triangles, but the proofs are not presented here. We will use these theorems along with slope-intercept form of an equation to determine if two equations represent parallel or perpendicular lines.

> **THEOREM 9.3**
> Two distinct lines with slopes m_1 and m_2 are parallel if and only if $m_1 = m_2$.

> **THEOREM 9.4**
> Two lines with slopes m_1 and m_2 are perpendicular if and only if $m_1 m_2 = -1$.

Technology Connection

A graphing calculator or graphing software will be needed.

1. Graph the lines $y = 3x$, $y = 3x + 2$, and $y = 3x + 4$ on one screen.
2. Make an observation about the lines.
3. What theorem does this illustrate?
4. From this observation, what is the equation of a line parallel to $y = 3x$ and with a y-intercept of 10?
5. Clear the screen.
6. Graph the lines $y = \frac{3}{4}x$ and $y = -\frac{4}{3}x$ on one screen.
7. Make an observation about the lines.
8. What theorem does this illustrate?
9. From this observation, what is the equation of a line through the origin and perpendicular to $y = \frac{2}{3}x$?

EXAMPLE 8   Determine whether the lines with equations $3x - 2y + 7 = 0$ and $2x + 3y - 6 = 0$ are parallel, perpendicular, or neither.

We solve each equation for y (write each in slope-intercept form) to determine the slope.

$$3x - 2y + 7 = 0 \qquad\qquad 2x + 3y - 6 = 0$$
$$-2y = -3x - 7 \qquad\qquad 3y = -2x + 6$$
$$y = \frac{3}{2}x + \frac{7}{2} \qquad\qquad y = -\frac{2}{3}x + 2$$
$$\downarrow \qquad\qquad\qquad\qquad \downarrow$$
$$m_1 = \frac{3}{2} \qquad\qquad\qquad m_2 = -\frac{2}{3}$$

Because $m_1 = \frac{3}{2}$ and $m_2 = -\frac{2}{3}$ are negative reciprocals or

$m_1 m_2 = \left(\frac{3}{2}\right)\left(-\frac{2}{3}\right) = -1$, the two lines are perpendicular.

EXAMPLE 9   Determine whether the lines with equations $x - 4y + 2 = 0$ and $3x - 12y + 6 = 0$ are parallel, perpendicular, or neither.

Solve each equation for y.

$$x - 4y + 2 = 0 \qquad\qquad 3x - 12y + 6 = 0$$
$$-4y = -x - 2 \qquad\qquad -12y = -3x - 6$$
$$y = \frac{1}{4}x + \frac{2}{4} \qquad\qquad y = \frac{3}{12}x + \frac{6}{12}$$
$$y = \frac{1}{4}x + \frac{1}{2} \qquad\qquad y = \frac{1}{4}x + \frac{1}{2}$$
$$\downarrow \qquad\qquad\qquad\qquad \downarrow$$
$$m_1 = \frac{1}{4} \qquad\qquad\qquad m_2 = \frac{1}{4}$$

Because $m_1 = m_2$, we are tempted to conclude that the two lines are

parallel. However, the y-intercepts are also equal $\left(\text{both are }\left(0, \frac{1}{2}\right)\right)$ so the

equations determine the same line. In situations like this, we say that the lines **coincide**.

Answers to Practice Exercises

1. $-\dfrac{7}{3}$ **2.** $3x + y - 3 = 0$

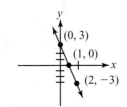

9.2 **Exercises**   *FOR EXTRA HELP:* Student's Solutions Manual   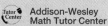 Addison-Wesley Math Tutor Center

In Exercises 1–6, find the slope of the line passing through the given pair of points.

1. $(-5, 2)$ and $(-1, -6)$ 2. $(7, -1)$ and $(3, 3)$ 3. $(-2, 7)$ and $(-5, 7)$
4. $(4, 2)$ and $(4, -2)$ 5. $(a, 0)$ and $(0, b)$ 6. (a, b) and (c, d)

In Exercises 7–24, find the general form of the equation of the line satisfying the conditions given and graph the line.

7. Through $(3, -1)$ with slope -2.
8. Through $(-2, -4)$ with slope -3.
9. Through $(2, -4)$ parallel to a line with slope $\frac{1}{3}$.
10. Through $(-3, 5)$ parallel to a line with slope -4.
11. Through $(7, -1)$ and $(5, 3)$.
12. Through $(-7, 1)$ and $(3, -5)$.
13. Through $(2, 3)$ and is a vertical line.
14. Through $(-1, 7)$ and is a vertical line.
15. Through $(2, 3)$ with slope 0.
16. Through $(5, -3)$ with slope 0.
17. With x-intercept $(-5, 0)$ and slope 2.
18. With x-intercept $(3, 0)$ and slope $\frac{1}{2}$.
19. With y-intercept $(0, 4)$ and slope -3.
20. With y-intercept $(0, 2)$ and slope -5.
21. Through $(4, -1)$ and $\left(7, \frac{2}{3}\right)$.
22. Through $\left(2, \frac{1}{2}\right)$ and $(-3, 8)$.
23. With x-intercept $(-2, 0)$ and y-intercept $(0, 5)$.
24. With x-intercept $(4, 0)$ and y-intercept $(0, -3)$.

In Exercises 25–30, find the slope and y-intercept of the line by writing the equation in slope-intercept form.

25. $5x - 2y + 4 = 0$ 26. $6x + 2y - 10 = 0$ 27. $5x + 4 = 0$
28. $x + 7 = 0$ 29. $-2y + 4 = 0$ 30. $3y + 9 = 0$
31. Find the slope-intercept form of the equation of the line passing through the points $(-2, 3)$ and $(1, -3)$. What is the slope? the y-intercept?
32. Find the slope-intercept form of the equation of the line passing through the points $(4, -5)$ and $(2, 8)$. What is the slope? the y-intercept?

In Exercises 33–38, determine whether the lines with the given equations are parallel, perpendicular, or neither.

33. $5x - 3y + 8 = 0$
    $3x + 5y - 7 = 0$
34. $2x - y + 3 = 0$
    $x + 2y - 5 = 0$
35. $2x - y + 7 = 0$
    $-6x + 3y - 1 = 0$
36. $5x - 2y + 3 = 0$
    $-10x + 4y + 3 = 0$
37. $2x + 1 = 0$
    $-3y - 4 = 0$
38. $4x - y + 7 = 0$
    $3x - 1 = 0$
39. Find the general form of the equation of the line passing through $(-1, 4)$ and parallel to the line with equation $-3x - y + 4 = 0$.
40. Find the general form of the equation of the line passing through $(-2, 5)$ and parallel to the line with equation $4x + 2y - 9 = 0$.
41. Find the general form of the equation of the line passing through $(-1, 4)$ and perpendicular to the line with equation $-3x - y + 4 = 0$.
42. Find the general form of the equation of the line passing through $(4, -7)$ and perpendicular to the line with equation $x - 5y + 7 = 0$.

In Exercises 43–46, find the general form of the equation of the line perpendicular to the line that contains the given points and that passes through the point midway between them.

43. $(-3, 2)$ and $(3, -6)$ **44.** $(-5, 1)$ and $(3, -1)$

45. $(5, -3)$ and $(7, 1)$ **46.** $(2, -7)$ and $(6, 1)$

47. A line is tangent to a circle, whose center is $(0, 0)$, at the point $(3, 5)$.
    a. Find the slope of the tangent line.
    b. Find the equation of the tangent line in slope-intercept form.
      [Hint: What is known about a radius to a tangent?]

48. If quadrilateral $ABCD$ is a rhombus and A is the point $(-1, 3)$ while C has coordinates $(1, -1)$, find the slope of diagonal $\overline{BD}$.
    [Hint: What is known about the diagonals of a rhombus?]

49. Given the vertices of a trapezoid are $(7, -3), (-1, 4), (-4, -3),$ and $(4, 4)$.
    a. Find the midpoints of the legs.
    b. Using slopes, show the median is parallel to the bases as stated in Theorem 4.22.

50. Given the coordinates of a triangle are $A(1, 1), B(7, 3),$ and $C(5, 9)$
    a. Find the midpoint of $\overline{AC}$ and call it point E.
    b. Find the midpoint of $\overline{BC}$ and call it point F.
    c. Using slopes, show $\overline{EF} \parallel \overline{AB}$ as stated in Theorem 4.19.

 51. Explain why $y - y_1 = m(x - x_1)$ is called point-slope form and $y = mx + b$ is called slope-intercept form.

9.3   Proofs Involving Polygons

OBJECTIVES

1. Prepare to do analytic geometry proofs.
2. Learn to do analytic geometry proofs.

OBJECTIVE 1   Prepare to do analytic geometry proofs.   Analytic geometry allows us to use algebra to prove certain geometric theorems. When thinking about an analytic geometry proof, the position of the figure on the Cartesian plane can make the proof easy or difficult. Before making a drawing for an analytic proof, there are several things to consider.

1. Look to the hypothesis of the statement for the type of figure needed. Do not add any additional attributes not given in the hypothesis. For example, if the statement describes an isosceles triangle, draw and label an isosceles triangle, *not* a right isosceles triangle, nor an equilateral triangle.
2. Remember, the coordinates of the vertices must be general coordinates like (c, d), not $(2, 5)$.
3. For simplicity in the calculations, make as many 0 coordinates as possible.

If a right triangle is given, the most convenient place to put the legs of the triangle is on the x-axis and y-axis. See Figure 9.18(a). If the triangle is not a right triangle, there are two possible convenient locations. See Figure 9.18(b) and 9.18(c).

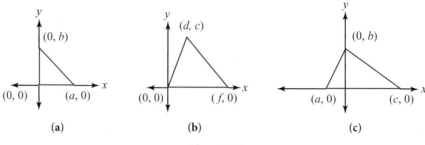

Figure 9.18

If the figure is a quadrilateral, it is usually convenient to place one vertex at the origin and one side along the x-axis. See Figure 9.19.

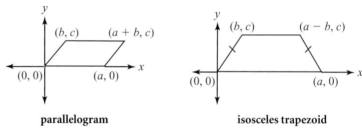

Figure 9.19

OBJECTIVE 2   **Learn to do analytic geometry proofs.**   We will illustrate analytic geometry proofs by proving earlier theorems. Theorems are assigned the original numbers given to them earlier in the text.

> **THEOREM 5.16**
> The median from the right angle in a right triangle is one-half the length of the hypotenuse.

PROOF: The hypothesis states there is a right triangle with a median from the right angle. Recall a median is a segment from an angle to the midpoint of the opposite side. The legs of the right triangle are placed on the x-axis and y-axis. The midpoint has coordinates $\left(\dfrac{a}{2}, \dfrac{b}{2}\right)$ as shown in Figure 9.20.

Using the distance formula, $AB = \sqrt{(a - 0)^2 + (0 - (-b))^2} = \sqrt{a^2 + b^2}$.

From the conclusion of the theorem, we can show $DC = AD = \dfrac{AB}{2}$ or $\dfrac{1}{2}AB$.

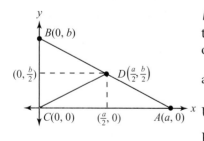

Figure 9.20

$$AD = \sqrt{\left(a - \frac{a}{2}\right)^2 + \left(0 - \frac{b}{2}\right)^2} = \sqrt{\left(\frac{a}{2}\right)^2 + \left(-\frac{b}{2}\right)^2} = \sqrt{\frac{a^2}{4} + \frac{b^2}{4}} = \frac{\sqrt{a^2 + b^2}}{2}$$

$$DC = \sqrt{\left(\frac{a}{2} - 0\right)^2 + \left(\frac{b}{2} - 0\right)^2} = \sqrt{\left(\frac{a}{2}\right)^2 + \left(\frac{b}{2}\right)^2} = \sqrt{\frac{a^2}{4} + \frac{b^2}{4}} = \frac{\sqrt{a^2 + b^2}}{2}$$

Thus, $DC = AD = \frac{1}{2}AB$, and the theorem is proved.

THEOREM 4.4

The diagonals of a parallelogram bisect each other.

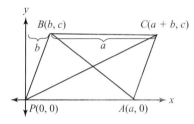

Figure 9.21

PROOF: The hypothesis states there is a parallelogram with diagonals. Place one vertex on the origin. A parallelogram has opposite sides parallel ($\overline{BC} \parallel \overline{PA}$) so B and C have the same y-coordinate (see Figure 9.21). Opposite sides are also congruent, therefore, their measures are the same ($BC = PA = a$). The x-coordinate of C is ($a + b, c$). From the conclusion, we must prove the diagonals bisect each other. If we can prove their midpoints are the same, the theorem will be proven.

$$\text{midpoint of } \overline{PC} = \left(\frac{a + b + 0}{2}, \frac{c + 0}{2}\right) = \left(\frac{a + b}{2}, \frac{c}{2}\right)$$

$$\text{midpoint of } \overline{AB} = \left(\frac{a + b}{2}, \frac{0 + c}{2}\right) = \left(\frac{a + b}{2}, \frac{c}{2}\right)$$

The midpoints of the segments are the same, $\left(\frac{a + b}{2}, \frac{c}{2}\right)$. It is the point of intersection of $\overline{PC}$ and $\overline{AB}$. Thus, the diagonals of a parallelogram bisect each other.

THEOREM 4.10

The diagonals of a rhombus are perpendicular.

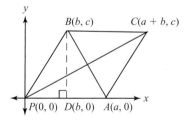

Figure 9.22

PROOF: Because a rhombus is a parallelogram, we label B and C in Figure 9.22 with the same y-coordinate. Also, as in the proof of Theorem 4.4 above, C can be labeled $C(a + b, c)$. To prove the theorem, we refer to Theorem 9.4 and prove that the product of the slopes of $\overline{PC}$ and $\overline{AB}$ is -1.

$$\text{slope of } \overline{PC} = \frac{c - 0}{a + b - 0} = \frac{c}{a + b}$$

$$\text{slope of } \overline{AB} = \frac{0 - c}{a - b} = \frac{-c}{a - b}$$

$$\text{product of slopes} = \left(\frac{c}{a + b}\right)\left(\frac{-c}{a - b}\right) = \frac{-c^2}{a^2 - b^2}$$

Draw auxillary segment $\overline{BD}$ where $\overline{BD} \perp \overline{PA}$. The coordinates of D are $(b, 0)$.

In right $\triangle BDP$ (Figure 9.22), $BD = c$, $PD = b$, and $BP = a$ because $ABCD$ is a rhombus and $BP = AP$. Thus, by the Pythagorean theorem

$$a^2 = b^2 + c^2$$
$$a^2 - b^2 = c^2.$$

Therefore, using substitution, the product of slopes $= \dfrac{-c^2}{a^2 - b^2} = \dfrac{-c^2}{c^2} = -1.$

By Theorem 9.4, the diagonals of a rhombus are perpendicular.

THEOREM 4.19

The segment joining the midpoints of two sides of a triangle is parallel to the third side and its length is one half the length of the third side.

A triangle is given. Draw any triangle ABP. It is a generic triangle. *Do not* draw it as an isosceles, right, or equilateral triangle. This would add additional attributes to the figure. The second part of the hypothesis states there is a segment joining the midpoints of two sides of the triangle. Let C and D be those midpoints as shown in Figure 9.23. Find the coordinates of C and D by using the midpoint formula.

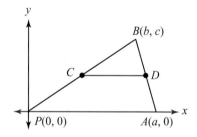

Figure 9.23

$$\text{midpoint of } \overline{BP} = \left(\frac{b + 0}{2}, \frac{c + 0}{2} \right) = \left(\frac{b}{2}, \frac{c}{2} \right) = \left(\frac{b}{2}, \frac{c}{2} \right)$$

$$\text{midpoint of } \overline{AB} = \left(\frac{a + b}{2}, \frac{0 + c}{2} \right) = \left(\frac{a + b}{2}, \frac{c}{2} \right) = \left(\frac{a + b}{2}, \frac{c}{2} \right)$$

The conclusion states we must show $\overline{CD} \parallel \overline{PA}$ and $CD = \frac{1}{2}PA$.

One way to prove lines are parallel is to show the slopes are equal by Theorem 9.3.

The slopes of $\overline{PA}$ and $\overline{CD}$ are:

$$\text{slope of } \overline{PA} = \frac{0 - 0}{a - 0} = \frac{0}{a} = 0$$

$$\text{slope of } \overline{CD} = \frac{\dfrac{c}{2} - \dfrac{c}{2}}{\dfrac{a + b}{2} - \dfrac{b}{2}} = \frac{0}{\dfrac{a}{2}} = 0$$

Thus, $\overline{CD} \parallel \overline{PA}$ by Theorem 9.3.

Because both $\overline{PA}$ and $\overline{CD}$ are horizontal, the length of each is the difference in the x-coordinates of their endpoints. Because the coordinates of A are $(a, 0)$ and the coordinates of P are $(0, 0)$, then

$$\text{length of } \overline{PA} = PA = a - 0 = a.$$

Also, because D is $\left(\dfrac{a + b}{2}, \dfrac{c}{2} \right)$ and C is $\left(\dfrac{b}{2}, \dfrac{c}{2} \right)$,

$$\text{length of } \overline{CD} = CD = \frac{a + b}{2} - \frac{b}{2} = \frac{a + b - b}{2} = \frac{a}{2}.$$

Thus, $CD = \dfrac{1}{2}PA$, and the theorem is proved.

 Student Activity

Groups of three people will be needed for this activity. Each member of the group will do one of the following:

1. Read the problem out loud to the group and identify the hypothesis and conclusion.
2. Draw the figure giving the coordinates of the polygon.
3. Identify the algebraic statements necessary to form the key deductive steps of the proof.

Problem: Use the slope formula to prove the following statement: if $ABCD$ is a parallelogram, M is the midpoint of $\overline{BC}$ and N is the midpoint of $\overline{AD}$, prove $AMCN$ is a parallelogram.

9.3 Exercises

FOR EXTRA HELP: Student's Solutions Manual   Tutor Center   Addison-Wesley Math Tutor Center

Complete the missing coordinates using only the variable given and the properties of the given figure.

1. Given $ABCD$ is a rectangle

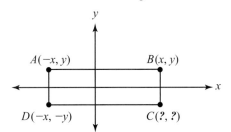

Exercise 1

2. Given $ABCD$ is a square

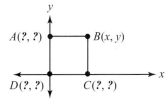

Exercise 2

3. Given $\triangle ABC$ is equilateral

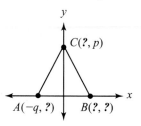

Exercise 3

4. Given $WXYZ$ is a parallelogram
Also name the coordinates of the midpoint of $\overline{WZ}$ and $\overline{XY}$.

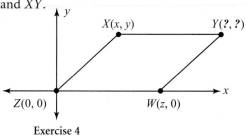

Exercise 4

5. Given $JKLM$ is a trapezoid
Also name the coordinates of the midpoints of $\overline{JK}$ and $\overline{ML}$.

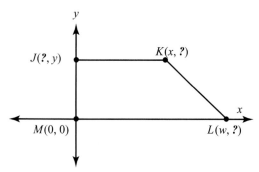

Exercise 5

6. Given $\triangle MNO$ is an isosceles right triangle

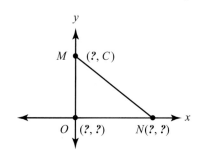

Exercise 6

7. Given $ABCD$ is an isosceles trapezoid

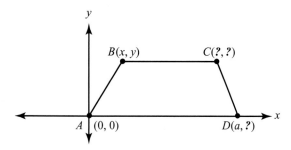

Exercise 7

8. Given $MNOP$ is a rectangle

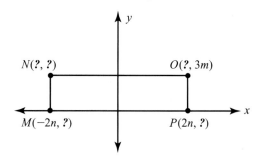

Exercise 8

9. Given isosceles trapezoid $ABCD$, where $c = a + b$, prove the diagonals are congruent.

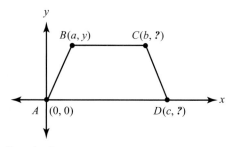

Exercise 9

10. Prove a quadrilateral formed by joining, in order, the midpoints of the sides of an isosceles trapezoid is a rhombus.

In Exercises 11–18, use analytic geometry to prove each theorem. Draw a figure using the hypothesis of each statement.

11. The diagonals of a rectangle are congruent.
12. The opposite sides of a parallelogram are congruent.
13. The diagonals of an isosceles trapezoid are congruent.
14. The medians drawn to the congruent sides of an isosceles triangle are congruent.
15. The diagonals of a square are perpendicular bisectors of each other.
16. The perpendicular bisector of the base of an isosceles triangle passes through the vertex of the triangle.
17. The median of a trapezoid is parallel to the bases.
18. The triangle formed by joining the midpoints of the sides of an isosceles triangle is isosceles.

Chapter (9)   Review

Key Terms and Symbols

9.1   Cartesian (rectangular)
      coordinate system
  coordinate plane
  x-axis
  y-axis
  origin
  quadrant
  ordered pair
  x-coordinate

y-coordinate
plot
linear (first-degree) equation
general form
solution
graph
table of values
intercept
x-intercept

y-intercept
distance formula
midpoint formula

9.2   slope
  rise of a line
  run of a line
  point-slope form
  slope-intercept form

Chapter 9 Proof Techniques

To Prove:
Two Lines Are Parallel

1. Show that their slopes are equal. (Theorem 9.3)

2. Write the equations in slope-intercept form and compare the slopes.

Two Lines Are Perpendicular

1. Show that the product of their slopes is -1. (Theorem 9.4)

2. Write equations in slope-intercept form and compare the slopes.

Chapter Formulas

distance formula $\quad d = \sqrt{x_2 - x_1)^2 + (y_2 - y_1)^2}$

midpoint formula $\quad \left(\dfrac{x_1 + x_2}{2}, \dfrac{y_1 + y_2}{2} \right)$

slope formula $\quad m = \dfrac{y_2 - y_1}{x_2 - x_1}$

general form $\quad ax + by + c = 0$

point-slope form $\quad y - y_1 = m(x - x_1)$

slope-intercept form $\quad y = mx + b$

Review Exercises

Section 9.1

1. What is the x-coordinate of the ordered pair $(-2, 3)$? What is the y-coordinate? In which quadrant is the point associated with this ordered pair located?

2. What are the perpendicular lines that form a Cartesian coordinate system called?

3. A linear equation of the form $y = b$ (b is a constant) has as its graph a straight line parallel to which axis?

4. A linear equation of the form $x = a$ (a is a constant) has as its graph a straight line parallel to which axis?

5. Give the coordinates of the points A, B, C, and D in the figure. In which quadrant is each point located?

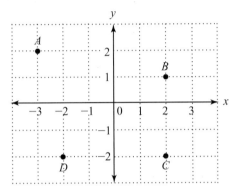

Exercise 5

6. Plot the points $(0, 4)$, $(-1, 2)$, $(2, -3)$, $(-3, 0)$, and $(-1, -4)$.

In Exercises 7–12, determine the intercepts and graph each linear equation.

7. $x + y - 2 = 0$

8. $3x - 2y - 6 = 0$

9. $3x - y = 0$

10. $x - 2y = 0$

11. $3x + 6 = 0$

12. $4y - 4 = 0$

Section 9.2

In Exercises 13–14, find the slope of the line through each pair of points.

13. $(1, -2)$ and $(3, 7)$

14. $(-5, -4)$ and $(-6, -2)$

15. If the slope of a line is 5, what is the slope of a line parallel to it? What is the slope of a line perpendicular to it?

16. Find the slope of the line passing through $(3, -1)$ and $(-2, 7)$. What is the slope of a line perpendicular to this line?

17. What is the distance between the points $(4, -1)$ and $(-2, -1)$?

18. The distance between $(3, 2)$ and $(x, 2)$ is 5. Find all possible values of x.

19. Find the midpoint of the line segment joining $(-1, 0)$ and $(5, -2)$.

20. What is the midpoint of the line segment joining $(1, -6)$ and $(3, 2)$?

21. Given the isosceles triangle in the figure, find the area and perimeter.

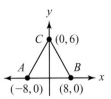

Exercise 21

22. Find the general form of the equation of the line with slope -4 and passing through $(-1, 3)$.

23. Find the slope-intercept form of the equation of the line $4x + 2y - 10 = 0$. What is the slope? the y-intercept?

24. Find the general form of the equation of a line passing through the points $(-2, 5)$ and $(6, 9)$.

25. Find the general form of the equation of the line that passes through the point $(-3, 1)$ and is perpendicular to the line $2x + y - 3 = 0$.

26. What are the intercepts and the slope of the line $3y - 15 = 0$?

27. Do the following equations represent lines that are parallel, perpendicular, or neither?
$$3x - y + 2 = 0$$
$$x + 3y - 7 = 0$$

28. Find the slope-intercept form of the equation of the line perpendicular to the line containing $(-2, -6)$ and $(8, -4)$ and passes through the point midway between them.

Section 9.3

29. If $ABCD$ is a parallelogram, what are the coordinates of C?

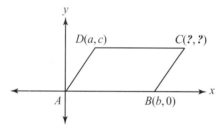

30. Using the figure from the previous exercise, if M is the midpoint of $\overline{DC}$ and N is the midpoint of $\overline{AB}$, use coordinate geometry and the slope formula to prove $ANMD$ is a parallelogram.

31. Supply the missing coordinates in the square, then prove the diagonals of a square are congruent using analytic geometry.

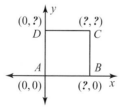

32. Prove using analytic geometry: The midpoint of the hypotenuse of a right triangle is equidistant from the three vertices of the triangle.

Chapter (9) Practice Test

For Exercises 1–5, use the equation $3x + 2y = 12$

1. Write the equation in slope-intercept form.

2. Name the x-intercept.

3. Name the y-intercept.

4. What is the slope of the line?

5. Graph the line.

6. Find the missing coordinate so that the ordered pair $(10, ?)$ is a solution to $y = 4x + 30$

7. Graph $y = x + 3$

8. Graph $y - 4 = 0$

9. Find the intercepts of the linear equation $x + y + 5 = 0$

For Exercises 9–13, two points $P(-2, 7)$ *and* $Q(4, -5)$ *are given.*

10. What is the slope of the line passing through P and Q?

11. What is the slope of a line perpendicular to $\overline{PQ}$?

12. What is the distance between P and Q?

13. What are the coordinates of the midpoint of the line segment joining P and Q?

14. Do the following equations represent lines that are parallel, perpendicular or neither?
$$2x - 5y + 7 = 0$$
$$-2x + 5y + 7 = 0$$

15. Find the general form of the equation of the line with slope -3 and passing through $(-2, 4)$.

16. Find the general form of the equation of the line passing through $(-1, 7)$ and $(4, -3)$.

17. Find the general form of the equation of the line passing through $(4, 5)$ and parallel to the line $2x + y - 4 = 0$.

18. Suppose the origin is the midpoint of a segment and one endpoint of the segment is $(5, -6)$. Find the coordinates of the other endpoint.

19. Complete the missing coordinates using only the variables given and the properties of the given isosceles trapezoid.

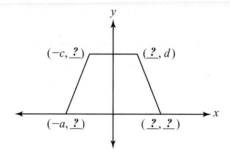

20. Using analytic geometry, prove: If the midpoint of one side of a rectangle is joined to the endpoints of the opposite side, an isosceles triangle is formed.

10

Introduction to Trigonometry

The word *trigonometry* is derived from Greek words that mean *three-angle measurement*, or *triangle measurement*. Historically, trigonometry was developed as a tool for finding the measurements of parts (sides and angles) of a triangle when other parts were known. As a result, it became indispensable in areas such as navigation, surveying, architecture, and astronomy. More recently, the study of trigonometry has expanded even further, but we will restrict our work here to triangle trigonometry. The following application is one we can solve using this basic trigonometry. Its solution appears in Section 10.4, Example 2.

The use of a scientific calculator is assumed.

AN APPLICATION

A small forest fire is sighted due south of a fire lookout tower on Woody Mountain. From a second tower, located 11.0 mi due east of the first, a ranger observes that the fire is 51.2° west of due south. How far is the fire from the tower on Woody Mountain?

10.1 Sine and Cosine Ratios

OBJECTIVES

1. Define the sine and cosine of an acute angle.
2. Use sine and cosine in special right triangles.
3. Find the measures of missing sides in a right triangle.

In Chapter 2, we defined a right triangle as a triangle that contains a right angle (90°); and we said the side opposite the right angle is called the hypotenuse of the triangle. Because the sum of the measures of the angles of a triangle is 180°, we determined that the remaining two angles are complementary acute angles. The sides opposite these angles are called the *legs* of the triangle.

From now on, we will label the angles of a triangle with capital letters. The sides of the triangle will be indicated by lowercase letters corresponding to the capital letter of the angle *opposite* the side. For example, the right triangle with angles *A*, *B*, and *C* and sides *a*, *b*, and *c* is shown in Figure 10.1. As in Chapter 5, we will identify the right angle with the letter *C*. The leg **opposite** *A* is *a* and the leg **adjacent** to *A* is *b*.

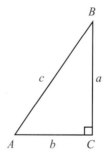

$m\angle A + m\angle B = 90°$
$m\angle C = 90°$
c is the hypotenuse
a is opposite $\angle A$
b is adjacent to $\angle A$
b is opposite $\angle B$
a is adjacent to $\angle B$

Figure 10.1 Right Triangle

In Chapter 5, we learned that similar triangles are triangles that have two congruent corresponding angles (AA theorem). As a result, two right triangles are similar if an acute angle of one is congruent to an acute angle of the other. Also, we know that corresponding sides of similar triangles are proportional. Consider the two similar right triangles in Figure 10.2 where $\triangle ABC \sim \triangle A'B'C'$. It follows that

$$\frac{a}{c} = \frac{a'}{c'} \text{ and } \frac{b}{c} = \frac{b'}{c'}.$$

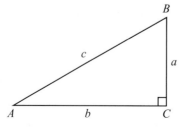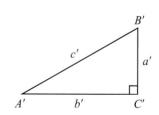

Figure 10.2 Similar Right Triangles

OBJECTIVE 1 **Define the sine and cosine of an acute angle.** Because these two ratios are independent of the size of the right triangle containing $\angle A$, we can use them to define two trigonometric ratios.

DEFINITION: Sine and Cosine Ratios

Let $\angle A$ be an acute angle of the right triangle.

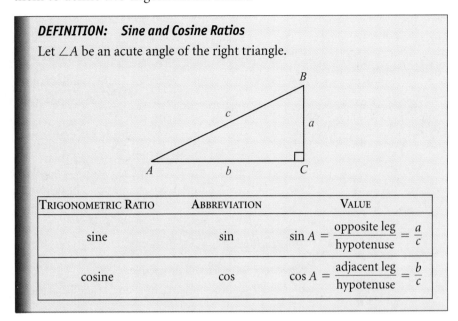

TRIGONOMETRIC RATIO	ABBREVIATION	VALUE
sine	sin	$\sin A = \dfrac{\text{opposite leg}}{\text{hypotenuse}} = \dfrac{a}{c}$
cosine	cos	$\cos A = \dfrac{\text{adjacent leg}}{\text{hypotenuse}} = \dfrac{b}{c}$

┌**Note** Sine and its abbreviation, sin, are pronounced like the word "sign."

Figure 10.3 is an extension of the above definition.

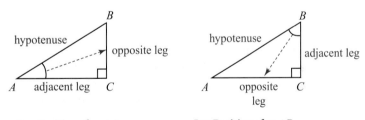

Leg Positions for $\angle A$ **Leg Positions for $\angle B$**

Figure 10.3

EXAMPLE 1 Determine $\sin A$, $\cos A$, $\sin B$, and $\cos B$ in Figure 10.4 given the legs of the right triangle are 3 and 4. Leave the answers in fraction form.

We must first find the value of c. We can use the Pythagorean Theorem.

$$c^2 = a^2 + b^2$$

In this example, $a = 3$ and $b = 4$.

$$c^2 = (3)^2 + (4)^2$$
$$= 9 + 16 = 25$$
$$c = 5$$

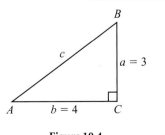

Figure 10.4

⌐**Note** The negative value was rejected because a distance cannot be negative.
└ Throughout the remainder of this chapter, the negative value will be omitted.

We now know the values of a, b, and c and can write the trigonometric ratios.

$$\sin A = \frac{\text{opposite leg}}{\text{hypotenuse}} = \frac{a}{c} = \frac{3}{5}$$

$$\cos A = \frac{\text{adjacent leg}}{\text{hypotenuse}} = \frac{b}{c} = \frac{4}{5}$$

$$\sin B = \frac{\text{opposite leg}}{\text{hypotenuse}} = \frac{b}{c} = \frac{4}{5}$$

$$\cos B = \frac{\text{adjacent leg}}{\text{hypotenuse}} = \frac{a}{c} = \frac{3}{5}$$

 Technology Connection

Geometry software will be needed.

1. Construct two similar right triangles and label them as in the figure.

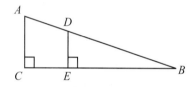

2. Measure $\angle ABC$, AB, AC, DB, and DE.
3. In $\triangle ACB$, measure the following ratio *in relation to* $\angle ABC$:

$$\frac{\text{opposite leg}}{\text{hypotenuse}} = \frac{AC}{AB}$$

4. In $\triangle DEB$, measure the following ratio *in relation to* $\angle ABC$:

$$\frac{\text{opposite leg}}{\text{hypotenuse}} = \frac{DE}{DB}$$

5. What do you notice about these ratios? [Note: Rounding by the computer may cause some unavoidable discrepancies.]
6. Drag point A vertically to change the measure of $\angle ABC$. [Note: $\triangle ACB$ and $\triangle DEB$ are still right triangles.] What happens to the ratios from steps 3 and 4?

Do the above results support the fact that the size of the triangle does not affect the

$\frac{\text{opposite leg}}{\text{hypotenuse}}$ ratio of $\angle ABC$? Notice the ratios stayed the same as point A was dragged.

Will the same results occur for the $\frac{\text{adjacent leg}}{\text{hypotenuse}}$ ratio in relation to $\angle ABC$?

OBJECTIVE 2   Use sine and cosine in special right triangles.

EXAMPLE 2   If one leg of a 45°-45°-90° triangle measures 1 inch, use Theorem 5.19 to find the lengths of the two remaining sides. Using these measurements, find sin 45° and cos 45°.

From Theorem 5.19, the other leg is 1 inch and the hypotenuse is $\sqrt{2}$ inches. See Figure 10.5.

$$\sin 45° = \frac{\text{opposite leg}}{\text{hypotenuse}} = \frac{1}{\sqrt{2}} = \frac{\sqrt{2}}{2};$$

$$\cos 45° = \frac{\text{adjacent leg}}{\text{hypotenuse}} = \frac{1}{\sqrt{2}} = \frac{\sqrt{2}}{2}$$

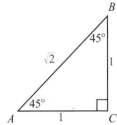

Figure 10.5

When a football player kicks a field goal, the angle of the trajectory and the velocity of the ball determine the distance the ball will travel. Trigonometry can be used to determine the best angle for the ball to be kicked at a constant velocity to achieve the maximum distance.

Note Before calculators were available, tables of values were used to find trigonometric ratios for angles. Now we can use calculators to determine the ratio. To find a trigonometric ratio with a calculator, first make sure the calculator is in degree mode. There may be a key that says DRG or a key that says MODE (if pressed, a menu will give you options). Read your instruction manual about the operation of your calculator. If sin 45° is to be evaluated on a *scientific* calculator, press

4 5 sin .

If sin 45° is to be evaluated on a *graphing calculator,* press

sin 4 5 Enter .

The order of the keystrokes depends on the individual calculator.

From Example 2, sin 45° = $\frac{\sqrt{2}}{2}$. On the calculator, sin 45° ≈ 0.707106781.

This is an irrational number that has been rounded. Confirm $\frac{\sqrt{2}}{2}$ ≈ 0.707106781 on your calculator.

Student Activity

The following equipment will be needed for this activity: a protractor, ruler, and a copy of the table in step 2 from your instructor.
1. Draw $\triangle ABC$, where $m\angle A = 30°$, $m\angle B = 60°$, and $m\angle C = 90°$. Make the triangle very large and use the protractor to ensure the accuracy of the angle measures.   *continued*

2. Insert the appropriate ratio $\left(\dfrac{BC}{AB} \text{ or } \dfrac{AC}{AB} \right)$ into the following table.

	(STEP 2) RATIO	(STEP 4) QUOTIENT	(STEP 5) CALCULATOR VALUE
sin 30°			
cos 30°			
sin 60°			
cos 60°			

3. Measure the sides of the triangle with the ruler to the nearest millimeter. Measure carefully.

4. Using the measurements from step 3, calculate the ratios and put them into the table. Write the answer as a decimal approximation rounded to three decimal places.

5. Find sin 30°, cos 30°, sin 60°, and cos 60° using a calculator. Insert those values into the table.

Compare the last two columns of the table. They should match because of the definition of sine and cosine of an angle. If the numbers do not match exactly, can you explain why?

EXAMPLE 3

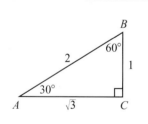

Figure 10.6

If the short leg of a 30°-60°-90° triangle measures 1 inch, use Theorem 5.20 to find the lengths of the two remaining sides. Using these measurements, (not a calculator), find sin 30°, cos 30°, sin 60°, and cos 60°.

From Theorem 5.20, if the short leg is 1 inch, the longer leg is $\sqrt{3}$ inches and the hypotenuse is 2 inches. See Figure 10.6.

$$\sin 30° = \frac{\text{opposite leg}}{\text{hypotenuse}} = \frac{1}{2}; \cos 30° = \frac{\text{adjacent leg}}{\text{hypotenuse}} = \frac{\sqrt{3}}{2}$$

$$\sin 60° = \frac{\text{opposite leg}}{\text{hypotenuse}} = \frac{\sqrt{3}}{2}; \cos 60° = \frac{\text{adjacent leg}}{\text{hypotenuse}} = \frac{1}{2}$$

PRACTICE EXERCISE 1

Using Figure 10.7, express the sin A, cos A, sin B, and cos B as fractions.

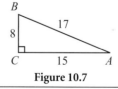

Figure 10.7

ANSWERS ON PAGE 463

OBJECTIVE 3 **Find the measures of missing sides in a right triangle.** Suppose the measure of an acute angle of a right triangle and the measure of one side of the triangle are known. By using trigonometry, the measure of the missing side of the triangle can be found.

EXAMPLE 4   State two different equations that could be used to find x, correct to one decimal place given the facts in Figure 10.8.

The first possible equation is

$$\sin 29° = \frac{x}{10}$$

$10 \sin 29° = x$ Multiply both sides by 10

$\qquad x \approx 4.8$ Rounded to one decimal place

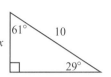

The second possible equation is

$$\cos 61° = \frac{x}{10}$$ Multiply both sides by 10

$10 \cos 61° = x$

$\qquad x \approx 4.8$ Rounded to one decimal place

Figure 10.8

Answers to Practice Exercise

1. $\sin A = \dfrac{\text{opposite leg}}{\text{hypotenuse}} = \dfrac{8}{17}$ $\cos A = \dfrac{\text{adjacent leg}}{\text{hypotenuse}} = \dfrac{15}{17}$

$\sin B = \dfrac{\text{opposite leg}}{\text{hypotenuse}} = \dfrac{15}{17}$ $\cos B = \dfrac{\text{adjacent leg}}{\text{hypotenuse}} = \dfrac{8}{17}$

 **10.1 Exercises**

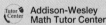 FOR EXTRA HELP: Student's Solutions Manual   Tutor Center   Addison-Wesley Math Tutor Center

In Exercises 1–6, two sides of right $\triangle ABC$ ($\angle C$ is the right angle) are given. Find the sin A, cos A, sin B, and cos B. Leave the answer in fraction form.

1. $a = 5$ and $b = 12$ 2. $a = 3$ and $b = 10$
3. $a = 7$ and $c = 15$ 4. $b = 7$ and $c = 20$
5. $a = 0.6$ and $b = 0.8$ 6. $a = \sqrt{5}$ and $b = \sqrt{11}$

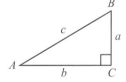

In Exercises 7–10, $\angle A$ is an acute angle in a right triangle. (See the figure for Exercises 1–6.) Use the given trigonometric ratio to find either the sine or cosine of $\angle A$.

7. $\sin A = \dfrac{5}{7}$

8. $\cos A = \dfrac{1}{3}$

9. $\cos A = \dfrac{\sqrt{33}}{7}$

10. $\sin A = \dfrac{1}{4}$

In Exercises 11–16, use your calculator to find each trigonometric ratio, correct to four decimal places.

11. $\sin 65°$ **12.** $\cos 5°$ **13.** $\cos 38°$

14. $\sin 12.4°$ **15.** $\cos 83.8°$ **16.** $\cos 57.1°$

In Exercises 17–20, state two different equations that could be used to find x.
Find x to one decimal place.

17.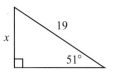

18.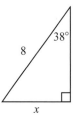

19.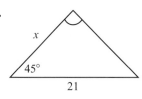

20.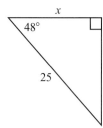

In Exercises 21–24, find x and y correct to one decimal place.

21.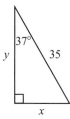

22.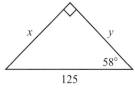

23.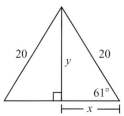

24.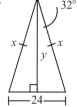

25. Suppose, in a right triangle, the length of one leg and the measure of one acute angle are given. If you need to find the length of the hypotenuse, how do you decide which trigonometric ratio to use?

(10.2) Tangent Ratio

OBJECTIVES

1. Define the tangent of an acute angle.
2. Use tangents in special right triangles.
3. Find an angle measure given a trigonometric ratio.

OBJECTIVE 1   **Define the tangent of an acute angle.**   Another trigonometric ratio is the **tangent ratio**. As in Section 10.1, we define this ratio for an acute angle in a right triangle.

DEFINITION: Tangent Ratio

Let $\angle A$ be an acute angle of the right triangle.

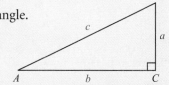

TRIGONOMETRIC RATIO	ABBREVIATION	VALUE
tangent	tan	$\tan A = \dfrac{\text{opposite leg}}{\text{adjacent leg}} = \dfrac{a}{b}$

EXAMPLE 1 Determine tan A and tan B using the information in Figure 10.9. Leave the answer in fraction form.

First we must find the value of b by using the Pythagorean Theorem.

$$a^2 + b^2 = c^2$$
$$9 + b^2 = 25$$
$$b^2 = 16$$
$$b = 4$$

Thus, the tangent ratios are

$$\tan A = \frac{\text{opposite leg}}{\text{adjacent leg}} = \frac{3}{4}$$

$$\tan B = \frac{\text{opposite leg}}{\text{adjacent leg}} = \frac{4}{3}$$

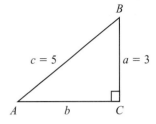

Figure 10.9

PRACTICE EXERCISE 🔟

Using the information in the given figure, find tan A and tan B. Express the answer as a ratio.

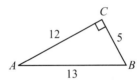

ANSWER ON PAGE 468

 Student Activity

A sheet of graph paper and a straightedge are needed for this activity.
1. Using the lines on the graph paper as a guide, draw a large square. Label it as in the figure.
2. Label the lengths of the four sides.
3. Draw $\overline{AC}$.
4. From the properties of a square, state $m\angle ACD$. What special right triangle is $\triangle ACD$?
5. Using your measurements, state the ratio $\frac{AD}{CD}$.
6. Find tan $45°$ on a calculator.

Compare steps 5 and 6. Compare your square to others in the room. Does the size of the square change the value of tan $45°$?

OBJECTIVE 2 **Use tangents in special right triangles.** Following a similar idea from Section 10.1, we will find the tangent of $45°$, $30°$, and $60°$ angles. Use Figures 10.10 and 10.11 on page 466 (taken from Section 10.1) to find the tangent values of $45°$, $30°$, and $60°$.

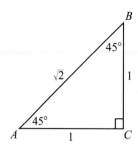

Figure 10.10 45°-45°-90° Triangle

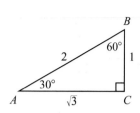

Figure 10.11 30°-60°-90° Triangle

$$\tan 45° = \frac{\text{opposite leg}}{\text{adjacent leg}} = \frac{1}{1} = 1 \qquad \tan 30° = \frac{\text{opposite leg}}{\text{adjacent leg}} = \frac{1}{\sqrt{3}} = \frac{\sqrt{3}}{3}$$

$$\tan 60° = \frac{\text{opposite leg}}{\text{adjacent leg}} = \frac{\sqrt{3}}{1} = \sqrt{3}$$

SUMMARY

x	45°	30°	60°
$\sin x$	$\dfrac{\sqrt{2}}{2}$	$\dfrac{1}{2}$	$\dfrac{\sqrt{3}}{2}$
$\cos x$	$\dfrac{\sqrt{2}}{2}$	$\dfrac{\sqrt{3}}{2}$	$\dfrac{1}{2}$
$\tan x$	1	$\dfrac{\sqrt{3}}{3}$	$\sqrt{3}$

Similar to Section 10.1, the calculator can be used to find the tangent of an acute angle. Make sure the calculator is in degree mode.

EXAMPLE 2 Use a calculator to find tan 35° correct to four decimal places.

$$\tan 35° \approx 0.700207538 \approx 0.7002 \qquad \text{correct to four decimal places}$$

Student Activity

A protractor and ruler are needed for this activity.
To ensure understanding of the tangent ratio when it is done on a calculator as in Example 2, do the following:
1. Use a protractor to draw a 35° angle as accurately as possible. This will be one angle of a right triangle.
2. What will be the measures of the other two angles? You are correct if you said 55° and 90°.

3. Complete the drawing of the right triangle as accurately as possible.
4. Carefully and accurately measure the lengths of the two legs to the nearest millimeter.
5. Use those measurements to write the tangent ratio of 35° as a fraction, then convert it to a decimal.
6. Compare the answer in step 5 to the value of tan 35° found in Example 2.
7. Both values should have been close to 0.7002 because the value found in Example 2 is the same as the ratio of the opposite leg to the adjacent leg of *any* right triangle containing a 35° angle.

OBJECTIVE 3   **Find an angle measure given a trigonometric ratio.**   We have been finding a trigonometric ratio for a given angle. Suppose we wanted to find an angle, given the value of one of the trigonometric ratios. This process is the reverse of what was shown in the previous examples. To accomplish this using the calculator, the $\boxed{\sin^{-1}}$, $\boxed{\cos^{-1}}$ or $\boxed{\tan^{-1}}$ key will be needed. Some scientific calculators have an $\boxed{\text{INV}}$ key that is used in combination with the $\boxed{\sin}$, $\boxed{\cos}$, or $\boxed{\tan}$ key, thus, sin⁻¹ means inverse sine.

EXAMPLE 3   (a) Find A if sin A = 0.5948 to the nearest tenth of a degree.

Set the calculator to degree mode and for a scientific calculator use the following sequence of keystrokes:

$\boxed{0.5948}$ $\boxed{\sin^{-1}}$ The answer is 36.498375 . . . ° ≈ 36.5°

For a graphing calculator, use the following keystrokes:

$\boxed{\sin^{-1}}$ $\boxed{0.5948}$ $\boxed{\text{Enter}}$ The answer is the same.

Note Calculators work differently, so be sure to understand the proper keystrokes for your calculator.

(b) Find A if tan A = 0.1263, to the nearest tenth of a degree.
$\tan^{-1} 0.1263 \approx 7.2°$

PRACTICE EXERCISE 2

Find A if cos A ≈ 0.90630779 to the nearest degree.

ANSWER ON PAGE 468

EXAMPLE 4 For Figure 10.12, name the ratio that should be used to find:

(a) a, if $\angle A$ and c are known

(b) $\angle B$, if a and c are known

(c) $\angle A$, if a and b are known

(d) b, if a and $\angle B$ are known

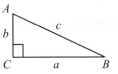

Figure 10.12

Answers

(a) sine, since $\sin A = \dfrac{a}{c}$

(b) cosine, since $\cos B = \dfrac{a}{c}$

(c) tangent, since $\tan A = \dfrac{a}{b}$

(d) tangent, since $\tan B = \dfrac{b}{a}$

Technology Connection

A spreadsheet program, graphing calculator, or scientific calculator will be needed.

What happens when you try to find tan 90° on your calculator? Why does this happen? Investigate the question by using a spreadsheet or the table function on a graphing calculator using $y = \tan x$ as the equation.

Another way to investigate is to copy and complete this table using a scientific calculator.

x	87°	88°	89°	89.5°	89.75°	89.8°	89.9°	90°
$\tan x$								

Answers to Practice Exercise

1. $\tan A = \dfrac{\text{opposite leg}}{\text{adjacent leg}} = \dfrac{5}{12}$ **2.** 25°

$\tan B = \dfrac{\text{opposite leg}}{\text{adjacent leg}} = \dfrac{12}{5}$

10.2 Exercises

FOR EXTRA HELP: Student's Solutions Manual 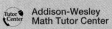 Tutor Center Addison-Wesley Math Tutor Center

In Exercises 1–6, two sides of a right triangle (see the figure) are given. Find tan A and tan B. Leave the answer in fraction form.

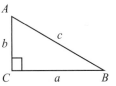

1. $a = 5$ and $b = 12$

2. $a = 3$ and $b = 10$

3. $a = 7$ and $c = 15$

4. $b = 7$ and $c = 20$

5. $a = 0.6$ and $b = 0.8$

6. $a = \sqrt{5}$ and $b = \sqrt{11}$

In Exercises 7–10, using the figure from Exercises 1–6 and the given trigonometric ratio, find the other two ratios.

7. $\tan A = \dfrac{\sqrt{5}}{2}$

8. $\cos A = \dfrac{40}{41}$

9. $\tan A = 7$

10. $\sin A = \dfrac{3}{\sqrt{13}}$

In Exercises 11–14, use a calculator to find each ratio correct to four decimal places.

11. $\tan 38°$ **12.** $\tan 57.1°$ **13.** $\tan 81.43°$ **14.** $\sin 27.5°$

In Exercises 15–20, use a calculator to find the measure of $\angle A$, to the nearest tenth of a degree.

15. $\sin A = 0.1258$ **16.** $\cos A = 0.8018$ **17.** $\tan A = 0.9301$
18. $\sin A = 0.9511$ **19.** $\cos A = 0.1022$ **20.** $\tan A = 2.5053$

In Exercises 21–22, use the tangent ratio to find the lengths of the indicated sides to the nearest tenth of a unit.

21.

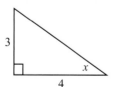

22.

In Exercises 23–28, use the sine, cosine, or tangent ratio to find the measures of the indicated angles to the nearest tenth of a degree.

23.

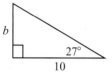

24.

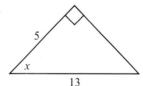

25.

26.

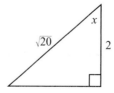

27.

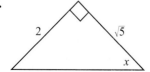

28.

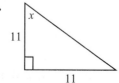

29.

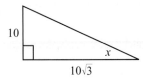

30.

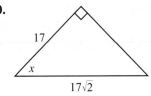

31.

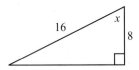

32.

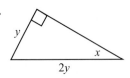

33. In the given circle, the radius is 14 inches. Find the length of chord $\overline{BC}$ to the nearest tenth of an inch.

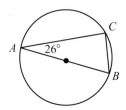

34. Find tan x, then find x to the nearest degree.

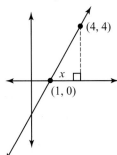

35.

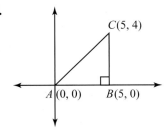

a. Find the slope of $\overline{AC}$

b. Find tan A in right $\triangle ABC$

 c. Make a conjecture about slope of $\overline{AC}$ and tan A. Will this always be true? Explain.

10.3 **Solving Right Triangles**

OBJECTIVES

1. Define solving a right triangle.
2. Solve right triangles.
3. Determine the area and perimeter of figures using trigonometry.

OBJECTIVE 1   Define solving a right triangle. The six parts of any triangle (three angles and three sides) can often be found when the measures of individual parts are given. This process is called **solving the triangle**. A right triangle can be solved when two of its sides are known or when one side and one acute angle are known. Simply knowing the angle measures of a triangle is not enough to solve the triangle because any similar triangle will have the same angles but, more than likely, different sides.

OBJECTIVE 2   Solve right triangles.

EXAMPLE 1

Solve the right triangle with $m\angle A = 26°$ and $b = 11$. See Figure 10.13. Round the answers to the nearest whole number.

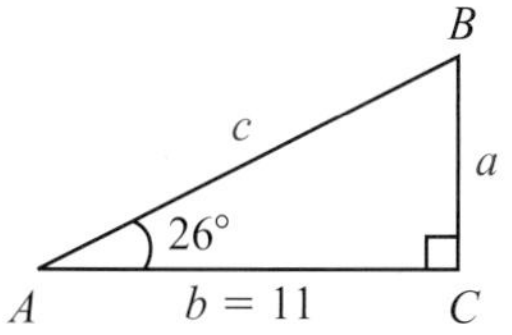

$\angle A$ and $\angle B$ are complementary. $m\angle B = 90° - 26° = 64°$
To find the missing sides, use trigonometry.

$$\tan A = \frac{\text{opposite leg}}{\text{adjacent leg}}$$

$$\tan 26° = \frac{a}{11} \qquad \text{Multiply both sides by 11}$$

$$11 \tan 26° = a$$

$$a \approx 5 \qquad \text{Rounded to nearest whole number}$$

To find c, we could use the Pythagorean theorem, but instead we will use the cosine ratio. *The most accurate answer is obtained by using only the given parts rather than approximated calculated parts.*

$$\cos A = \frac{\text{adjacent leg}}{\text{hypotenuse}}$$

$$\cos 26° = \frac{11}{c} \qquad \text{Multiply both sides by c}$$

$$c(\cos 26°) = 11$$

$$c = \frac{11}{\cos 26°}$$

$$c \approx 12 \qquad \text{Rounded to nearest whole number}$$

Thus, $m\angle B = 64°$, $a \approx 5$, and $c \approx 12$.

Figure 10.13

(Figure: right triangle with vertices A, B, C; angle $26°$ at A; side $b = 11$ along AC; side a along CB; hypotenuse c from A to B; right angle at C.)

EXAMPLE 2   Solve the right triangle with $a = 3$ and $b = 7$. Round answers to the nearest whole number.

Figure 10.14 gives a reasonably accurate sketch of the triangle. We must find $m\angle A$, $m\angle B$, and c.

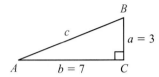

Figure 10.14

To find $m\angle A$: $\tan A = \dfrac{3}{7}$

$$\tan^{-1}\left(\dfrac{3}{7}\right) \approx 23° \qquad \text{Rounded to the nearest whole number}$$

Thus, to the nearest degree, $m\angle A \approx 23°$.

To find $m\angle B$:   Because $\angle A$ and $\angle B$ are complementary,

$$m\angle B = 90° - m\angle A \approx 90° - 23° \approx 67°.$$

$m\angle B$ can also be found by using, for example, $\tan B = \dfrac{7}{3}$, which also gives $m\angle B \approx 67°$.

To find c: We can find c with the Pythagorean Theorem. Remember, for the most accurate answer, we want to use the given parts.

$$c^2 = a^2 + b^2$$
$$= 3^2 + 7^2$$
$$= 58$$
$$c = \sqrt{58} \approx 8$$

Thus, $m\angle A \approx 23°$, $m\angle B \approx 67°$, and $c \approx 8$.

PRACTICE EXERCISE ❶

Solve the right triangle in which $b = 10$ and $c = 16$. Round answers to the nearest whole number.

ANSWER ON PAGE 473

CAUTION   To achieve the most accurate answer, round only in the *final* calculation.

EXAMPLE 3   Solve the right triangle in which $c = 8.1$ cm and $m\angle B = 72.5°$. Also find the area of the triangle. Round answers to the nearest tenth.

To solve the triangle, find a, b, and $m\angle A$. See Figure 10.15.

To find $m\angle A$: $m\angle A = 90° - 72.5° = 17.5°$

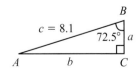

Figure 10.15

To find a: Remember to use the given parts.

$$\cos B = \frac{a}{c}; \cos 72.5° = \frac{a}{8.1}; a = (8.1)(\cos 72.5°) \approx 2.4 \qquad \text{Rounded to nearest tenth}$$

To find b: Remember to use the given parts.

$$\sin 72.5° = \frac{b}{8.1}; b = (8.1)(\sin 72.5°) \approx 7.7 \qquad \text{Rounded to nearest tenth}$$

To find the area: $A = \frac{1}{2}(base)(height)$, where $base = (8.1)(\cos 72.5°)$ and $a = (8.1)(\sin 72.5°)$

$$A = \frac{1}{2}[(8.1)(\cos 72.5°)][(8.1)(\sin 72.5°)]$$

$$A \approx 9.4 \qquad \text{Rounded to nearest tenth}$$

Note The rounded values for a and b were not used since they would not give the most accurate answer.

Thus, $m\angle A = 17.5°$, $a \approx 2.4$ cm, $b \approx 7.7$ cm, $A \approx 9.4$ cm².

OBJECTIVE 3   **Determine the area and perimeter of figures using trigonometry.**   The next example shows how to use trigonometry to find the area of a regular pentagon.

EXAMPLE 4   Find the area of a regular pentagon with a side measuring 6 inches. Round the answer to the nearest whole number. See Figure 10.16.

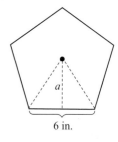

The formula for area of a regular polygon is $A = \frac{1}{2}ap$. The perimeter of the pentagon is $6(5) = 30$; therefore, $p = 30$. Each central angle measures $\frac{360}{5} = 72°$. The triangle drawn in the figure is isosceles, therefore, the other two angles of the triangle measure $\frac{180 - 72}{2} = \frac{108}{2} = 54°$.

Figure 10.16

Let $a =$ apothem in Figure 10.16. By Theorem 7.9, the apothem of a regular polygon bisects its respective side; thus, the leg adjacent to the 54° angle is 3 inches. See the figure to the right.

Use the tangent relationship to find the apothem.

$$\tan 54° = \frac{a}{3}; 3\tan 54° = a$$

Remember, don't round until the final calculation.

$$A = \frac{1}{2}(30)(3\tan 54°) \approx 62 \qquad \text{Rounded to the nearest whole number}$$

Thus, the area of the pentagon is about 62 in.².

Answers to Practice Exercise

1. $a \approx 12, m\angle A \approx 51°, m\angle B \approx 39°$

10.3   Exercises

Solve each right triangle in Exercises 1–18. Remember that $m\angle C = 90°$.
Round to the same number of decimal places as in the given information.

1. $a = 9$ and $m\angle A = 60°$ **2.** $b = 4$ and $m\angle B = 45°$ **3.** $c = 12$ and $m\angle A = 20°$

4. $c = 29$ and $m\angle B = 55°$ **5.** $a = 12$ and $c = 13$ **6.** $a = 6$ and $b = 8$

7. $b = 7$ and $c = 15$ **8.** $a = 19$ and $b = 48$ **9.** $a = 9.2$ and $c = 10.1$

10. $a = 4.3$ and $b = 6.1$ **11.** $m\angle A = 26.7°$ and $c = 12.0$ **12.** $m\angle A = 62.2°$ and $b = 7.3$

13. $b = 3.2$ and $m\angle B = 10.8°$ **14.** $c = 16.6$ and $m\angle B = 12.8°$ **15.** $m\angle A = 22.5°$ and $c = 28.7$

16. $m\angle B = 41.3°$ and $a = 0.8$ **17.** $c = 21.9$ and $m\angle B = 81.7°$ **18.** $a = 19.3$ and $m\angle A = 26.6°$

For Exercises 19–24, round the answer to the nearest tenth of a unit.

19. Find the perimeter of the given rectangle.

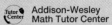

Exercise 19

20. Find the area of the given parallelogram.

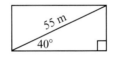

Exercise 20

21. Find the area of the right triangle.

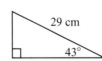

Exercise 21

22. Given a right circular cone with slant height 13 inches. The angle formed by the slant height and the radius is 46°. Find the height and lateral area of the cone.

23. Find the length of an apothem in a regular pentagon whose radius is 12 inches.

24. Find the area of the regular pentagon described in Exercise 23.

(10.4) Applications Involving Right Triangles

OBJECTIVES

1. Apply sine, cosine, and tangent ratios.
2. Define angle of elevation and angle of depression.
3. Apply angle of elevation and angle of depression.

OBJECTIVE 1 Apply sine, cosine, and tangent ratios. In this section we will investigate a variety of applied problems, all of which depend in some way on finding unknown parts of a right triangle. For the most part, we will use a process of *indirect measurement:* finding a particular length, distance, or angle without applying a measuring instrument such as a tape measure or protractor.

EXAMPLE 1 During harvest time, farmers will often make a cone-shaped pile of grain near their silos. The angle the grain makes with the ground depends on the type of grain because some have more or less friction "grain on grain" than others. This farmer knows that an angle of 27° is formed with the ground. He also approximated the radius of the pile of grain to be 20 feet. How high is the pile of grain? See Figure 10.17.

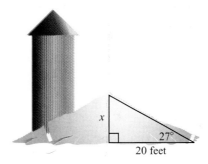

Figure 10.17

$$\tan 27° = \frac{x}{20}$$
$$20 \tan 27° = x$$
$$x \approx 10 \text{ feet}$$

The height of the pile of grain is about 10 feet. The answer is rounded to a whole number because the radius was estimated as whole number.

Forest rangers can use right-triangle trigonometry to locate forest fires. The following example solves the applied problem given in the introduction.

EXAMPLE 2   A small forest fire is sighted due south of a fire lookout tower on Woody Mountain. From a second tower, located 11.0 mi due east of the first, a ranger observes that the fire is 51.2° west of due south. How far is the fire from the tower on Woody Mountain? Round the answer to the nearest tenth of a mile.

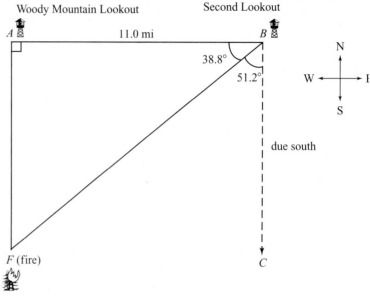

Figure 10.18

In Figure 10.18, $m\angle CBF$ is 51.2° and $\angle FAB$ is a right angle. In $\triangle ABF$, $m\angle ABF = 38.8°$ because it is the complement of $\angle CBF$. AF is the distance from the fire to the lookout on Woody Mountain.

$$\tan 38.8° = \frac{AF}{11.0}$$
$$AF = (11.0)\tan 38.8°$$
$$AF \approx 8.844227$$

Thus, the fire is approximately 8.8 mi from the lookout tower on Woody Mountain.

PRACTICE EXERCISE 1

Find the distance of the fire in Example 2 from the second lookout tower. Round the answer to the nearest tenth of a mile.

ANSWER ON PAGE 478

EXAMPLE 3   An airplane leaves the East Coast of the United States and flies for two hours at 350 km/hr in a direction that is 32.7° to the east of due north. Assuming that the East Coast is a straight north-south line, how far is the airplane from the coastline? Round the answer to the nearest whole number.

Figure 10.19

The sketch in Figure 10.19 shows the essential elements of the problem. AB is 700 km because 350 km/hr for 2 hr gives

$$\text{distance} = (\text{rate})(\text{time})$$
$$= (350)(2)$$
$$= 700.$$

We must find BC. Let $x = BC$.

$$\sin 32.7° = \frac{x}{700}$$
$$x = (700)\sin 32.7°$$
$$x \approx 378.16822$$

The airplane is approximately 378 km from the coast.

OBJECTIVE 2   **Define angle of elevation and angle of depression.**   Suppose a search-and-rescue mission is taking place using a helicopter and a rescue ship. If a crew member on the ship looks up at the helicopter, then the angle from the horizontal *up* to the line of sight is called the **angle of elevation**. An angle of elevation is an acute angle formed by the line of sight and a horizontal line when looking up. If the pilot in the helicopter is looking down to the crew member on the ship, then the angle from the horizontal *down* to the line of sight is called the **angle of depression**. The angle of depression is an acute angle formed by the line of sight and a horizontal line when looking down. See Figure 10.20.

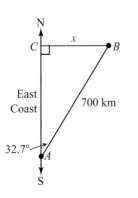

Archimedes (287–212 B.C.)

Archimedes used mathematics to solve many practical problems of his day. He frequently used geometry in his scientific studies as well as in the design of weapons. His approximation of π as being between $3\frac{10}{71}$ and $3\frac{1}{7}$ was particularly accurate for his time.

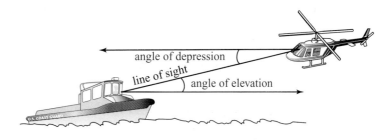

Figure 10.20

Notice that the angle of elevation from the crew member of the ship looking up to the helicopter is the same as the angle of depression from the pilot of the helicopter looking down to the crew member on the ship. The line of sight can form the hypotenuse of a right triangle.

OBJECTIVE 3   Apply angle of elevation and angle of depression.

EXAMPLE 4   You are in the rescue ship and it is about 1000 feet from the stranded boat. The helicopter is hovering about 170 feet directly above the stranded boat. What is the angle of elevation from the rescue ship to the helicopter? Round the answer to the nearest degree.

Start by drawing a picture of the facts. (See Figure 10.21.)

$$\tan x = \frac{170}{1000}$$

$$x = \tan^{-1}(170/1000)$$

$$x \approx 10° \qquad \text{Rounded to nearest degree}$$

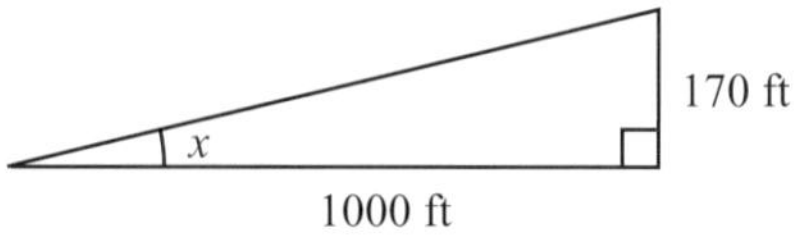

170 ft

1000 ft

Figure 10.21

The angle of depression is about 10°.

Answer to Practice Exercise

1. 14.1 mi

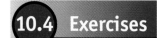

10.4 Exercises   *FOR EXTRA HELP:*   📖 Student's Solutions Manual   🎧 Addison-Wesley Math Tutor Center

For Exercises 1–3, use the given figure, where *ACEF* is a rectangle. Give the number of the angle described and whether it is an angle of elevation or an angle of depression.

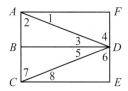

1. You are at point *D* looking at point *A*.
2. You are at point *D* looking at point *C*.
3. You are at point *A* looking at point *D*.

For Exercises 4–16, round the answers to the nearest tenth of a unit.

4. Sue is flying a kite with 150 feet of string. The kite flies at an angle of elevation of 50°. Find the kite's height.

5. A lighthouse stands atop a 200-foot cliff (see figure). The lighthouse is 50 feet tall. Suppose the angle of elevation from the ship to the top of the lighthouse is 10°. What is the distance from the base of the cliff to the ship?

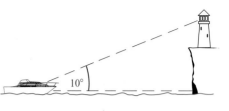

Exercise 5

6. An Olympic-size swimming pool is 50 meters long. Suppose there is a 10-meter-high diving platform at one end of the pool. If you are sitting on the deck looking at the platform from the other end of the pool, what is the angle of elevation?

7. The height of the Washington Monument is 555 feet. If Teaha is standing 100 feet from the monument's base, find the angle of elevation.

8. A machine part is in the shape of an equilateral triangle with an altitude of length 10.8 cm. Find (a) its perimeter and (b) its area using trigonometry.

9. When the sun is at an elevation of 49° above level ground (see the figure below) a pole casts a shadow 40 ft in length. What is the height of the pole?

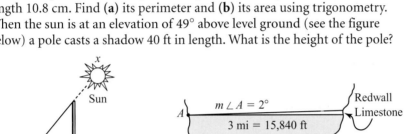

Exercise 9   Exercise 10

10. From a point in the Grand Canyon 3 mi from the base of the Redwall Limestone but on the same horizontal plane, a geologist sights the top of the Redwall at an angle of 2° (see the figure above). How many feet thick is the Redwall Limestone formation at this point?

11. Devil's Tower is a rock formation. Jeremiah is standing 1200.0 feet from the base of the formation. The angle with the ground to the top of the rock is 35.8°. How high is the tower?

12. A forest ranger in a 100-foot-tall fire tower sees a fire on the forest floor. The angle of depression from the tower to the fire is 84°. How far from the base of the tower is the forest fire?

13. Fire tower Bravo is located 8.0 miles due south of fire tower Spirit. A fire is located due east of Spirit and 31.7° east of due north from Bravo. How far is the fire from Bravo?

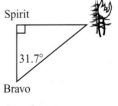

Exercise 13

14. Phoenix is 140.0 miles due south of Flagstaff, and Winslow is 60.0 miles due east of Flagstaff. If a pilot were to fly from Phoenix to Winslow, how many degrees east of due north should she head?

15. A chairlift at a ski resort makes an average angle of 17.5° with the horizontal ground at its base. If the vertical rise is 800 meters, what is the approximate length of the ride to the top of the lift?

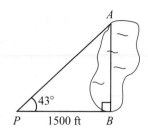

Exercise 16

16. A surveyor wants to know the distance across the lake from points A to B. He is standing at point P, 1500 feet from point B. He has measured $m\angle APB = 43°$ as shown in the figure. How far is it from A to B?

17. A playground is in the shape of an isosceles triangle with vertex angle 90° and equal sides 82 feet. Rounding the answers to the nearest foot, find the perimeter and area of the playground.

18. A holding area for cattle is in the shape of an isosceles triangle with a vertex angle of 90° and a base of 22 meters. Rounding the answers to the nearest meter, find the perimeter and area of the holding area.

19. A large plot of land is in the shape of an equilateral triangle with sides measuring 6.2 miles. Rounding the answers to the nearest mile, find the perimeter and area of the plot of land.

20. A ship sails 30.0 miles in a direction of 37.7° west of due north. How far north has it sailed? How far west has it sailed? Round the answers to the nearest tenth of a mile.

Chapter (10) Review

Key Terms and Symbols

10.1 opposite side
adjacent side
sine ratio (sin)
cosine ratio (cos)

10.2 tangent ratio (tan)

10.3 solving a triangle

10.4 angle of elevation
angle of depression

Chapter Formulas

$$\text{sine ratio} = \frac{\text{opposite leg}}{\text{hypotenuse}}$$

$$\text{cosine ratio} = \frac{\text{adjacent leg}}{\text{hypotenuse}}$$

$$\text{tangent ratio} = \frac{\text{opposite leg}}{\text{adjacent leg}}$$

Review Exercises

Section 10.1

In Exercises 1 and 2, two sides of a right triangle are given, find the sin A, cos A, sin B, and cos B. Leave the answers in fraction form.

1. $a = 16, b = 12$ **2.** $a = 15, c = 20$

3. If $\angle B$ is an acute angle in a right triangle and $\cos B = \frac{1}{3}$, find $\sin B$.

In Exercises 4 and 5, use your calculator to find each trigonometric ratio correct to four decimal places.

4. $\sin 21°$ **5.** $\cos 65.8°$

6. Find x and y correct to one decimal place.

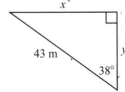

Section 10.2

7. If $\angle A$ is an acute angle in a right triangle and $\tan A = \frac{20}{15}$, find $\sin A$ and $\cos A$. Leave the answers in fraction form.

8. Use your calculator to find $\tan 82.1°$ correct to four decimal places.

In Exercises 9–11, use your calculator to find $m\angle A$ to the nearest tenth of a degree.

9. $\sin A = 0.2196$ **10.** $\cos A = 0.9108$ **11.** $\tan A = 1.6432$

In Exercises 12 and 13, use the sine, cosine, or tangent ratio to find the measures of the indicated angles to the nearest degree.

12.

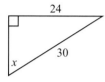

13.

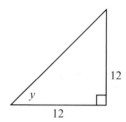

Section 10.3

Solve each right triangle in Exercises 14–19. Remember $m\angle C = 90°$. Round the answers to the same number of decimal places as the given information.

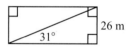

14. $b = 6$ and $m\angle A = 30°$ **15.** $c = 12$ and $m\angle B = 42°$ **16.** $a = 2.4$ and $m\angle A = 72.3°$

17. $a = 7$ and $b = 11$ **18.** $b = 12$ and $c = 20$ **19.** $a = 6.2$ and $c = 9.6$

Section 10.4

For Exercises 20–25 round the answers to the nearest tenth of a unit.

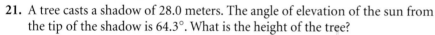

20. Find the area of the given figure.

Exercise 20

21. A tree casts a shadow of 28.0 meters. The angle of elevation of the sun from the tip of the shadow is 64.3°. What is the height of the tree?

22. A forest fire is sighted due west of outpost Alpha. From outpost Bravo, 5.2 miles due south of Alpha the angle of the fire is 51.7° west of due north. How far is the fire from outpost Bravo? From outpost Alpha?

23. Jon is in Florida watching the launch of a rocket. He is at the visitor's site, 3 miles from the launch pad. The announcer at the visitor's site says the rocket is about 2.3 miles high. What is the angle of elevation from Jon to the rocket?

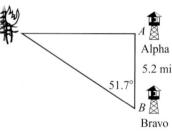

Exercise 22

24. A guy wire is stretched from the top of a tower to 10 yards from the base of the tower. The wire makes a 65° angle with the ground. Find the length of the wire.

25. The distance between the two levels in the MetroCenter Shopping Mall is 16.8 feet, and the angle that the escalator makes with the horizontal is 32.2°. How far would a person travel while riding from level 1 to level 2?

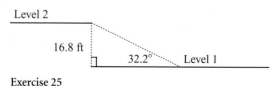

Exercise 25

Chapter **10** Practice Test

1. Find the three trigonometric ratios of $\angle A$ in the right triangle with $a = 5$, $c = 9$, and $m\angle C = 90°$. Leave answers in fraction form.

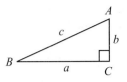

Exercise 1

2. Using the figure from Exercise 1, find the three trigonometric ratios of $\angle B$ in the right triangle with $a = 5$, $c = 9$, and $m\angle C = 90°$. Leave answers in fraction form.

3. If $\angle A$ is an acute angle in a right triangle and $\sin A = \dfrac{2}{5}$, find $\cos A$ and $\tan A$. Leave answers in fraction form.

4. Find the three trigonometric ratios of $68.9°$, correct to four decimal places.

5. Using the figure, find x and y, correct to the nearest tenth of an inch.

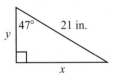

Exercise 5

6. Using the figure, find x and y, correct to the nearest tenth of a millimeter.

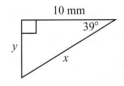

Exercise 6

7. If $\cos A = 0.1096$, find $m\angle A$ to the nearest tenth of a degree.

8. Solve the right triangle with $a = 1.1$ and $c = 4.5$. Round the answers to the nearest tenth of a unit.

9. Solve the right triangle with $b = 15$ and $m\angle A = 38°$. Round the answers to the nearest whole unit.

10. Find the area of the given rectangle, rounded to the nearest tenth of a centimeter.

11. At an altitude of 2500 feet, the engine on a plane suddenly fails. What angle of glide with the horizontal (see the figure) is needed to reach an airport runway 3 miles away? Remember, there are 5280 feet in a mile. Round the answer to the nearest degree.

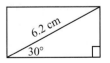

Exercise 10

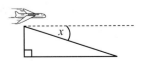

Exercise 11

12. Brian's dog, Charlie, is very old and has a hard time getting into the car. Brian is planning to build a ramp for Charlie. Brian wants the ramp to slant no more than $15°$. If the vertical distance from the ground into the car is 9.5 inches, how long should the ramp be? Round the answer to the nearest tenth of an inch.

Chapters (8-10) Cumulative Review

Chapter 8

1. Determine the surface area and volume of a right prism with a rectangular base measuring 5.4 cm by 2.3 cm and a height of 3.4 cm.

2. Determine the surface area and volume of a regular square pyramid, each edge measuring 6 ft and the height of the pyramid is 4 ft.

3. A can of lemonade has a diameter of 2.5 inches and a height of 5 inches. How much lemonade is in the can? Round the answer to the nearest tenth of a unit.

4. Find the surface area of a square pyramid used for a store display. An edge of the square measures 12 inches and the height of the pyramid is 8 inches.

5. A perfectly spherical scoop of ice cream with a radius of 1 inch is placed on top of an empty ice cream cone with diameter of 2 inches and height of 4.5 inches. When the ice cream melts, is the cone big enough to hold all the ice cream? Justify the answer.

Chapter 9

6. Determine the intercepts and graph the line $x + 3y = 9$. Find the distance between the intercepts.

7. Find the slope-intercept form of the equation of a line that contains the points $(-2, 5)$ and $(0, -3)$.

8. Use analytic geometry to prove any point on the perpendicular bisector of a segment is equidistant from the endpoints of the segment.

Chapter 10

9. Find a and b correct to one decimal place.

10. Find $m\angle ABC$ to the nearest degree.

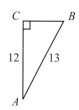

11. Solve the right triangle where $a = 2.7$, $b = 3.4$ and $\angle C$ is a right angle. Round the answer to the nearest tenth.

12. An escalator at Plaza mall makes a $32°$ angle with the floor. It carries people to the second floor that is 30 feet above the first floor. What is the length of the escalator to the nearest tenth of a foot?

Answers to Odd-Numbered Exercises

Chapter 1

1.1 Section Exercises

1. 28 **3.** 243 **5.** −32 **7.** 34 **9.** S **11.** U **13.** an even number **15.** A postulate is a statement that is assumed true without proof. A definition is a statement that gives us a new term to be used in the system. **17.** A postulate is a statement that is assumed true, and a theorem is a statement that is proved. **19.** The hypothesis follows "if"; the conclusion follows "then." **21.** postulate **23.** There will probably be disagreement in their negotiations. **25.** The conclusion does not follow; this is a fallacy in reasoning. **27.** The conclusion follows logically; deductive reasoning. **29.** The conclusion does not follow; inductive reasoning. **31.** The conclusion follows logically; deductive reasoning. **33.** The conclusion follows logically; deductive reasoning. **35.** We cannot conclude it is necessarily green, only that *if* it hops or is a frog, *then* it is green. **37.** The older dog is the younger dog's father. **39.** 12 **41.** It would have been impossible in 300 B.C. to know that that was the date as we now know it. **43.** 9 minutes **45.** Once **47.** Identify the sacks as sack 1, sack 2, and sack 3. Take one coin from sack 1, two coins from sack 2, and three coins from sack 3. Place the coins on the scale. The reading will be 6 lb 1 oz, 6 lb 2 oz, or 6 lb 3 oz. The number of ounces identifies the sack with the counterfeit coins.

49. The shortest route for the bug to take is shown in the figure to the right. The fact that it is the shortest is easy to see if we "unfold" the top face of the cube and draw a straight line between 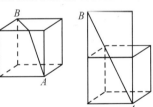 points A and B. (The shortest distance between two points is the distance along the straight line joining them.)

51. The four pieces could be arranged as shown to the right. The fallacy here is that the pieces do not actually fit to form the diagonal line shown. Some parts overlap covering part of the area and giving rise to the "missing" square.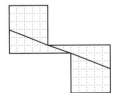

53. Take away the seven matches forming the F and the E, which leaves IV, the Roman numeral for four. **55.** Smith is the engineer, Jones is the brakeman, and Rodriquez is the fireman. Passenger Smith lives in Denver, Passenger Jones lives in New York, Passenger Rodriquez lives in San Francisco.

1.2 Section Exercises

1. Exactly one **3.** C is in plane $\mathcal{P}$ **5.** Postulate 1.1 **7.** Postulate 1.4 **9.**

$$-1.5 \quad -\tfrac{1}{3} \quad \tfrac{7}{8} \quad \sqrt{3} \quad 3$$
$$-2 \;-1 \;\; 0 \;\; 1 \;\; 2 \;\; 3$$

11. A:2; B:−3; C:3.5; D: $-\dfrac{1}{2}$ **13.** 5 **15.** 7 **17.** a **19.** False; the statement is true by the transitive law. **21.** False: it's true by the multiplication-division law. **23.** 2. Add-Subt Post 4. Add-Subt Post 6. Mult-Div Post 8. Symmetric law 9. Substitution law **25.** Yes, if the given point is a common point where two lines meet. If five lines meet at one point, then the point exists on all five lines. **27.** Yes, the line is where the two planes meet. If all five planes meet in one common place, this will be the described line. **29.** By Postulate 1.2, three distinct points (the ends of the legs) all lie in the same plane (the floor on which the stool stands). However, four points (the ends of a 4-legged stool) need not all lie in the same plane (the floor on which that stool stands).

1.3 Section Exercises

1. True **3.** True **5.** True **7.** False **9.** True **11.** False **13.** True **15.** False **17.** True **19.** True **21.** 65° **23.** 90° **25.** 115° **27.** 72° **29.** 53°20′ **31.** 106° **33.** 122°25′ **35.** 45° **37.** 40° **39.** 25° **41.** 60° **43.** Acute **45.** Straight **47.** Acute **49.** Acute **51.** Right **53.** 107°10′ **55.** 139°20′16″ **57.** Two **59.** Two **61.** It depends on the length of $\overline{AB}$. If $AB < 5$ there is only one. If $AB \geq 5$ there are two. **63.** 45° **65.** $x = 9$ **67.** $y = 20$

1.4 Section Exercises

1. *Given:* The president gets his budget passed.
Prove: He will be voted out of office.
  1. Given
  2. Taxes will rise.
  3. Premise 1
  4. They will go to the polls.
  5. He will be voted out of office.
  ∴ If the President gets his budget passed, then he will be voted out of office.

A-1

3. *Given:* I watch TV.
Prove: My parents will be upset.
Proof:

STATEMENTS	REASONS
1. I watch TV.	1. Given
2. I will not do my homework.	2. Premise 1
3. I will fail geology.	3. Premise 3
4. I will lose my scholarship.	4. Premise 2
5. My parents will be upset.	5. Premise 4

∴ If I watch TV, then my parents will be upset.
5. No. What we proved is *if* I watch TV, *then* we know that my parents will be upset.   **7.** No, the theorem could not be proved.
9. If I am watching football, then it is Sunday.   **11.** If you go to the airport, then you will fly in an airplane.   **13.** The city is not large.   **15.** Joe ran in the race.   **17.** If this animal is not a bird, then it does not have two legs.   **19.** If it isn't gold, then it doesn't glitter.   **21.** If a figure is not a parallelogram, then it is not a rectangle.   **23.** If we are not the conference champions, then we lost to Central State excluding a tie.   **25.** Hypothesis: Two lines have slopes m_1 and m_2 where $m_1 = m_2$. Conclusion: The lines are parallel.   **27.** Hypothesis: Two angles are vertical angles. Conclusion: The angles are congruent.

1.5 Section Exercises

1. 1. Given
   2. Given
   3. Given
   4. Given
   5. Addition-subtraction law (Post. 1.9)
   6. Segment addition postulate (Post. 1.13)
   7. Substitution law (Post. 1.11)
3. 1. $\angle ABC$ is a right angle
   2. Def. of right angle
   3. Angle addition postulate (Post 1.14)
   4. Transitive law (Post 1.8)
   5. Given
   6. Def. of complementary angles
   7. Symmetric and transitive laws using 4 and 6
   8. $m\angle ABD = m\angle 1$
5. $\angle 1$ and $\angle 2$; $\angle 2$ and $\angle 3$; $\angle 3$ and $\angle 4$; $\angle 4$ and $\angle 1$ **7.** 150°
9. 150°   **11.** 15°   **13.** Yes, by Theorem 1.11   **15.** Yes, if they are the same measure, which must be 90°.   **17.** No, obtuse means greater than 90°.   **19.** Yes, obtuse means greater than 90°.   **21.** Yes, the lines must be perpendicular   **23.** No, vertical angles must be congruent.   **25.** Yes, adjacent means they have a common vertex and a common side. These angles can be supplementary if their sum is 180°.

Note to students about proofs: Proofs are not unique. Your proof may differ slightly from the solutions manual. This does not mean your proof is necessarily incorrect. Consult with your instructor.

27. *Given:* $\angle A$ and $\angle C$ are supplementary
   $\angle B$ and $\angle D$ are supplementary
   $m\angle C = m\angle D$
Prove: $m\angle A = m\angle B$
Proof:

STATEMENTS	REASONS
1. $\angle A$ and $\angle C$ are supp.	1. Given
2. $m\angle A + m\angle C = 180°$	2. Def. of supp. $\angle$
3. $\angle B$ and $\angle D$ are supp.	3. Given
4. $m\angle B + m\angle D = 180°$	4. Def. of supp. $\angle$
5. $m\angle A + m\angle C = m\angle B + m\angle D$	5. Sym. and trans. laws
6. $m\angle C = m\angle D$	6. Given
7. $m\angle A = m\angle B$	7. Add.-subt. law

29. *Given:* $PR = QS$
Prove: $PQ = RS$
Proof:

STATEMENTS	REASONS
1. $PR = QS$	1. Given
2. $QR = QR$	2. Ref. law
3. $PR - QR = QS - QR$	3. Add.-subt. law
4. $PQ = PR - QR$	4. Seg.-add. post.
5. $RS = QS - QR$	5. Seg.-add. post.
6. $PQ = RS$	6. Substitution law

31. *Given:* $\angle 3$ and $\angle 4$ are complementary
   $\angle 1$ and $\angle 3$ are supplementary
   $\angle 2$ and $\angle 4$ are supplementary
Prove: $m\angle 1 + m\angle 2 = 270°$
Proof:

STATEMENTS	REASONS
1. $\angle 3$ and $\angle 4$ are complementary	1. Given
2. $m\angle 3 + m\angle 4 = 90°$	2. Def. of comp. $\angle$
3. $\angle 1$ and $\angle 3$ are supplementary	3. Given
4. $m\angle 1 + m\angle 3 = 180°$	4. Def. of supp. $\angle$
5. $\angle 2$ and $\angle 4$ are supplementary	5. Given

6. $m\angle 2 + m\angle 4 = 180°$ 6. Def. of supp. $\angle$
7. $m\angle 1 + m\angle 2 + m\angle 3 + m\angle 4 = 360°$ 7. Add.-subt. law using statements 4 and 6
8. $m\angle 1 + m\angle 2 + 90° = 360°$ 8. Substitution law using statement 2
9. $m\angle 1 + m\angle 2 = 270°$ 9. Add.-subt. law

1.6 Section Exercises

1. Use Construction 1.1.   **3.** Use Construction 1.2.   **5.** Use Construction 1.2 twice.   **7.** Use Construction 1.3.   **9.** Use Construction 1.3.   **11.** Use Construction 1.4.   **13.** Use Construction 1.6.   **15.** Use Construction 1.6 three times.   **17.** A ruler is used to measure lengths, but a straightedge is used only to draw a straight line between two points.   **19.** Exactly one   **21.** Use Construction 1.5.   **23.** Construct the bisector of the angle with vertex at the bridge and sides containing the ranger station and the cabin. Locate the point of intersection of this bisector (ray) and the edge of the forest.   **25.** 1. B is the midpoint of $\overline{AC}$ 2. Def. of midpoint 3. Segment addition postulate (Post. 1.13) 4. Substitution law (Post. 1.11) 5. $2AB = AC$ 6. Multiplication-division law (Post. 1.10)   **27.** 1. B is the midpoint of $\overline{AC}$ 3. Given 4. $PQ = \dfrac{PR}{2}$ 5. $AC = PR$ 6. Multiplication-division law (Post. 1.10) 7. Substitution law (Post. 1.11)   **29.** The angles are complementary. Adjacent angles have a common vertex and a common side. 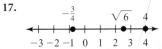 If the nonshared sides are perpendicular, this means the sum of the adjacent angles is 90°, thus the angles are complementary.

Chapter 1 Review Exercises

1. The four parts to any axiomatic system are undefined terms, definitions, axioms or postulates, and theorems.   **2.** They serve as a starting point when building the system.   **3.** 17   **4.** K
5. $\dfrac{1}{32}$ **6.** The conclusion follows logically; deductive reasoning.
7. The conclusion does not follow; this is a fallacy of reasoning.
8. The conclusion does not follow; inductive reasoning.
9. The conclusion follows logically; deductive reasoning.
10. We cannot conclude that tomorrow is a holiday, only that *if* it is Sunday, *then* tomorrow is a holiday.   **11.** 3 minutes
12. 3   **13.** True   **14.** False   **15.** True   **16.** True
17.

$$-\tfrac{3}{4} \qquad \sqrt{6} \qquad 4$$

$$-3 \ -2 \ -1 \ \ 0 \ \ 1 \ \ 2 \ \ 3 \ \ 4$$

18. Reflexive law   **19.** Symmetric law   **20.** Transitive law
21. Multiplication-division law   **22.** Substitution law
23. False   **24.** True   **25.** True   **26.** True   **27.** False
28. True   **29.** False   **30.** False   **31.** True   **32.** False
33. True   **34.** False   **35.** 65°29′15″   **36.** 43°17′9″
37. 20°   **38.** No   **39.** Exactly one   **40.** Two

41. *Given:* It is Gleep.
  Prove: It is Grob.
  Proof:

STATEMENTS	REASONS
1. It is Gleep.	1. Given
2. It is Glop.	2. Premise 1
3. It is Gunk.	3. Premise 3
4. It is Grob.	4. Premise 2

∴ If it is Gleep, then it is Grob.

42. *Given:* It is a tree.
  Prove: It can be destroyed by fire.
  Proof:

STATEMENTS	REASONS
1. It is a tree.	1. Given
2. It is made of wood	2. Premise 2
3. It will burn.	3. Premise 3
4. It can be destroyed by fire.	4. Premise 1

∴ If it is a tree, it can be destroyed by fire.

43. If it is cold, then it is ice.   **44.** If you want the best for someone, then you love that person.   **45.** My car is not red.
46. The moon is made of green cheese.   **47.** If the runner does not win the race, then she is not in excellent condition.
48. If a painting is not a Picasso, then it is not valuable.
49. If I do not climb the mountain, then the weather is not good.   **50.** If I drive, then I do not drink.   **51.** 42°   **52.** 138°
53. 17°   **54.** 163°   **55.** 1. Given 2. $m\angle 1 + m\angle 2 = 90°$
3. $\angle 1$ and $\angle 3$ are vertical angles 4. Vertical angles are equal in measure 5. $\angle 2$ and $\angle 4$ are vertical angles 6. Vertical angles are equal in measure 8. $\angle 3$ and $\angle 4$ are complementary.
56. 1. Given 2. Reflexive law 3. Addition-subtraction law
4. Segment addition postulate 5. Segment addition postulate
6. Substitution law   **57.** First bisect $\overline{PQ}$ determining midpoint T. Then use Construction 1.1 to construct segment $\overline{CD}$ on m with length PT.   **58.** Use Construction 1.6.   **59.** Use Construction 1.3.   **60.** Infinitely many   **61.** Exactly one
62. Use Construction 1.2.

Chapter 1 Practice Test

1. A postulate is a statement that is assumed true without proof; and a theorem is a statement that is proved using deductive reasoning.   **2.** Inductive reasoning involves reaching a general conclusion based on specific observations. Deductive reasoning involves reaching a specific conclusion based on assumed general conditions.   **3.** Deductive reasoning   **4.** No. The specific observations that lead to the conclusion may not be sufficient in number.   **5.** $\dfrac{1}{5}$ **6.** No. This inductive reasoning does not necessarily lead to the same conclusion.   **7.** Infinitely many
8. $y + 2$ **9.** True   **10.** True   **11.** False   **12.** True   **13.** True
14. False   **15.** True   **16.** False   **17.** 56°20′

18. *Given:* I rob a bank.
Prove: I will go to jail.
Proof:

STATEMENTS	REASONS
1. I rob a bank.	1. Given
2. I will be arrested.	2. Premise 4
3. I will go on trial.	3. Premise 2
4. I will be convicted.	4. Premise 1
5. I will go to jail.	5. Premise 3

∴ If I rob a bank, then I will go to jail.
19. If it is white, then it is milk.   **20.** The road to success is not difficult.   **21.** If it is not a collie, then it is not a dog.
22. If the fruit is not ripe, then it is not picked.   **23.** 1. Given 2. Adj. ∠'s whose noncommon sides are in line are supp. 3. Same as 2 4. Supp's of = ∠'s are = in measure (Theorem 1.8)
24. Use Construction 1.3 to find midpoint C; then Construction 1.2 to construct the desired angle.   **25.** Use Construction 1.6   **26.** 1. Given 2. Angle-addition postulate. 3. Angle-addition postulate. 4. Substitution 5. Reflexive law 6. Addition-subtraction law   **27.** (a) Midpoint (b) Bisector (c) Perpendicular bisector

Chapter 2

2.1 Section Exercises

1. D, E, and F   **3.** Obtuse triangle   **5.** $\overline{DE}$ **7.** ∠F   **9.** $\overline{DE}$
11. $\overline{AB}$, $\overline{BC}$, and $\overline{AC}$ **13.** Equilateral triangle   **15.** $\overline{BC}$
17. ∠A   **19.** It has no hypotenuse since it is not a right triangle.   **21.** 90 cm   **23.** 33 in.   **25.** 23 ft   **27.** Base: 15 cm, sides: 45 cm   **29.** △AED and △AEB   **31.** △ECD and △ACD   **33.** True   **35.** False   **37.** False   **39.** True
41. Yes, a triangle can be scalene and obtuse. Scalene means all sides have different lengths and obtuse means the triangle has one obtuse angle.

43. Yes, a right triangle can be isosceles. A right triangle contains one right angle while an isosceles triangle has two congruent sides.

45. No, an equilateral triangle has three angles that measure 60°.

2.2 Section Exercises

1. Congruent by SSS   **3.** Since right angles are congruent, the triangles are congruent by SAS.   **5.** Congruent by ASA
7. 1. Given 2. ∠3 ≅ ∠4 3. $\overline{BD}$ ≅ $\overline{BD}$ 4. ASA   **9.** 1. Given 2. ⊥ lines form rt. ∠'s 3. ∠ABD ≅ ∠CDB 4. $\overline{DB}$ ≅ $\overline{BD}$ 5. Given 6. SAS
Note to students about proofs: Proofs are not unique. Your proof may differ slightly from the solutions manual. This does not mean your proof is necessarily incorrect. Consult with your instructor.

11. *Proof:*

STATEMENTS	REASONS
1. $\overline{AD}$ bisects $\overline{BE}$	1. Given
2. $\overline{BC}$ ≅ $\overline{EC}$	2. Def. of bisector
3. $\overline{BE}$ bisects $\overline{AD}$	3. Given
4. $\overline{AC}$ ≅ $\overline{DC}$	4. Def. of bisector
5. ∠ACB ≅ ∠DCE	5. Vertical angles are ≅
6. △ABC ≅ △DEC	6. SAS

13. *Proof:*

STATEMENTS	REASONS
1. $\overline{AD}$ ≅ $\overline{BD}$ so AD = BD	1. Given, def. ≅ seg.
2. $\overline{AE}$ ≅ $\overline{BC}$ so AE = BC	2. Given, def. ≅ seg.
3. $AD - AE = BD - BC$	3. Addition-subtraction law
4. $ED = AD - AE$ and $CD = BD - BC$	4. Segment addition postulate
5. $ED = CD$ so $\overline{ED}$ ≅ $\overline{CD}$	5. Substitution, def. ≅ seg.
6. ∠D ≅ ∠D	6. Reflexive law
7. △ACD ≅ △BED	7. SAS

15. (a) Not correct because $\overline{BA}$ ≇ $\overline{ED}$ (b) Correct
(c) Not correct because $\overline{ED}$ ≇ $\overline{AB}$ (d) Not correct, △BAC ≅ △FED   **17.** (b) Same   (c) Match   (d) Match; yes
(e) Yes   (f) No   **19.** 3   **21.** 4.   **23.** 120°   **25.** $\overline{PR}$ **27.** $\overline{QR}$
29. ∠P

2.3 Section Exercises

1. 9 cm   **3.** 46°
5. 1. Given
2. ∠1 ≅ ∠2
3. Adj. ∠'s whose noncommon sides are in a line are supp. ∠'s
4. ∠2 and ∠ECB are supplementary
5. Supp. of ≅ ∠'s are ≅
6. Reflexive law
7. SAS
8. ∠G ≅ ∠E
7. 1. Given
2. Given
3. $\overline{AB}$ ≅ $\overline{CD}$; AB = CD
4. Segment addition postulate
5. BD = CD + BC
7. Substitution law and Def. ≅ seg.
8. △AFC ≅ △DEB
9. CPCTC
9. *Proof:*

STATEMENTS	REASONS
1. $\overline{AC}$ ≅ $\overline{CE}$	1. Given
2. $\overline{DC}$ ≅ $\overline{CB}$	2. Given
3. ∠DCE and ∠BCA are vertical angles	3. Def. of vert. ∠'s
4. ∠DCE ≅ ∠BCA	4. Vert. ∠'s are ≅
5. △DCE ≅ △BCA	5. SAS
6. ∠A ≅ ∠E	6. CPCTC

11. *Proof:*

STATEMENTS	REASONS
1. $\overline{BC} \cong \overline{CD}$	1. Given
2. $\angle 1 \cong \angle 2$	2. Given
3. $\overline{AC} \cong \overline{AC}$	3. Reflexive law
4. $\triangle ABC \cong \triangle ADC$	4. SAS
5. $\overline{AB} \cong \overline{AD}$	5. CPCTC

13. *Proof:*

STATEMENTS	REASONS
1. $\angle 1 \cong \angle 2$	1. Given
2. $\angle 3 \cong \angle 4$	2. Given
3. $\overline{DB} \cong \overline{BD}$	3. Reflexive law
4. $\triangle ABD \cong \triangle CDB$	4. ASA
5. $\angle A \cong \angle C$	5. CPCTC

15. *Proof:*

STATEMENTS	REASONS
1. $\overline{GB} \perp \overline{AF}$ and $\overline{FD} \perp \overline{GE}$	1. Given
2. $\angle 1$ and $\angle 2$ are right angles	2. $\perp$ lines form rt. $\angle$'s
3. $\angle 1 \cong \angle 2$	3. Rt. $\angle$'s are $\cong$
4. $\overline{GD} \cong \overline{FB}$ and $\overline{GB} \cong \overline{FD}$	4. Given
5. $\triangle BGF \cong \triangle DFG$	5. SAS
6. $\angle BGF \cong \angle DFG$; $m\angle BGF = m\angle DFG$	6. CPCTC; def. $\cong \angle$'s.
7. $\angle 3 \cong \angle 4$; $m\angle 3 = m\angle 4$	7. CPCTC; def. $\cong \angle$'s.
8. $m\angle DFG - m\angle 3 = m\angle BGF - m\angle 4$	8. Add-Subt. law
9. $m\angle 5 = m\angle DFG - m\angle 3$	9. Angle add. post.
10. $m\angle 6 = m\angle BGF - m\angle 4$	10. Angle add. post.
11. $m\angle 5 = m\angle 6$ so $\angle 5 \cong \angle 6$	11. Substitution, def. $\cong \angle$'s.
12. $\triangle BGC \cong \triangle DFC$	12. ASA
13. $\overline{BC} \cong \overline{DC}$	13. CPCTC

17. Since $\overline{AB}$ and $\overline{DE}$ are corresponding parts of congruent triangles $\triangle ABC$ and $\triangle EDC$ (by SAS), $AB = 105$ yd (the same as DE).   **19.** Because the triangle is a rigid figure that cannot be distorted like a four-sided figure. This is a result of SSS since there is only one triangle possible with three given sides. The bridge cannot change shape without breaking.   **21.** $\overline{AD} \cong \overline{CD}$ and $\angle ADB \cong \angle CDB$ are given. $\overline{DB} \cong \overline{DB}$ by reflexive law, thus $\triangle ADB \cong \triangle CDB$ by SAS and $\overline{AB} \cong \overline{CB}$ by CPCTC.

$$8x - 3 = 6x + 1$$
$$2x = 4$$
$$x = 2$$

23. $\angle ZWY \cong \angle XYW$ and $\angle ZYW \cong \angle XWY$ are given. $\overline{WY} \cong \overline{YW}$ by reflexive law, thus $\triangle ZYW \cong \triangle XWY$ by ASA and $\overline{WZ} \cong \overline{YX}$ by CPCTC

$$3x - 2 = 2x + 1$$
$$x = 3$$

25. $\angle BAC \cong \angle DCA$ and $\angle CAD \cong \angle ACB$ is given information. $\overline{AC} \cong \overline{CA}$ by reflexive law, thus $\triangle ABC \cong \triangle CDA$ by ASA. $\overline{AB} \cong \overline{CD}$ because CPCTC.

2.4 Section Exercises

1. $\overline{AB} \cong \overline{CB}$ **3.** $\overline{WY} \cong \overline{XY}$ **5.** $\overline{RS} \cong \overline{TR}$

7.
1. Given
2. $\overline{AC}$ bisects $\angle BAD$
3. Def. of $\angle$ bisector
4. Reflexive law
5. $\triangle BAE \cong \triangle DAE$
6. CPCTC
7. $\triangle BDE$ is isosceles

9.
1. Given
2. $\overline{AC} \cong \overline{AD}$ so $AC = AD$
3. Given
4. Def. of midpoint and def. $\cong$ seg.
5. E is the midpoint of $\overline{AD}$
6. Def. midpoint and def. $\cong$ seg.
7. Segment addition postulate
8. Substitution law
10. Multiplication-division law
13. $\angle 1 \cong \angle 2$

11. *Proof:*

STATEMENTS	REASONS
1. $\angle 1 \cong \angle 2$	1. Given
2. $\overline{BE} \cong \overline{CE}$	2. Sides opp $\cong \angle$'s are $\cong$
3. $\angle 3 \cong \angle 4$	3. Given
4. $\angle AEB$ and $\angle DEC$ are vertical angles	4. Def. of vert. $\angle$'s
5. $\angle AEB \cong \angle DEC$	5. Vert. $\angle$'s are $\cong$
6. $\triangle ABE \cong \triangle DCE$	6. ASA
7. $\angle A \cong \angle D$	7. CPCTC

13. An altitude is a line segment from a vertex of a triangle, perpendicular to the side opposite the vertex (or an extension of that side). What distinguishes this from the other two is an altitude of a triangle may be outside the triangle and an altitude is perpendicular to a side of the triangle.

An angle bisector is a ray that separates an angle into two congruent adjacent angles. What distinguishes this from the other two is all angle bisectors of a triangle are inside the triangle and cut the angle into two congruent angles.

A median of a triangle is a line segment joining a vertex with the midpoint of the opposite side of the triangle. What distinguishes this from the other two is all medians of a triangle are inside the triangle and the midpoint of a side of the triangle is needed.   **15.** True   **17.** True   **19.** False   **21.** The orthocenter of a triangle is the intersection of the altitudes of the triangle. From the figure we see this point is D.   **23.** FI = 3.5 inches   **25.** AI = 20.4 m   **27.** E   **29.** No, the centroid is on a median of a triangle. The median joins a vertex with the midpoint of the opposite side. A median is always inside a triangle.

2.5 Section Exercises

1. LA **3.** LA **5.** $\triangle ABT$ **7.** Given **9.** Given
11. Def. rt. $\triangle$ **13.** Given **15.** Reflexive law
17. *Proof:*

STATEMENTS	REASONS
1. $\overline{XY} \cong \overline{YZ}$	1. Given
2. $\overline{WY} \cong \overline{WY}$	2. Reflexive law
3. $\overline{WY} \perp \overline{XZ}$	3. Given
4. $\angle XYW$ and $\angle ZYW$ are rt. $\angle$'s	4. $\perp$ lines form rt. $\angle$'s
5. $\triangle XYW$ and $\triangle ZYW$ are rt. $\triangle$'s	5. Def. rt. $\triangle$
6. $\triangle XYW \cong \triangle ZYW$	6. LL

19. *Proof:*

STATEMENTS	REASONS
1. $\overline{AB} \cong \overline{DE}$	1. Given
2. $\angle ACB \cong \angle ECD$	2. Vertical $\angle$'s $\cong$
3. $\overline{AB} \perp \overline{BD}, \overline{DE} \perp \overline{BD}$	3. Given
4. $\angle ABC$ and $\angle EDC$ are rt. $\angle$'s	4. $\perp$ lines form rt. $\angle$'s
5. $\triangle ABC$ and $\triangle EDC$ are rt. $\triangle$'s	5. Def. rt. $\triangle$
6. $\triangle ABC \cong \triangle EDC$	6. LA
7. $\overline{AC} \cong \overline{EC}$	7. CPCTC

2.6 Section Exercises

1. Use Construction 2.1. **3.** Use Construction 2.2.
5. Use Construction 2.3. **7.** Any two obtuse angles can be tried since their noncommon sides will not intersect. **9.** Use Construction 2.2. **11.** The sides do not intersect to form a triangle. **13.** First construct a right angle using Construction 1.4, then use Construction 2.2. **15.** There are two triangles with these parts. **17.** Use Construction 2.5 three times. The point of concurrency is the orthocenter. **19.** Use Construction 2.6 three times. The point of concurrency is the centroid. **21.** Use Construction 1.6 three times. The point of concurrency is the incenter. **23.** No, the orthocenter is the point of concurrency of the altitudes of a triangle. Some altitudes in an obtuse triangle will be outside the triangle, thus, so will the orthocenter. **25.** Yes, the incenter is the point of concurrency of the angle bisectors. **27.** They intersect at the circumcenter. **29.** Use Construction 2.4.

Chapter 2 Review Exercises

1. A, B, and C **2.** $\overline{AB}$, $\overline{BC}$, and $\overline{CA}$ **3.** Scalene triangle
4. Obtuse triangle **5.** $\overline{AC}$ **6.** $\angle B$ **7.** $\overline{BC}$ **8.** $\angle C$
9. 44 cm **10.** Base: 13 ft, sides: 26 ft **11.** True **12.** True
13. False **14.** True **15.** Yes **16.** SSS **17.** ASA
18. 1. Given 2. $\overline{AC} \cong \overline{AE}$ 3. $\angle A \cong \angle A$ 4. $\triangle ACF \cong \triangle AEB$
19. 1. Given 2. $\overline{AC}$ bisects $\angle BAD$ 3. $\angle BAC \cong \angle EAC$
4. Reflexive law 5. $\triangle ABC \cong \triangle AEC$

20. *Proof:*

STATEMENTS	REASONS
1. $\overline{AD} \cong \overline{CB}$	1. Given
2. $\overline{AB} \cong \overline{CD}$	2. Given
3. $\overline{DB} \cong \overline{BD}$	3. Reflexive law
4. $\triangle ADB \cong \triangle CBD$	4. SSS

21. *Proof:*

STATEMENTS	REASONS
1. $\overline{AC} \perp \overline{BD}$	1. Given
2. $\angle 3 \cong \angle 4$	2. $\perp$ lines form $\cong$ adj. $\angle$'s
3. $\angle 1 \cong \angle 2$	3. Given
4. $\overline{EC} \cong \overline{EC}$	4. Reflexive law
5. $\triangle DEC \cong \triangle BEC$	5. ASA
6. $\overline{DC} \cong \overline{BC}$	6. CPCTC
7. $\overline{AC} \cong \overline{AC}$	7. Reflexive law
8. $\triangle ABC \cong \triangle ADC$	8. SAS
9. $\overline{AB} \cong \overline{AD}$	9. CPCTC

22. 2 **23.** 25 **24.** 4 **25.** 35° **26.** $\overline{BC}$ **27.** $\angle D$
28. 1. Given
2. $\overline{AC}$ bisects $\angle DAB$
3. Def. of $\angle$ bisector; Def. $\cong$ $\angle$'s
4. Given
5. $\angle EBA \cong \angle DBE$ so $m\angle EBA = m\angle DBE$
6. Angle addition postulate
7. Substitution law
9. Symmetric and transitive laws
10. Multiplication-division law and def. $\cong$ $\angle$'s
11. Reflexive law
12. ASA
13. $\overline{AC} \cong \overline{BE}$

29. *Proof:*

STATEMENTS	REASONS
1. E is the midpoint of $\overline{AC}$	1. Given
2. $\overline{AE} \cong \overline{CE}$	2. Def. of midpoint
3. E is the midpoint of $\overline{BD}$	3. Given
4. $\overline{BE} \cong \overline{DE}$	4. Def. of midpoint
5. $\angle AEB$ and $\angle CED$ are vertical angles	5. Def. of vert. $\angle$'s
6. $\angle AEB \cong \angle CED$	6. Vert. $\angle$'s are $\cong$
7. $\triangle AEB \cong \triangle CED$	7. SAS
8. $\overline{AB} \cong \overline{CD}$	8. CPCTC

30. *Proof:*

STATEMENTS	REASONS
1. $\overline{AB} \cong \overline{CD}$	1. Given
2. $\overline{BC} \cong \overline{DE}$	2. Given
3. $\angle CAE \cong \angle CEA$	3. Given
4. $\overline{AC} \cong \overline{CE}$	4. Sides opp. $\cong$ $\angle$'s are $\cong$
5. $\triangle ABC \cong \triangle CDE$	5. SSS
6. $\angle B \cong \angle D$	6. CPCTC

31. *Proof:*

STATEMENTS	REASONS
1. △ABD is isosceles with base $\overline{BD}$	1. Given
2. $\overline{AB} \cong \overline{AD}$	2. Def. of isosceles △
3. △BDE is isosceles with base $\overline{BD}$	3. Given
4. $\overline{BE} \cong \overline{DE}$	4. Def. of isosceles △
5. $\overline{AE} \cong \overline{AE}$	5. Reflexive law
6. △ABE ≅ △ADE	6. SSS
7. ∠3 ≅ ∠4	7. CPCTC
8. ∠3 and ∠1 are supplementary and ∠4 and ∠2 are supplementary	8. Adj. ∠'s whose noncommon sides are in line are supp.
9. ∠1 ≅ ∠2	9. Supp. of ≅ ∠'s are ≅

32. *Proof:*

STATEMENTS	REASONS
1. $\overline{AB} \cong \overline{CB}$; $\overline{AD} \cong \overline{CD}$	1. Given
2. $\overline{BD} \cong \overline{BD}$	2. Reflexive law
3. △ABD ≅ △CBD	3. SSS
4. ∠ABD ≅ ∠CBD	4. CPCTC
5. $\overline{BD}$ is bisector of ∠ABC	5. Def. ∠ bisector

33. True **34.** False **35.** True **36.** True **37.** False **38.** False **39.** 8.5 m = EF, AC = 24 m

40. *Proof:*

STATEMENTS	REASONS
1. ∠C ≅ ∠E	1. Given
2. $\overline{AD} \perp \overline{BC}$; $\overline{AD} \perp \overline{DE}$	2. Given
3. ∠ABC and ∠BDE are rt. ∠'s	3. ⊥ lines form rt. ∠'s
4. △ABC and △BDE are rt. △'s	4. Def. rt. △'s
5. B is midpoint $\overline{AD}$	5. Given
6. $\overline{AB} \cong \overline{DB}$	6. Def. of midpoint
7. △ABC ≅ △BDE	7. LA

41. *Proof:*

STATEMENTS	REASONS
1. △WXY is isosceles triangle with $\overline{WX} \cong \overline{YX}$	1. Given
2. ∠W ≅ ∠Y	2. Sides opp. ≅ ∠'s are ≅.
3. $\overline{XZ} \cong \overline{XZ}$	3. Reflexive law
4. $\overline{WY} \perp \overline{XZ}$	4. Given
5. ∠XZW, ∠XZY are rt. ∠'s	5. ⊥ lines form rt. ∠'s.
6. △WXZ and △YXZ are rt. △'s.	6. Def. rt. △'s
7. △WXZ ≅ △YXZ	7. LA

42. Use Construction 2.1. **43.** First construct a right angle using Construction 1.4. Then use the right angle, the segment, and the acute angle with Construction 2.3. **44.** Use Construction 2.6. **45.** Use Construction 2.2. **46.** Use Construction 2.5.

Chapter 2 Test

1. Acute triangle **2.** Isosceles triangle **3.** ∠A **4.** ∠B ≅ ∠C **5.** ∠B **6.** $\overline{BC}$ **7.** Yes **8.** Yes **9.** ∠ACD **10.** 13 cm **11.** 33 in. **12.** 28°

13. *Proof:*

STATEMENTS	REASONS
1. ∠1 ≅ ∠2	1. Given
2. D is the midpoint of $\overline{CE}$	2. Given
3. $\overline{CD} \cong \overline{DE}$	3. Def. of midpoint
4. $\overline{AC} \cong \overline{AE}$	4. Given
5. ∠E ≅ ∠C	5. Angles opp. ≅ sides are ≅
6. △BCD ≅ △FED	6. ASA
7. $\overline{BD} \cong \overline{FD}$	7. CPCTC

14. *Proof:*

STATEMENTS	REASONS
1. ∠3 ≅ ∠4	1. Given
2. $\overline{AC} \cong \overline{AD}$	2. Sides opp. ≅ ∠'s are ≅
3. ∠1 and ∠3 are supplementary and ∠2 and ∠4 are supplementary	3. Adj. ∠'s whose noncommon sides are in line are supp.
4. ∠1 ≅ ∠2	4. Supp. of ≅ ∠'s are ≅
5. $\overline{BC} \cong \overline{ED}$	5. Given
6. △ABC ≅ △AED	6. SAS
7. ∠5 ≅ ∠6	7. CPCTC

15. *Proof:*

STATEMENTS	REASONS
1. $\overrightarrow{AB}$ bisects $\overline{CD}$	1. Given
2. $\overline{DE} \cong \overline{CE}$	2. Def. bisector
3. ∠DEB ≅ ∠CEA	3. Vertical ∠'s ≅
4. ∠C and ∠D are rt. ∠'s	4. Given
5. △ACE and △BDE are rt. △'s	5. Def. rt. △
6. △ACE ≅ △BDE	6. LA

16. Use Construction 2.2 followed by Construction 2.5 and Construction 2.6. **17.** It is both an altitude and a perpendicular bisector. **18.** Perpendicular bisector **19.** Neither **20.** (a) centroid (b) 14.5 m = WX (c) MX = 9 m

Chapter 3

3.1 Section Exercises

1. 2. Assumption 3. Premise 1 4. Premise 2 5. Premise 3
3. *Given:* The weather is nice.
Prove: I am healthy.
Proof:

STATEMENTS	REASONS
1. The weather is nice.	1. Given
2. Assume I am unhealthy.	2. Assumption
3. I don't exercises.	3. Premise 2
4. I'm not playing tennis regularly.	4. Premise 1
5. The weather is bad.	5. Premise 3

But this is a contradiction of Statement 1 "the weather is nice." Thus, we must conclude that our assumption in Statement 2 was incorrect. ∴ If the weather is nice, then I am healthy.
5. Yes **7.** Yes **9.** No **11.** No **13.** One; only line ℓ
15. Use the fact that if two lines are parallel to a third line and they intersect, then there are two lines through a point that are parallel to a line not containing the point, a contradiction.
17. Yes **19.** No **21.** No **23.** Many examples can be given.
25. The most obvious answer is the yard lines and the side lines.
27. No. If a triangle has two right angles, the uncommon sides of the angles could not intersect to form the triangle. **29.** Answers will vary.

3.2 Technology Connection
4. $\angle GBA$ and $\angle BED$; $\angle ABE$ and $\angle DEH$; $\angle GBC$ and $\angle BEF$; $\angle CBE$ and $\angle FEH$ **5.** $\angle ABE$ and $\angle BEF$; $\angle CBE$ and $\angle BED$
6. $\angle GBA$ and $\angle FEH$; $\angle GBC$ and $\angle DEH$ **7.** $\angle ABE$ and $\angle BED$; $\angle CBE$ and $\angle BEF$

3.2 Section Exercises
1. $\angle 3$ and $\angle 5$; $\angle 4$ and $\angle 6$ **3.** $\angle 1$ and $\angle 5$; $\angle 2$ and $\angle 6$; $\angle 3$ and $\angle 7$; $\angle 4$ and $\angle 8$ **5.** $\angle 1, \angle 3, \angle 5, \angle 7$ **7.** $\angle 4, \angle 6, \angle 8$
9. $m\angle 2 = m\angle 4 = m\angle 8 = 135°$, $m\angle 1 = m\angle 3 = m\angle 5 = m\angle 7 = 45°$ **11.** $m\angle 1 = m\angle 3 = m\angle 5 = m\angle 7 = 55°$; $m\angle 2 = m\angle 4 = m\angle 6 = m\angle 8 = 125°$ **13.** 69° **15.** 128° **17.** 45°
19. No; Thm 3.3 says $\overleftrightarrow{BC} \parallel \overleftrightarrow{DE}$ **21.** Yes $\overleftrightarrow{BC} \nparallel \overleftrightarrow{DE}$ **23.** No; Thm 2.5, transitive law, and Thm 3.3 says $\overleftrightarrow{BC} \parallel \overleftrightarrow{DE}$ **25.** Yes by Thm 2.5, transitive law, Thm 3.3
27. *Proof:*

STATEMENTS	REASONS
1. $\angle 3$ is supplementary to $\angle 4$	1. Given
2. $\angle 3$ is supplementary to $\angle 5$	2. Adj. $\angle$'s whose non-common sides are in line are supp.
3. $\angle 4 \cong \angle 5$	3. Supp. of $\cong$ $\angle$'s are $\cong$
4. $m \parallel n$	4. If corr. $\angle$'s are $\cong$, then lines are $\parallel$.

29. *Proof:*

STATEMENTS	REASONS
1. $m \parallel n$	1. Given
2. $\angle ABC$ is supplementary to $\angle 5$ and $\angle BAC$ is supplementary to $\angle 1$	2. Int. $\angle$'s same side transv. are supp.
3. $\angle 5 \cong \angle 1$	3. Given
4. $\angle ABC \cong \angle BAC$	4. Supp. of $\cong$ $\angle$'s are $\cong$
5. $\overline{BC} \cong \overline{AC}$	5. Sides opp. $\cong$ $\angle$'s are $\cong$
6. $\triangle ABC$ is isosceles	6. Def. of isosc. $\triangle$

31. *Proof:*

STATEMENTS	REASONS
1. $m \parallel n$	1. Given
2. $\overline{CD} \cong \overline{CE}$	2. Given
3. $\angle CDE \cong \angle CED$	3. $\angle$'s opp. $\cong$ sides are $\cong$
4. $\angle ABC \cong \angle CDE$ and $\angle BAC \cong \angle CED$	4. If lines are $\parallel$ then alt. int. $\angle$'s are $\cong$.
5. $\angle ABC \cong \angle BAC$	5. Transitive law
6. $\overline{AC} \cong \overline{BC}$	6. Sides opp. $\cong$ $\angle$'s are $\cong$

33. *Proof:*

STATEMENTS	REASONS
1. $\overline{AB} \cong \overline{DE}$	1. Given
2. $\overline{AD} \cong \overline{BE}$ (Draw $\overline{AD}$ and $\overline{BE}$)	2. Given
3. $\overline{BD} \cong \overline{BD}$	3. Reflexive law
4. $\triangle ABD \cong \triangle EDB$	4. SSS
5. $\angle ABD \cong \angle EDB$	5. CPCTC
6. $m \parallel n$	6. If alt. int. $\angle$'s are $\cong$, then lines are $\parallel$.

35. *Given:* $m \parallel n$ and m and n are cut by transversal ℓ
Prove: $\angle 2 \cong \angle 1$ and $\angle 5 \cong \angle 6$
Proof:

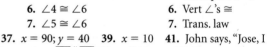

STATEMENTS	REASONS
1. $m \parallel n$	1. Given
2. $\angle 2 \cong \angle 3$	2. If lines are $\parallel$, corresp. $\angle$'s $\cong$.
3. $\angle 3 \cong \angle 1$	3. Vert $\angle$'s $\cong$
4. $\angle 2 \cong \angle 1$	4. Trans. law

Prove $\angle 5 \cong \angle 6$ in a similar manner.

5. $\angle 5 \cong \angle 4$	5. If lines are $\parallel$, corresp. $\angle$'s $\cong$.
6. $\angle 4 \cong \angle 6$	6. Vert $\angle$'s $\cong$
7. $\angle 5 \cong \angle 6$	7. Trans. law

37. $x = 90$; $y = 40$ **39.** $x = 10$ **41.** John says, "Jose, I know since $\overline{AX} \parallel \overline{BZ}$ that $\angle XAB \cong \angle ZBC$ because they are corresponding angles. Along with the other given information that shows $\triangle AXC \cong \triangle BZD$. By CPCTC, $\angle YCB \cong \angle ZDC$. This proves $\overline{CX} \parallel \overline{DZ}$ because using $\overline{AD}$ as the transversal if corresponding angles are congruent, lines are parallel."

3.3 Section Exercises
1. (a) triangle (b) 180° (c) 60° (d) 360° (e) 120° (f) 15 cm
3. (a) pentagon (b) 540° (c) 108° (d) 360° (e) 72° (f) 25 cm

5. **(a)** heptagon
   (b) 900°
   (c) about 128.6°
   (d) 360°
   (e) about 51.4°
   (f) 35 cm

7. **(a)** nonagon
   (b) 1260°
   (c) 140°
   (d) 360°
   (e) 40°
   (f) 45 cm

9. 11   **11.** there is no polygon satisfying these conditions.
13. 16   **15.** There is no polygon satisfying these conditions.
17. 160°   **19.** Decrease   **21.** 60°   **23.** 6   **25.** 4
27. The sides are 7 inches, 7 inches, and 10 inches.   **29.** $x = 20$
$m\angle A = 3(20 + 16) = 108°$ $m\angle B = 108°$ $m\angle C = 6(20 - 2) = 108°$ $m\angle D = 5(20) + 8 = 108°$ $m\angle E = 4(20 + 7) = 108°$ The figure is not regular since only the angles are congruent, not the sides.   **31.** Consider any regular polygon with n sides and n congruent angles. By Theorem 3.17, the sum of the measures of the exterior angles of any polygon is 360°. In Problem 30, it was proven that the exterior angles of a regular polygon are congruent. Thus one exterior angle of a regular polygon with n sides and n angles measures $\dfrac{360°}{n}$.   **33.** The desired point is the point of intersection of the bisectors of $\angle DAB$ and $\angle ABC$, the incenter.
35.

Polygon	Drawing	Number of Sides	Number of Diagonals from One Vertex
Triangle		3	0
Quadrilateral		4	1
Pentagon		5	2
Hexagon		6	3
Octagon		8	5
Decagon		10	7
Dodecagon		12	9
n-gon		n	$n - 3$

3.4 Section Exercises

1. LA   **3.** LL   **5.** HL   **7.** $\angle ACB \cong \angle ECD$ by vertical $\angle$'s $\cong$; AAS   **9.** $\overline{NL} \cong \overline{NL}$ by reflexive law; HL

11.

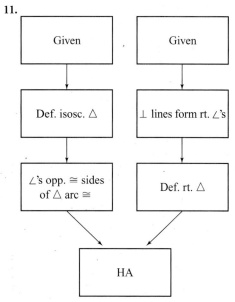

13. STATEMENTS

1. $\overline{AE} \cong \overline{BE}$; $\overline{AC} \cong \overline{BD}$
   $\overline{AB} \perp \overline{CD}$
2. $\angle AEC$ and $\angle BED$ are rt. $\angle$'s
3. $\triangle AEC$ and $\triangle BED$ are rt. $\triangle$'s
4. $\triangle AEC \cong \triangle BED$

REASONS

1. Given
2. $\perp$ lines form rt. $\angle$'s
3. Def. rt. $\triangle$'s
4. HL

15. STATEMENTS

1. $\overline{WX} \cong \overline{ZX}$
   $\angle V$ and $\angle Y$ are rt. $\angle$'s
2. $\triangle YXZ$ and $\triangle VXW$ are rt. $\triangle$'s
3. $\angle YXZ \cong \angle VXW$
4. $\triangle YXZ \cong \triangle VXW$

REASONS

1. Given
2. Def. rt. $\triangle$
3. Vertical $\angle$'s $\cong$
4. HA

17. STATEMENTS

1. Isosc. $\triangle VWX$ with base $\overline{VX}$
2. $\overline{VW} \cong \overline{XW}$
3. $\overline{WY}$ is perp. bisector of $\overline{VX}$
4. $\overline{VY} \cong \overline{XY}$
5. $\angle VYW$ and $\angle XYW$ are rt. $\angle$'s
6. $\triangle VWY$ and $\triangle XWY$ are rt. $\triangle$'s
7. $\triangle VWY \cong \triangle XWY$

REASONS

1. Given
2. Def. isos. $\triangle$
3. Given
4. Def. seg. bisector
5. $\perp$ lines form rt. $\angle$'s
6. Def. rt. $\triangle$
7. HL

Note Using different statements, the $\triangle$'s could be $\cong$ by HA or LL or LA or SAS or AAS or SSS or ASA.

19.

STATEMENTS	REASONS
1. $\overline{AB} \cong \overline{AD}$ $\angle ACB$ and $\angle ACD$ are rt. $\angle$'s	1. Given
2. $\triangle ACB$ and $\triangle ACD$ are rt. $\triangle$'s	2. Def. rt. $\triangle$
3. $\overline{AC} \cong \overline{AC}$	3. Reflexive law
4. $\triangle ACB \cong \triangle ACD$	4. HL
5. $\angle BAC \cong \angle DAC$	5. CPCTC
6. $\overline{AC}$ bisects $\angle BAD$	6. Def. $\angle$ bisector

21.

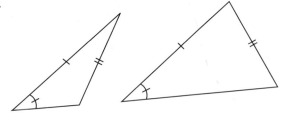

Chapter 3 Review Exercises

1. The first statement in a direct proof is P, and we form successive statements arriving at Q. The first statement in an indirect proof is $\sim Q$ and we form successive statements arriving at a contradiction (usually $\sim P$). Thus, our assumption of $\sim Q$ is wrong so we have Q, and therefore $P \longrightarrow Q$.

2. *Given:* I have the money.
Prove: I will buy my wife a present.
Proof:

STATEMENTS	REASONS
1. I have the money	1. Given
2. Assume I don't buy my wife a present.	2. Assumption
3. My wife will be unhappy.	3. Premise 3
4. She won't wash my shirts.	4. Premise 4
5. I can't go to work.	5. Premise 1
6. I won't have any money.	6. Premise 2

But this is a contradiction of Statement 1 "I have money." Thus, the assumption that "I don't buy my wife a present" is incorrect so we must conclude that I bought her one.
$\therefore$ If I have the money, then I will buy my wife a present.
3. For a given line ℓ and a point P not on ℓ, one and only one line through P is parallel to ℓ.   **4.** Yes, m and n are both $\perp$ to a third line, ℓ, therefore by Thm. 3.1 $m \parallel n$.   **5.** No, because the Parallel Postulate says there can only be one line through P parallel to ℓ **6.** Yes, since $\ell \parallel m$, alt. int. $\angle$'s $\cong$.   **7.** No
8. Yes; $\angle 1 \cong \angle 2$ by Statement 6 and $\angle 2 \cong \angle 3$ by vertical $\angle$'s thus $\angle 1 \cong \angle 3$ by transitive law   **9.** Yes, since $\ell \parallel m$, same side interior $\angle$'s are supp.   **10.** $\angle 1, \angle 2,$ and $\angle 3$ **11.** 30   **12.** 80

13. *Proof:*

STATEMENTS	REASONS
1. $\ell \parallel m$ and $\angle 1 \cong \angle 2$	1. Given
2. $\angle 3 \cong \angle 1$ and $\angle 4 \cong \angle 2$	2. Alt. int. $\angle$'s are $\cong$
3. $\angle 3 \cong \angle 4$	3. Trans. laws
4. $\overline{AB} \cong \overline{AC}$	4. Sides opp. $\cong \angle$'s are $\cong$
5. $\triangle ABC$ is isosceles	5. Def. of isosc. $\triangle$

14. $y = 140$ and $x = -60$ **15.** $720°$ **16.** $120°$ **17.** $360°$
18. $60°$ **19.** 15   **20.** 22   **21.** $35°$ **22.** No, the interior and exterior $\angle$ at the same vertex are supplementary, so interior could be less than corresponding exterior.   **23.** No, if there were two right $\angle$'s, their sum is $180°$ but the sum of all three $\angle$'s in $\triangle = 180°$.   **24.** $m\angle 1 = 60°$ $m\angle 2 = 120°$ $m\angle 3 = 120°$
25. $x = 19$

$$m\angle 1 = 2(19) + 1 = 39°$$
$$m\angle 2 = 3(19 - 2) = 51°$$
$$m\angle 3 = 39 + 90 = 129°$$

26. $\angle B \cong \angle D$ is needed.   **27.** $\overline{AB} \cong \overline{CB}$ is needed.
28. $\overline{AB} \cong \overline{DC}$ is needed.
29.

STATEMENTS	REASONS
1. $\overline{DB} \cong \overline{AC}; \overline{AB} \perp \overline{BC};$ $\overline{CD} \perp \overline{BC}$	1. Given
2. $\angle ABC$ and $\angle DCB$ are rt. $\angle$'s	2. $\perp$ lines form rt. $\angle$'s
3. $\triangle ABC$ and $\triangle DCB$ are rt. $\triangle$'s	3. Def. rt. $\triangle$'s
4. $\overline{BC} \cong \overline{BC}$	4. Reflexive law
5. $\triangle ABC \cong \triangle DCB$	5. HL

Chapter 3 Practice Test

1. Exactly one by the Parallel Postulate.   **2.** True, since $\ell \parallel m$, the $40°$ angle and $\angle 1$ are corresponding $\angle$'s so their measures are equal.   **3.** False, $m\angle 2 = 140°$ since it's supplementary to $\angle 1$.
4. True, since $\angle 1$ and $\angle 3$ are vertical angles, their measures are $=$.
5. False, the $\angle$'s are $\cong$ because they are corresponding $\angle$'s but not supplementary. The angles may appear to be right $\angle$'s but this is not known for sure since $\perp$ are not given.   **6.** $1080°$
7. $360°$ **8.** $135°$ **9.** $m\angle 1 = 47°$ **10.** $m\angle 3 = 144°$
11.

STATEMENTS	REASONS
1. $\overline{AB} \cong \overline{CB}$, $\angle A \cong \angle C$ $\angle B$ is a right angle	1. Given
2. $\triangle ABE$ and $\triangle CBD$ are rt. $\triangle$'s	2. Def. rt. $\triangle$
3. $\triangle ABE \cong \triangle CBD$	3. $\angle A$

12. 17 cm   **13.** 20
14.

STATEMENTS	REASONS
1. $\angle B \cong \angle D; \overline{AB} \parallel \overline{CD}$	1. Given
2. $\angle BAC \cong \angle DCA$	2. If lines are $\parallel$, alt. int. $\angle$'s $\cong$
3. $\overline{AC} \cong \overline{AC}$	3. Reflexive law
4. $\triangle ABC \cong \triangle CDA$	4. AAS

15. 90°, since the sum of the measures of ∠'s of a △ = 180° and one angle measures 90°, the sum of remaining two angles is 90°. 180° − 90° = 90°.

Chapters 1–3 Cumulative Review

1. True   **2.** False   **3.** True   **4.** False   **5.** True   **6.** True
7. True   **8.** False   **9.** True   **10.** False   **11.** True   **12.** True
13. False   **14.** False   **15.** 63°

16.

17. Scalene triangle   **18.** Obtuse triangle   **19.** ∠A (or ∠1)
20. ∠4   **21.** $\overline{AB}$ **22.** 22.5 in (7.5 × 3)

23.

STATEMENTS	REASONS
1. $\overline{AB} \cong \overline{BC}$	1. Given
2. ∠A ≅ ∠BCA	2. If two sides of △ ≅, ∠'s opp. them are ≅.
3. $\overline{DE} \cong \overline{EC}$	3. Given
4. ∠ECD ≅ ∠D	4. If two sides of △ ≅, ∠'s opp. them are ≅.
5. ∠BCA ≅ ∠ECD	5. Vertical ∠'s ≅.
6. ∠A ≅ ∠D	6. Transitive law

24. $\overline{AD}$ is an altitude
$\overleftrightarrow{EH}$ is a perpendicular bisector
$\overline{CF}$ is a median

25.

STATEMENTS	REASONS
1. $\overline{WY}$ bisects ∠XYZ	1. Given
2. ∠XYW ≅ ∠ZYW	2. Def. ∠ bisector
3. $\overline{YW}$ bisects ∠XWZ	3. Given
4. ∠XWY ≅ ∠ZWY	4. Def. ∠ bisector
5. $\overline{WY} \cong \overline{WY}$	5. Reflexive law
6. △WXY ≅ △WZY	6. AAS

26.

STATEMENTS	REASONS
1. $\overline{AD}$ and $\overline{BC}$ bisect each other	1. Given
2. $\overline{AE} \cong \overline{DE}$; $\overline{BE} \cong \overline{CE}$	2. Def. of bisector
3. ∠AEB ≅ ∠DEC	3. Vertical ∠'s ≅
4. △AEB ≅ △DEC	4. SAS
5. $\overline{AB} \cong \overline{DC}$	5. CPCTC

27. 54 mm
28. (a) 29°   (b) 92°   (c) 122°   (d) 58°   (e) 122°
(f) 58°   (g) 97°   (h) 54°   (i) 59°   (j) 63°   (k) 23°
29. $m\angle 1 = m\angle 4 = m\angle 5 = m\angle 8 = 133°$ **30.** 1080°
$m\angle 2 = m\angle 3 = m\angle 6 = m\angle 7 = 47°$
31. 360°

32.

STATEMENTS	REASONS
1. $\overline{AB} \perp \overline{BE}$; $\overline{DE} \perp \overline{BE}$; ∠A ≅ ∠D	1. Given
2. ∠ABE and ∠DEB are rt. ∠'s.	2. ⊥ lines form rt. ∠'s.
3. △ABE and △DEB are rt. △'s	3. Def. rt. △'s
4. $\overline{BE} \cong \overline{EB}$	4. Reflexive law
5. △ABE ≅ △DEB	5. ∠A

33. △ABC ≅ △ETP by ASA   **34.** △FOR ≅ △QDE by SSS
35. △HOP ≅ △FIT by AAS   **36.** Not enough information
37. △ABC ≅ △EMC by ASA or LA   **38.** △WXZ ≅ △YXZ by
AAS or HA   **39.** Not enough information
40. △SUM ≅ △FHG by SAS or LL

Chapter 4

4.1 Section Exercises

1. True   **3.** True   **5.** True   **7.** False   **9.** False   **11.** True
13. True   **15.** True   **17.** False   **19.** True   **21.** 1. Sum of ∠'s
of quadralateral = 360° 2. Given; Def. ≅ ∠'s 3. Substitution law
5. Mult.-div. law 6. ∠A and ∠B are supplementary 7. $\overline{AD} \| \overline{BC}$
9. Def. supp. ∠'s. 10. $\overline{AB} \| \overline{DC}$ 11. ABCD is a parallelogram
23. No, the four sides of two parallelograms could be congruent
without the corresponding angles being congruent. [▱   ▱]

25.

STATEMENTS	REASONS
1. $\overline{VZ} \| \overline{WY}$; $\overline{VZ} \cong \overline{WX}$; $\overline{WY} \cong \overline{WX}$	1. Given
2. $\overline{VZ} \cong \overline{WY}$	2. Transitive law
3. VWYZ is ▱	3. If one pair of opp. sides of quad. are ≅ and |, the quad is ▱.

27. 60° and 120°
29. *Proof:*

STATEMENTS	REASONS
1. $\overline{AP} \cong \overline{QC}$	1. Given
2. ABCD is a parallelogram	2. Given
3. ∠1 ≅ ∠2	3. If lines |, alt. int. ∠'s are ≅
4. $\overline{AB} \cong \overline{DC}$	4. Opp. sides of ▱ are ≅
5. △ABP ≅ △CDQ	5. SAS
6. $\overline{PB} \cong \overline{QD}$	6. CPCTC
7. ∠3 ≅ ∠4	7. If lines |, alt. int. ∠'s are ≅
8. $\overline{AD} \cong \overline{CB}$	8. Opp. sides of ▱ are ≅
9. △APD ≅ △CQB	9. SAS
10. $\overline{PD} \cong \overline{QB}$	10. CPCTC
11. PBQD is a parallelogram	11. If opp. sides are ≅, then ▱

31. ▱   **33.** approximately 72 pounds

4.2 Section Exercises

1. (a) Isosceles △ because △ABC has 2 ≅ sides since $ABCD$ is a rhombus which has 4 ≅ sides (b) Right △ because the diagonals of a rhombus are ⊥. (c) Yes, one reason is SSS. $\overline{AD} \cong \overline{AB}$ because they are sides of a rhombus. $\overline{BE} \cong \overline{DE}$ because the diagonals of a rhombus bisect each other. Finally, $\overline{AE} \cong \overline{AE}$ by the reflexive law. **3.** A rhombus and a kite are similar in that they are both quadrilaterals, diagonals are ⊥, have 2 ≅ adj. sides, and one diagonal bisects the other diagonal. The differences are a rhombus is a parallelogram, so it has all the properties of a parallelogram. A kite is not a parallelogram. **5.** 50.4 in **7.** 4.2 in **9.** $m\angle 1 = 60°$; $m\angle 2 = 60°$ **11.** $m\angle 2 = 121°$ $m\angle 1 = 59°$

13. *Proof:*

STATEMENTS	REASONS
1. $\square ABCD$ is a rhombus	1. Given
2. $\overline{AC} \perp \overline{BD}$	2. Diag. of rhombus are ⊥
3. $m\angle DPC = 90°$	3. ⊥ lines form 90° angles
4. $m\angle DPC + m\angle 1 +$ $m\angle 2 = 180°$	4. Sum of ∠'s of △ = 180°
5. $90° + m\angle 1 + m\angle 2$ $= 180°$	5. Substitution law
6. $m\angle 1 + m\angle 2 = 90°$	6. Add.-subt. law
7. $m\angle 1$ and $m\angle 2$ are complementary	7. Def. of comp. ∠'s

4.3 Section Exercises

1. Parallelogram, rhombus, rectangle, square **3.** Parallelogram, rhombus, rectangle, square **5.** Parallelogram, rhombus, rectangle, square **7.** Rectangle, square **9.** Rhombus, kite, square **11.** Rhombus, kite, square **13.** Rhombus, square **15.** 8 **17.** 8 **19.** 60 inches (10 + 24 + 26) because opposites sides of a rectangle are ≅ and diagonals of a rectangle bisect each other. **21.** 50 inches (13 + 13 + 24) because opposites sides of a rectangle are ≅ and diagonals of a rectangle bisect each other. **23.** Use Construction 4.2 **25.** 112 ft.
27. (a) 9 cm
(b) 5 cm
(c) 7 cm
(d) 42 cm
(e) 21 cm
(f) The perimeter of the △ formed by joining the midpoints is one-half the perimeter of △ABC. Yes, the perimeter will always be one-half the perimeter of the original △ because the lengths of the sides of the first △ are always $\frac{1}{2}$ the lengths of the original sides by Theorem 4.19.

29. False,

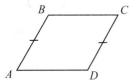

31. True, because all the ∠'s of a rectangle are rt. ∠'s.
33. 204 squares **35.** $m\angle 1 = 59°$ $m\angle 2 = 121°$ $m\angle 3 = 121°$ $m\angle 4 = 31°$ $m\angle 5 = 39°$ $m\angle 6 = 20°$ $m\angle 7 = 90°$ $m\angle 8 = 20°$ $m\angle 9 = 62°$ $m\angle 10 = 28°$ $m\angle 11 = 90°$ $m\angle 12 = 90°$ $m\angle 13 = 31°$ $m\angle 14 = 59°$ $m\angle 15 = 62°$ $m\angle 16 = 59°$ $m\angle 17 = 59°$ $m\angle 18 = 121°$ $m\angle 19 = 62°$ $m\angle 20 = 28°$

4.4 Section Exercises

1. True **3.** True **5.** True **7.** True **9.** True **11.** False **13.** True **15.** True **17.** True **19.** False **21.** 4.5 cm by Thm. 4.23 **23.** 81° by Thm. 3.8 **25.** 63° by Thm. 3.8 **27.** 99° by Thm. 3.11 **29.** 9 **31.** $4x + 2 = AB$ **33.** 1. By construction 2. $ABCD$ is a trapezoid with median $\overline{EF}$ 3. Def. of median of trapezoid 4. $\overline{AE} \cong \overline{ED}$ and $\overline{BF} \cong \overline{FC}$ 5. Vert. ∠'s are ≅ 6. $\overline{AB} \parallel \overline{CD}$ 7. If lines are ∥, alt. int. ∠'s are ≅ 8. AAS 9. CPCTC 11. Seg. Add. post. 12. Substitution law 15. Line ∥ to one of two ∥ lines is ∥ to other.

35.

A B

D C

Given: Isosceles trapezoid $ABCD$ with equal sides $\overline{AD}$ and $\overline{BC}$ and diagonals $\overline{AC}$ and $\overline{BD}$
Prove: $\overline{AC} \cong \overline{BD}$

Proof:

STATEMENTS	REASONS
1. $ABCD$ is an isosceles trapezoid with $\overline{AD} \cong \overline{BC}$	1. Given
2. $\angle ADC \cong \angle BCD$	2. Base ∠'s of isosc. trapezoid are ≅
3. $\overline{DC} \cong \overline{DC}$	3. Reflexive law
4. $\triangle ADC \cong \triangle BCD$	4. SAS
5. $\overline{AC} \cong \overline{BD}$	5. CPCTC

37.

A P B
S Q
D R C

Given: $\square ABCD$ with $P, Q, R,$ and S midpoints of adjacent sides
Prove: $PQRS$ is a rhombus

Proof

STATEMENTS	REASONS
1. *ABCD* is a rectangle with *P, Q, R,* and *S* midpoints of sides	1. Given
2. $\angle A, \angle B, \angle C,$ and $\angle D$ are right angles	2. All $\angle$'s of $\square$ are rt. $\angle$'s
3. $\angle A \cong \angle B \cong \angle C \cong \angle D$	3. All rt. $\angle$'s are $\cong$
4. $\overline{AB} \cong \overline{DC}$ and $\overline{AD} \cong \overline{BC}$	4. Opp. sides of $\square$ are $\cong$
5. $\overline{AP} \cong \overline{PB}, \overline{BQ} \cong \overline{QC}, \overline{CR} \cong \overline{DR},$ and $\overline{DS} \cong \overline{AS}$	5. Def. of midpoint
6. $\triangle APS \cong \triangle BPQ \cong \triangle CRQ \cong \triangle DRS$	6. SAS
7. $\overline{SP} \cong \overline{QP} \cong \overline{QR} \cong \overline{SR}$	7. CPCTC
8. *PQRS* is a parallelogram	8. Both pairs of opp. sides are $\cong$
9. *PQRS* is a rhombus	9. Two adj. sides are $\cong$ in a $\square$

Chapter 4 Review Exercises

1. True **2.** True **3.** False **4.** True **5.** True **6.** False
7. False **8.** True **9.** True **10.** True

11. *Proof:*

STATEMENTS	REASONS
1. $\angle 1 \cong \angle 2$	1. Given
2. $\overline{NO} \cong \overline{NP}$	2. If two $\angle$'s in a $\triangle \cong$, sides opp. them are $\cong$.
3. $\overline{MQ} \cong \overline{NO}; \overline{MN} = \overline{QP}$	3. Given
4. $\overline{MQ} \cong \overline{NP}$	4. Transitive law
5. *MNPQ* is $\square$	5. If opp. sides of quad. $\cong$, it's a $\square$.

12.

STATEMENTS	REASONS
1. $\overline{AB} \cong \overline{CD}; \angle 1 \cong \angle 2$	1. Given
2. $\overline{AB} \parallel \overline{CD}$	2. If alt. int. $\angle$'s $\cong$, lines $\parallel$
3. *ABCD* is parallelogram	3. If two opp. sides of quad. $\cong$ and $\parallel$, then $\square$.

13. *Proof:*

STATEMENTS	REASONS
1. $\angle 1 \cong \angle 2; \angle 3 \cong \angle 4$	1. Given
2. $\overline{NL} \cong \overline{NL}$	2. Reflexive law
3. $\triangle MNL \cong \triangle ONL$	3. ASA
4. $\overline{MN} \cong \overline{ON}$ and $\overline{ML} \cong \overline{OL}$	4. CPCTC
5. *LMNO* is kite	5. Def. kite (2 pr. $\cong$ consec. sides)

14. *Proof:*

STATEMENTS	REASONS
1. $\square ABCD; \square DEFG$	1. Given
2. $\angle B \cong \angle CDA$	2. Opp. $\angle$'s $\square \cong$
3. $\angle CDA \cong \angle EDG$	3. Vertical $\angle$'s $\cong$
4. $\angle EDG \cong \angle F$	4. Opp. $\angle$'s $\square \cong$
5. $\angle B \cong \angle F$	5. Transitive law

15. $m\angle 1 = 108°$ by Thm. 4.13, one pair of opp. $\angle$'s $\cong$ in a kite. The sum of the $\angle$'s of quadrilateral is 360° by Thm 3.15. $m\angle 2 = 360 - (108 + 108 + 43) = 101°$

16.

STATEMENTS	REASONS
1. $\triangle EBF \cong \triangle GDH$ $\triangle AEH \cong \triangle CGF$	1. Given
2. $\overline{EF} \cong \overline{GH}; \overline{EH} \cong \overline{GF}$	2. CPCTC
3. *EFGH* $\square$	3. If both pr. opp. sides of quad $\cong$, then $\square$.

17. False **18.** True **19.** True **20.** True **21.** True
22. False **23.** False **24.** 50.8 cm **25.** True **26.** False
27. True **28.** False **29.** True **30.** True **31.** False
32. True **33.** $X = 3$ $CD = 9$ **34.** $X = 1$ $DE = 5; BC = 10$

35.

STATEMENTS	REASONS
1. Rectangle *ABCD* with diagonals $\overline{AC}$ and $\overline{BD}$	1. Given
2. $\overline{AB} \cong \overline{DC}$	2. Opp. sides rect. $\cong$
3. $\overline{BD} \cong \overline{CA}$	3. Diagonals of rect. $\cong$
4. $\overline{AD} \cong \overline{AD}$	4. Reflexive
5. $\triangle ABD \cong \triangle DCA$	5. SSS
6. $\angle 1 \cong \angle 2$	6. CPCTC

36. Use Construction 1.3 to construct the two diagonals; they are $\perp$ bisectors of each other. Draw the square by connecting endpoints of diagonals.

37. *Proof:*

STATEMENTS	REASONS
1. *ABCE* is a rectangle	1. Given
2. *BCDE* is a parallelogram	2. Given
3. $\overline{AC} \cong \overline{BE}$	3. Diag. of rect. are $\cong$
4. $\overline{BE} \cong \overline{CD}$	4. Opp. sides of $\square$ are $\cong$
5. $\overline{AC} \cong \overline{CD}$	5. Transitive law
6. $\triangle ACD$ is isosceles	6. Def. of isosc. $\triangle$

38. False **39.** True **40.** True **41.** False **42.** True
43. No **44.** Use Construction 4.3. **45.** 36 ft **46.** $AD = 22.2$
47. $m\angle B = 113°, m\angle C = 113°, m\angle D = 67°$

Chapter 4 Practice Test

1. True **2.** True **3.** False **4.** True **5.** $m\angle ADC = 70°$

6.

STATEMENTS	REASONS
1. *ABCD* is rectangle; $\overline{AE} \cong \overline{FC}$	1. Given
2. $\angle A$ and $\angle C$ are rt. $\angle$'s	2. All $\angle$'s of $\square$ are rt. $\angle$'s
3. $\triangle ADE$ and $\triangle CBF$ are rt. $\triangle$'s	3. Def. rt. $\triangle$
4. $\overline{AD} \cong \overline{CB}$	4. Opp. sides of $\square \cong$
5. $\triangle ADE \cong \triangle CBF$	5. LL
6. $\overline{DE} \cong \overline{BF}$	6. CPCTC

7.

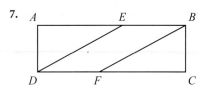

Given: $ABCD$ is a rectangle with $\overline{AE} \cong \overline{FC}$
Prove: $DEBF$ is a parallelogram
Proof:

STATEMENTS	REASONS
1. $ABCD$ is a rectangle with $\overline{AE} \cong \overline{FC}$ therefore $AE = FC$	1. Given; def. $\cong$ seg.
2. $\overline{EB} \parallel \overline{DF}$	2. $\overline{AB} \parallel \overline{DC}$ since they are opp. sides of a $\square$
3. $\overline{AB} \cong \overline{DC}$ therefore $AB = DC$	3. Opp. sides of $\square$ are $\cong$; def. $\cong$ seg.
4. $AB = AE + EB$ and $DC = DF + FC$	4. Seg. add. post.
5. $AE + EB = DF + FC$	5. Substitution law (Steps 3 and 4)
6. $AE + EB = DF + AE$	6. Substitution (Steps 1 and 5)
7. $EB = DF$	7. Add.-subtr. law
8. $DEBF$ is a parallelogram	8. Opp. sides are $\cong$ and $\parallel$

8. Use Construction 1.2 and a construction similar to 4.2
9. 88 cm   **10.** $AC = 30$ **11.** $DC = 14.9$ in   **12.** $m\angle B = 113°$. $m\angle C = 113°$ **13.** $MN = 24$ cm   **14.** $14 = x$ **15.** Yes, the two parallelograms with $\cong$ diagonals are a square and a rectangle. Since a square is a rectangle by definition, the figure must be a rectangle.

16.

STATEMENTS	REASONS
1. $\square$ $ABCD$; $\overline{AD} \cong \overline{CN}$; $\angle 1 \cong \angle 2$	1. Given
2. $\overline{CN} \cong \overline{CD}$	2. If 2 $\angle$'s $\triangle$ are $\cong$, sides opp. those $\angle$'s $\cong$
3. $\overline{AD} \cong \overline{CD}$	3. Transitive law
4. $ABCD$ is rhombus	4. Def. rhombus ($\square$ with 2 $\cong$ adj. sides)

17. Square   **18.** Square or rhombus   **19.** Trapezoid
20. Kite, square, or rhombus

Chapter 5

5.1 Section Exercises

1. $\frac{4}{7}$ **3.** $\frac{1}{4}$ **5.** 50 mph   **7.** $\frac{1}{6}$ **9.** $\frac{4}{3}$ **11.** a = 2   **13.** $y = \frac{7}{4}$
15. $x = 5$ or $x = -5$.   **17.** $5 = a$ **19.** $4 = y$ **21.** 12 or -12
23. 24 or -24 **25.** 6500 votes.   **27.** 130 mi.   **29.** 805 lbs.
31. Take the appropriate measurements and approximate the various ratios.   **33.** 22.5 cm, 30 cm, and 37.5 cm   **35.** 40° and 140°   **37.** A ratio compares numbers by using division. A ratio is written as a fraction. A proportion is an equation that shows two ratios are equal.

5.2 Section Exercises

1. 15 cm   **3.** $z = 12$ cm   **5.** $m\angle C' = 103°$ **7.** Always
9. Sometimes   **11.** Always   **13.** Sometimes   **15.** Never
17. Sometimes   **19.** 10 ft   **21.** $AD = 16$ yd and $BD = 6$ yd
23. $AD = 10$ ft and $BD = 5$ ft   **25.** $DC = 3$ ft   **27.** $BD = 18$ in and $DC = 27$ in   **29.** 120 ft
31. *Proof*

STATEMENTS	REASONS
1. ABC is a triangle and $\angle 1 \cong \angle 2$	1. Given
2. $\angle A \cong \angle A$	2. Reflexive law
3. $\triangle ABC \sim \triangle ADE$	3. AA
4. $\dfrac{AD}{AB} = \dfrac{DE}{BC}$	4. Def. $\sim \triangle$'s

33. *Proof*

STATEMENTS	REASONS
1. ABC is a triangle and $\overline{DE} \parallel \overline{BC}$	1. Given
2. $\angle ADP \cong \angle ABF$ and $\angle AEP \cong \angle ACF$	2. If lines $\parallel$ corr. $\angle$'s are $\cong$
3. $\angle DAP \cong \angle BAF$ and $\angle PAE \cong \angle FAC$	3. Reflexive law
4. $\triangle ADP \sim \triangle ABF$ and $\triangle APE \sim \triangle AFC$	4. AA
5. $\dfrac{DP}{BF} = \dfrac{AP}{AF}$ and $\dfrac{AP}{AF} = \dfrac{PE}{FC}$	5. Def. $\sim \triangle$'s
6. $\dfrac{DP}{BF} = \dfrac{PE}{FC}$	6. Trans. law

35. Use Construction 5.1.
37.

STATEMENTS	REASONS
1. Trapezoid $WXYZ$ with $\overline{XY} \parallel \overline{WZ}$	1. Given
2. $\angle YXZ \cong \angle XZW$ and $\angle XYW \cong \angle YWZ$	2. If lines $\parallel$, alt. int. $\angle$'s $\cong$
3. $\triangle XAY \sim \triangle ZAW$	3. AA

39. Answers will vary.
41. *Given:* $\triangle ABC \sim \triangle DEF$ with $\angle A \cong \angle D$. $\overline{AP}$ and $\overline{DQ}$ are altitudes
Prove: $\dfrac{AP}{DQ} = \dfrac{AB}{DE} = \dfrac{BC}{EF} = \dfrac{AC}{DF}$

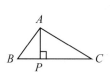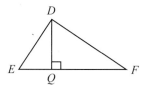

Proof

STATEMENTS	REASONS
1. $\triangle ABC \sim \triangle DEF$ with $\angle A \cong \angle D$	1. Given
2. $\overline{AP}$ and $\overline{DQ}$ are altitudes	2. Given
3. $\overline{AP} \perp \overline{BC}$ and $\overline{DQ} \perp \overline{EF}$	3. Def. of altitude
4. $\angle APB$ and $\angle DQE$ are right angles	4. $\perp$ lines form right $\angle$'s
5. $\angle APB \cong \angle DQE$	5. All rt. $\angle$'s are $\cong$
6. $\angle B \cong \angle E$	6. Corresponding $\angle$'s in $\sim$ $\triangle$'s are $\cong$ (def. $\sim\triangle$'s)
7. $\triangle ABP \sim \triangle DEQ$	7. AA
8. $\dfrac{AP}{DQ} = \dfrac{AB}{DE}$	8. Corr. sides of $\sim$ $\triangle$'s are prop. (Def. $\sim\triangle$'s)

Note: Since $\dfrac{AB}{DE} = \dfrac{BC}{EF} = \dfrac{AC}{DF}$, the ratio of corresponding altitudes is equal to the ratio of *any* two corresponding sides.

43. Use Construction 5.2.

45. (a) By AA

(b) $\dfrac{1}{3}$

(c) $\dfrac{1}{3}$

(d) They are 3 times longer; that is, this pantograph will enlarge a drawing to one 3 times larger than the original.

(e) Place the pen at D, trace the original at E and a reduced copy will be made at D.

5.3 Section Exercises

1. $x = 4\ y = \sqrt{20}$ or $2\sqrt{5}$ **3.** $x = 8\ y = \sqrt{128}$ or $8\sqrt{2}$

5. $x = \dfrac{27}{5}\ y = \dfrac{48}{5}$ **7.** $x = 9$ **9.** $10 = x$ **11.** $CE = 10$ cm

13. $CD = 6$ cm   **15.** $AB = 10$ ft   **17.** $BD = 27$ cm   **19.** $AC = 8$ yd   **21.** 5.48 ft   **23.** 4.06 cm   **25.** 8.94 yd   **27.** An altitude of a triangle is a line segment from a vertex perpendicular to the opposite side. A median is also a line segment from a vertex but it connects the vertex to the midpoint of the opposite side. The similarities are they are both segments from a vertex of a triangle. The difference is the altitude is always perpendicular to the opposite side and the median always bisects the opposite side.

5.4 Section Exercises

1. $c = 5$ cm   **3.** $a = 60$ ft   **5.** $b = \sqrt{85}$ yd   **7.** $b = 18$ cm

9. Yes   **11.** Yes   **13.** No   **15.** $c = 10\sqrt{2}$ ft

17. $b = 3\sqrt{2}$ yd   **19.** $c = 6$ cm   **21.** $c = 3\sqrt{6}$ ft

23. $a = 3\sqrt{2}$ yd   **25.** $b = \dfrac{1}{2}$ cm   **27.** $c = 20$ ft

29. $b = 8$ yd   **31.** $a = 7\sqrt{3}$ cm   **33.** $b = 2$ ft

35. $a = \dfrac{3}{2}$ yd   **37.** $c = 2$ cm   **39.** $4\sqrt{2}$ inches   **41.** 17.0 ft

43. 230.9 ft   **45.** $5\sqrt{3}$ ft   **47.** $25\sqrt{3}$ ft^2 **49.** $\sqrt{3}x = d$

51. $\sqrt{3}$ times the length of the original segment. The process can be continued to find a segment $\sqrt{n}$ times the length of the original segment for $n = 2, 3, 4, 5, \ldots$ **53.** The quadrilateral is clearly a rhombus, because all sides have length c. Show it contains a right angle by showing the angle is the supplement of the angle formed by adding the two acute angles of the right triangle.

55. (a) Show $\angle DAB$ is a right angle by an argument similar to that in Exercise 53.

(b) $\dfrac{1}{2}ab; \dfrac{1}{2}ab; \dfrac{1}{2}c^2$

(c) Show that $\overline{ED} \parallel \overline{CB}$ since each is perpendicular to the same line.

(d) $\dfrac{1}{2}(a + b)(a + b)$

(e) When the expressions from parts b and d are equated, the result simplifies to $a^2 + b^2 = c^2$.

5.5 Section Exercises

1. Transitive law   **3.** Division property of inequalities   **5.** Subtraction property of inequalities   **7.** The whole is greater than its parts   **9.** Transitive law   **11.** $x < 4$ **13.** $x > -\dfrac{7}{3}$ **15.** $x < 8$ **17.** $x < -18$ **19.** False, an exterior angle of a triangle is greater than (not equal to) each remote interior angle.   **21.** True   **23.** True   **25.** False, $BC > 16$ yd since sides opposite equal angles are unequal in the same order.   **27.** True   **29.** True   **31.** $AC > DE$ **33.** $m\angle C > m\angle D$ **35.** The Triangle Inequality Theorem prevents this from happening since $3 + 4 = 7 < 8$ not greater than 8.

37. *Proof*

STATEMENTS	REASONS
1. $BQ = ED$ thus $\overline{BQ} \cong \overline{ED}$	1. By Construction; Def. $\cong$ seg.
2. $BC = EF$ thus $\overline{BC} \cong \overline{EF}$	2. Given; def. $\cong$ seg.
3. $m\angle QBC = m\angle DEF$ thus $\angle QBC \cong \angle DEF$	3. Given; def. $\cong \angle$'s.
4. $\triangle QBC \cong \triangle DEF$	4. SAS
5. $\overline{QC} \cong \overline{DF}$ thus $QC = DF$	5. CPCTC; def. $\cong$ seg.
6. $AC = QC + AQ$	6. Seg. add. post.
7. $AC > QC$	7. Whole $>$ any part
8. $AC > DF$	8. Substitution law

39. *Proof*

STATEMENTS	REASONS
1. $AB = DE$ and $BC = EF$ thus $\overline{AB} \cong \overline{DE}$; $\overline{BC} \cong \overline{EF}$	1. Given; def. $\cong$ seg.
2. $BQ = ED$ thus $\overline{BQ} \cong \overline{ED}$	2. By construction; def. $\cong$ seg.
3. $\overline{AB} \cong \overline{BQ}$	3. Trans. laws and Statements 1 and 2
4. $\overline{BR} \cong \overline{BR}$	4. Reflexive law
5. $\overline{BR}$ bisects $\angle ABQ$	5. By construction
6. $\angle ABR \cong \angle RBQ$	6. Def. of $\angle$ bisector
7. $\triangle ABR \cong \triangle RBQ$	7. SAS
8. $\overline{AR} \cong \overline{RQ}$ thus $AR = RQ$	8. CPCTC; def. $\cong$ seg.
9. $AC = AR + RC$	9. Seg. add. post.
10. $AC = RQ + RC$	10. Substitution law
11. $RQ + RC > QC$	11. Triangle Inequality Theorem
12. $AC > QC$	12. Substitution law
13. $\angle QBC \cong \angle DEF$	13. By construction
14. $\triangle QBC \cong \triangle DEF$	14. SAS
15. $\overline{QC} \cong \overline{DF}$ thus $QC = DF$	15. CPCTC; def. $\cong$ seg.
16. $AC > DF$	16. Substitution law

41. *Proof*

STATEMENTS	REASONS
1. $m\angle A > m\angle B$	1. Given
2. $BC > AC$	2. Sides opp. $\neq$ $\angle$'s are $\neq$ in same order
3. $m\angle D > m\angle E$	3. Given
4. $CE > CD$	4. Same as Statement 2
5. $BC + CE > AC + CD$	5. Add. prop. of inequalities
6. $BE = BC + CE$ and $AD = AC + CD$	6. Seg. add. post.
7. $BE > AD$	7. Substitution law

43. *Given:* $\triangle ABC$

Prove: $AC - AB < BC$

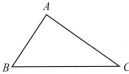

Proof

STATEMENTS	REASONS
1. ABC is a triangle	1. Given
2. $AC < AB + BC$	2. Triangle Inequality
3. $AB = AB$	3. Reflexive law
4. $AC - AB < BC$	4. Subtraction prop. of inequalities using Statements 2 and 3

45. *Given:* $\triangle ABC$ with P inside the triangle

Prove: $PA + PB + PC > \frac{1}{2}(AB + BC + AC)$

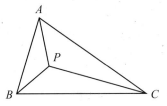

Proof

STATEMENTS	REASONS
1. ABC is a triangle with P inside it	1. Given
2. $PA + PB > AB$, $PA + PC > AC$, and $PB + PC > BC$	2. Triangle inequality applied to $\triangle APB$, $\triangle APC$, and $\triangle BPC$
3. $PA + PB + PA + PC + PB + PC > AB + AC + BC$	3. Add. prop. of inequalities
4. $2PA + 2PB + 2PC > AB + AC + BC$	4. Distributive law
5. $2(PA + PB + PC) > AB + AC + BC$	5. Distributive law
6. $PA + PB + PC > \frac{1}{2}(AB + AC + BC)$	6. Div. prop. of inequalities

47. *Proof*

STATEMENTS	REASONS
1. $PA = AD$	1. D is $\perp$ reflection of P in mirror
2. $AB = AB$, $AC = AC$	2. Reflexive law
3. $\angle DAC$ and $\angle PAC$ are rt. $\angle$'s	3. $\perp$ lines form rt. $\angle$'s
4. $\triangle DAC$, $\triangle PAC$, $\triangle DAB$, and $\triangle PAB$ are rt. $\triangle$'s	4. Def. of rt. $\triangle$
5. $\triangle DAB \cong \triangle PAB$ and $\triangle DAC \cong \triangle PAC$	5. LL
6. $\overline{DB} \cong \overline{PB}$ and $\overline{DC} \cong \overline{PC}$ thus $DB = PB$ and $DC = PC$	6. CPCTC; def. $\cong$ seg.
7. $DQ = DB + BQ$	7. Seg. add. post.
8. $DQ < DC + CQ$	8. Triangle ineq.
9. $DB + BQ < DC + CQ$	9. Substitution law
10. $PB + BQ < PC + CQ$	10. Substitution law using Statements 6 and 9

49. (a) The minimum distance is 62 miles and the maximum distance is 1240 miles. (b) The minimum distance is 217 miles and the maximum distance is 1519 miles. (c) The minimum distance is 279 miles and the maximum distance is 1457 miles.

Chapter 5 Review Exercises

1. $\dfrac{4}{5}$ **2.** 60 mph **3.** $a = 3$ **4.** $x = 40$ **5.** $2 = y$ **6.** $5 = a$

7. $x = 20$ **8.** x is 20 or x is -20. **9.** $2\dfrac{1}{4}$ inches on the map

represents an actual distance of 180 miles. **10.** False
11. True **12.** True **13.** $y = 3.5$ cm **14.** $x = 2$ cm
15. $z = 8$ cm **16.** $m\angle C' = 95°$ **17.** Yes **18.** No
19. $EC = 21$ ft **20.** $AE = 37.5$ ft and $EC = 12.5$ ft
21. $FC = 20$ cm **22.** $BF = 21$ ft and $FC = 56$ ft
23. *Proof*

STATEMENTS	REASONS
1. $\triangle ABC$ is isosceles with base $\overline{BC}$	1. Given
2. $\angle C$ is supplementary to $\angle 1$	2. Given
3. $m\angle C + m\angle 1 = 180°$	3. Def. supp. $\angle$'s
4. $\angle C \cong \angle B$; thus $m\angle C = m\angle B$	4. Base angles of isos. $\triangle \cong$; def. $\cong \angle$'s
5. $m\angle B + m\angle 1 = 180°$	5. Substitution
6. $\angle B$ supp to $\angle 1$	6. Def. supp. $\angle$'s
7. $\overline{DE} \parallel \overline{BC}$	7. If interior $\angle$'s on same side of transversal supp., then lines $\parallel$.
8. $\dfrac{AD}{DB} = \dfrac{AE}{EC}$	8. Line $\parallel$ to third side of $\triangle$ divides other sides into prop. segments (Theorem 5.11)

24. *Proof*

STATEMENTS	REASONS
1. ABC is a triangle with $\overline{BD}$ bisecting $\angle B$	1. Given
2. $\overline{ED} \parallel \overline{BC}$	2. Given
3. $\dfrac{AE}{EB} = \dfrac{AD}{DC}$	3. Line $\parallel$ to 3rd side of $\triangle$ divides other sides in prop. seg. (Thm 5.11)
4. $\dfrac{AB}{BC} = \dfrac{AD}{DC}$	4. Bisector of $\angle$ divides side into seg. prop. to other two sides (Thm 5.12)
5. $\dfrac{AE}{EB} = \dfrac{AB}{BC}$	5. Trans. laws

25. Use Construction 5.1. **26.** 22 cm **27.** 20 ft
28. $AC = 8\sqrt{7}$ cm **29.** $AD = 4$ yd **30.** 915 ft²
31. *Given:* $ABCD$ is a rhombus with diagonals $\overline{AC}$ and $\overline{BD}$, and $\overline{PQ} \perp \overline{BC}$
Prove: $(PE)^2 = (BP)(PC)$
Proof

STATEMENTS	REASONS
1. $ABCD$ is a rhombus with diagonals $\overline{AC}$ and $\overline{BD}$	1. Given
2. $\overline{AC} \perp \overline{BD}$	2. Diag. of rhombus are $\perp$
3. $\angle BEC$ is a right angle	3. $\perp$ lines form rt. $\angle$'s
4. $\triangle BEC$ is a right triangle	4. Def. of rt. $\triangle$
5. $\overline{PQ} \perp \overline{BC}$	5. Given
6. $\overline{EP}$ is an altitude of $\triangle BEC$ from right angle $\angle BEC$	6. Def. of altitude
7. $\dfrac{BP}{PE} = \dfrac{PE}{PC}$	7. Corollary 5.14
8. $(PE)^2 = (BP)(PC)$	8. Means-extremes prop.

32. *Given:* $ABCD$ is a rhombus with diagonals $\overline{AC}$ and $\overline{BD}$, and $\angle EQA$ is a right angle
Prove: $(AE)^2 = (AQ)(AD)$
Proof

STATEMENTS	REASONS
1. $ABCD$ is a rhombus with diagonals $\overline{AC}$ and $\overline{BD}$	1. Given
2. $\angle EQA$ is a right angle	2. Given
3. $\overline{EQ} \perp \overline{AD}$	3. Lines meeting in rt. $\angle$ are $\perp$
4. $\overline{BD} \perp \overline{AC}$	4. Diag. of rhombus are $\perp$
5. $\angle AED$ is a right angle	5. $\perp$ lines form rt. $\angle$'s
6. $\triangle AED$ is a right triangle	6. Def. of rt. $\triangle$
7. $\overline{EQ}$ is an altitude of $\triangle AED$ from right angle $\angle AED$	7. Def. of altitude
8. $\dfrac{AD}{AE} = \dfrac{AE}{AQ}$	8. Corollary 5.15
9. $(AE)^2 = (AD)(AQ)$	9. Means-extremes prop.

33. 2400 yd² **34.** $c = \sqrt{185}$ cm **35.** $b = 22$ ft **36.** No
37. 60 cm **38.** $20\sqrt{3}$ yd **39.** 4 ft

40. $6\sqrt{3}$ cm **41.** 3 yd **42.** $\dfrac{14\sqrt{3}}{3}$ ft **43.** $\dfrac{1}{5}$ yd

44. 10 cm **45.** $3\sqrt{2}$ ft **46.** 19.3 ft **47.** 1 mile
48. Consider the equilateral triangle with sides x shown below. Construct an altitude. Since the altitude bisects the vertex angle, $60°$, $m\angle 1 = 30°$, $m\angle B = 60°$, and $m\angle ADB = 90°$. By

Thm 5.20 $AD = \dfrac{\sqrt{3}}{2}AB = \dfrac{\sqrt{3}}{2}x$. Then the area of $\triangle ABC$

is $A = \dfrac{1}{2}bh = \dfrac{1}{2}(BC)(AD) = \dfrac{1}{2}(x)\left(\dfrac{\sqrt{3}}{2}x\right) = \dfrac{\sqrt{3}}{4}x^2.$

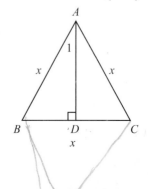

49. Transitive law   **50.** Subtraction property of inequalities
51. Division property of inequalities   **52.** Trichotomy law
53. The whole is greater than its parts.   **54.** $x < -17$
55. $y > -7$ **56.** False   **57.** True   **58.** True   **59.** False, since
$m\angle C = 53°$ and $m\angle 2 = 64°$, $m\angle B = 180° - (53° + 64°) = 63°$.
Thus, $m\angle C < m\angle B < m\angle 2$ so by Thm 5.23 $AB < AC < BC$.
60. True   **61.** $AC < DE$ **62.** No
63. *Given:* $\triangle ABC$ with median $\overline{CP}$ and $m\angle APC > m\angle BPC$
  Prove: $AC > BC$
  Proof

STATEMENTS	REASONS
1. ABC is a triangle with median $\overline{CP}$	1. Given
2. P is the midpoint of $\overline{AB}$	2. Def. of median
3. $AP = PB$	3. Def. of midpoint
4. $PC = PC$	4. Reflexive law
5. $m\angle APC > m\angle BPC$	5. Given
6. $AC > BC$	6. SAS Inequality Theorem

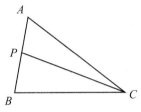

Chapter 5 Practice Test

1. 40 mph   **2.** $a = 7$ **3.** $x = 12$ or $x = -12$ **4.** 91 pounds
5. True   **6.** $B'C' = 2$ ft   **7.** yes   **8.** $x = 7.5$ **9.** 8.75
10. Sometimes   **11.** Never   **12.** 80 feet
13.

STATEMENTS	REASONS
1. $\overline{AB} \parallel \overline{ED}$	1. Given
2. $\angle B \cong \angle D$; $\angle A \cong \angle E$	2. If $\parallel$, alt. int. $\angle$'s are $\cong$
3. $\triangle ABC \sim \triangle EDC$	3. AA

14. $10\sqrt{3}$ and $-10\sqrt{3}$ **15.** $WZ = 8$ inches $XW = 8$ inches
16. $BD = 2\sqrt{15}$ cm   **17.** $CD = 13.5$ yd   **18.** 4 ft or 16 ft
19. $3\sqrt{2}$ cm   **20.** 7 yd   **21.** 17.0 yd   **22.** 5 cm   **23.** $\dfrac{\sqrt{3}}{4}x^2$
24. Subtraction property of inequalities   **25.** $x < -2$
26. $m\angle E > m\angle B$ **27.** $m\angle C < m\angle F$ **28.** AC is less than
18 ft and more than 2 ft

Chapter 6

6.1 Section Exercises

1. 22 in   **3.** $\dfrac{3}{2}$ ft   **5.** 8 in   **7.** $\dfrac{1}{3}$ ft   **9.** a   **11.** b   **13.** d   **15.** h
17. j   **19.** e   **21.** $m\widehat{XY} = 100°$, $m\widehat{YZ} = 120°$, $m\widehat{ZX} = 140°$
23. $m\angle 4 = 30°$
25. (a) central angle
  (b) inscribed angle

(c) $m\widehat{AC} = 120°$
(d) 240°
(e) 60°
(f) 120°
27. (a) 64°
  (b) 296°
  (c) 64°
  (d) 32°
29. *Proof*

STATEMENTS	REASONS
1. $\overline{AB} \perp \overline{BC}$	1. Given
2. $\angle ABC$ is a right angle	2. $\perp$ lines form rt. $\angle$'s
3. $m\angle ABC = 90°$	3. Def. of rt. $\angle$'s
4. $\frac{1}{2}m\widehat{ADC} = m\angle ABC$	4. Measure of inscribed angle $= \frac{1}{2}$ measure intercepted arc.
5. $m\widehat{ADC} = 2m\angle ABC$	5. Mult.-div. law
6. $m\widehat{ADC} = 2(90°) = 180°$	6. Substitution law

31. *Construction:* Draw radius $\overline{OC}$
  Proof

STATEMENTS	REASONS
1. $\overline{AB}$ is a diameter	1. Given
2. $\overline{AC} \parallel \overline{OD}$	2. Given
3. $\overline{OC}$ is a radius	3. By construction
4. $\angle CAO \cong \angle DOB$; $m\angle CAO = m\angle DOB$	4. If $\parallel$ lines corr. $\angle$'s are $\cong$; def. $\cong \angle$'s
5. $\overline{AO} \cong \overline{OC}$	5. Radii are $\cong$
6. $\angle CAO \cong \angle ACO$	6. $\angle$'s opp. $\cong$ sides of $\triangle$ are $\cong$
7. $\angle ACO \cong \angle COD$	7. If $\parallel$ lines alt. int. $\angle$'s are $\cong$
8. $\angle CAO \cong \angle COD$; $m\angle CAO = m\angle COD$	8. Transitive law; def. $\cong \angle$'s
9. $m\angle DOB = m\angle COD$	9. Substitution law
10. $m\widehat{BD} = m\angle DOB$ and $m\widehat{DC} = m\angle COD$	10. Def. of measure of arc
11. $m\widehat{BD} = m\widehat{DC}$	11. Trans. laws

33. Yes, by the definition of a semicircle.

6.2 Section Exercises

1. 94°   **3.** 60°   **5.** 12 in   **7.** 90°   **9.** 10 cm   **11.** 11 ft
13. 4 cm   **15.** 15 cm   **17.** 40°   **19.** 30°   **21.** 4 ft
23. 9 cm   **25.** $m\widehat{AC} = 20°$, $m\angle ABC = 10°$
27. *Proof*

STATEMENTS	REASONS
1. $m\widehat{BC} + m\widehat{AD} = 180°$	1. Given
2. $m\angle AED = 90°$	2. $\angle$ formed by chords is $\frac{1}{2}$ sum intercepted arc and arc intercepted by vertical $\angle$ (Thm 6.5)

3. $m\overarc{AB} + m\overarc{CD} +$ 3. A circle is an arc of $360°$.
$m\overarc{BC} + m\overarc{AD} = 360°$

4. $m\overarc{AB} + m\overarc{CD} = 180°$ 4. Add.-subt. law

5. $m\angle AEB = 90°$ 5. Same as 2

6. $\overline{AC} \perp \overline{BD}$ 6. Def. of $\perp$ lines

29. *Proof*

STATEMENTS	REASONS
1. $\overline{AB} \cong \overline{BC}$	1. Given
2. $m\angle P = \frac{1}{2}(m\overarc{BC} - m\overarc{AD})$	2. If secants intersect outside $\odot$, measure of $\angle = \frac{1}{2}$ difference of intercepted arcs. (Thm 6.14)
3. $\overarc{AB} \cong \overarc{BC}$ thus $m\overarc{AB} = m\overarc{BC}$	3. $\cong$ chords have $\cong$ arcs and def. $\cong$ arcs
4. $m\angle P = \frac{1}{2}(m\overarc{AB} - m\overarc{AD})$	4. Substitution law

31. *Given:* $\overarc{AB} \cong \overarc{CD}$
Prove: $\overline{AB} \cong \overline{CD}$
Construction: Draw radii $\overline{AO}, \overline{BO}, \overline{CO},$ and $\overline{DO}$

Proof

STATEMENTS	REASONS
1. $\overarc{AB} \cong \overarc{CD}$	1. Given
2. $\overline{AO}, \overline{BO}, \overline{CO},$ and $\overline{DO}$ are radii	2. By construction
3. $\overline{AO} \cong \overline{BO} \cong \overline{CO} \cong \overline{DO}$	3. Radii are $\cong$
4. $m\angle BOA = m\angle COD$ thus $\angle BOA \cong \angle COD$	4. Def. of measure of an arc; def. $\cong \angle$'s
5. $\triangle BOA \cong \triangle COD$	5. SAS
6. $AB = CD$	6. CPCTC

33. *Given:* $\overline{AB}$ and $\overline{CD}$ are chords with $\overline{OE} \perp \overline{AB}, \overline{OF} \perp \overline{CD},$ and $OE = OF$
Prove: $\overline{AB} \cong \overline{CD}$
Construction: Draw radii $\overline{AO}, \overline{BO}, \overline{CO},$ and $\overline{DO}$

Proof

STATEMENTS	REASONS
1. $\overline{AB}$ and $\overline{CD}$ are chords	1. Given
2. $\overline{OE} \perp \overline{AB}$ and $\overline{OF} \perp \overline{CD}$	2. Given
3. $\angle AEO, \angle BEO, \angle CFO,$ and $\angle DFO$ are right angles	3. $\perp$ lines form rt. $\angle$'s
4. $OE = OF$	4. Given
5. $\overline{AO} \cong \overline{BO} \cong \overline{CO} \cong \overline{DO}$	5. Radii are $\cong$
6. $\triangle AEO, \triangle BEO, \triangle CFO,$ and $\triangle DFO$ are right triangles	6. Def. of rt. $\triangle$
7. $\triangle AEO \cong \triangle CFO$ and $\triangle BEO \cong \triangle DFO$	7. HL

8. $\overline{AE} \cong \overline{CF}$ and $\overline{BE} \cong \overline{DF}$ 8. CPCTC; def. $\cong$ seg.
thus $AE = CF$ and $BE = DF$

9. $AE + BE = CF + DF$ 9. Add.-subt. law

10. $AB = AE + BE$ and   10. Seg. add. post.
$CD = CF + DF$

11. $AB = CD$ thus $\overline{AB} \cong \overline{CD}$ 11. Substitution law and def. $\cong$ seg.

35. 13 inches

6.3 Section Exercises

1. $35°$ 3. $36°$ 5. $40°$ 7. $35°$ 9. $34°$ 11. $29°$ 13. $50°$
15. $50°$ 17. 17 cm 19. 8 cm 21. 12.5 yd

23. There are two internal tangents, m and n, to circles in the given positions as shown in the figure to the right.

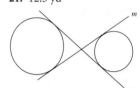

25. *Proof*

STATEMENTS	REASONS
1. $\overleftrightarrow{PA}$ and $\overleftrightarrow{PB}$ are tangents to the circle	1. Given
2. $m\angle P = \frac{1}{2}(m\overarc{ADB} - m\overarc{ACB})$	2. Thm 6.18
3. $m\overarc{ADB} + m\overarc{ACB} = 360°$	3. Circle measures $360°$
4. $m\overarc{ADB} = 360° - m\overarc{ACB}$	4. Arc add. post.
5. $m\angle P = \frac{1}{2}([360° - m\overarc{ACB}] - m\overarc{ACB})$	5. Substitution law
6. $m\angle P = \frac{1}{2}(360° - 2m\overarc{ACB})$	6. Distributive law
7. $m\angle P = 180° - m\overarc{ACB}$	7. Distributive law
8. $m\angle P + m\overarc{ACB} = 180°$	8. Add.-subt. law

27. *Construction:* Extend $\overleftrightarrow{AB}$ and $\overleftrightarrow{CD}$ to meet at point P
Proof

STATEMENTS	REASONS
1. $\overleftrightarrow{AB}$ and $\overleftrightarrow{CD}$ are common external tangents to circles that are not congruent	1. Given
2. $\overleftrightarrow{AB}$ and $\overleftrightarrow{CD}$ meet at P	2. By construction since circles are not $\cong$
3. $AP = CP$ and $BP = DP$	3. Thm 6.19
4. $AB = AP - BP$ and $CD = CP - DP$	4. Seg. add. post.
5. $AP - BP = CP - DP$	5. Add.-subt. law
6. $AB = CD$	6. Substitution law

29. *Given:* $\overleftrightarrow{AP}$ and $\overleftrightarrow{BP}$ are tangents

Prove: $m\angle APB = \frac{1}{2}(m\widehat{ADB} - m\widehat{ACB})$

Construction: Choose point D on major arc AB and draw secant $\overrightarrow{PD}$ intersecting the circle at C and D.

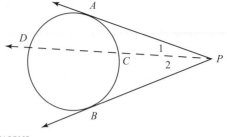

Proof

STATEMENTS	REASONS
1. $\overleftrightarrow{AP}$ and $\overleftrightarrow{BP}$ are tangents	1. Given
2. $\overrightarrow{PD}$ is a secant	2. By construction
3. $m\angle 1 = \frac{1}{2}(m\widehat{AD} - m\widehat{AC})$ and $m\angle 2 = \frac{1}{2}(m\widehat{DB} - m\widehat{BC})$	3. Thm 6.17
4. $m\angle APB = m\angle 1 + m\angle 2$	4. $\angle$ add. post.
5. $m\angle 1 + m\angle 2 = \frac{1}{2}(m\widehat{AD} - m\widehat{AC}) + \frac{1}{2}(m\widehat{DB} - m\widehat{BC})$	5. Add.-subt. law
6. $m\angle 1 + m\angle 2 = \frac{1}{2}[m\widehat{AD} + m\widehat{DB} - (m\widehat{AC} + m\widehat{BC})]$	6. Simplify
7. $m\widehat{AD} + m\widehat{DB} = m\widehat{ADB}$ and $m\widehat{AC} + m\widehat{BC} = m\widehat{ACB}$	7. Arc add. post.
8. $m\angle APB = \frac{1}{2}(m\widehat{ADB} - m\widehat{ACB})$	8. Substitution law

31. Use Construction 6.1.   **33.** Use Construction 6.3.
35. Use Construction 6.4.   **37.** 89.4 miles   **39.** 3

6.4 Section Exercises

1. 94°   **3.** 96°   **5.** 90°   **7.** 90°   **9.** The endpoints of a diameter and the points of intersection of the circle with the perpendicular bisector of the diameter give four points that divide the circle into four equal arcs. By Thm 6.25, the chords formed by these arcs form a regular 4-gon, that is, form a square. **11.** (g) Yes, the polygon is equilateral by Thm 6.7. The sides of the polygon are chords of the circle. Chords formed by $\cong$ arcs are $\cong$. (h) Yes, the polygon is equiangular. First construct all radii of the polygon. They are $\cong$ to each other and the chords of the circle thus forming six equilateral $\triangle$'s. Equilateral $\triangle$'s are equiangular. Thus the polygon is equiangular.   **13.** Since the length of a side of a regular inscribed hexagon is equal to the radius, by marking off six points on the circle with a compass set at the length of the radius, and joining every other point we form an equilateral triangle that is inscribed in the circle (by Thm 6.25).   **15.** Use Construction 4.2 and Construction 6.6. **17.** Use Construction 2.4 and Construction 6.5.   **19.** 120° **21.** 72°   **23.** 40°   **25.** 8 sides   **27.** 18 sides **29.** *Given:* $ABCD$ is a parallelogram inscribed in a circle

Prove: $ABCD$ is a rectangle

Proof

STATEMENTS	REASONS
1. $ABCD$ is a parallelogram	1. Given
2. $\overline{AB} \cong \overline{DC}$ and $\overline{AD} \cong \overline{BC}$	2. Opp. sides of $\square$ are $\cong$.
3. $\widehat{AB} \cong \widehat{DC}$ and $\widehat{AD} \cong \widehat{BC}$ thus $m\widehat{AB} = m\widehat{DC}$ and $m\widehat{AD} = m\widehat{BC}$	3. $\cong$ chords form $\cong$ arcs; def. $\cong$ arcs
4. $m\widehat{AB} + m\widehat{AD} = m\widehat{DC} + m\widehat{BC}$	4. Add.-subt. law
5. $m\widehat{BAD} = m\widehat{AB} + m\widehat{AD}$ and $m\widehat{BCD} = m\widehat{DC} + m\widehat{BC}$	5. Arc add. post.
6. $m\widehat{BAD} = m\widehat{BCD}$	6. Substitution law
7. $m\angle A = \frac{1}{2}m\widehat{BAD}$ and $m\angle C = \frac{1}{2}m\widehat{BCD}$	7. Measure inscribed $\angle = \frac{1}{2}$ measure intercepted arc
8. $m\angle A = m\angle C$	8. Substitution
9. $\angle A$ and $\angle C$ are supplementary	9. Opp. $\angle$'s of inscribed quad. are supp.
10. $m\angle A + m\angle C = 180°$	10. Def. of supp. $\angle$'s
11. $m\angle A + m\angle A = 180°$	11. Substitution law
12. $2m\angle A = 180°$	12. Distributive law
13. $m\angle A = 90°$	13. Mult.-div. law
14. $\angle A$ is a right angle	14. Def. of rt. $\angle$
15. $ABCD$ is a rectangle	15. Def. of $\square$

31. The proof follows from the fact that each side is an equal chord of the *circumscribed* circle; hence, the same distance from the center of the circle so that all radii of the inscribed circle must be equal in length.   **33.** The proof follows by circumscribing a circle around the regular polygon and using the fact that ≅ chords (the equal sides of the regular polygon) have ≅ central angles.

35. Consider an equilateral triangle inscribed in a circle with radius 4 cm as shown to the right. Draw altitude $\overline{OP}$ to side $\overline{AB}$. In right triangle $\triangle PBO$, $m\angle PBO = 30°$, $m\angle POB = 60°$, and hypotenuse $\overline{BO}$(radius $\overline{BO}$) = 4 cm. Thus, by

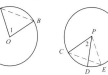

Thm 5.20, $PB = \dfrac{\sqrt{3}}{2}\, BO = \dfrac{\sqrt{3}}{2}(4) = 2\sqrt{3}$. Since $AB = 2PB$, $AB = 2(2\sqrt{3}) = 4\sqrt{3}$ cm.

6.5 Section Exercises

1. True **3.** True **5.** False **7.** True **9.** $m\overparen{AB} > m\overparen{CD}$
11. $m\overparen{AB} > m\overparen{CD}$ **13.** $AB > CD$ **15.** $\overline{CD}$
17. *Given:* Congruent circles with centers O and P, $m\angle 1 > m\angle 2$
Prove: $m\overparen{AB} > m\overparen{CD}$
Construction: Construct $\overparen{CDE} \cong \overparen{AB}$ and draw chords $\overline{AB}$ and $\overline{CE}$

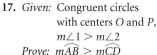

Proof

STATEMENTS	REASONS
1. The circles centered at O and P are congruent	1. Given
2. $\overline{AO}, \overline{BO}, \overline{CP},$ and $\overline{EP}$ are radii	2. Def. of radii
3. $\overline{AO} \cong \overline{BO} \cong \overline{CP} \cong \overline{EP}$	3. Radii of same and ≅ circles are ≅
4. $\angle 1 \cong \angle CPE$ thus $m\angle 1 = m\angle CPE$	4. By construction; def. ≅ $\angle$s
5. $\triangle AOB \cong \triangle CPE$	5. SAS
6. $\overline{AB} \cong \overline{CE}$	6. CPCTC
7. $m\overparen{AB} = m\angle 1$ and $m\overparen{CE} = m\angle CPE$	7. Def. of arc measure
8. $m\overparen{AB} = m\overparen{CE}$	8. Def. of arc measure
9. $m\angle 1 > m\angle 2$	9. Given
10. $m\angle CPE > m\angle 2$	10. Substitution law
11. $m\overparen{CE} > m\overparen{CD}$	11. In same circle, > of 2 cent. $\angle$'s has > arc
12. $m\overparen{AB} > m\overparen{CD}$	12. Substitution law

19. *Given:* Congruent circles with centers O and P, $m\overparen{AB} > m\overparen{CD}$
Prove: $m\angle 1 > m\angle 2$

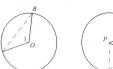

Construction: Construct $\overparen{CDE} \cong \overparen{AB}$ and draw chords $\overline{AB}, \overline{CD},$ and $\overline{CE}$.

Proof

STATEMENTS	REASONS
1. The circles with centers O and P congruent	1. Given
2. $\overline{AO}, \overline{BO}, \overline{CP}, \overline{DP},$ and $\overline{EP}$ are radii	2. Def. of radii
3. $\overline{AO} \cong \overline{BO} \cong \overline{CP} \cong \overline{DP} \cong \overline{EP}$	3. Radii of same and ≅ circles are ≅
4. $\overparen{CDE} \cong \overparen{AB}$	4. By construction
5. $\overline{AB} \cong \overline{CE}$	5. Chords formed by ≅ arc are ≅
6. $\triangle AOB \cong \triangle CPE$	6. SSS
7. $\overline{AB} \cong \overline{CE}$	7. CPCTC
8. $\overparen{AB} \cong \overparen{CE}$ thus $m\overparen{AB} = m\overparen{CE}$	8. ≅ chords have ≅ arcs; def. arcs
9. $m\overparen{AB} > m\overparen{CD}$	9. Given
10. $m\overparen{CE} > m\overparen{CD}$	10. Substitution
11. $m\angle CPE > m\angle 2$	11. Exercise 18
12. $m\angle 1 > m\angle 2$	12. Substitution law

21. *Given:* Congruent circles O and P, arcs $\overparen{AB}$ and $\overparen{CD}$ with $m\overparen{AB} > m\overparen{CD}$
Prove: $AB > CD$

Construction: Draw radii $\overline{AO}, \overline{BO}, \overline{CP},$ and $\overline{DP}$ and chords $\overline{AB}$ and $\overline{CD}$

Proof

STATEMENTS	REASONS
1. Congruent circles O and P with arcs $\overparen{AB}$ and $\overparen{CD}$	1. Given
2. $m\overparen{AB} > m\overparen{CD}$	2. Given
3. $m\overparen{AB} = m\angle AOB$ and $m\overparen{CD} = m\angle CPD$	3. Def. of arc measure
4. $m\angle AOB > m\angle CPD$	4. Substitution law
5. $\overline{AO} \cong \overline{BO} \cong \overline{CP} \cong \overline{DP}$	5. Radii of same or ≅ circles are ≅
6. $AB > CD$	6. SAS Inequality Theorem

23. The proof is in two parts: The case for the same circle first, then the case for congruent circles. We present the first case, the second follows from it after using a technique similar to that given in the proof of Thm 8.11.

Given: The circle centered at O, chords $\overline{AB}$ and $\overline{CD}$ with $\overline{OE} \perp \overline{AB}, \overline{OF} \perp \overline{CD},$ and $OF > OE$.

Prove: $AB > CD$

Construction: Construct chord $\overline{DE}$ such that $DE = AB$.
Draw $\overline{OG}$ such that $\overline{OG} \perp \overline{DE}$.
Draw segment $\overline{FG}$.

[Note: The case we are showing here has O on the interior of inscribed angle $\angle CDE$. There are two other possibilities: O is on $\angle CDE$ and O is exterior to $\angle CDE$. These two cases can be verified separately. The first uses Thm 6.36 and the second can be converted to the case shown below by drawing $\overline{DE}$ so O is inside $\angle CDE$.]

Proof

STATEMENTS	REASONS
1. The circle centered at O with chords $\overline{AB}$ and $\overline{CD}$	1. Given
2. $DE = AB$	2. By construction
3. $\overline{OG} \perp \overline{DE}$	3. By construction
4. $\overline{OE} \perp \overline{AB}$ and $\overline{OF} \perp \overline{CD}$	4. Given
5. $OG = OE$	5. $=$ chords are equidistant from center.
6. $OF > OE$	6. Given
7. $OF > OG$	7. Substitution law
8. $m\angle 3 > m\angle 4$	8. $\angle$'s opp. $\neq$ sides of $\triangle$ are $\neq$ in same order
9. $m\angle 3 + m\angle 2 = m\angle OGD$ and $m\angle 4 + m\angle 1 = m\angle OFD$	9. $\angle$ add. post.
10. $\angle OGD$ and $\angle OFD$ are right angles	10. $\perp$ lines form rt. $\angle$'s
11. $m\angle OGD = m\angle OFD$	11. All rt. $\angle$'s are $=$ in measure
12. $m\angle 3 + m\angle 2 = m\angle 4 + m\angle 1$	12. Substitution law
13. $m\angle 2 < m\angle 1$	13. Subtraction property of inequalities with statements 8 and 12
14. $FD < DG$	14. Sides opp. $\neq$ $\angle$'s are $\neq$ in same order
15. $2FD < 2DG$	15. Mult. prop. of ineq's
16. $CF = FD$ and $DG = GD$	16. Line through $O \perp$ to chord bisects chord
17. $CD = CF + FD$ and $DE = DG + GE$	17. Seg. add. post.
18. $CD = FD + FD$ and $DE = DG + DG$	18. Substitution law
19. $CD = 2FD$ and $DE = 2DG$	19. Distributive law
20. $CD < DE$	20. Substitution law
21. $CD < AB$	21. Substitution using statements 2 and 20

25. *Given:* $\triangle ABC$ inscribed in a circle centered at O with $m\angle A > m\angle B$.
Prove: $m\overset{\frown}{BC} > m\overset{\frown}{AC}$

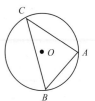

Proof

STATEMENTS	REASONS
1. $\triangle ABC$ is inscribed in the circle centered at O	1. Given
2. $m\angle A > m\angle B$	2. Given
3. $BC > AC$	3. Two $\angle$'s of $\triangle$ are $\neq$, sides opp. $\angle$'s are $\neq$ in same order
4. $m\overset{\frown}{BC} > m\overset{\frown}{AC}$	4. The $>$ of two chords has the $>$ arc (Thm 6.32)

27. First prove that $RS = CD$ by drawing perpendiculars from O to $\overline{RS}$ and $\overline{CD}$ at points W and V respectively. Two right triangles formed where $\overline{OP} \cong \overline{OP}$ by reflexive law and $\angle 2 \cong \angle RPO$ by construction thus $\triangle OWP \cong \triangle OVP$ by HA. $m\angle 1 + m\angle RPA = m\angle RPO$ by angle addition postulate. $m\angle 1 + m\angle RPA = m\angle 2$ by substitution thus $m\angle 1 > m\angle 2$.

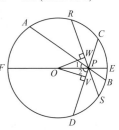

6.6 Section Exercises

1. The locus is a circle with center P and radius 5 units.
3. The locus is two circles concentric with the given circle, one with radius 1 unit and the other has radius 7 units.   **5.** The locus is a line between the given lines parallel to each of them at a distance 1.5 units from each.   **7.** There are three possibilities: (1) There is no locus if m does not intersect the circle. (2) The locus is one point, the point of tangency, if m is tangent to the circle. (3) The locus is two points if m intersects the circle in two points.   **9.** The locus is a diameter (excluding its endpoints) that is perpendicular to the given chords.   **11.** The locus of all points that are centers of circles tangent to both of two parallel lines is a line between the given parallel lines that is parallel to each. Thus, the desired locus is one point that is the intersection of the above line and the given intersecting line.
13. The proof consists of two parts. First, we prove that the center of a circle tangent to both of two parallel lines is equidistant from the two given lines.

Given: P is the center of a circle tangent to m and n with $m \parallel n$
Prove: P is equidistant from m and n
Construction: Draw radii $\overline{PE}$ and $\overline{PF}$ to points of tangency E and F

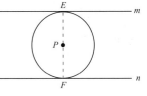

Proof

STATEMENTS	REASONS
1. P is the center of a circle tangent to m and n with $m \parallel n$	1. Given
2. $\overline{PE}$ and $\overline{PF}$ are radii to points of tangency E and F	2. By construction
3. $\overline{PE} \cong \overline{PF}$	3. Radii are $\cong$
4. $\overline{PE} \perp m$ and $\overline{PF} \perp n$	4. Radii to pts. of tangency are $\perp$ to tangents
5. PE is distance from P to m and PF is distance from P to n	5. Def. of distance pt. to line
6. P is equidistant from m and n	6. By statements 3 and 5

Next, we prove that a point on the line equidistant from two parallel lines is the center of a circle tangent to each line.

Given: P is on line l that is equidistant from $\parallel$ lines m and n.

Prove: P is the center of a circle tangent to both m and n.

Construction: Draw $\overline{PE} \perp m$ and $\overline{PF} \perp n$

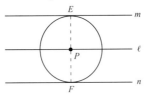

Proof

STATEMENTS	REASONS
1. P is on l that is equidistant from parallel lines m and n.	1. Given
2. $\overline{PE} \perp m$ and $\overline{PF} \perp n$	2. By construction
3. $PE = PF$	3. Def. of equidistant
4. Construct circle centered at P with radius PE ($= PF$).	4. Circle can be constructed with given radius and center.
5. m and n are tangent to circle in Step 4	5. Lines $\perp$ to radii are tangent to circle.

15. The locus consists of the point of intersection of the diagonals of the rectangle.   **17.** The locus of points is a line segment joining the center of the floor and ceiling.

Chapter 6 Review Exercises

1. 5.2 cm   **2.** $\frac{2}{5}$ yd   **3.** $m\angle P = 55°$ **4.** $m\angle PQR = 90°$

5. (a) central angle
(b) inscribed angle
(c) 50°
(d) 310°
(e) 25°

6. (a) 90°
(b) 55°
(c) 27.5°
(d) 60°
(e) 120°
(f) 120°
(g) 30°
(h) 125°

7. 94°   **8.** 135°   **9.** 17 cm   **10.** 10 ft   **11.** By Theorem 6.10, the distances to the center are equal.   **12.** 20°   **13.** 78°   **14.** 9 ft

15. *Proof*

STATEMENTS	REASONS
1. $ABCD$ is a rectangle	1. Given
2. $\overline{AB} \cong \overline{DC}$	2. Opp. sides of rect. are $\cong$
3. $\overparen{AB} \cong \overparen{CD}$	3. $\cong$ chords have $\cong$ arcs

16. *Proof*

STATEMENTS	REASONS
1. $\overline{AD} \perp \overline{BC}$ and $\overline{AD}$ contains the center O	1. Given
2. $\overline{AD}$ bisects $\overline{BC}$	2. Line through center $\perp$ chord bisects the chord
3. $\overline{BD} \cong \overline{CD}$	3. Def. of bisector
4. $\angle ADB$ and $\angle ADC$ are right angles	4. $\perp$ lines form rt. $\angle$'s
5. $\triangle ADB$ and $\triangle ADC$ are right triangles	5. Def. of rt. $\triangle$
6. $\overline{AD} \cong \overline{AD}$	6. Reflexive law
7. $\triangle ADB \cong \triangle ADC$	7. LL
8. $\overline{AB} \cong \overline{AC}$	8. CPCTC
9. $\overparen{AB} \cong \overparen{AC}$	9. $\cong$ chords have $\cong$ arcs

17. 33°   **18.** 35°   **19.** 33°   **20.** 50°   **21.** 38 cm   **22.** $13.\overline{3}$ ft

23. *Proof*

STATEMENTS	REASONS
1. $m\overparen{BD} = 2m\overparen{AC}$	1. Given
2. $m\angle B = \frac{1}{2}m\overparen{AC}$	2. Insc. $\angle$ measures $\frac{1}{2}$ measure of intercepted arc
3. $m\angle P = \frac{1}{2}(m\overparen{BD} - m\overparen{AC})$	3. Thm 6.14
4. $m\angle P = \frac{1}{2}(2m\overparen{AC} - m\overparen{AC})$	4. Substitution law
5. $m\angle P = \frac{1}{2}m\overparen{AC}$	5. Distributive law
6. $m\angle P = m\angle B$ thus $\angle P \cong \angle B$	6. Trans. laws; def. $\cong \angle$'s
7. $\overline{PC} \cong \overline{BC}$	7. Sides opp. $\cong \angle$'s are $\cong$

24. *Proof*

STATEMENTS	REASONS
1. $\overline{AD}$ and $\overline{DC}$ are tangents	1. Given
2. $AD = DC$	2. Thm 6.19
3. $ABCD$ is a parallelogram	3. Given
4. $ABCD$ is a rhombus	4. Def. of rhombus

25. *Proof*

STATEMENTS	REASONS
1. $\overline{AB}$ is a common internal tangent	1. Given
2. $\overleftrightarrow{OP}$ is the line of centers	2. Given
3. $\overline{OA} \perp \overline{AB}$ and $\overline{PB} \perp \overline{AB}$	3. Radius to pt. of tangency is $\perp$ to tangent
4. $\angle OAC$ and $\angle CBP$ are right angles	4. $\perp$ lines form rt. $\angle$'s
5. $\angle OAC \cong \angle PBC$	5. All rt. $\angle$'s are $\cong$
6. $\angle OCA \cong \angle PCB$	6. Vert. $\angle$'s are $\cong$
7. $\triangle OAC \sim \triangle PBC$	7. AA
8. $\angle O \cong \angle P$	8. Corr. $\angle$'s in ~ $\triangle$'s are $\cong$

26. *Given:* Circles centered at O and P are tangent externally at D, n is their common internal tangent, A is on n, $\overline{AC}$ is tangent to circle centered at O, and $\overline{AB}$ is tangent to circle centered at P.

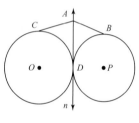

Prove: $AC = AB$

Proof

STATEMENTS	REASONS
1. n is common internal tangent to circles at D with A on n	1. Given
2. $\overleftrightarrow{AC}$ is tangent to circle centered at O and $\overleftrightarrow{AB}$ is tangent to circle centered at P	2. Given
3. $AC = AD$ and $AD = AB$	3. Tangents to circle from same pt. are $=$
4. $AC = AB$	4. Trans. law

27. Use Construction 6.1.

28. (a) $100°$
(b) $60°$
(c) $10°$
(d) $5°$
(e) $65°$
(f) 7.5 cm
(g) 5.2 cm
(h) 5.2 cm

29. $92°$ **30.** $90°$ **31.** Divide the circle into four equal arcs using the endpoints of a diameter and the points of intersection of the perpendicular bisector of the diameter and the circle. The perpendicular bisector of a side of the square determines the bisector of the arc formed by the side that is used to mark off 8 equal arcs to form the regular octagon. **32.** Construct the regular hexagon by inscribing it in a circle marking off arcs using the radius of the circle. Then use Construction 6.6 to inscribe a circle in the hexagon. **33.** $20°$ **34.** 36 sides **35.** 14 cm
36. The following paragraph-style proof shows a regular hexagon but the same proof can be applied to any regular polygon with n sides where $n > 2$.

Proof
Given: Regular polygon $ABCDEF$ where $\overline{OE}$ is a radius of the polygon
Prove: $\overline{OE}$ bisects $\angle DEF$

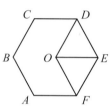

$\overline{OD} \cong \overline{OE} \cong \overline{OF}$ because all radii of a regular polygon are congruent. $\overline{DE} \cong \overline{EF}$ by the definition of a regular polygon, all sides are congruent. Thus $\triangle ODE \cong \triangle OFE$ by SSS. $\angle DEO \cong \angle FEO$ by CPCTC thus $\overline{OE}$ bisects $\angle DEF$ by the definition of an angle bisector.
37. True **38.** False **39.** False **40.** False **41.** True
42. The locus is two circles concentric with the given circle and with radii 8 units and 12 units. **43.** The locus is one circle concentric with the given circle with radius 10 units. **44.** The locus is one point at the intersection of the bisector of $\angle ABC$ and the circle centered at B. **45.** There are four possibilities: (1) There is no locus if m does not intersect the square. (2) The locus is one point if m passes through a vertex of the square only. (3) The locus is two points if m intersects two sides of the square. (4) The locus is a side of the square if m is collinear with that side.

Chapter 6 Practice Test

1. 0.65 ft **2.** $65°$ **3.** $35°$ **4.** 3 cm **5.** $95°$ **6.** 4 cm
7. 8 cm **8.** 15 ft **9.** Use Construction 6.2. **10.** $80°$
11. Use Constructions 2.4 and 6.6. **12.** $45°$
13. $m\angle 1 > m\angle 2$ **14.** $\overline{UV}$ **15.** $XY > UV$
16. *Given:* Rhombus $ABCD$ with $m\angle A > 90°$ (not a square), and diagonals $\overline{AC}$ and $\overline{BD}$.
Prove: $AC \neq BD$

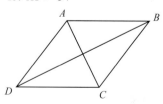

Proof

STATEMENTS	REASONS
1. $ABCD$ is a rhombus with $m\angle A > 90°$	1. Given
2. $m\angle A + m\angle D = 180°$	2. Adj. $\angle$'s in $\square$ are supp.
3. $m\angle D < 90°$	3. Subtraction property of inequalities
4. $m\angle A > m\angle D$	4. Transitive law
5. $\overline{AB} \cong \overline{AD} \cong \overline{DC}$	5. Sides of rhombus are $\cong$
6. $BD > AC$ (thus, $AC \neq BD$)	6. SAS Inequality Theorem

17. The locus is a line parallel to and between the given parallel lines and 3 units from each line. **18.** There are three possibilities: (1) There is no locus if the circles do not intersect. (2) The locus is one point (the point of tangency) if the centers of the two circles are 2 units or 10 units apart. (3) The locus is two points if the circles intersect in two points. **19.** The locus of the midpoints of all the chords parallel to a diameter is the perpendicular bisector of the diameter. **20.** The locus is all points interior to a semicircle with radius 8 ft and centered at the midpoint of the fence.

Chapter 7

7.1 Section Exercises

1. $A = 12 \text{ cm}^2, P = 14 \text{ cm}$ **3.** $A = 50 \text{ yd}^2, P = 32 \text{ yd}$
5. $A = 121 \text{ cm}^2, P = 44 \text{ cm}$, **7.** $A = 36 \text{ yd}^2, P = 30 \text{ yd}$
9. $A = 32 \text{ cm}^2, P = 34 \text{ cm}$, **11.** $A = 337.5 \text{ yd}^2, P = 79 \text{ yd}$
13. $A = 15.075 \text{ in}^2, P = 16.0 \text{ in}$ **15.** $A = 29.25 \text{ cm}^2, P = 27.5 \text{ cm}$
17. $A = 80 \text{ yd}^2, P = 36 \text{ yd}$ **19.** $\left(12 + \dfrac{25\sqrt{3}}{4}\right) \text{ft}^2$
21. 1. Given 2. $\overline{AC} \perp \overline{BD}$ 3. Add. prop. of area 4. Form. for area of $\triangle$ 5. Area of $\triangle ABC = \dfrac{1}{2}(d)(BE)$ 6. Substitution law
7. Distributive law 8. $DE + BE = BD = d'$ 9. $A = \dfrac{1}{2}dd'$
23. 72.3 in^2 **25.** 216 ft^2 **27.** 9.36 ft^2 **29.** 64.08 in^2
31. 7 cm **33.** 2 gallons **35.** \$1,949.44 **37.** \$40.83
39. Answers will vary. **41.** 39 ft^2 **43.** height is 10 ft, base is 12 ft **45.** 80 cm^2 **47.** 16 cm^2

7.2 Section Exercises

1. False **3.** False **5.** True **7.** $C = 12.4\pi \text{ mi}, A = 38.44\pi \text{ mi}^2$ **9.** $C = 12.78\pi \text{ ft}, A = 40.832\pi \text{ ft}^2$ **11.** $C \approx 4.71 \text{ cm}$, $A \approx 1.77 \text{ cm}^2$ **13.** $C \approx 57.68 \text{ yd}, A \approx 264.75 \text{ yd}^2$ **15.** 59.69 m^2
17. \$645.24 **19.** 25.68 in^2 **21.** 1257.84 cm^2 **23.** $\dfrac{5}{8} \text{ in}$
25. 11.8 cm^2 **27.** little over 8 in **29.** 1,600,000 miles

7.3 Section Exercises

1. $A = \dfrac{25}{4}\pi \text{ cm}^2, P = \left(\dfrac{5}{2}\pi + 10\right) \text{cm}$
3. $A = 2\pi \text{ cm}^2, P = (\pi + 8) \text{ cm}$ **5.** 4.6 yd^2 **7.** 13.0 ft^2
9. 45.1 cm^2 **11.** 4.4 cm^2 **13.** 24 yd **15.** 34.6 in^2
17. 24 cm^2 **19.** $P = (5\pi + 20) \text{ cm}, A = (25\pi - 50) \text{ cm}^2$

21. 9.1 yd^2

23. $\dfrac{9.4}{2\pi} = r$ or $r \approx 1.5 \text{ in}$

The circumference around the top of the cone corresponds to the arc length of the sector formed by the flattened cone.

7.4 Section Exercises

1. 400 yd^2 **3.** apothem is $4\sqrt{3}$ inches, side = 8 inches
5. apothem = 3 cm, radius = $3\sqrt{2}$ cm **7.** $384\sqrt{3} \text{ ft}^2$
9. 30 cm **11.** 100 cm^2 **13.** 138.24 cm^2 **15.** 452.4 ft^2
17. 86.4 m **19.** 5 in **21.** 87 mm^2 **23.** Answers will vary

Chapter 7 Review Exercises

1. $P = 46 \text{ cm}, A \approx 89.98 \text{ cm}^2$ **2.** $P = 37 \text{ in}, A = 43.5 \text{ in}^2$
3. $A = 36\sqrt{3} \text{ in}^2$ **4.** 4 times the original area **5.** 3 gallons
6. 126 cm^2 **7.** $A = 2535 \text{ cm}^2, P = 253 \text{ cm}$ **8.** 5.2 cm
9. $\dfrac{2}{5}\text{yd}$ **10.** $A = \dfrac{25\pi}{256} \text{ m}^2, C = \dfrac{5\pi}{8} \text{ m}$ **11.** $A \approx 66.48 \text{ ft}^2$,
$C \approx 28.90 \text{ ft}$ **12.** 36.2 in^2 **13.** 111.3 cm^2 **14.** 20.4 in^2
15. 9.1 cm^2 **16.** 3.6 ft **17.** $A = 6\pi \text{ cm}^2, P = (2\pi + 12)\text{cm}$
18. $112\pi \text{ in}^2$ **19.** 144 cm^2 **20.** 420 ft^2 **21.** 750 in^2
22. $2,338.3 \text{ yd}^2$ **23.** 754.8 ft^2

Chapter 7 Practice Test

1. 340 ft^2 **2.** 8 yd **3.** \$1,279.60 **4.** 14.7 mm^2 **5.** 96 in^2
6. 4 inches **7.** The areas of the two figures are equal but the figures are not congruent. Congruent figures are figures that can be made to coincide. These figures are not the exact same shape therefore will not coincide even though the areas are equal.
8. 471.6 in^2 **9.** $C = \pi(12.6) \approx 39.58 \text{ cm}, A = \pi(39.69) \approx$ 124.69 cm^2 **10.** 30.9 in^2 **11.** $10\pi \text{ cm}$ **12.** 218.2 cm^2
13. $A = 54\pi \text{ m}^2, P = (9\pi + 24)\text{m}$ **14.** $12\pi \text{ cm}$ **15.** 196 yd^2
16. 5.8 cm^2 **17.** $384\sqrt{3} \text{ ft}^2$ **18.** 467.6 cm^2

Chapters 4–7 Cumulative Review

1. False **2.** True **3.** False **4.** True **5.** True **6.** True
7. $m\angle A = 68° = m\angle C, m\angle B = 112° = m\angle D$ **8.** $x = 5$
9. *Given:* $\triangle ABC$ and $\overline{AB} \parallel \overline{DE}$

Prove: $\dfrac{EC}{BC} = \dfrac{DC}{AC}$

Proof

STATEMENTS	REASONS
1. $\triangle ABC$ and $\overline{AB} \parallel \overline{DE}$	1. Given
2. $\angle ABC \cong \angle DEC$	2. If lines $\parallel$, corresp. $\angle$'s$\cong$
3. $\angle BAC \cong \angle EDC$	3. If lines $\parallel$, corresp. $\angle$'s$\cong$
4. $\triangle ABC \sim \triangle DEC$	4. AA
5. $\dfrac{EC}{BC} = \dfrac{DC}{AC}$	5. Def $\sim\triangle$'s

10. 16 cm   **11.** 8 ft   **12.** 30 in   **13.** hypotenuse is 20 mm, long leg is $10\sqrt{3}$ mm   **14.** 32 m²   **15.** False; If the measure of two angles of a triangle are unequal, the measures of the sides opposite those angles are unequal in the same order (Theorem 5.23). Since $m\angle A > m\angle B$ then $BC > AC$.   **16.** 40°   **17.** 20°
18. 140°   **19.** 90°   **20.** 30°   **21.** 60°   **22.** 120°   **23.** 70°
24. 120°   **25.** By Theorem 6.33 the greater of two arcs has the greater chord.   **26.** By Corollary 6.36, a diameter is greater than any other chord.   **27.** 10   **28.** 45°   **29.** True
30. There are no points in the locus if line m does not intersect the pentagon. There is one point in the locus if line m only goes through one vertex of the pentagon. There are two points in the locus if line m intersects 2 sides of the pentagon. If line m coincides with one side of the pentagon, the locus consists of all the points on that side of the pentagon.

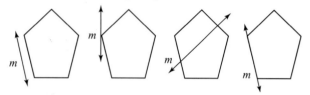

31. 8 yd   **32.** 50π in.   **33.** 36.2 mm²   **34.** $600\sqrt{3}$ yd²
35. 51.5 in.²   **36.** $16\sqrt{3}$ in.²   **37.** 10.5 or $\dfrac{21}{2}$ in.²

Chapter 8

8.1 Section Exercises

1. True   **3.** False   **5.** True   **7.** False   **9.** True   **11.** True
13. $f = 7, v = 6, e = 11; 7 + 6 - 11 = 2$ **15.** $f = 6, v = 6, e = 10;$
$6 + 6 - 10 = 2$ **17.** $f = 8, v = 6, e = 12; 8 + 6 - 12 = 2$ **19.** 20
21.

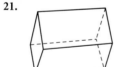

23.

25.

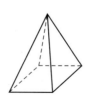

27. $2,160.00   **29.** $e\sqrt{3}$ **31.** Answers will vary.

8.2 Technology Connection

EDGE (cm)	SURFACE AREA $S = 6s^2$ (cm²)	VOLUME $V = s^3$ (cm³)
2	24	8
4	96	64
6	216	216
8	384	512
10	600	1000
12	864	1728

The graphs cross at $x = 6$. This is where the volume and surface area are equal. Before $x = 6$, the suface area increases faster but after it the volume increases faster.

8.2 Section Exercises

1. True   **3.** True   **5.** 13.5 m²   **7.** 1083.16 cm²
9. $(168 + 11\sqrt{72})$cm² or $(168 + 66\sqrt{2})$cm²
11. $(506 + 31\sqrt{115})$mm²   **13.** 1061.208 cm³   **15.** 282.625 in³
17. $384\sqrt{3}$ cm³   **19.** $93.312\sqrt{3}$ yd³   **21.** 7 cm, $V = 343$ cm³
23. 1560 ft³   **25.** 10,648 g   **27.** $504.00
29. $d = \sqrt{l^2 + w^2 + h^2}$ **31.** Answers will vary.

8.3 Section Exercises

1. 924 m²   **3.** $248\sqrt{11240}$ cm²   **5.** 61.3 ft²   **7.** 648.1 mm²
9. 2420 yd³   **11.** 4344.5 cm²   **13.** 28.2 ft³   **15.** 1018.4 mm³
17. 50.8 or about 51 gallons of paint   **19.** 500 cm³
21. 4 faces; They all appear congruent.

8.4 Section Exercises

1. True   **3.** 414.7 cm²   **5.** 3234.8 in²   **7.** 361.7 m²
9. 1231.5 cm³   **11.** 23448.5 in³   **13.** 2020.0 m³   **15.** 75.4 yd²
17. 6334.7 in²   **19.** 2212.6 ft²   **21.** 37.7 in³   **23.** 708.0 cm³
25. 14510.4 m³   **27.** 53.6 cm³, 107.2 cm³   **29.** $SA = 3515.6$ in²,
$V \approx 14476.5$ in³   **31.** 1.6 cans   **33.** 2.2 gallons   **35.** 1.8 gallons
37. Cylinder holds more cereal.   **39.** 113.1 in²

8.5 Section Exercises

1. 1809.6 m²   **3.** 7.1 ft²   **5.** 394.1 yd²   **7.** 1436.8 ft³
9. 3.1 m³   **11.** 11008.4 yd³   **13.** The sphere has the greater volume.   **15.** $r = 3$ **17.** 12   **19.** 307.9 in²   **21.** 137,258.3 ft³
23. 320.4 cm³   **25.** $479.41   **27.** 11.5 in³   **29.** $35.36

Chapter 8 Review Exercises

1. True   **2.** True   **3.** True   **4.** True   **5.** False   **6.** False
7. $f = 6, v = 8, e = 12; 6 + 8 - 12 = 2$ **8.** $f = 12, v = 8; e = 18;$
$12 + 8 - 18 = 2$ **9.** 24   **10.** $49.50   **11.** 942.6 cm²
12. 191.44 ft²   **13.** 6434.856 m³   **14.** 519.9 m³   **15.** 15.6 ft³
16. 118,221 cm²   **17.** 327.2 ft²   **18.** 1038.8 in³
19. 34835.6 m³   **20.** 2.9 in³   **21.** 127.2 ft²   **22.** 5421.4 cm²
23. 18,303,390.0 in³   **24.** 89,509.4 m³   **25.** 305.8 cm²

26. 3069.4 ft^2 **27.** 3294.2 m^3 **28.** 20,572.0 in^3 **29.** 377.0 in^3
30. 78,539.8 cm^3 **31.** 191.1 ft^2 **32.** 9025.7 m^2 **33.** 3.3 cm^3
34. 8992.0 in^3 **35.** 286.5 in^2 **36.** 94.0 m^2 **37.** 14,869.7 g
38. 100.5 cm^3 **39.** 88,555.2 kg   **40.** $321.69

Chapter 8 Practice Test

1. True   **2.** True   **3.** True   **4.** True   **5.** $1394.11
6. 447.5 cm^2 **7.** 207.46 m^3 **8.** 26,927.9 in^3 **9.** 1569.7 m^2
10. 6324.724 m^3 **11.** 59.6 ft^2 **12.** 30.3 ft^3 **13.** 615.8 cm^2
14. 1995.3 cm^3 **15.** 549.8 in^2 **16.** 16,272.7 m^3 **17.** 75.4 in^2
18. 332.4 in^3

Chapter 9

9.1 Section Exercises

1.

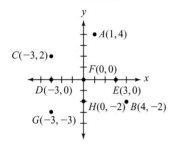

3. $A(3, 5)$, $B(4, 1)$, $C(0, 0)$, $D(-3, 2)$, $E(-5, 0)$, $F(0, -4)$, $G(-4, -6)$, $H(5, -4)$

5.

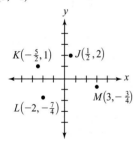

7. J:I, K:II, L:III, M:IV   **9.** (a) -2 (b) -2 (c) -3 (d) 0
11. (a) impossible (b) -3 (c) impossible (d) -3 **13.** (a) $-\dfrac{1}{3}$
(b) impossible (c) $-\dfrac{1}{3}$ (d) impossible

15. x-intercept $(-2, 0)$;
y-intercept $(0, -2)$

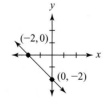

17. x-intercept $(2, 0)$;
y-intercept $(0, 6)$

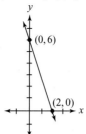

19. x-intercept $(2, 0)$;
y-intercept $(0, -2)$

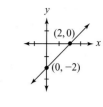

21. x-intercept $(0, 0)$;
y-intercept $(0, 0)$

23. x-intercept $\left(\dfrac{7}{3}, 0\right)$:
no y-intercept

25. no x-intercept;
y-intercept $(0, -1)$

27. x-intercept $(0, 0)$;
y-intercept $(0, 0)$

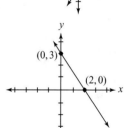

29. x-intercept $(2, 0)$;
y-intercept $(0, 3)$

31. x-intercept $(0, 0)$; every point on the y-axis is a y-intercept; the graph is the y-axis.
33. All have the same y-intercept, $(0, 1)$.

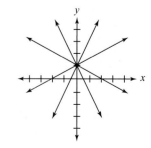

35. 13   **37.** $\sqrt{13}$ **39.** 5   **41.** 13   **43.** $(1, 0)$ **45.** $\left(\dfrac{11}{2}, 4\right)$

47. $(k + p, n + q)$ **49.** $(-2, 4)$ **51.** (a) $(3, 4)$ (b) $AM = 5$; $MC = 5$; $MB = 5$ **53.** (a) $AB = 4$; $BC = \sqrt{32}$; $AC = 4$ (b) right isosceles $\triangle$ **55.** Answers will vary.

9.2 Section Exercises

1. -2 **3.** 0   **5.** $-\dfrac{b}{a}$

7. $2x + y - 5 = 0$

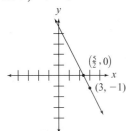

9. $x - 3y - 14 = 0$

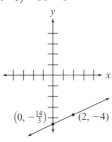

11. $2x + y - 13 = 0$

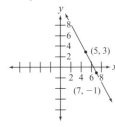

13. $x = 2$ or $x - 2 = 0$

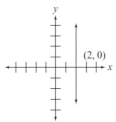

15. $y = 3$ or $y - 3 = 0$

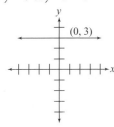

17. $2x - y + 10 = 0$

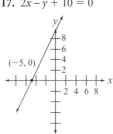

19. $3x + y - 4 = 0$

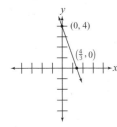

21. $5x - 9y - 29 = 0$

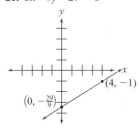

23. $5x - 2y + 10 = 0$

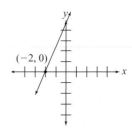

25. Then the slope is $\dfrac{5}{2}$, and the y-intercept is $(0, 2)$.   **27.** It has no y-intercept and the slope is undefined.   **29.** The slope is 0, and the y-intercept is $(0, 2)$.   **31.** The slope is -2, and the y-intercept is $(0, -1)$.   **33.** perpendicular   **35.** parallel   **37.** perpendicular   **39.** $3x + y - 1 = 0$ **41.** $x - 3y + 13 = 0$ **43.** $3x - 4y - 8 = 0$

45. $x + 2y - 4 = 0$ **47.** a) $m = -\dfrac{3}{5}$ b) $y = -\dfrac{3}{5}x + \dfrac{34}{5}$

49. a) $\left(-\dfrac{5}{2}, \dfrac{1}{2}\right); \left(\dfrac{11}{2}, \dfrac{1}{2}\right)$ b) Since the slopes of the two bases and the median are 0, the lines are parallel.   **51.** To use the form $y - y_1 = m(x - x_1)$ the slope (m) and a point (x_1, y_1) are needed, hence the name point-slope. To use the form $y = mx + b$, the slope (m) and the y-intercept (b) are needed, hence the name slope-intercept.

9.3 Section Exercises

1. $C(x, -y)$ **3.** $A(-q, 0); B(q, 0); C(0, p)$ **5.** $J(0, y); K(x, y);$ $L(w, 0)$ Midpoint of $\overline{JK}$ is $\left(\dfrac{x}{2}, y\right)$; midpoint of $\overline{ML}$ is $\left(\dfrac{w}{2}, 0\right)$

7. $C(a - x, y); D(a, 0)$ **9.** The coordinates of C are (b, y) and the coordinates of D are $(c, 0)$. Use the distance formula, $d = \sqrt{(x_2 - x_1)^2 + (y_2 - y_1)^2}$ to find BD and AC.

$$BD = \sqrt{(c - a)^2 + (0 - y)^2}$$
$$= \sqrt{(c - a)^2 + y^2}$$

substitute $a + b = c$

$$= \sqrt{(a + b - a)^2 + y^2}$$
$$= \sqrt{b^2 + y^2}$$
$$AC = \sqrt{(b - 0)^2 + (y - 0)^2}$$
$$= \sqrt{b^2 + y^2}$$

Thus, the diagonals of an isosceles trapezoid are congruent.

11. Draw a figure.
Use the distance formula to prove $AC = BD$.

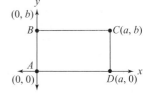

$$AC = \sqrt{(a-0)^2 + (b-0)^2} = \sqrt{a^2 + b^2}$$
$$BD = \sqrt{(a-0)^2 + (0-b)^2} = \sqrt{a^2 + (-b)^2} =$$
$$\sqrt{a^2 + b^2}$$

Thus, the diagonals of a rectangle are $\cong$.

13. Draw an isosceles trapezoid. Because the figure is isosceles, $a - b = d$. Use the distance formula to show $AC = BD$.

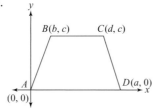

$$AC = \sqrt{(d-0)^2 + (c-0)^2} = \sqrt{d^2 + c^2}$$
$$BD = \sqrt{(a-b)^2 + (0-c)^2} = \sqrt{(a-b)^2 + (-c)^2} =$$
$$\sqrt{(a-b)^2 + c^2}$$

Substitute $a - b = d$

$$BD = \sqrt{d^2 + c^2}$$

Thus, $AC = BD$ and the diagonals of an isosceles trapezoid are $\cong$.

15. Draw a square. Use the midpoint formula to show the diagonals bisect each other.

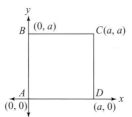

midpoint of $\overline{AC} = \left(\dfrac{a+0}{2}, \dfrac{a+0}{2}\right) = \left(\dfrac{a}{2}, \dfrac{a}{2}\right)$

midpoint of $\overline{BD} = \left(\dfrac{0+a}{2}, \dfrac{a+0}{2}\right) = \left(\dfrac{a}{2}, \dfrac{a}{2}\right)$

Thus, $\left(\dfrac{a}{2}, \dfrac{a}{2}\right)$ is the common midpoint of the diagonals; the diagonals bisect each other. To show the diagonals are perpendicular, use the slope formula.

slope $\overline{AC} = \dfrac{a-0}{a-0} = 1$

slope $\overline{BD} = \dfrac{a-0}{0-a} = \dfrac{a}{-a} = -1$

The diagonals are $\perp$ because the product of the slopes is -1.

17. Draw a figure. Find the midpoints of $\overline{AB}$ and $\overline{CD}$ using the midpoint formula.

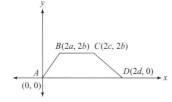

midpoint $\overline{AB} = \left(\dfrac{2a+0}{2}, \dfrac{2b+0}{2}\right) = (a, b)$

Let $M = (a, b)$

midpoint $\overline{CD} = \left(\dfrac{2c+2d}{2}, \dfrac{2b+0}{2}\right) = ((c+d), b)$

Let $N = ((c+d), b)$

The median is $\overline{MN}$. To show the median is parallel to the bases, use the slope formula.

slope $\overline{BC} = \dfrac{2b-2b}{2a-2c} = \dfrac{0}{2a-2c} = 0$

slope $\overline{MN} = \dfrac{b-b}{c+d-a} = \dfrac{0}{c+d-a} = 0$

slope $\overline{AD} = \dfrac{0-0}{2d-0} = \dfrac{0}{2d} = 0$

Since the slopes are equal, the three lines are parallel. The median of a trapezoid is parallel to the bases.

Chapter 9 Review Exercises

1. x-coordinate: -2; y-coordinate: 3 ; II   **2.** x-axis and y-axis
3. x-axis   **4.** y-axis
5. $A(-3, 2)$; $B(2, 1)$; $C(2, -2)$; $D(-2, -2)$

   II   I   IV   III

6.

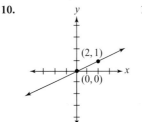

7.

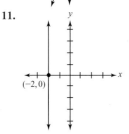

8.

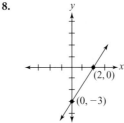

9.

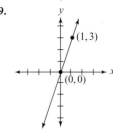

10.

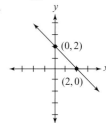

11.

12.

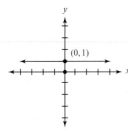

13. $\dfrac{9}{2}$ **14.** -2 **15.** 5; A line perpendicular to a line with slope 5 has slope $-\dfrac{1}{5}$ **16.** $-\dfrac{8}{5}$; A line perpendicular to a line with slope $-\dfrac{8}{5}$ has slope $\dfrac{5}{8}$ **17.** 6   **18.** 8   **19.** $(2,-1)$ **20.** $(2,-2)$

21. $P = 36$; $A = 48$ **22.** $4x + y + 1 = 0$ **23.** The slope is -2, and the y-intercept is $(0, 5)$.   **24.** $x - 2y + 12 = 0$ **25.** $x - 2y + 5 = 0$ **26.** The slope is 0, and the y-intercept is $(0, 5)$. There is no x-intercept.   **27.** perpendicular   **28.** $y = -5x + 10$

29. $(a + b, c)$ **30.** $\left(\dfrac{2a + b}{2}, c\right) = M$; $\left(\dfrac{b}{2}, 0\right) = N$; slope $\overline{AD} = \dfrac{c}{a}$; slope $\overline{MN} = \dfrac{c}{a}$; thus $\overline{AD} \parallel \overline{MN}$; slope $\overline{DM} = 0$

slope $\overline{AN} = 0$; thus $\overline{DM} \parallel \overline{AN}$.
Therefore $ANMD$ is a parallelogram because opposite sides are parallel.

31.

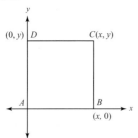

Prove $AC = BD$ using the distance formula.
$AC = \sqrt{(x - 0)^2 + (y - 0)^2} = \sqrt{x^2 + y^2}$
$BD = \sqrt{(x - 0)^2 + (0 - y)^2} = \sqrt{x^2 + y^2}$
Thus, $AC = BD$, therefore $\overline{AC} \cong \overline{BD}$.

32. For convenience, a factor of 2 will be used in the coordinates of B and C.
    First find the midpoint of $\overline{BC}$.

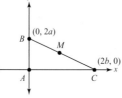

midpoint $= \left(\dfrac{x_1 + x_2}{2}, \dfrac{y_1 + y_2}{2}\right) = \left(\dfrac{0 + 2b}{2}, \dfrac{2a + 0}{2}\right) = (b, a)$
Next show $MB = AM = MC$ using the distance formula.

$MB = \sqrt{(b - 0)^2 + (a - 2a)^2} = \sqrt{b^2 + a^2}$
$AM = \sqrt{(b - 0)^2 + (a - 0)^2} = \sqrt{b^2 + a^2}$
$MC = \sqrt{(2b - b)^2 + (0 - a)^2} = \sqrt{b^2 + a^2}$
Since the three distances are equal, the midpoint of the hypotenuse is equidistant from the three vertices of the triangle.

Chapter 9 Practice Test

1. $y = -\dfrac{3}{2}x + 6$ **2.** $(4, 0)$ **3.** $(0, 6)$ **4.** $-\dfrac{3}{2}$

5.

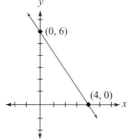

6. $(10, 70)$

7.

8.

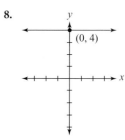

9. x-intercept: $(-5, 0)$ y-intercept: $(0, -5)$ **10.** -2

11. $\dfrac{1}{2}$ **12.** $= \sqrt{180}$ or $6\sqrt{5}$ **13.** $(1, 1)$ **14.** parallel

15. $3x + y + 2 = 0$ **16.** $2x + y - 5 = 0$ **17.** $2x + y - 13 = 0$
18. $(-5, 6)$ **19.** $(-c, d)$; (c, d); $(-a, 0)$; $(a, 0)$
20. Let $M = $ midpoint of $\overline{AD}$.

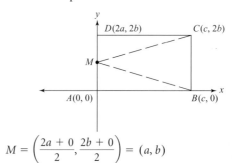

$M = \left(\dfrac{2a + 0}{2}, \dfrac{2b + 0}{2}\right) = (a, b)$

Use the distance formula to show $MB = MC$.

$$MC = \sqrt{(a-c)^2 + (b-2b)^2} = \sqrt{(a-c)^2 + b^2}$$
$$MB = \sqrt{(a-c)^2 + (b-0)^2} = \sqrt{(a-c)^2 + b^2}$$

Since $MB = MC$, $\triangle BMC$ is isosceles.

Chapter 10

10.1 Section Exercises

1. $\sin A = \dfrac{5}{13}$; $\cos A = \dfrac{12}{13}$; $\sin B = \dfrac{12}{13}$; $\cos B = \dfrac{5}{13}$

3. $\sin A = \dfrac{7}{15}$; $\cos A = \dfrac{4\sqrt{11}}{15}$; $\sin B = \dfrac{4\sqrt{11}}{15}$; $\cos B = \dfrac{7}{15}$

5. $\sin A = 0.6$; $\cos A = 0.8$; $\sin B = 0.8$; $\cos B = 0.6$

7. $\cos A = \dfrac{2\sqrt{6}}{7}$ 9. $\sin A = \dfrac{4}{7}$ 11. 0.9063 13. 0.7880

15. 0.1080 17. $\sin 51° = \dfrac{x}{19}$; $\cos 39° = \dfrac{x}{19}$; $x \approx 14.8$

19. $\cos 45° = \dfrac{x}{21}$; $\sin 45° = \dfrac{x}{21}$; $x \approx 14.8$

21. $x \approx 21.1$; $y \approx 28.0$ 23. $x \approx 9.7$; $y \approx 17.5$ 25. If the given leg is adjacent to the given acute angle, use the cosine ratio. If the given leg is opposite the given acute angle, use the sine ratio.

10.2 Section Exercises

1. $\tan A = \dfrac{5}{12}$; $\tan B = \dfrac{12}{5}$

3. $\tan A = \dfrac{7}{4\sqrt{11}}$ or $\dfrac{7\sqrt{11}}{44}$; $\tan B = \dfrac{4\sqrt{11}}{7}$

5. $\tan A = \dfrac{3}{4}$; $\tan B = \dfrac{4}{3}$ 7. $\sin A = \dfrac{\sqrt{5}}{3}$; $\cos A = \dfrac{2}{3}$

9. $\sin A = \dfrac{7}{5\sqrt{2}}$ or $\dfrac{7\sqrt{2}}{10}$; $\cos A = \dfrac{1}{5\sqrt{2}}$ or $\dfrac{\sqrt{2}}{10}$

11. 0.7813 13. 6.6357 15. 7.2° 17. 42.9° 19. 84.1°

21. 5.1 23. 36.9° 25. 22.6° 27. 41.8° 29. 30° 31. 60°

33. 12.3 in 35. (a) $\dfrac{4}{5}$, (b) $\tan A = \dfrac{4}{5}$, (c) Yes, it will always be true. The slope of $\overline{AC} = \tan A$ because the slope of a line is the change in y divided by the change in x: $\dfrac{y_2 - y_1}{x_2 - x_1}$. The length of the opposite leg is the change in y, while the length of the adjacent leg is the change in x. Thus, $\dfrac{y_2 - y_1}{x_2 - x_1} = \dfrac{\text{opposite leg}}{\text{adjacent leg}}$.

10.3 Section Exercises

1. $m\angle B = 30°$, $b \approx 5$, and $c \approx 10$ 3. $m\angle B = 70°$, $a \approx 4$, and $b \approx 11$.   5. $b = 5$, $m\angle A \approx 67°$, and $m\angle B \approx 23°$.   7. $a \approx 13$, $m\angle B \approx 28°$, $m\angle A \approx 62°$ 9. $b \approx 4.2$, $m\angle A \approx 65.6°$, $m\angle B \approx 24.4°$ 11. $a \approx 5.4$, $b \approx 10.7$, $m\angle B \approx 63.3°$ 13. $a \approx 16.8$, $c \approx 17.1$, $m\angle A \approx 79.2°$ 15. $a \approx 11.0$, $b \approx 26.5$, $m\angle B \approx 67.5°$ 17. $a \approx 3.2$, $b \approx 21.7$, $m\angle A = 8.3°$ 19. $P \approx 155.0$ m   21. $A \approx 209.7$ cm   23. 9.7 in.

10.4 Section Exercises

1. $\angle 3$ is an angle of elevation.   3. $\angle 1$ is an angle of depression.   5. 1417.8 ft   7. 79.8°   9. 46.01 ft   11. 865.5 ft   13. 9.4 mi   15. 2660.4 m   17. (a) $P \approx 280$ ft (b) $A \approx 3362$ ft^2 19. (a) $P = 18.6$ mi; $A \approx 16.6$ mi^2

Chapter 10 Review Exercises

1. $\sin A = \dfrac{4}{5}$; $\cos A = \dfrac{3}{5}$; $\sin B = \dfrac{3}{5}$; $\cos B = \dfrac{4}{5}$

2. $\sin A = \dfrac{3}{4}$; $\cos A = \dfrac{\sqrt{7}}{4}$; $\sin B = \dfrac{\sqrt{7}}{4}$; $\cos B = \dfrac{3}{4}$ 3. $\dfrac{2\sqrt{2}}{3}$

4. 0.3584   5. 0.4099   6. $x = 26.5$; $y \approx 33.9$ 7. $\sin A = \dfrac{4}{5}$; $\cos A = \dfrac{3}{5}$ 8. 7.2066   9. 12.7°   10. 24.4°   11. 58.7°

12. 53°   13. 45°   14. $m\angle B = 60°$, $a \approx 3$, $c \approx 7$ 15. $m\angle A = 48°$, $a \approx 9$, $b \approx 8$ 16. $m\angle B = 17.7°$, $b \approx 0.8$, $c \approx 2.5$ 17. $c \approx 13$, $m\angle B \approx 58°$, $m\angle A \approx 32°$ 18. $16 = a$, $m\angle A \approx 53°$, $m\angle B \approx 37°$ 19. $b \approx 7.3$, $m\angle A \approx 40.2°$, $m\angle B \approx 49.8°$ 20. 1125.1 m^2 21. 58.2 m   22. 8.4 miles from outpost Bravo; 6.6 miles from outpost Alpha   23. 37.5°   24. 23.7 yards   25. 31.5 ft

Chapter 10 Practice Test

1. $\sin A = \dfrac{5}{9}$; $\cos A = \dfrac{2\sqrt{14}}{9}$; $\tan A = \dfrac{5\sqrt{14}}{28}$

2. $\sin B = \dfrac{2\sqrt{14}}{9}$; $\cos B = \dfrac{5}{9}$; $\tan B = \dfrac{2\sqrt{14}}{5}$

3. $\cos A = \dfrac{\sqrt{21}}{5}$; $\tan A = \dfrac{2\sqrt{21}}{21}$ 4. 0.9330; 0.3600; 2.5916

5. $x \approx 15.4$ in.; $y \approx 14.3$ in.   6. $x \approx 12.9$ mm; $y \approx 8.1$ mm

7. 83.7°   8. $b \approx 4.4$; $m\angle A \approx 14.1°$ $m\angle B \approx 75.9°$

9. $m\angle B = 52°$; $a \approx 12$; $c \approx 19$ 10. $A \approx 16.6$ cm^2 11. 9°

12. 36.7 inches

Chapters 8–10 Cumulative Review

1. $SA = 77.2$ cm^2 $V = 42.228$ cm^3 2. $SA = 96$ ft^2 $V = 48$ ft^3

3. 24.5 in^3 4. 384 in^2 5. Since the volume of the cone is larger than the volume of the ice cream, the cone will hold all the ice cream when it melts.

6. x-intercept is $(9, 0)$
   y-intercept is $(0, 3)$
   $d = \sqrt{90}$ or $3\sqrt{10}$

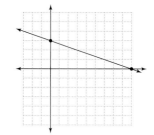

7. $y = -4x - 3$

8. Let $\overline{AB}$ be the given segment with endpoints $(-a, 0)$ and $(a, 0)$, then $(0, 0)$ is the midpoint of the segment. The perpendicular bisector is the y-axis. Choose any point C on the y-axis, call it $(0, b)$. Prove $AC = BC$ (See the figure.)

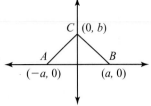

$AC = \sqrt{(-a - 0)^2 + (0 - b)^2}$

$AC = \sqrt{a^2 + b^2}$

$BC = \sqrt{(a - 0)^2 + (0 - b)^2}$

$BC = \sqrt{a^2 + b^2}$

Thus $AC = BC$, and any point on the perpendicular bisector of a segment is equidistant from the endpoints of the segment.

9. $a \approx 13.5$ and $b \approx 8.9$ **10.** $m\angle ABC \approx 67°$ **11.** $c \approx 4.3$;

$m\angle A \approx 38.5°$; $m\angle B \approx 51.5°$ **12.** 56.6 feet

Glossary

acute angle	An acute angle measures between 0 and 90 degrees.
acute triangle	An acute triangle contains all acute angles.
adjacent angles	Adjacent angles have a common vertex, share a common side, and have no interior points in common.
alternate exterior angles	Alternate exterior angles are nonadjacent angles, on opposite sides of the transversal, and on the exterior of two lines.
alternate interior angles	Alternate interior angles are nonadjacent angles, on opposite sides of the transversal, and on the interior of the two lines.
altitude of a parallelogram	The altitude of a parallelogram is the line segment from a vertex of the parallelogram perpendicular to the nonadjacent side (possibly extended).
altitude of a pyramid	The altitude of a pyramid is the line segment from the vertex of the pyramid perpendicular to the plane of the base.
altitude of a trapezoid	The altitude of a trapezoid is the line segment from a vertex of the trapezoid perpendicular to the nonadjacent base.
altitude of a triangle	The altitude of a triangle is the line segment from a vertex perpendicular to the side opposite that vertex (possibly extended).
angle	An angle is a geometric figure consisting of two rays that share a common endpoint.
angle bisector of a triangle	An angle bisector of a triangle is a line segment that separates a given angle into two congruent adjacent angles.
angle of depression	An angle of depression is an acute angle formed, when looking down, by the line of sight and a horizontal line.
angle of elevation	An angle of elevation is an acute angle formed, when looking up, by the line of sight and a horizontal line.
apothem	An apothem is a line segment from the center of a polygon perpendicular to one of its sides.
arc	An arc is the part of a circle determined by two points on the circle and all of the points between them.
area	An area is the number of square units needed to cover the surface of a polygonal region.
auxiliary line	An auxiliary line is the line added to a drawing to help complete a proof.
axiomatic system	An axiomatic system consists of four parts: undefined terms, definitions, axioms or postulates, and theorems.

axiom	An axiom is a statement about undefined terms and definitions that is accepted as true without proof.
base angles of an isosceles trapezoid	The base angles of an isosceles trapezoid is a pair of angles formed by one base and the legs of a trapezoid.
base angles of isosceles triangle	The base angles of an isosceles triangle are the angles opposite the two congruent sides.
base of an isosceles triangle	The base of an isosceles triangle is the side of the triangle that is not equal to the other two sides.
base of a parallelogram	The base of a parallelogram is the side to which the altitude, or height, is drawn.
bases of a trapezoid	The bases of a trapezoid are the parallel sides of the trapezoid.
bisect a line segment	To bisect a line segment is to separate it into two equal parts.
bisector of an angle	The bisector of an angle is a ray separating the angle into two congruent angles.
bisector of an arc	The bisector of an arc is a line that divides the arc into two arcs with the same measure.
center of a circle	The center of a circle is the fixed point from which all points on the circle are equidistant.
center of a regular polygon	The center of a regular polygon is the center of the circle circumscribed around the polygon.
central angle	A central angle is the angle with sides that are the radii of a circle and the vertex at the center of the circle.
central angle of a regular polygon	The central angle of a regular polygon is the angle formed by two radii to two adjacent vertices.
centroid	A centroid is the point determined by the intersection of the three medians of a triangle.
chord	A chord is the line segment joining two distinct points on a circle.
circle	A circle is the set of all points in a plane that are located a fixed distance from a fixed point, the center.
circumcenter	The circumcenter is the point determined by the intersection of the perpendicular bisectors of the three sides of a triangle.
circumference	The circumference is the distance around the circle.
circumscribed circle	A circumscribed circle is a circle containing all the vertices of a polygon. The sides of the polygon are chords of the circle.
circumscribed polygon	A circumscribed polygon is a polygon whose sides are all tangent to a circle. The circle is in the interior of the polygon.
collinear points	Collinear points are points that lie on the same line.
common tangent	A common tangent is a line that intersects two nonintersecting circles in one and only one place. If the circles are on the same side of the tangent, then the tangent is called a *common external tangent*. If the circles are on opposite sides of the tangent, then it is called a *common internal tangent*.
compass	A compass is a tool used to draw circles or parts of circles.
complementary angles	Complementary angles are two angles whose measures add up to 90 degrees.
concave polygon	A concave polygon is a polygon in which a line segment joining two points in the polygon may include points not in the interior of the polygon.
concentric circles	Concentric circles are circles with the same center but different radii.
conclusion	A conclusion is the part of a statement providing the claim to be verified. It is the "then" phrase of an "if, then" statement.
concurrent lines	Concurrent lines are lines that intersect in one and only one point.
conditional statement	A conditional statement is a statement written in the "if, then" format.

cone	A cone is a solid figure formed by connecting a circle with a point, or vertex, not in the plane of the circle.
congruent	The term *congruent* describes figures that can be made to coincide.
congruent angles	Congruent angles are angles with the same measure.
congruent arcs	Congruent arcs are arcs with the same measure.
congruent segments	Congruent segments are segments with the same length, or measure.
congruent triangles	Congruent triangles are triangles whose corresponding parts have the same measure.
contrapositive	A contrapositive is the opposite of a conditional. For example, given the conditional $P \longrightarrow Q$, the contrapositive is $\sim Q \longrightarrow \sim P$.
converse	The converse of a conditional statement is formed by interchanging the hypothesis and conclusion. If given the conditional $P \longrightarrow Q$, the converse is $Q \longrightarrow P$.
convex polygon	A convex polygon is a polygon whose angles all measure between 0 and 180 degrees.
coordinate	A coordinate is the number corresponding to a given point on a number line.
coplanar	Coplanar refers to points that lie on the same plane.
corollary	A corollary is a statement whose proof is a direct result of a previously proved theorem.
corresponding angles	Corresponding angles are nonadjacent angles on the same side of the transversal and in the same corresponding positions with respect to the two lines in question.
corresponding parts	Corresponding parts are parts of congruent triangles that coincide when one is placed on top of the other.
cosine	In a right triangle, cosine is the ratio of the leg adjacent to the acute angle to the hypotenuse.
cylinder	A cylinder is a solid figure formed by joining two congruent circles in parallel planes.
decagon	A decagon is a ten-sided polygon.
deductive reasoning	Deductive reasoning is a form of reasoning in which a specific conclusion is reached based on a collection of generally accepted assumptions.
definition	A definition is a statement that gives meaning to new terms.
degree	A degree is an angle unit of measure.
diagonal	A diagonal is a segment that joins two nonadjacent vertices.
diameter	A diameter is a line segment that passes through the center of a circle and has endpoints on that circle.
direct proof	A direct proof is one that shows a series of statements, starting with the hypothesis, followed by a statement that follows from the preceding one, and so on.
distance	Distance is the length of a line segment.
dodecahedron	A dodecahedron is a polyhedron with twelve congruent faces that are regular pentagons.
edge of a polyhedron	The edge of a polyhedron is a line segment between two consecutive vertices.
endpoint	An endpoint is part of a line segment. (See *line segment* or *ray*.)
equal angles	Equal angles are angles with the same measure.
equiangular triangle	An equiangular triangle is a triangle in which all angles are equal in measure.
equilateral triangle	An equilateral triangle is a triangle in which all sides are equal in length.
extended ratio	An extended ratio compares more than two quantities and is written $a:b:c$.
exterior of an angle	The exterior of an angle is all points that lie outside the sides of the angle.
exterior angle	An exterior angle is an angle formed by one side of a figure and an extension of another side.

extremes of a proportion	The extremes of a proportion are the first and last terms in a proportion.
face of a polyhedron	The face of a polyhedron is the flat surface on a polyhedron formed by the polygons and their interiors.
fallacy	A fallacy is a statement that is made when a conclusion is reached that does not necessarily follow from the premises.
geometric construction	A geometric construction is a geometric drawing made with only a straightedge and a compass.
geometric figure	A geometric figure is any set of points, lines, or planes in space.
geometric mean	In the proportion $a/b = b/c$, b is the geometric mean between a and c.
height	Height is the length of an altitude.
hemisphere	A hemisphere is one-half of a sphere.
heptagon	A heptagon is a seven-sided polygon.
hexagon	A hexagon is a six-sided polygon.
hexahedron	A hexahedron is a polyhedron with six congruent square faces.
hypotenuse	A hypotenuse is the side of a right triangle opposite the right angle.
hypothesis	A hypothesis is the part of a statement providing the given information. It is the "if" phrase of an "if, then" statement.
icosahedron	An icosahedron is a polyhedron with twenty congruent faces that are equilateral triangles.
incenter	The incenter is the point determined by the intersection of the three angle bisectors of a triangle.
indirect proof	An indirect proof is a proof in which you assume the conclusion is not true and then deduce a contradiction.
inductive reasoning	Inductive reasoning is a form of reasoning in which a limited collection of specific observations are used to reach a general conclusion.
inscribed angle	An inscribed angle is an angle whose vertex is on a circle and whose sides intersect the circle in two other points.
inscribed circle	An inscribed circle is a circle that lies inside a polygon such that the sides of the polygon are tangents of the circle.
inscribed polygon	An inscribed polygon is a polygon that has its vertices on a circle such that the sides of the polygon are chords of the circle.
interior of an angle	The interior of an angle is all of the points that lie inside the sides of the angle.
interior angle of a polygon	The interior angle of a polygon is an angle formed by two sides of the polygon such that the angle lies in the interior of the polygon.
inverse	The inverse of the conditional $P \longrightarrow Q$ is $\sim\!P \longrightarrow \sim\!Q$
isosceles trapezoid	An isosceles trapezoid is a trapezoid with congruent legs.
isosceles triangle	An isosceles triangle is a triangle with two sides equal in length.
kite	A kite is a quadrilateral with exactly two distinct pairs of congruent consecutive sides.
lateral area	The lateral area is the sum of the areas of the lateral faces of a solid figure.
legs of a right triangle	The legs of a right triangle are the two sides that form the right angle.
legs of trapezoid	The legs of a trapezoid are the nonparallel sides of the trapezoid.
length	The length is the distance between the endpoints of a segment.

length of an arc	The length of an arc is the distance along a circle between the endpoints of an arc; it is part of the circumference.
line	A line is an undefined term that can be thought of as a set of points in a one-dimensional straight figure extending in opposite directions without ending.
line of centers	The line of centers is the line passing through the centers of two circles.
line segment	A line segment is the part of a line containing two endpoints and all the points between them.
locus	The locus is the set of points and only those points that satisfy one or more conditions.
major arc of a circle	The major arc of a circle is an arc that is longer than a semicircle.
mean proportional	See *geometric mean.*
mean of a proportion	The mean of a proportion is the middle terms in a proportion.
measure of an arc	The measure of an arc is the number of degrees in the central angle that intercepts the arc.
median of a trapezoid	The median of a trapezoid is the segment joining the midpoints of the legs of the trapezoid.
median of a triangle	The median of a triangle is the segment joining a vertex to the midpoint of the side opposite that vertex.
midpoint	The midpoint is a point on a line segment that separates it into two equal parts.
minor arc of a circle	The minor arc of a circle is an arc that is shorter than a semicircle.
negation	Negation refers to the negative of a statement $(\sim P)$.
nonagon	A nonagon is a nine-sided polygon.
number line	The number line is a line labeled in equal units marked with integers.
oblique planes	Oblique planes are two planes or a line and a plane that intersect but are not perpendicular.
oblique prism	An oblique prism is a nonright prism.
obtuse angle	An obtuse angle is one measuring between 90 and 180 degrees.
obtuse triangle	An obtuse triangle is a triangle with one obtuse angle.
octagon	An octagon is an eight-sided polygon.
octahedron	An octahedron is a polyhedron with eight congruent faces that are equilateral triangles.
ordered pair	An ordered pair refers to the coordinates (x, y) that give the position of a point on a two dimensional coordinate plane.
origin	On a number line, the origin is the starting point selected to mark off equal units in both directions. On a coordinate plane, the origin is the intersection point of the axes.
orthocenter	The orthocenter is the point determined by the intersection of the three altitudes of a triangle.
parallel lines	Parallel lines are two lines in the same plane that do not intersect.
parallel planes	Parallel planes are two planes that do not intersect.
parallelogram	A parallelogram is a quadrilateral whose opposite sides are parallel.
pentagon	A pentagon is a five-sided polygon.
perimeter	The perimeter is the sum of the lengths of the sides of a figure.
perpendicular bisector	A perpendicular bisector is a line that both bisects and is perpendicular to a given line segment.
perpendicular lines	Perpendicular lines are intersecting lines that form equal adjacent right angles.
plane	A plane is an undefined term that can be thought of as a set of points in a flat surface with two dimensions extending without boundary.

plot	To plot is to identify a point on a number line with a given real number.
point	A point is an undefined term that can be thought of as an object that determines a position but has no dimension.
point of tangency	A point of tangency is the point of intersection between a tangent line and a circle.
point-slope form	The point-slope form refers to the equation of a line with slope m that passes through the point (x_1, y_1) where $y - y_1 = m(x - x_1)$.
polygon	A polygon is a closed figure in a plane.
polyhedron	A polyhedron is a solid figure formed by the intersection of planes.
postulate	A postulate is a statement about undefined terms and definitions that is accepted as true without proof.
premise	A premise is a collection of undefined terms, definitions, postulates, or previously proved theorems.
prism	A prism is a solid figure formed by joining two congruent polygonal regions in parallel planes.
proportion	A proportion is an equation showing that two ratios are equal.
proportional segments	Proportional segments are segments whose lengths are proportional to one another.
pyramid	A pyramid is a solid figure that is formed by connecting a polygon with a point not in the plane of the polygon.
quadrant	A quadrant is one of four sections formed by the axes on a coordinate plane.
quadrilateral	A quadrilateral is a four-sided polygon.
radius of a circle	The radius of a circle is the line segment joining the center of a circle to one of its points. (Its plural is *radii*.)
radius of a regular polygon	The radius of a regular polygon is the segment joining the center of a polygon to one of its vertices.
ratio	A ratio is a comparison of two quantities, a and b, which is written as a fraction a/b or as $a:b$.
ray	A ray is the part of a line that begins at an endpoint and extends infinitely in one direction.
rectangle	A rectangle is a parallelogram with one right angle.
regular polygon	A regular polygon is a polygon in which all sides are congruent and all angles are congruent.
regular polyhedron	A regular polyhedron is a solid figure in which all faces are congruent regular polygons.
regular prism	A regular prism is one whose base is a regular polygon.
regular pyramid	A regular pyramid is one that has a regular polygon for its base, congruent isosceles triangles for its lateral surfaces, and an altitude passing through the center of its base.
remote interior angles	Remote interior angles are the angles of a triangle that are not adjacent to a given exterior angle of the triangle.
rhombus	A rhombus is a parallelogram that has two congruent adjacent sides.
right angle	A right angle is an angle measuring 90 degrees.
right circular cone	A right circular cone is a cone in which the line segment joining the vertex and the center of the base is perpendicular to the base.
right circular cylinder	A right circular cylinder is a cylinder in which the line segment joining the centers of the bases is perpendicular to the plane of both bases.
right prism	A right prism is a prixm in which the lateral faces are rectangles.

right triangle	A right triangle is a triangle in which one angle is a right angle.
scalene triangle	A scalene triangle is a triangle in which no two sides are equal in length.
secant	A secant is a line that intersects a circle in two points.
sector of a circle	The sector of a circle is the region bounded by two radii of the circle and the arc of the circle determined by the radii.
segment of a circle	A segment of a circle is a region bounded by a chord of a circle and the arc formed by the chord.
semicircle	A semicircle is the arc of a circle whose endpoints are the endpoints of a diameter of the circle.
set	A set is a collection of objects.
sides of an angle	The sides of an angle are the two rays that form the angle.
sides of a polygon	The sides of a polygon are the segments of the polygon that only intersect at their endpoints.
sides of a triangle	The sides of a triangle are the segments of a triangle.
similar polygons	Similar polygons are polygons whose vertices can be paired in such a way that the corresponding angles are congruent and the corresponding sides are proportional.
sine	In a right triangle, a sine is the ratio of the leg opposite the acute angle to the hypotenuse.
slant height of a cone	The slant height of a cone is a line segment from the vertex to a point on the edge of the circular base.
slant height of a pyramid	The slant height of a pyramid is a line segment joining the vertex of the pyramid on the perpendicular to a base edge of the pyramid.
slope	The slope is the measure of the steepness of a line. It is the ratio of the vertical change to the horizontal change, or the rise over the run.
slope-intercept form	The slope-intercept form refers to the equation of a line with slope m and y-intercept $(0, b)$ where $y = m(x + b)$.
space	Space is the set of all points.
sphere	A sphere is a set of all points in space a given distance, or radius, from a given point, the center.
square	A square is a rhombus with one right angle.
straight angle	A straight angle measures 180 degrees; the sides form a line.
straightedge	A straightedge is a ruler with no marks of scale.
supplementary angles	Supplementary angles are two angles whose measures total 180 degrees.
surface area	The surface area is the sum of the areas of the surfaces of a solid figure, or the lateral area plus the base area.
tangent	In a right triangle, tangent is the ratio of the leg opposite the acute angle to the leg adjacent to the acute angle.
tangent line	A tangent line is a line that intersects a circle in one and only one point, the point of tangency.
tetrahedron	A tetrahedron is a polyhedron with four congruent faces that are equilateral triangles.
theorem	A theorem is a statement that can be proved using definitions, postulates, and rules of deduction and logic.
transversal	The transversal is a line that intersects two or more distinct lines in different points.
trapezoid	A trapezoid is a quadrilateral with exactly one pair of parallel sides.

triangle	A triangle is a closed figure formed by three segments.
undefined terms	Undefined terms are terms whose meanings are assumed.
vertex angle of an isosceles triangle	The vertex angle of an isosceles triangle is the angle formed by the congruent sides of the triangle.
vertex of an angle	The vertex of an angle is the point at which the two sides of an angle meet.
vertex of a polygon	The vertex of a polygon is the intersection point of the sides of a polygon.
vertex of a triangle	The vertex of a triangle is the intersection point of the sides of a triangle.
vertical angles	Vertical angles are two nonadjacent angles formed by two intersecting lines.
volume	Volume is the measurement of the amount of space within a solid figure.
x-coordinate	The x-coordinate is the first number in an ordered pair.
y-coordinate	The y-coordinate is the second number in an ordered pair.
x-intercept	The x-intercept is the x-coordinate of the point where the graph crosses the x axis.
y-intercept	The y-intercept is the y-coordinate of the point where the graph crosses the y axis.

P1.1 Given any two distinct points in space, there is exactly one line that passes through them.

P1.2 Given any three distinct points in space not on the same line, there is exactly one plane that passes through them.

P1.3 The line determined by any two distinct points in a plane is also contained in the plane.

P1.4 No plane contains all points in space.

P1.5 (Ruler Postulate) There is a one-to-one correspondence between the set of all points on a line and the set of all real numbers.

P1.6 (Reflexive Law) Any quantity is equal to itself. ($x = x$)

P1.7 (Symmetric Law) If x and y are any two quantities and $x = y$, then $y = x$.

P1.8 (Transitive Law) If x, y, and z are any three quantities with $x = y$ and $y = z$, then $x = z$.

P1.9 (Addition-Subtraction Law) If w, x, y, and z are any four quantities with $w = x$ and $y = z$, then $w + y = x + z$ and $w - y = x - z$.

P1.10 (Multiplication-Division Law) If w, x, y, and z are any four quantities with $w = x$ and $y = z$, then $wy = xz$ and $\frac{w}{y} = \frac{x}{z}$ (provided $y \neq 0$ and $z \neq 0$).

P1.11 (Substitution Law) If x and y are any two quantities with $x = y$, then x can be substituted for y in any expression containing y.

P1.12 (Distributive Law) If x, y, and z are any three quantities, then $x(y + z) = xy + xz$.

P1.13 (Segment Addition Postulate) Let A, B, and C be three points on the same line with B between A and C. Then $AC = AB + BC$, $BC = AC - AB$, and $AB = AC - BC$.

P1.14 (Angle Addition Postulate) Let A, B, and C be points that determine $\angle ABC$ with P a point in the interior of the angle. Then $m\angle ABC = m\angle ABP + m\angle PBC$, $m\angle PBC = m\angle ABC - m\angle ABP$, and $m\angle ABP = m\angle ABC - m\angle PBC$.

P1.15 (Midpoint Postulate) Each line segment has exactly one midpoint.

P1.16 (Perpendicular Bisector Postulate) Each given line segment has exactly one perpendicular bisector.

P1.17 There is exactly one line perpendicular to a given line passing through a given point on the line.

P1.18 There is exactly one line perpendicular to a given line passing through a given point not on that line.

P1.19 (Angle Bisector Postulate) Each angle has exactly one bisector.

P2.1 (SAS) If two sides and the included angle of one triangle are congruent to two sides and the included angle of a second triangle, then the triangles are congruent.

P2.2 (ASA) If two angles and the included side of one triangle are congruent to two angles and the included side of a second triangle, then the triangles are congruent.

P2.3 (SSS) If three sides of one triangle are congruent to three sides of a second triangle, then the triangles are congruent.

P3.1 (Parallel Postulate) For a given line ℓ and a point P not on ℓ, one and only one line through P is parallel to ℓ.

P3.2 A polygon has the same number of angles as sides.

P5.1 (AAA) Two triangles are similar if three angles of one are congruent to the corresponding three angles of the other triangle.

P5.2 (Trichotomy Law) If a and b are real numbers, exactly one of the following is true: $a < b$, $a = b$, or $a > b$.

P5.3 (Transitive Law) If a, b, and c are real numbers with $a < b$ and $b < c$, then $a < c$.

P5.4 (Addition Properties of Inequalities) If a, b, c, and d are real numbers with $a < b$ and $c < d$, then $a + c < b + c$ and $a + c < b + d$.

P5.5 (Subtraction Properties of Inequalities) If a, b, c, and d are real numbers with $a < b$ and $c = d$, then $a - c < b - c$, $a - c < b - d$, and $c - a > d - b$.

P5.6 (Multiplication Properties of Inequalities) If a, b, and c are real numbers with $a < b$, then $ac < bc$ if $c > 0$ and $ac > bc$ if $c < 0$.

P5.7 (Division Properties of Inequalities) If a, b, and c are real numbers with $a < b$ then $\frac{a}{c} < \frac{b}{c}$ if $c > 0$ and $\frac{a}{c} > \frac{b}{c}$ if $c < 0$.

P5.8 (The Whole Is Greater Than Its Parts) If a, b, and c are real numbers with $c = a + b$ and $b > 0$, then $c > a$.

P5.9 The measure of each angle of a triangle is greater than $0°$.

P6.1 (Congruent Circles) If two circles are congruent then their radii and diameters are congruent. Conversely, if the radii or diameters are congruent, then two circles are congruent.

P6.2 (Arc Addition Postulate) Let A, B, and C be three points on the same circle with B between A and C. Then $m\widehat{AC} = m\widehat{AB} + m\widehat{BC}$, $m\widehat{BC} = m\widehat{AC} - m\widehat{AB}$, and $m\widehat{AB} = m\widehat{AC} - m\widehat{BC}$.

P6.3 If a line is perpendicular to a radius of a circle and passes through the point where the radius intersects the circle, then the line is a tangent.

P6.4 A radius drawn to the point of tangency of a tangent is perpendicular to the tangent.

P7.1   (Area of a Rectangle) The area of a rectangle with base b and height h is determined with the formula $A = bh$.

P7.2   (Additive Property of Areas) If lines divide a given area into several smaller nonoverlapping areas, the given area is the sum of the smaller areas.

P7.3   Two congruent polygons have the same area.

P7.4   (Circumference of a Circle) The circumference C of any circle with radius r and diameter d is determined with the formula $C = 2\pi r = \pi d$.

P7.5   (Area of a Circle) The area of a circle with radius r is determined with the formula $A = \pi r^2$.

P7.6   (Area of a Sector) The area of a sector of a circle with radius r whose arc has measure $m°$ is determined with the formula $A = \dfrac{m}{360}\pi r^2$.

P7.7   (Arc Length) The length of an arc measuring $m°$ in a circle with radius r is determined with the formula $L = \dfrac{m}{360}2\pi r = \dfrac{m}{180}\pi r$.

P8.1   The intersection of two distinct planes is a line.

P8.2   (Surface Area and Volume of a Sphere) For a sphere with radius r, the surface area is $SA = 4\pi r^2$ and the volume is $V = \dfrac{4}{3}\pi r^3$.

P9.1   Associated with each point in the plane there is one and only one ordered pair of numbers.

T1.1 (Addition Theorem for Segments) If B is a point between A and C on segment $\overline{AC}$, Q is a point between P and R on segment $\overline{PR}$, $AB = PQ$ and $BC = QR$, then $AC = PR$.

T1.2 (Subtraction Theorem for Segments) If B is a point between A and C on segment $\overline{AC}$, Q is a point between P and R on segment $\overline{PR}$, $AC = PR$, and $AB = PQ$, then $BC = QR$.

T1.3 (Addition Theorem for Angles) If D is a point in the interior of $\angle ABC$, S is a point in the interior of $\angle PQR$, $m\angle ABD = m\angle PQS$, and $m\angle DBC = m\angle SQR$, then $m\angle ABC = m\angle PQR$.

T1.4 (Subtraction Theorem for Angles) If D is a point in the interior of $\angle ABC$, S is a point in the interior of $\angle PQR$, $m\angle ABC = m\angle PQR$, and $m\angle DBC = m\angle SQR$, then $m\angle ABD = m\angle PQS$.

T1.5 Two equal supplementary angles are right angles.

T1.6 Complements of equal angles are equal in measure.

C1.7 Complements of the same angle are equal in measure.

T1.8 Supplements of equal angles are equal in measure.

C1.9 Supplements of the same angle are equal in measure.

T1.10 If A, B, and C are three points on a line, with B between A and C, and $\angle ABD$ and $\angle DBC$ are adjacent angles, then $\angle ABD$ and $\angle DBC$ are supplementary angles.

T1.11 Vertical angles are equal in measure.

T1.12 All right angles are equal in measure.

T2.1 (Transitive Law for Congruent Triangles) If $\triangle ABC \cong \triangle DEF$ and $\triangle DEF \cong \triangle GHI$, then $\triangle ABC \cong \triangle GHI$.

T2.2 (Segment Bisector Theorem) Construction 1.3 gives the perpendicular bisector of a given line segment.

T2.3 Every point on the perpendicular bisector of a segment is equidistant from the two endpoints.

T2.4 (Angle Bisector Theorem) Construction 1.6 gives the bisector of a given angle.

T2.5 If two sides of a triangle are congruent, then the angles opposite these sides are also congruent.

C2.6 If a triangle is equilateral, then it is equiangular.

T2.7 If two angles of a triangle are congruent, then the sides opposite these angles are also congruent.

C2.8 If a triangle is equiangular, then it is equilateral.

T2.9 The perpendicular bisectors of the sides of a triangle are concurrent.

T2.10 The medians of a triangle are concurrent and meet at a point that is two-thirds the distance from the vertex to the midpoint of the opposite side.

T2.11 The altitudes of a triangle are concurrent.

T2.12 The bisectors of the angles of a triangle are concurrent and meet at a point equidistant from the sides of the triangle.

T2.13 (LA) If a leg and acute angle of one right triangle are congruent to a leg and the corresponding acute angle of another right triangle, then the two right triangles are congruent.

T2.14 (LL) If the two legs of one right triangle are congruent to the two legs of another right triangle, then the two right triangles are congruent.

T3.1 If two lines in a plane are both perpendicular to a third line, then they are parallel.

T3.2 If two lines are cut by a transversal and a pair of alternate interior angles are congruent, then the lines are parallel.

T3.3 If two lines are cut by a transversal and a pair of corresponding angles are congruent, then the lines are parallel.

T3.4 If two lines are cut by a transversal and a pair of alternate exterior angles are congruent, then the lines are parallel.

T3.5 If two lines are cut by a transversal and two interior angles on the same side of the transversal are supplementary, then the lines are parallel.

T3.6 (Converse of Theorem 3.2) If two parallel lines are cut by a transversal, then all pairs of alternate interior angles are congruent.

T3.7 (Converse of Theorem 3.1) If two lines are parallel and a third line is perpendicular to one of them, then it is also perpendicular to the other.

T3.8 (Converse of Theorem 3.3) If two parallel lines are cut by a transversal, then all pairs of corresponding angles are congruent.

T3.9 (Converse of Theorem 3.4) If two parallel lines are cut by a transversal, then all pairs of alternate exterior angles are congruent.

T3.10 (Converse of Theorem 3.5) If two parallel lines are cut by a transversal, then all pairs of interior angles on the same side of the transversal are supplementary.

T3.11 The sum of the measures of the angles of a triangle is $180°$.

C3.12 Any triangle can have at most one right angle or at most one obtuse angle.

C3.13 If two angles of one triangle are congruent to two angles of another triangle, then the third angles are also congruent.

C3.14 The measure of an exterior angle of a triangle is equal to the sum of the measures of the nonadjacent interior angles.

T3.15 The sum of the measures of the angles of a polygon with n sides is given by the formula $S = (n-2)180°$.

C3.16   The measure of each angle of a regular polygon with n sides is given by the formula $a = \dfrac{(n-2)180°}{n}$.

T3.17   The sum of the measures of the exterior angles of a polygon, one at each vertex, is 360°.

C3.18   The measure of each exterior angle of a regular polygon with n sides is determined with the formula $e = \dfrac{360°}{n}$.

T3.19   (AAS) If two angles and any side of one triangle are congruent to two angles and the corresponding side of another triangle, then the triangles are congruent.

T3.20   (HA) If the hypotenuse and an acute angle of one right triangle are congruent to the hypotenuse and an acute angle of another right triangle, then the two right triangles are congruent.

T3.21   (HL) If the hypotenuse and a leg of one right triangle are congruent to the hypotenuse and a leg of another right triangle, then the two right triangles are congruent.

T4.1   Each diagonal divides a parallelogram into two congruent triangles.

C4.2   The opposite sides and opposite angles of a parallelogram are congruent.

T4.3   Consecutive angles of a parallelogram are supplementary.

T4.4   The diagonals of a parallelogram bisect each other.

T4.5   If both pairs of opposite sides of a quadrilateral are congruent, then the quadrilateral is a parallelogram.

T4.6   If both pairs of opposite angles of a quadrilateral are congruent, then the quadrilateral is a parallelogram.

T4.7   If two opposite sides of a quadrilateral are congruent and parallel, then the quadrilateral is a parallelogram.

T4.8   If the diagonals of a quadrilateral bisect each other, then the quadrilateral is a parallelogram.

T4.9   All four sides of a rhombus are congruent.

T4.10   The diagonals of a rhombus are perpendicular.

T4.11   If the diagonals of a parallelogram are perpendicular, then the parallelogram is a rhombus.

T4.12   The diagonals of a rhombus bisect the angles of the rhombus.

T4.13   If a quadrilateral is a kite, one pair of opposite angles is congruent.

T4.14   If a quadrilateral is a kite, one diagonal is the perpendicular bisector of the other diagonal.

T4.15   All angles of a rectangle are right angles.

T4.16   The diagonals of a rectangle are congruent.

T4.17   If the diagonals of a parallelogram are congruent, then the parallelogram is a rectangle.

T4.18   Two parallel lines are always the same distance apart.

T4.19   The segment joining the midpoints of two sides of a triangle is parallel to the third side and its length is one-half the length of the third side.

T4.20   The base angles of an isosceles trapezoid are congruent.

T4.21   The diagonals of an isosceles trapezoid are congruent.

T4.22   The median of a trapezoid is parallel to the bases and equal to one-half their sum.

T4.23   If three or more parallel lines intercept congruent segments on one transversal, then they intercept congruent segments on all transversals.

T5.1   (Means-Extremes Property) In any proportion, the product of the means is equal to the product of the extremes. That is, if $\dfrac{a}{b} = \dfrac{c}{d}$, then $ad = bc$.

T5.2   (Reciprocal Property of Proportions) The reciprocals of both sides of a proportion are also proportional. That is, if $\dfrac{a}{b} = \dfrac{c}{d}$, then $\dfrac{b}{a} = \dfrac{d}{c}$.

T5.3   (Means Property of Proportions) If the means are interchanged in a proportion, a new proportion is formed. That is, if $\dfrac{a}{b} = \dfrac{c}{d}$, then $\dfrac{a}{c} = \dfrac{b}{d}$.

T5.4   (Extremes Property of Proportions) If the extremes are interchanged in a proportion, a new proportion is formed. That is, if $\dfrac{a}{b} = \dfrac{c}{d}$, then $\dfrac{d}{b} = \dfrac{c}{a}$.

T5.5   (Addition Property of Proportions) If the denominators in a proportion are added to their respective numerators, a new proportion is formed. That is, if $\dfrac{a}{b} = \dfrac{c}{d}$, then $\dfrac{a + b}{b} = \dfrac{c + d}{d}$.

T5.6   (Subtraction Property of Proportions) If the denominators in a proportion are subtracted from their respective numerators, a new proportion is formed. That is, if $\dfrac{a}{b} = \dfrac{c}{d}$, then $\dfrac{a - b}{b} = \dfrac{c - d}{d}$.

T5.7   If a, b, c, d, e, and f are numbers satisfying $\dfrac{a}{b} = \dfrac{c}{d} = \dfrac{e}{f}$, then $\dfrac{a + c + e}{b + d + f} = \dfrac{a}{b}$.

T5.8   (AA) Two triangles are similar if two angles of one triangle are congruent to the corresponding two angles of the other triangle.

T5.9   If $\triangle ABC \cong \triangle DEF$, then $\triangle ABC \sim \triangle DEF$.

T5.10   (Transitive Law for Similar Triangles) If $\triangle ABC \sim \triangle DEF$ and $\triangle DEF \sim \triangle GHI$, then $\triangle ABC \sim \triangle GHI$.

T5.11   (Triangle Proportionality Theorem) A line parallel to one side of a triangle that intersects the other two sides divides the two sides into proportional segments.

T5.12   (Triangle Angle-Bisector Theorem) The bisector of one angle of a triangle divides the opposite side into segments that are proportional to the other two sides.

T5.13   The altitude from the right angle to the hypotenuse in a right triangle forms two right triangles that are similar to each other and to the original triangle.

C5.14   The altitude from the right angle to the hypotenuse in a right triangle is the geometric mean or mean proportional between the segments of the hypotenuse.

C5.15   If the altitude is drawn from the right angle to the hypotenuse in a right triangle, then each leg is the geometric mean or mean proportional between the hypotenuse and the segment of the hypotenuse adjacent to the leg.

T5.16   The median from the right angle in a right triangle is one-half the length of the hypotenuse.

T5.17   (The Pythagorean Theorem) In a right triangle, the square of the length of the hypotenuse is equal to the sum of the squares of the lengths of the legs.

T5.18   (Converse of the Pythagorean Theorem) If the sides of a triangle have lengths a, b, and c, and $a^2 + b^2 = c^2$, then the triangle is a right triangle.

T5.19   ($45°$-$45°$-$90°$ Triangle Theorem) In a $45°$-$45°$-$90°$ triangle, the hypotenuse is $\sqrt{2}$ times as long as each (congruent) leg.

T5.20   ($30°$-$60°$-$90°$ Triangle Theorem) In a $30°$-$60°$-$90°$ triangle, the length of the hypotenuse is twice the length of the shorter leg, and the length of the longer leg is $\sqrt{3}$ times as long as the length of the shorter leg.

T5.21   The measure of an exterior angle of any triangle is greater than each remote interior angle.

T5.22   If the measures of two sides of a triangle are unequal, then the measures of the angles opposite those sides are unequal in the same order.

T5.23   If the measures of two angles of a triangle are unequal, then the measures of the sides opposite those angles are unequal in the same order.

T5.24   (The Triangle Inequality Theorem) The sum of the lengths of any two sides of a triangle is greater than the length of the third side.

T5.25   (SAS Inequality Theorem) If two sides of one triangle are equal in measure to two sides of another triangle, and the measure of the included angle of the first is greater than the measure of the included angle of the second, then the third side of the first triangle is greater than the third side of the second triangle.

T5.26   (SSS Inequality Theorem) If two sides of one triangle are equal in measure to two sides of another triangle, and the third side of the first is greater than the third side of the second, then the measure of the included angle of the first triangle is greater than the measure of the included angle of the second triangle.

T6.1   The diameter d of a circle is twice the radius r of the circle. That is, $d = 2r$.

T6.2   The measure of an inscribed angle is one-half the measure of its intercepted arc.

C6.3   Inscribed angles that intercept the same or equal arcs are congruent.

C6.4   Every angle inscribed in a semicircle is a right angle.

T6.5   When two chords of a circle intersect, the measure of each angle formed is one-half the sum of the measures of its intercepted arc and the arc intercepted by its vertical angle.

T6.6   In the same circle, the arcs formed by congruent chords are congruent.

T6.7   In the same circle, the chords formed by congruent arcs are congruent.

T6.8   A line drawn from the center of a circle perpendicular to a chord bisects the chord and the arc formed by the chord.

T6.9   A line drawn from the center of a circle to the midpoint of a chord (not a diameter) or to the midpoint of the arc formed by the chord is perpendicular to the chord.

T6.10   In the same circle, congruent chords are equidistant from the center of the circle.

T6.11   In the same circle, chords equidistant from the center of the circle are congruent.

T6.12   The perpendicular bisector of a chord passes through the center of the circle.

T6.13   If two chords intersect inside a circle, the product of the lengths of the segments of one chord is equal to the product of the lengths of the segments of the other.

T6.14   If two secants intersect forming an angle outside the circle, then the measure of this angle is one-half the difference of the measures of the intercepted arcs.

T6.15   If two secants are drawn to a circle from an external point, the product of the lengths of one secant segment and its external segment is equal to the product of the lengths of the other secant segment and its external segment.

T6.16   The angle formed by a tangent and a chord has a measure one-half the measure of its intercepted arc.

T6.17   The angle formed by the intersection of a tangent and a secant has a measure one-half the difference of the measures of the intercepted arcs.

T6.18   The angle formed by the intersection of two tangents has a measure one-half the difference of the measures of the intercepted arcs.

T6.19   Two tangent segments to a circle from the same point have equal lengths.

T6.20   If a secant and a tangent are drawn to a circle from an external point, the length of the tangent segment is the geometric mean between the length of the secant segment and its external segment.

T6.21   If two circles are tangent internally or externally, the point of tangency is on their line of centers.

T6.22   If two circles intersect in two points, then their line of centers is the perpendicular bisector of their common chord.

T6.23   If a quadrilateral is inscribed in a circle, the opposite angles are supplementary.

T6.24   If a parallelogram is inscribed in a circle then it is a rectangle.

T6.25 If a circle is divided into n equal arcs, $n > 2$, then the chords formed by these arcs form a regular n-gon.

T6.26 If a circle is divided into n equal arcs, $n > 2$, and tangents are constructed to the circle at the endpoints of each arc, then the figure formed by these tangents is a regular n-gon.

T6.27 All radii of a regular polygon are equal in length.

T6.28 All central angles of a regular polygon have the same measure.

T6.29 The measure a of each central angle in a regular n-gon is determined with the formula $a = \dfrac{360°}{n}$.

T6.30 In the same circle or in congruent circles, the greater of two central angles intercepts the greater of two arcs.

T6.31 In the same circle or in congruent circles, the greater of two arcs is intercepted by the greater of two central angles.

T6.32 In the same circle or in congruent circles, the greater of two chords forms the greater arc.

T6.33 In the same circle or in congruent circles, the greater of two arcs has the greater chord.

T6.34 In the same circle or in congruent circles, the greater of two unequal chords is nearer the center of the circle.

T6.35 In the same circle or in congruent circles, if two chords have unequal distances from the center of the circle, the chord nearer the center is greater.

C6.36 Every diameter of a circle is greater than any other chord that is not a diameter.

T6.37 The locus of all points equidistant from two given points A and B is the perpendicular bisector of $\overline{AB}$.

T6.38 The locus of all points equidistant from the sides of an angle is the angle bisector.

C7.1 The area of a square with sides of length s is determined with the formula $A = s^2$.

T7.2 The area of a parallelogram with length of base b and height h is determined with the formula $A = bh$.

T7.3 The area of a triangle with length of base b and height h is determined with the formula $A = \dfrac{1}{2}bh$.

T7.4 (Heron's Formula) If the three sides of a triangle have lengths a, b, and c, the area is
$$A = \sqrt{s(s - a)(s - b)(s - c)},$$
where $s = \dfrac{a + b + c}{2}$.

C7.5 The area of an equilateral triangle with sides length a is
$$A = \dfrac{a^2\sqrt{3}}{4}.$$

T7.6   The area of a trapezoid with length of bases b and b' and height h is determined with the formula $A = \frac{1}{2}(b + b')h$.

T7.7   The area of a rhombus with diagonals of length d and d' is determined with the formula $A = \frac{1}{2}dd'$.

T7.8   Every apothem of a regular polygon has the same length.

T7.9   The apothem of a regular polygon bisects its respective side.

T7.10 Every radius of a regular polygon bisects the angle at the vertex to which it is drawn.

T7.11 (Area of a Regular Polygon) The area of a regular polygon with apothem of length a and perimeter p is determined with the formula
$$A = \frac{1}{2}ap.$$

T8.1   (Lateral Area of a Right Prism) The lateral area of a right prism is determined with the formula $LA = ph$, where p is the perimeter of a base and h is the height of the prism.

T8.2   (Volume of a Right Prism) The volume of a right prism is determined with the formula $V = Bh$, where B is the area of a base and h is the height.

T8.3   (Lateral Area of a Regular Pyramid) The lateral area of a regular pyramid is determined with the formula $LA = \frac{1}{2}p\ell$, where p is the perimeter of the base and ℓ is the slant height.

T8.4   (Volume of a Regular Pyramid) The volume of a pyramid is determined with the formula $V = \frac{1}{3}Bh$, where B is the area of the base and h is the height.

T8.5   (Lateral Area of a Right Circular Cylinder) The lateral area of a right circular cylinder is determined with the formula $LA = 2\pi rh$, where r is the radius of a base and h is the height, the length of the altitude.

T8.6   (Volume of a Right Circular Cylinder) The volume of a right circular cylinder is determined with the formula $V = \pi r^2 h$, where r is the radius of a base and h is the height.

T8.7   (Lateral Area of a Right Circular Cone) The lateral area of a right circular cone is determined with the formula $LA = \pi r\ell$, where r is the radius of the base and ℓ is the slant height.

T8.8   (Volume of a Right Circular Cone) The volume of a right circular cone is determined with the formula
$$V = \frac{1}{3}Bh \text{ or } \frac{1}{3}\pi r^2 h,$$
where r is the radius of the base, B is the area of the base, and h is the height of the cone.

T9.1   (Distance Formula) The distance between two points with coordinates (x_1, y_1) and (x_2, y_2) is determined with the formula
$$d = \sqrt{(x_2 - x_1)^2 + (y_2 - y_1)^2}.$$

T9.2 (Midpoint Formula) The coordinates of the midpoint of the line segment joining (x_1, y_1) and (x_2, y_2) are determined with the midpoint formula $\left(\dfrac{x_1 + x_2}{2}, \dfrac{y_1 + y_2}{2} \right)$. That is, the x-coordinate of the midpoint is the average of the x-coordinates of the points, and the y-coordinate of the midpoint is the average of the y-coordinates of the points.

T9.3 Two distinct lines with slopes m_1 and m_2 are parallel if and only if $m_1 = m_2$.

T9.4 Two lines with slopes m_1 and m_2 are perpendicular if and only if $m_1 m_2 = -1$.

APPENDIX C

Constructions in Geometry

C1.1 Construct a line segment with the same length as a given line segment.

C1.2 Construct an angle with the same measure as a given angle.

C1.3 Construct a bisector of a given line segment.

C1.4 Construct a line perpendicular to a given line passing through a given point on the line.

C1.5 Construct a line perpendicular to a given line passing through a given point not on that line.

C1.6 Construct a bisector of a given angle.

C2.1 Construct a triangle that is congruent to a given triangle.

C2.2 Construct a triangle with two sides and the included angle given.

C2.3 Construct a triangle with two angles and the included side given.

C2.4 Construct an equilateral triangle when given a single side.

C2.5 Construct an altitude of a given triangle.

C2.6 Construct a median of a given triangle.

C3.1 Construct the line parallel to a given line that passes through a point not on the given line.

C4.1 Construct a rectangle when two adjacent sides are given.

C4.2 Construct a square when a side is given.

C4.3 Divide a given segment into a given number of congruent segments.

C5.1 Construct a segment proportional to three given line segments.

C5.2 Construct a polygon similar to a given polygon.

C6.1 Construct a tangent to a circle at a given point on the circle.

C6.2 Construct a tangent to a circle from a point outside the circle.

C6.3 Construct a common external tangent to two given circles.

C6.4 Construct a common internal tangent to two given circles.

C6.5 Construct a circle that is circumscribed around a given regular polygon.

C6.6 Construct a circle that is inscribed in a given regular polygon.

Index